Properties of Areas of Common Shapes

Rectangle

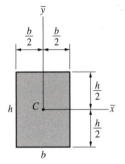

$A = bh$

$\overline{I}_x = \frac{1}{12}bh^3$

$\overline{I}_y = \frac{1}{12}hb^3$

$\overline{J} = \frac{1}{12}bh(h^2 + b^2)$

$\overline{r}_x = \frac{h}{\sqrt{12}}$

$\overline{r}_y = \frac{b}{\sqrt{12}}$

Triangle

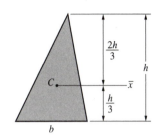

$A = \frac{1}{2}bh$

$\overline{I}_x = \frac{1}{36}bh^3$

$\overline{r}_x = \frac{h}{\sqrt{18}}$

Circle

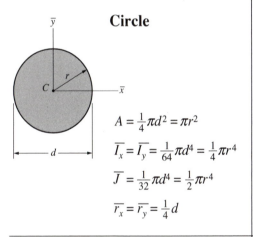

$A = \frac{1}{4}\pi d^2 = \pi r^2$

$\overline{I}_x = \overline{I}_y = \frac{1}{64}\pi d^4 = \frac{1}{4}\pi r^4$

$\overline{J} = \frac{1}{32}\pi d^4 = \frac{1}{2}\pi r^4$

$\overline{r}_x = \overline{r}_y = \frac{1}{4}d$

Circular Ring

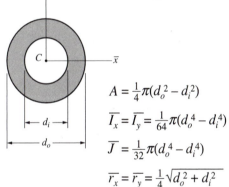

$A = \frac{1}{4}\pi(d_o^2 - d_i^2)$

$\overline{I}_x = \overline{I}_y = \frac{1}{64}\pi(d_o^4 - d_i^4)$

$\overline{J} = \frac{1}{32}\pi(d_o^4 - d_i^4)$

$\overline{r}_x = \overline{r}_y = \frac{1}{4}\sqrt{d_o^2 + d_i^2}$

Semicircle

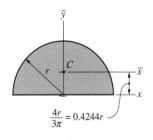

$\frac{4r}{3\pi} = 0.4244r$

$A = \frac{1}{2}\pi r^2$

$\overline{I}_x = 0.1098r^4$

$\overline{I}_y = \overline{I}_x = \frac{1}{8}\pi r^4$

$\overline{J} = 0.5025r^4$

$\overline{r}_x = 0.2644r$

$\overline{r}_y = r_x = \frac{1}{2}r$

Quarter-Circle

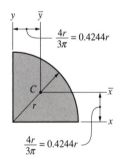

$\frac{4r}{3\pi} = 0.4244r$

$\frac{4r}{3\pi} = 0.4244r$

$A = \frac{1}{4}\pi r^2$

$\overline{I}_x = \overline{I}_y = 0.0549r^4$

$I_x = I_y = \frac{1}{16}\pi r^4$

$\overline{J} = 0.1098r^4$

$\overline{r}_x = \overline{r}_y = 0.2644r$

$r_x = r_y = \frac{1}{2}r$

Statics *and* Strength *of* Materials

Second Edition

Fa-Hwa Cheng, Ph.D., P.E.
Professor of Engineering and Industrial Technologies
Virginia Western Community College

GLENCOE
McGraw-Hill

New York, New York Columbus, Ohio Woodland Hills, California Peoria, Illinois

Cheng, Fa-Hwa.
    Statics and strength of materials/Fa-Hwa Cheng.—2nd ed.
        p.   cm.

Includes index.
ISBN 0-02-803067-2   ISBN 0-02-803068-0 (I.M.)
    1. Statics.   2. Strength of materials.   I. Title.
TA351.C52   1997
624.1'71—dc20   95–49620
                                                 CIP

Figure 10–2 photo courtesy of the Tinius Olsen Testing Machine Company. Part One photo, Zigy Kaluzny, Tony Stone Images (motorway overpass under construction). Part Two photo, Donovan Reese, Tony Stone Worldwide (a tow-angle view of office buildings seen through steel sculpture).

Cover: South Grand Island bridges in Tonawanda, New York, by Mike Dobel.

Send all inquiries to:
Glencoe/McGraw-Hill
8787 Orion Place
Columbus, Ohio 43240

11 12   DOW/DOW   0 9 8 7 6

ISBN-13: 978-0–02–803067–8

ISBN-10: 0–02–803067–2

Printed in the United States of America

The second edition of this book, *Statics and Strength of Materials,* presents an elementary, practical, and easy-to-understand text for the course in the subject. It is designed to be used primarily in two- and four-year technology programs in engineering, construction, and architecture. It provides sufficient material to be used for a two-semester course or for two separate, one-semester courses in statics and strength of materials. The book can also serve as a reference guide for practicing technologists and technicians. For those who are preparing for the examinations for engineering-in-training or the state license for professional engineers, this text is useful for reviewing statics as well as strength of materials.

This book requires a level of mathematics that does not include calculus. However, a working knowledge of algebra, geometry, and trigonometry is essential. A review of trigonometry and algebra is included in Chapter 1.

This revised text presents expanded coverage of the existing material plus many new topics, including important treatment of the plastic design approach. The following are among the new topics: accuracy and precision, solution of simultaneous equations, rolling resistance, mechanical properties of materials, composite beams, reinforced concrete beams, plastic analysis of beams, design of shear connectors (to resist shear flow), load and resistance factor design (LRFD, or ultimate strength design), the three-moment theorem (for continuous beams), and eccentrically loaded connections. These topics may be covered partially or omitted altogether without affecting the continuity of the text. Load and resistance factor design (LRFD) is included to meet the increasing trend toward the ultimate strength design approach in engineering practice.

The following features are included in this updated edition:

1. Organization has been greatly improved in the second edition. For example, area moments of inertia now follows the discussion on centroid; statically indeterminate, axially loaded members and thermal stresses are covered right after the discussion on axial deformation; stresses in thin-walled pressure vessels can be considered as simple stresses and they are placed in the chapter on simple stresses. A new chapter on mechanical properties of materials follows the chapters on simple stresses and strains. Design of beams for strength is an expanded chapter following the chapter on stresses in beams. Statically indeterminate beams is another expanded chapter placed right after the chapter on beam deflection.

2. Within each chapter, topics are carefully organized and clearly explained. Each section generally includes several subtopics denoted by boldface subtitles, to improve clarity and provide convenience for later reference and review.

3. An abundance of example problems with detailed step-by-step solutions illustrate each particular phase of the topic under consideration.

4. Understanding of the basic principles and their far-reaching applicability are emphasized so that students can apply these basic principles and concepts to a vast variety of problems.

5. Extensive and well-developed coverage of design topics (including the ultimate strength design approach) familiarize students with the general procedure involved in the design process and give them a clear perception of the design work preformed in the real world.

6. Homework problems of various levels of difficulty are furnished at the end of each chapter and are grouped by sections. Many of the homework problems in the second edition are brand new. Some first-edition problems have been revised to reflect better the practical aspects in engineering practice. The problems in each group are arranged in order of increasing difficulty. They begin with relatively simple, uncomplicated problems to help students gain confidence. The last few problems in the group are usually more involved than the others, to provide challenge and maintain interest. Answers to all odd-numbered problems are given at the back of the book.

7. Computer program assignments follow regular homework problems in most chapters. These computer problems require students to write computer programs to solve some general problems in the chapter topics. Any appropriate computer language may be used.

8. Because free-body diagrams are the foundation of mechanics, they are discussed in great detail in Chapter 3. The importance of free-body diagrams is emphasized throughout the book, and they are used extensively to determine external and internal forces.

9. Each chapter ends with a summary of the topics covered in that chapter. This summary section is particularly useful for review purposes.

10. Both U.S. customary and SI units (the international system of units) are used throughout the book. All the section property tables and mechanical property tables in the appendix pp. 761–779, are in both systems of units so that analysis and design problems can be given in either system. The corresponding tables in the two unit systems appear on facing pages for easy correlation.

The instructor's manual that accompanies *Statics and Strength of Materials* includes detailed solutions for all the homework problems in the book. Solutions to the computer program assignments, written in QBASIC, are also included. The instructor's manual also includes several test problems accompanied by detailed solutions for each chapter. These problems can be duplicated and handed out to students for review purposes or they can be administered to students as test problems.

My hearty thanks go to the following reviewers of the second and the first editions; their constructive suggestions have made this text a better book: Richard Ciocci, Harrisburg Area Community College; Anthony L. Brizendine, Fairmont State College; Daniel M. Hahn, Kirkwood Community College; Victor G. Forsnes, Ricks College; Ronald F. Amberger, Rochester Institute of Technology; David A. Pierce, Columbus Technical Institute; John O. Pautz, Middlesex County College; T. M. Brittain, The University of Akron; and John Keeley, Mt. Hood Community College. Special thanks are due to my colleague, David Webb, who spent countless hours proofreading the entire second-edition manuscript and offered many valuable suggestions for improvement. I appreciate expressively the superb job that the editorial and production staffs at Glencoe did for this book, with particular thanks to John Beck and Jan Hall for their continuing support and help. I am also indebted to my wife, Rosa, for her patience and understanding. In the time between the first and second editions, we have seen our two sons, Lincoln and Lindsay, grow up from teenager to adulthood. I hope that the quality of this text has grown the same amount in this new edition.

Fa-Hwa Cheng

CONTENTS

PART 1

Statics

CHAPTER 1

FUNDAMENTAL CONCEPTS AND PRINCIPLES

1-1
INTRODUCTION TO MECHANICS

Mechanics is the branch of physical science that deals with the state of rest or motion of bodies under the action of forces. Mechanics is the foundation for most engineering sciences and it is an indispensable prerequisite to most engineering or technical courses. The principles of mechanics are used in almost all technical analysis and design. A thorough understanding of these principles and their applications are of paramount importance for students in the engineering and technical programs.

Mechanics is divided into three branches: statics, dynamics, and strength of materials. *Statics* concerns the equilibrium of bodies under the action of *balanced forces*. *Dynamics* deals with the *motions* of bodies under the action of *unbalanced forces*. *Strength of materials* deals with the relationships among the external forces applied to the bodies, the resulting stresses (intensity of internal forces), and deformation (change of size or shape). The determination of the proper sizes of structural members to satisfy strength and deformation requirements are also important topics of strength of materials.

In the study of statics and dynamics, all bodies are assumed to be perfectly rigid. A *rigid body* is a solid in which the distance between any two points in the body remain unchanged. This is an idealization. In reality, deformations do occur in all bodies when they are subjected to forces. However, the deformations are usually very small and they can be neglected in the statics and dynamic analyses, without appreciable errors.

In the study of the strength of materials, deformation of structural members becomes very important because the concerns are the strength and stiffness of structural or machine members. Strength and stiffness are directly or indirectly related to the deformation, even if the deformation is very small.

This book concentrates on two major topics. The first eight chapters deal with statics, and the remaining twelve chapters deal with strength of materials. Dynamics will not be covered in this book.

1-2
THE NATURE OF A FORCE

A *force* is any effect that may change the state of rest or motion of a body. It represents the action of one body on another. The existence of a force can be observed by the effects that the force produces. Force is applied either by direct physical contact between bodies or by remote action. Gravitational, electrical, and magnetic forces are applied through remote action. Most other forces are applied by direct contact.

Forces Exerted by Direct Contact. The forces exerted on a rope pulled by someone's hands and the forces between a beam and its supports are examples of forces applied by direct contact. Less obvious cases of contact forces occur when a solid body comes in contact with a liquid or a gas. For example, forces exist between water and the hull of a boat, and similarly, between air and airplane wings.

Forces Exerted Through Remote Action. When a ball is thrown into the air, it falls to the ground. The pull of the earth's gravity, exerted through remote action, causes the ball to fall. The attraction force of the earth is usually referred to as the *weight* of the body. A satellite is kept in its orbit around the earth by the gravitational attraction from the earth. When a magnet attracts a small piece of iron through remote action, the magnetic force causes the iron to move without any physical contact.

Characteristics of a Force. A force can be defined completely by a magnitude, a direction, and a point of application. These are called the *characteristics of a force*. The magnitude of a force is described by a number with a proper unit. The direction of a force is indicated by a line (called the *line of action*) with an arrowhead. The *point of application* is the point at which the force is exerted. For example, the force applied to the block in Fig. 1–1 is described as a 100-lb (magnitude) force acting along a horizontal line (line of action) to the right (direction) through point A (point of application).

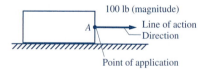

FIGURE 1–1

1-3
SCALAR AND VECTOR QUANTITIES

The quantities that we deal with in mechanics can be classified into two groups.

Scalar Quantities. Scalar quantities can be described completely by a magnitude. Examples of scalar quantities are length, area, volume, speed, mass, time, etc.

Vector Quantities. A vector quantity is characterized by its magnitude, direction, line of action, and sometimes point of application. Furthermore, vector quantities must be added by the *parallelogram law* (to be discussed in the next chapter); that is, vectors must be added geometrically rather than algebraically. As described in the preceding section, a force is a vector quantity. Other examples of vector quantities include moment, displacement, velocity, acceleration, etc.

1–4
TYPES OF FORCES

Forces can be classified into the following types.

Distributed and Concentrated Forces. A *distributed force* is exerted on a line, over an area, or throughout an entire volume. The weight of a beam can be treated as a distributed force over its length. A *concentrated force* is an idealization in which a force is assumed to act at a point. A force can be regarded as a concentrated force if the area of application is relatively small compared to the total surface area of the body.

External and Internal Forces. A force is called an *external force* if it is exerted on the body by another body. If a structure is formed by several connected components, the forces holding the component parts together are *internal forces* within the structure. In Fig. 1–2, the applied forces **P** and **Q**, together with the reactions A_x, A_y, and B_y at the supports, are external forces to the truss. The internal forces are developed in the truss members due to the applied loads and the reactions. These internal forces are responsible for holding the truss together.

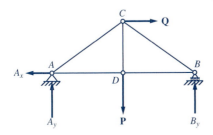

FIGURE 1–2

1–5
TYPES OF FORCE SYSTEMS

Forces treated as a group constitute a force system. Forces whose lines of action lie in the same plane are called *coplanar forces*. Forces whose lines of action act in a three-dimensional space are called *spatial forces*. If the lines of action of all the forces in a system pass through a common point, they are said to be *concurrent*. On the other hand, if there is no common point of intersection, the forces are said to be *nonconcurrent*. Depending on whether the forces are coplanar or spatial, concurrent or nonconcurrent, force systems may be classified into the following types.

Concurrent Coplanar Force System.   The lines of action of all the forces in the system pass through a common point and lie in the same plane, as shown in Fig. 1–3a.

Nonconcurrent Coplanar Force System.   The lines of action of all the forces in the system lie in the same plane but do not pass through a common point, as shown in Fig. 1–3b.

Spatial Force System.   The lines of action of all the forces in the system do not lie in the same plane. Spatial force systems can be either concurrent (Fig. 1–3c) or nonconcurrent (Fig. 1–3d).

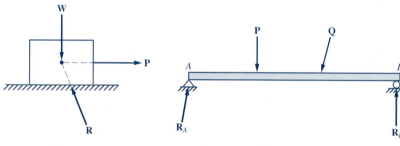

(a) Concurrent coplanar force system   (b) Nonconcurrent coplanar force system

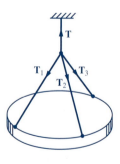   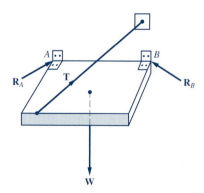

(c) concurrent spatial force system   (d) Nonconcurrent spatial force system

FIGURE 1–3

1–6
NEWTON'S LAWS

In the latter part of the seventeenth century, Sir Isaac Newton (1642–1727) formulated three laws governing the equilibrium and motion of a particle (a point mass). *Newton's laws* now form the foundation of *Newtonian mechanics.*

First Law. *A particle remains at rest or continues to move along a straight line with a constant velocity if the force acting on it is zero.*

Second Law. *If the force acting on a particle is not zero, the particle accelerates (changes velocity with respect to time) in the direction of the force, and the magnitude of the acceleration (the rate of change of velocity per unit time) is proportional to the magnitude of the force.*

Third Law. *The forces of action and reaction between interactive bodies always have the same magnitudes and opposite directions.*

The first law deals with the condition for equilibrium of a particle. It provides the foundation for the study of *statics.* The second law is the basis for the study of *dynamics* and it may be expressed in the vector equation as:

$$\mathbf{F} = m\mathbf{a} \tag{1–1}$$

where $\mathbf{F}$ = the force acting on the particle

m = the mass of the particle

$\mathbf{a}$ = the acceleration of the particle caused by the force

The third law is important for both statics and dynamics. It states that active and reactive forces always exist in equal and opposite pairs. For example, the downward weight of a body resting on a table is accompanied by an upward reaction of the same magnitude exerted by the table on the body. The third law applies equally well to forces exerted through remote action. For instance, two magnets always attract each other with equal and opposite forces.

1–7
THE PRINCIPLE OF TRANSMISSIBILITY

The *principle of transmissibility* states that the point of application of a force acting on a rigid body may be placed anywhere along its line of action without altering the conditions of equilibrium or motion of the rigid body. As an example, consider the disabled car shown in Fig. 1–4a. The car is pulled forward by a force $\mathbf{F}$ applied to the front bumper. Using the principle of transmissibility, the pulling force $\mathbf{F}$ may be replaced by an equivalent force $\mathbf{F}'$ acting on the rear bumper (Fig. 1–4b). We see that the state of motion of the car is unaffected, and all the other external forces, $\mathbf{W}$, $\mathbf{R}_1$, and

R_2, acting on the car remain unchanged whether the car is pulled by **F** or pushed by **F'**, as long as the two forces have the same magnitude, have the same direction, and act along the same line.

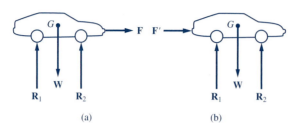

(a)   (b)

FIGURE 1–4

One must be aware, however, that the internal effect of a force on a body is dependent on its point of application. Hence, the principle of transmissibility does not apply if our concern is the internal force or deformation. For example, the force **F** applied at point *B* of the bracket in Fig. 1–5a cannot be placed at point *C* (Fig. 1–5b) if our concern is the internal force or the deflection in the part labeled *BC*.

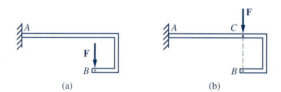

(a)   (b)

FIGURE 1–5

1–8
SYSTEMS OF UNITS

Equation 1–1 involves the measurements of length, time, mass, and force. Hence, the units of these quantities must satisfy Equation 1–1 and they cannot be chosen independently. If three *base units* are chosen, the fourth unit must be derived from Equation 1–1 in terms of the base units.

Currently, two systems of units are used in engineering practices in the United States: the *U.S. customary system of units* (the old English system) and the *International System* or *SI units* (from the French *Système International d'Unités*). The SI units have now been widely adopted throughout the world. In the United States, the U.S. customary units are gradually being replaced by the SI units. During the transition years, however, engineers must be familiar with both systems.

U.S. Customary Units. The three base units in this system are:

> Length: foot (ft)
> Force: pound (lb)
> Time: second (s)

The base unit pound is dependent on the gravitational attraction of the earth, which varies with location; hence, this system is also referred to as a *gravitational system.*

The unit of mass, called the *slug,* is a derived unit. From Equation 1–1, we have:

$$m = F/a$$

$$1 \text{ slug} = (1 \text{ lb})/(1 \text{ ft/s}^2) = 1 \text{ lb} \cdot \text{s}^2/\text{ft}$$

Thus, the slug is a derived unit for mass expressed in the base units as $\text{lb} \cdot \text{s}^2/\text{ft}$. The slug is commonly used in dynamics. In this book, it is rarely used.

The weight of a body (due to the gravitational attraction) causes the body to accelerate at 32.2 ft/s^2 on the surface of the earth. This quantity is usually denoted by g and is called the *gravitational acceleration.* Equation 1–1 may be rewritten as:

$$W = mg \qquad\qquad (1\text{–}2)$$

From this equation, the weight of a 1-slug mass on the surface of the earth is:

$$W = mg$$
$$= (1 \text{ slug})(32.2 \text{ ft/s}^2)$$
$$= (1 \text{ lb} \cdot \text{s}^2/ \text{ ft})(32.2 \text{ ft/s}^2)$$
$$= 32.2 \text{ lb}$$

Other U.S. customary units frequently encountered in engineering practice are:

$$1 \text{ mile (mi)} = 5280 \text{ ft}$$
$$1 \text{ inch (in.)} = \frac{1}{12} \text{ ft}$$
$$1 \text{ kilo-pound (kip)} = 1000 \text{ lb}$$
$$1 \text{ U.S. ton (ton)} = 2000 \text{ lb}$$
$$1 \text{ minute (min)} = 60 \text{ s}$$
$$1 \text{ hour (h)} = 60 \text{ min} = 3600 \text{ s}$$

SI Units. The three base SI units are:

> Length: meter (m)
> Mass: kilogram (kg)
> Time: second (s)

The SI units are an *absolute system* because the three base units chosen are independent of the location where the measurements are made.

The unit of force, called the *newton* (N), is a derived unit expressed in terms of the three base units. From Equation 1–1:

$$F = ma$$

$$1 \text{ N} = (1 \text{ kg}) \cdot (\text{m/s}^2) = 1 \text{ kg} \cdot \text{m/s}^2$$

Thus, the unit of force, newton, is equivalent to $\text{kg} \cdot \text{m/s}^2$. With gravitational acceleration, $g = 9.81 \text{ m/s}^2$, the weight of a 1-kg mass on the surface of the earth is:

$$W = mg$$

$$= (1 \text{ kg})(9.81 \text{ m/s}^2)$$

$$= 9.81 \text{ kg} \cdot \text{m/s}^2 = 9.81 \text{ N}$$

When a quantity is either very large or very small, the units may be modified by using a *prefix*. Commonly used SI prefixes are shown in Table 1–1. The following are some typical examples of the use of prefixes in mechanics:

$$1 \text{ kg} = 1000 \text{ g}$$

$$1 \text{ km} = 10^3 \text{ m} = 1000 \text{ m}$$

$$1 \text{ m} = 1000 \text{ mm}$$

$$1 \text{ kN} = 10^3 \text{ N} = 1000 \text{ N}$$

$$1 \text{ Mg (metric ton)} = 10^3 \text{ kg} = 1000 \text{ kg}$$

Unit reductions with quantities involving prefixes can be done simply by moving the decimal point three places to the right (multiply by 10^3) or to the left (multiply by 10^{-3}). For example:

TABLE 1–1   Commonly Used SI Prefixes

Prefix	Symbol	Multiplication Factor	
giga-	G	10^9	= 1 000 000 000[a]
mega-	M	10^6	= 1 000 000
kilo-	k	10^3	= 1 000
centi- [b]	c	10^{-2}	= 0.01
milli-	m	10^{-3}	= 0.001
micro-	µ	10^{-6}	= 0.000 001
nano-	n	10^{-9}	= 0.000 000 001

[a] Use a space rather than a comma to separate numbers in groups of three, counting from the decimal point in both directions. Spaces may be omitted for numbers of four digits. Commas are not used for this purpose because they are used as decimal points in some countries.
[b] The use of *centi* should be avoided in general except for certain measurements of areas and volumes.

$$1.59 \text{ km} = 1590 \text{ m} \quad = 1.59 \times 10^3 \text{ m}$$

$$4.83 \text{ Mg} = 4830 \text{ kg} \quad = 4.83 \times 10^3 \text{ kg}$$

$$75.4 \text{ mm} = 0.0754 \text{ m} = 75.4 \times 10^{-3} \text{ m}$$

With the exception of the base unit kg, in general a prefix must be used as the leading symbol in a derived unit. For example, a spring constant (force per unit length of deformation) can be expressed as kN/m, but not as N/mm. A moment (force multiplied by distance) can be expressed as $kN \cdot m$, but not as $N \cdot km$ or $N \cdot mm$.

1–9
UNIT CONVERSION

Changing units within a system is call *unit reduction*. Changing units from one system to another system is called *unit conversion*. In this book, problems are solved in the same system of units as in the given data. Therefore, unit conversion is not needed in problem solutions. In actual engineering applications, however, unit conversions are sometimes necessary. The following conversion factors are useful:

$$1 \text{ ft} = 0.3048 \text{ m}$$

$$1 \text{ slug} = 14.59 \text{ kg}$$

$$1 \text{ lb} = 4.448 \text{ N}$$

For a more extensive list of conversion factors, refer to Table 1–2, p. 12, which lists the U.S. customary units for quantities frequently used in mechanics and their SI equivalents.

──────── **EXAMPLE 1–1** ────────

Convert a velocity of 30 mph into its equivalent value in m/s.

Solution. The unit mile must be reduced to feet first before being converted to meters. Thus:

$$30 \text{ mph} = \left(\frac{30 \text{ mi}}{h}\right)\left(\frac{5280 \text{ ft}}{1 \text{ mi}}\right)\left(\frac{0.3048 \text{ m}}{1 \text{ ft}}\right)\left(\frac{1 \text{ h}}{3600 \text{ s}}\right)$$

$$= 13.41 \text{ m/s} \qquad\qquad \Leftarrow \textbf{Ans.}$$

The conversion factors used above are each equal to unity because the numerator and denominator of each factor represent the same measurement. The value of a quantity is not altered if it is multiplied by factors of unity.

Using the conversion factor for mph and m/s listed in Table 1–2 (1 mph = 0.4470 m/s), the result can be obtained more conveniently as

$$30 \text{ mph} = (30 \text{ mi/h})\left(\frac{0.4470 \text{ m/s}}{1 \text{ mi/h}}\right)$$

$$= 13.41 \text{ m/s}$$

───── **EXAMPLE 1–2** ─────

Convert a moment 10 N · m to an equivalent value in the appropriate U.S. customary units.

Solution. The given units are force multiplied by length. Thus, the corresponding U.S. customary units are lb · ft:

$$10 \text{ N} \cdot \text{m} = (10 \text{ N} \cdot \text{m})\left(\frac{1 \text{ lb}}{4.448 \text{ N}}\right)\left(\frac{1 \text{ ft}}{0.3048 \text{ m}}\right)$$

$$= 7.38 \text{ lb} \cdot \text{ft} \qquad\qquad \Leftarrow \textbf{Ans.}$$

Using the conversion factor for moment, N · m and lb · ft, listed in Table 1–2 (1 lb · ft = 1.356 N · m), we get

$$10 \text{ N} \cdot \text{m} = (10 \text{ N} \cdot \text{m})\left(\frac{1 \text{ lb} \cdot \text{ft}}{1.356 \text{ N} \cdot \text{m}}\right)$$

$$= 7.38 \text{ lb} \cdot \text{ft}$$

TABLE 1–2　U.S. Customary Units and Their SI Equivalents

Quantity	U.S. Customary Unit	SI Equivalent
Length	ft	0.3048 m
	in.	25.40 mm
	mi	1.609 km
Mass	slug	14.59 kg
Force	lb	4.448 N
	kip	4.448 kN
Area	ft^2	0.0929 m^2
	in.^{2}	0.6452×10^{-3} m^2
Volume	ft^3	0.02832 m^3
	in.^{3}	16.39×10^{-6} m^3
Velocity	ft/s	0.3048 m/s
	mi/h (mph)	0.4470 m/s
	mi/h (mph)	1.609 km/h
Acceleration	ft/s^2	0.3048 m/s^2
Moment of a force	lb · ft	1.356 N · m
	lb · in.	0.1130 N · m
Pressure or stress	lb/ft^2 (psf)	47.88 Pa (pascal or N/m^2)
	lb/in.^{2} (psi)	6.895 kPa (kN/m^2)
Spring constant	lb/ft	14.59 N/m
	lb/in.	175.1 N/m
Load intensity	lb/ft	14.59 N/m
	kip/ft	14.59 kN/m
Area moment of inertia	in.^{4}	0.4162×10^{-6} m^4
Work or energy	lb · ft	1.356 J (joule or N · m)
Power	lb · ft/s	1.356 W (watt or N · m/s)
	hp (1 horsepower = 550 ft · lb/s)	745.7 W (watt or N · m/s)

1-10
CONSISTENCY OF UNITS IN AN EQUATION

An equation relating several physical quantities is called a physical equation. Every term in a physical equation must be *dimensionally homogeneous*; that is, each term must be reduced to the same units.

When substituting the values of known quantities into an equation, it is extremely important that the units in the quantities be consistent. *Students should form the habit of carrying units with all quantities when substituting into an equation and making sure that the result is in the correct units.* The following examples illustrate the proper procedure.

────── **EXAMPLE 1-3** ──────

Using the equation $s = vt$, find the distance s in feet traveled by a car at a constant speed $v = 60$ mi/h for a period of $t = 30$ s.

Solution. Before the speed v and the period t are substituted in $s = vt$, we must first reduce the units of v from mi/h to ft/s. Thus:

$$v = \left(\frac{65 \text{ mi}}{h}\right)\left(\frac{5280 \text{ ft}}{1 \text{ mi}}\right)\left(\frac{1 \text{ h}}{3600 \text{ s}}\right)$$

$$= 95.33 \text{ ft/s}$$

$$s = vt$$

$$= \left(95.33 \frac{\text{ft}}{\text{s}}\right)(30 \text{ s})$$

$$= 2860 \text{ ft} \qquad \Leftarrow \textbf{Ans.}$$

────── **EXAMPLE 1-4** ──────

Solve the equation $mv = Ft$ for the velocity v and find its value if the force $F = 2.56$ N, the time $t = 16.8$ s, and the mass $m = 1.89$ kg.

Solution. Dividing both sides of the equation by m, we get:

$$v = \frac{Ft}{m}$$

Substituting the given quantities into the right-hand side of the equation, we get:

$$v = \frac{Ft}{m}$$

$$= \frac{(2.56 \text{ N})(16.8 \text{ s})}{(1.89 \text{ kg})}$$

At first glance, it seems that the units lead nowhere. But we note that the unit newton (N) is a derived unit that can be expressed in the base units as $kg \cdot m/s^2$. Thus, we write:

$$v = \frac{(2.56 \text{ kg} \cdot \text{m/s}^2)(16.8 \text{ s})}{(1.89 \text{ kg})}$$

$$= 22.8 \text{ m/s} \qquad \Leftarrow \textbf{Ans.}$$

Note that as long as consistent units (kg, m, s, and N) are used, the result for v is in the units of m/s.

EXAMPLE 1–5

The elongation e of a rod subjected to an axial load F can be computed from the equation

$$e = \frac{FL}{AE}$$

where F = axial load

L = length of the rod

A = cross-sectional area of the rod

E = modulus of elasticity of the material

Find the elongation of a steel rod subjected to an axial load of 15 kips. The rod has a length of 4 ft 6 in., and a diameter of 1 in., and the modulus of elasticity of steel is $E = 30 \times 10^6$ lb/in.^{2}.

Solution. For this problem, it is convenient to use lb and in. units. Before substituting the quantities into the equation, we must reduce each quantity to the lb and in. units:

$$F = 15 \text{ kips} = (15 \text{ kips})\left(\frac{1000 \text{ lb}}{1 \text{ kip}}\right) = 15\,000 \text{ lb}$$

$$L = 4 \text{ ft 6 in.} = (4.5 \text{ ft})\left(\frac{12 \text{ in.}}{1 \text{ ft}}\right) = 54.0 \text{ in.}$$

$$A = \frac{\pi}{4}(1 \text{ in.})^2 = 0.754 \text{ in.}^2$$

$$E = 30 \times 10^6 \text{ lb/in.}^2$$

Now these quantities, together with their units, are substituted into the equation. The reason for including the units in the substitution is to make sure that all the units are consistent and that the result is dimensionally correct.

$$e = \frac{FL}{AE}$$

$$= \frac{(15\,000 \text{ lb})(54.0 \text{ in.})}{(0.785 \text{ in.}^2)(30 \times 10^6 \text{ lb/in.}^2)}$$

$$= 0.0344 \text{ in.} \qquad \Leftarrow \textbf{Ans.}$$

1-11
RULES FOR NUMERICAL COMPUTATIONS

Approximate Numbers. Although some numbers that we encounter in engineering computations are exact numbers, most numbers are approximate. *Exact numbers* are either derived from definition or obtained by counting. For example, one hour has exactly 60 minutes by definition. One inch is defined to equal 25.4 mm exactly. An automobile has four wheels by counting. *Approximate numbers* are usually obtained through some kind of measurement. For example, the distance between two points on the ground is measured to be 235.7 ft; the voltage of a house current is measured to be 115 volts.

Approximate numbers are usually written with a decimal and often include zeros that serve as placeholders. For example, the zeros in the numbers 7400 and 0.0057 are used as placeholders. The zeros in the numbers 2005 and 0.708 indicate that the values at those digits are zero, so they are not used merely as placeholders.

Significant Digits. Except for the zeros used as placeholders, all the other digits in an approximate number are considered *significant digits.* For example, the numbers 176, 0.587, 1350, 3050, 0.00408 are of three significant digits each.

Accuracy and Precision. The *accuracy* of a number refers to its number of significant digits. For example, the numbers 1570, 60.9, and 0.0805 are all accurate to three significant digits. The *precision* of a number refers to the decimal position of the last significant digit. For example, the number 1.35 is precise to the nearest hundredths (two decimal places), and the number 0.745 is precise to the nearest thousandths (three decimal places).

──────── **EXAMPLE 1-6** ────────

Determine the number of significant digits in the following groups of numbers:
(a)   1240, 254 000, 0.348, 0.005 86
(b)   304.3, 28.06, 7003, 1.704
(c)   22.40, 2.890, 8.000, 4.560×10^8

Solution
(*a*) The numbers 1240, 254 000, 0.348, and 0.005 86 each have three significant digits. Note that we count only the nonzero digits as significant. All the zeros in these numbers are used as placeholders.

(*b*) The numbers 304.3, 28.06, 7003, and 1.704 each have four significant digits. Note that the zeros in these numbers are significant because they are not used here as placeholders. They are used to indicate that, in those digits, the values are zero.

(*c*) The numbers 22.40, 2.890, 8.000, and 4.560×10^8 each have four significant digits. The zeros at the end of each number after the decimal point are used to show that these digits are indeed zero. These zeros should not be written unless they are significant. Note that the number 34 000 has only two significant digits. If the number is known to have four significant digits, it should be written in scientific notation* as 3.400×10^4.

* The *scientific notation* of a number is expressed as a product of a number between 1 and 10 and an integer power of 10. For example, the numbers 43 200 and 0.000 068 3 can be expressed in scientific notation as 4.32×10^4 and 6.83×10^{-5}, respectively.

EXAMPLE 1–7

A steel plate 1.25 in. thick is coated with a thin layer of paint 0.014 in. thick. Of these two values of thickness, which one has a greater accuracy and which one has a greater precision?

Solution. The number 1.25 has three significant digits, while the number 0.014 has only two significant digits. Therefore, the thickness of the plate, 1.25 in., has a greater accuracy. On the other hand, the number 1.25 is precise to the nearest hundredths, and the number 0.014 is precise to the nearest thousandths; therefore, the thickness of the paint has a greater precision.

Rules for Numerical Computation. When calculations are performed on approximate numbers, the results must be expressed with the proper number of digits. It is a common mistake to express the final result in a greater accuracy or precision than is warranted by the given data. Therefore, it is essential that the following rules are observed.

Rule 1 When approximate numbers are *multiplied or divided,* the result is expressed with the *same accuracy as the least accurate number.*

Rule 2 When approximate numbers are *added or subtracted,* the result is expressed with the *same precision as the least precise number.*

EXAMPLE 1–8

Calculate the area of a circle from the following expression:

$$\frac{\pi (2.683)^2}{4}$$

Solution. The number 4 in the expression is an exact number, so it does not limit the accuracy of the result. The number π is a built-in feature in a calculator and accurate to at least 8 significant digits. Thus, the number 2.683, with four significant digits, is the least accurate number in the expression. Therefore, from rule 1, the final result should be expressed in four significant digits. Thus,

$$\frac{\pi\,(2.683)^2}{4} = 5.654 \qquad \Leftarrow \textbf{Ans.}$$

which is rounded off from 5.6536.

————— **EXAMPLE 1–9** —————

Find the sum of the following approximate numbers:

$$12.36 + 26.53 + 4.782 + 1.203 + 204.5$$

Solution. The least precise number in the expression is 204.5, which is precise to the tenths. According to rule 2, the sum should be rounded off to the tenths also. Thus,

$$\text{The sum} = 249.4 \qquad \Leftarrow \textbf{Ans.}$$

which is rounded off from 249.375.

————— **EXAMPLE 1–10** —————

Evaluate the following expression:

$$\frac{(27.83)(4.756)}{9.843 - 2.78}$$

Solution. After all the numbers are entered, the calculator displays a result of 18.7398. To see how many significant digits should be expressed in the result, we first note that in the denominator, the difference of the two numbers must be precise to the hundredths because the least precise number (2.78) is precise to the hundredths (rule 2). That number gives the denominator an accuracy of three significant digits and makes it the least accurate number. Then, according to rule 1, the final result should be accurate to three significant digits. Thus:

$$\text{The result} = 18.7 \qquad \Leftarrow \textbf{Ans.}$$

which is rounded off from 18.74.

The following rule of thumb is commonly used when the given data are of unknown accuracy: *Round off a result to four significant digits if its first significant digit is 1; round off a result to three significant digits if its first significant digit is other than 1.* For example, a result of 136.25 N is rounded off to 136.3 N, and a result of 2864 lb is rounded off to 2860 lb. However, all intermediate computations must be carried through with one more significant digit than in the final results. When working on a calculator, retain all digits and round only the final answer. Unless otherwise indicated, the data given in a problem is assumed to be accurate to three or four significant digits. For example, a force of 120 lb is to be read as 120.0 lb, and a length of 5.6 m is to be read as 5.60 m.

1–12
A BRIEF REVIEW OF MATHEMATICS

Mathematics Used in Mechanics. This book requires a knowledge of only basic mathematics, including arithmetic, algebra, geometry, and trigonometry. This section is designed to give students a brief review of some important mathematical skills useful in the study of mechanics.

Right Triangles. A right triangle is a closed, three-sided figure that has a right angle (angle equals 90°). The side opposite the right angle is called the *hypotenuse.* Figure 1–6 shows a right triangle, with the right angle at *C*. The sides opposite to angles *A, B,* and *C* are denoted by *a, b,* and *c,* respectively. Side *c* is the hypotenuse. With respect to angle *A, a* is the *opposite side* and *b* is the *adjacent side.* (With respect to angle *B, a* is the adjacent side and *b* is the opposite side.)

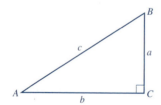

FIGURE 1–6

Since the sum of the three interior angles of a triangle is 180° and *C* = 90°, we have:

$$A + B = 90° \tag{1–3}$$

The Pythagorean theorem. The hypotenuse of a right triangle squared is equal to the sum of the other two sides squared; that is:

$$c^2 = a^2 + b^2 \tag{1–4}$$

Trigonometry of Right Triangles. Trigonometry relates the lengths of the sides of a right triangle by means of trigonometric functions. The three trigonometric functions used in this book are the sine, cosine, and tangent functions. These functions are commonly abbreviated as sin, cos, and tan, respectively. The three trigonometric functions for angle *A* are as follows:

$$\sin A = \frac{\text{opposite side}}{\text{hypotenuse}} = \frac{a}{c} \qquad (1\text{–}5)$$

$$\cos A = \frac{\text{adjacent side}}{\text{hypotenuse}} = \frac{b}{c} \qquad (1\text{–}6)$$

$$\tan A = \frac{\text{opposite side}}{\text{adjacent side}} = \frac{a}{b} \qquad (1\text{–}7)$$

Since only the ratio of sides is necessary to define the trigonometric functions, the values of trigonometric functions are constant for a given angle, regardless of the size of the triangle. The trigonometric functions of an angle can be obtained by using a scientific calculator.

If two sides of a right triangle or one side and an *acute angle* (angle less than 90°) of a right triangle are known, the other unknown elements can be determined by using the Pythagorean theorem and the trigonometric functions. These processes are illustrated in the following examples.

EXAMPLE 1–11

A 16-ft ladder leans against a wall, forming an angle of 75° with the floor (see Fig. E1–11). Determine the height h that the ladder reaches on the wall.

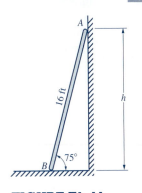

Solution. Using the definition of the sine function, we write:

$$\sin 75^\circ = \frac{h}{16 \text{ ft}}$$

From which

FIGURE E1–11

$$h = (16 \text{ ft}) \sin 75^\circ$$
$$= 15.45 \text{ ft} \qquad \Leftarrow \textbf{Ans.}$$

EXAMPLE 1–12

A symmetrical roof has the dimensions indicated in Fig. E1–12(1). Determine the height h and the angle of inclination A of the roof.

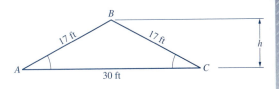

FIGURE E1–12(1)

Solution. From B draw a line BD perpendicular to AC. [see Fig. E1–12(2)]. Line BD bisects AC; thus, $AD = AC/2$ $= 15$ ft. Triangle ADB is a right triangle. By the Pythagorean theorem:

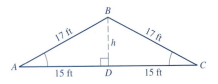

FIGURE E1–12(2)

$$h^2 = 17^2 - 15^2 = 64$$

$$h = 8 \text{ ft} \qquad \Leftarrow \textbf{Ans.}$$

By the definition of the cosine function:

$$\cos A = \frac{15}{17}$$

Thus,

$$A = \cos^{-1}\left(\frac{15}{17}\right)$$
$$= 28.1° \qquad \Leftarrow \textbf{Ans.}$$

where $\cos^{-1}(15/17)$ stands for the inverse cosine function of the ratio 15/17.

Oblique Triangles. An *oblique triangle* is a triangle in which none of the interior angles is equal to 90°. Figure 1–7a shows an oblique triangle with three acute angles. Figure 1–7b shows an oblique triangle with an *obtuse angle B,* which is greater than 90°. In both cases, *a, b,* and *c* represent the three sides of the oblique triangle opposite to the angles *A, B,* and *C,* respectively.

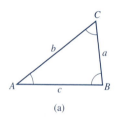
(a)

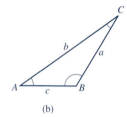
(b)

FIGURE 1–7

The sum of the three interior angles in a triangle is 180°; that is:

$$A + B + C = 180° \qquad (1\text{–}8)$$

The Law of Sines. *The ratio of any side of a triangle to the sine function of its opposite angle is a constant:*

$$\frac{a}{\sin A} = \frac{b}{\sin B} = \frac{c}{\sin C} \qquad (1\text{–}9)$$

The Law of Cosines. *The square of any side of a triangle is equal to the sum of the squares of the other two sides minus twice the product of the two sides multiplied by the cosine function of the angle between them:*

$$a^2 = b^2 + c^2 - 2bc \cos A \qquad (1\text{–}10a)$$

$$b^2 = c^2 + a^2 - 2ca \cos B \qquad (1\text{–}10b)$$

$$c^2 = a^2 + b^2 - 2ab \cos C \qquad (1\text{–}10c)$$

The following examples demonstrate the use of these laws in solving problems that involve oblique triangles.

———————— **EXAMPLE 1–13** ————————————————————————————————

A 20-ft ladder AB leaning on a wall makes an angle of 25° with the wall, as shown in Fig. E1–13. The ground is 15° from the horizontal. Determine the height h that the ladder reaches on the wall.

Solution. In triangle ABC, the interior angle at B is 25°. The other two interior angles at A and C are:

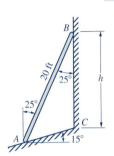

$$A = 90° - 25° - 15°$$

$$= 50°$$

$$C = 180° - 25° - 50°$$

$$= 105°$$

FIGURE E1–13

Applying the law of sines to the triangle ABC, we write:

$$\frac{h}{\sin 50°} = \frac{20 \text{ ft}}{\sin 105°}$$

From which we get:

$$h = \frac{(20 \text{ ft}) \sin 50°}{\sin 105°}$$

$$= 15.86 \text{ ft} \qquad\qquad \Leftarrow \textbf{Ans.}$$

———————— **EXAMPLE 1–14** ————————————————————————————————

For the bracket support shown in Fig. E1–14, rod AB is 500 mm long and makes an angle of 60° with the vertical. The vertical distance between the supports is 600 mm. Compute the length of rod BC.

Solution. In the oblique triangle ABC, the interior angle at A is

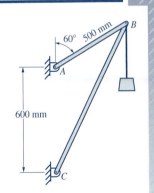

$$A = 180° - 60°$$

$$= 120°$$

FIGURE E1–14

Now the oblique triangle ABC has two known sides and a known angle between the two sides. Applying the law of cosines to the triangle we write

$$(BC)^2 = (500)^2 + (600)^2 - 2(500)(600) \cos 120°$$

From which, we obtain:

$$BC = 954 \text{ mm} \qquad \Leftarrow \textbf{Ans.}$$

EXAMPLE 1–15

A lamp is suspended between two walls with cables AB and AC, as shown in Fig. E1–15(1). Compute the angle α between the two cables.

Solution. Refer to Fig. E1–15(2). To find the distance BC, we connect two points, B and C, and draw line CD perpendicular to the wall. Triangle BCD is a right triangle, as shown in Fig. E1–15(2). From the Pythagorean theorem, we get:

$$BC = \sqrt{(0.5 \text{ m})^2 + (2.5 \text{ m})^2}$$
$$= 2.55 \text{ m}$$

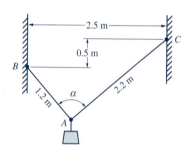

FIGURE E1–15(1)

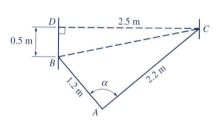

FIGURE E1–15(2)

The oblique triangle ABC has three known sides. Applying the law of cosines to the triangle, we write:

$$(2.55 \text{ m})^2 = (1.2 \text{ m})^2 + (2.2 \text{ m})^2 - 2(1.2 \text{ m})(2.2 \text{ m}) \cos \alpha$$

From which, we get

$$\cos \alpha = -0.04214$$
$$\alpha = \cos^{-1}(-0.04214)$$
$$= 92.4° \qquad \Leftarrow \textbf{Ans.}$$

Simultaneous Equations. In solving two-dimensional equilibrium problems, sometimes we need to solve two simultaneous linear equations containing two unknowns. Solving three simultaneous linear equations containing three unknowns occurs in the solutions of three-dimensional equilibrium problems. The following examples illustrate the solution of two and three simultaneous linear equations by the methods of substitution and addition or subtraction.

──────── **EXAMPLE 1–16** ──

In solving a two-dimensional equilibrium problem, the following equations are obtained:

$$P \cos 20° - Q \cos 40° = 0 \qquad\qquad\text{(a)}$$

$$P \sin 20° + Q \sin 40° = 100 \text{ lb} \qquad\qquad\text{(b)}$$

where P and Q are two unknown forces. Solve the equations for the two unknowns.

Solution. Two methods are presented in the following.

Method 1: Elimination by Substitution. Solving Equation (a) for Q, we get:

$$Q = P \cos 20° / \cos 40° = 1.227P \qquad\qquad\text{(c)}$$

Substituting into Equation (b) gives:

$$P (\sin 20° + 1.227 \sin 40°) = 100 \text{ lb}$$

$$P = 88.5 \text{ lb} \qquad\qquad \Leftarrow \textbf{Ans.}$$

Substituting into Equation (c), we obtain:

$$Q = 108.5 \text{ lb} \qquad\qquad \Leftarrow \textbf{Ans.}$$

Method 2: Elimination by Addition and Subtraction. The unknown Q can be eliminated by adding the two equations if the coefficients of Q are reduced to the same value but opposite in sign. We can reduce the coefficient of Q in Equation (a) to -1 if *every term* of the equation is divided by $\cos 40°$:

$$(a) \div \cos 40°: \qquad P \cos 20° / \cos 40° - Q = 0$$

$$1.227P - Q = 0 \qquad\qquad\text{(d)}$$

Similarly, the coefficient of Q in (b) becomes 1 if *every term* of the equation is divided by $\sin 40°$:

$$(b) \div \sin 40°: \qquad P \sin 20° / \sin 40° + Q = (100 \text{ lb}) / \sin 40°$$

$$0.5321P + Q = 155.6 \text{ lb} \qquad\qquad\text{(e)}$$

If we add Equations (d) and (e), Q can be eliminated:

$$(d) + (e): \qquad 1.759P = 155.6 \text{ lb}$$

$$P = 88.5 \text{ lb} \qquad\qquad \Leftarrow \textbf{Ans.}$$

Substituting in Equation (d), we obtain:

$$Q = 108.5 \text{ lb} \qquad\qquad \Leftarrow \textbf{Ans.}$$

━━━ **EXAMPLE 1–17** ━━━

In solving a three-dimensional equilibrium problem, the following equations are obtained:

$$0.429F_1 - 0.549F_2 - 0.570F_3 = 0 \qquad\qquad \text{(a)}$$

$$0.857F_1 - 0.824F_2 + 0.684F_3 = 400 \text{ N} \qquad\qquad \text{(b)}$$

$$0.286F_1 + 0.1374F_2 - 0.456F_3 = 0 \qquad\qquad \text{(c)}$$

Solve the equations for the three unknown forces: F_1, F_2, and F_3.

Solution. Note that all three equations contain the three unknowns. We must eliminate one of the three unknowns (say, F_3) from the three equations. First, the coefficients of F_3 in the three equations are reduced to either $+1$ or -1 as follows:

$$\text{(a)} \div 0.570: \quad 0.753F_1 - 0.963F_2 - F_3 = 0 \qquad\qquad \text{(d)}$$

$$\text{(b)} \div 0.684: \quad 1.253F_1 - 1.205F_2 + F_3 = 585 \text{ N} \qquad\qquad \text{(e)}$$

$$\text{(c)} \div 0.456: \quad 0.627F_1 + 0.301F_2 - F_3 = 0 \qquad\qquad \text{(f)}$$

The unknown F_3 can be eliminated by addition or subtraction:

$$\text{(d)} + \text{(e)}: \quad 2.006F_1 - 2.168F_2 = 585 \text{ N} \qquad\qquad \text{(g)}$$

$$\text{(d)} - \text{(f)}: \quad 0.126F_1 - 1.264F_2 = 0 \qquad\qquad \text{(h)}$$

The unknown F_2 can be eliminated from the above two equations by:

$$\text{(g)} \div 2.168: \quad 0.9253F_1 - F_2 = 269.8 \text{ N} \qquad\qquad \text{(i)}$$

$$\text{(h)} \div 1.264: \quad 0.0997F_1 - F_2 = 0 \qquad\qquad \text{(j)}$$

$$\text{(i)} - \text{(j)}: \qquad\quad 0.8256F_1 = 269.8 \text{ N} \qquad\qquad \text{(k)}$$

From Equation (k), we get:

$$F_1 = 327 \text{ N} \qquad\qquad\qquad \Leftarrow \textbf{Ans.}$$

Substituting in Equation (j) yields:

$$F_2 = 32.6 \text{ N} \qquad\qquad\qquad \Leftarrow \textbf{Ans.}$$

Substituting in Equation (d) yields:

$$F_3 = 215 \text{ N} \qquad\qquad\qquad \Leftarrow \textbf{Ans.}$$

Cramer's Rule for Two Equations. The solution to two linear equations containing two unknowns can be expressed in terms of *determinants of the second order,* defined as follows:

$$\begin{vmatrix} a_1 & b_1 \\ a_2 & b_2 \end{vmatrix} = a_1b_2 - a_2b_1 \qquad (1\text{--}11)$$

Equation 1–11 is a square array of numbers with two rows and two columns, enclosed by vertical bars. Each number in the array is called an *element*. Elements a_1 and b_1 are in the *first row*; elements a_2 and b_2 are in the *second row*; elements a_1 and a_2 are in the *first column*; elements b_1 and b_2 are in the second column. The *principal diagonal* is the direction along elements a_1 and b_2. The *secondary diagonal* is the direction along elements a_2 and b_1. The value of a determinant of the second order is *the product of the two elements along the principal diagonal subtracting the product of the two elements along the secondary diagonal.*

Using second-order determinants, the solution to a system of two linear equations with two unknowns in the form

$$a_1x + b_1y = k_1 \qquad (1\text{--}12)$$

$$a_2x + b_2y = k_2 \qquad (1\text{--}13)$$

can be written as

$$x = \frac{D_x}{D} \qquad y = \frac{D_y}{D} \qquad (1\text{--}14)$$

where

$$D = \begin{vmatrix} a_1 & b_1 \\ a_2 & b_2 \end{vmatrix} \qquad D_x = \begin{vmatrix} k_1 & b_1 \\ k_2 & b_2 \end{vmatrix} \qquad D_y = \begin{vmatrix} a_1 & k_1 \\ a_2 & k_2 \end{vmatrix} \qquad (1\text{--}15)$$

In Equation 1–15, the elements of determinant D are made up of the coefficients of x and y in the two equations. The elements in determinant D_x are essentially the same as those of determinant D, except that the elements in the first column (the coefficients of x) are replaced by the right-hand constant k's. Similarly, the elements in determinant D_y are essentially the same as those of determinant D, except that the elements in the second column (the coefficients of y) are replaced by the right-hand constant k's. The solution given in Equations 1–14 and 1–15 is known as *Cramer's rule.*

──────── EXAMPLE 1–18 ────────

Solve the following equations for x and y by using Cramer's rule.

$$4x + y = 10$$

$$3x - 5y = 19$$

Solution. First, we set up and evaluate determinant D using the coefficients of x and y as elements:

$$D = \begin{vmatrix} 4 & 1 \\ 3 & -5 \end{vmatrix} = 4(-5) - 3(1) = -23$$

Then set up D_x by replacing the elements in the first column of D (the coefficients of x) with the right-hand constants:

$$D_x = \begin{vmatrix} 10 & 1 \\ 19 & -5 \end{vmatrix} = 10(-5) - 19(1) = -69$$

The determinant D_y can be set up by replacing the elements in the second column of D (the coefficients of y) with the right-hand constants:

$$D_y = \begin{vmatrix} 4 & 10 \\ 3 & 19 \end{vmatrix} = 4(19) - 3(10) = 46$$

By Cramer's rule, the solution is

$$x = \frac{D_x}{D} = \frac{-69}{-23} = 3 \qquad \Leftarrow \textbf{Ans.}$$

$$y = \frac{D_y}{D} = \frac{46}{-23} = -2 \qquad \Leftarrow \textbf{Ans.}$$

Cramer's Rule for Three Equations. Cramer's rule can also be applied to three linear equations with three unknowns in the form

$$a_1 x + b_1 y + c_1 z = k_1 \tag{1-16}$$

$$a_2 x + b_2 y + c_2 z = k_2 \tag{1-17}$$

$$a_3 x + b_3 y + c_3 z = k_3 \tag{1-18}$$

The solution to the equations is

$$x = \frac{D_x}{D} \qquad y = \frac{D_y}{D} \qquad z = \frac{D_z}{D} \tag{1-19}$$

where

$$D = \begin{vmatrix} a_1 & b_1 & c_1 \\ a_2 & b_2 & c_2 \\ a_3 & b_3 & c_3 \end{vmatrix} \tag{1-20}$$

$$D_x = \begin{vmatrix} k_1 & b_1 & c_1 \\ k_2 & b_2 & c_2 \\ k_3 & b_3 & c_3 \end{vmatrix} \tag{1-21}$$

$$D_y = \begin{vmatrix} a_1 & k_1 & c_1 \\ a_2 & k_2 & c_2 \\ a_3 & k_3 & c_3 \end{vmatrix} \tag{1-22}$$

$$D_z = \begin{vmatrix} a_1 & b_1 & k_1 \\ a_2 & b_2 & k_2 \\ a_3 & b_3 & k_3 \end{vmatrix} \tag{1-23}$$

These determinants, having three rows and three columns, are called *determinants of the third order.* In Equation 1–20, the elements of the determinant D are made up of the coefficients of x, y, and z in the three equations. In Equation 1–21, the elements in the determinant D_x are essentially the same as those of the determinant D, except that the elements in the first column (the coefficients of x) are replaced by the right-hand constant k's.

Similarly, in Equation 1–22, the elements in the determinant D_y are essentially the same as those of determinant D, except that the elements in the second column (the coefficients of y) are replaced by the right-hand constant k's. And in Equation 1–23, the elements in the determinant D_z are essentially the same as those of determinant D, except that the elements in the third column (the coefficients of z) are replaced by the right-hand constant k's.

The value of a third-order determinant can be computed from

$$\begin{vmatrix} a_1 & b_1 & c_1 \\ a_2 & b_2 & c_2 \\ a_3 & b_3 & c_3 \end{vmatrix} = a_1 b_2 c_3 + a_3 b_1 c_2 + a_2 b_3 c_1 - a_3 b_2 c_1 - a_1 b_3 c_2 - a_2 b_1 c_3 \quad (1\text{–}24)$$

The computation of the right-hand side of Equation 1–24 can be simplified by using the following steps:

1. Duplicate the first and the second columns and place them to the right of the determinant, as shown in the following:

$$\begin{vmatrix} a_1 & b_1 & c_1 \\ a_2 & b_2 & c_2 \\ a_3 & b_3 & c_3 \end{vmatrix} \quad \begin{matrix} a_1 & b_1 \\ a_2 & b_2 \\ a_3 & b_3 \end{matrix}$$

2. Find the product of the three elements along the principal diagonal, and the products of the three elements along the parallel diagonals to the right. The sum of the three products gives the first three terms of the expression in Equation 1–24, as shown in the following:

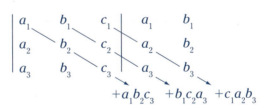

$$+a_1 b_2 c_3 \quad +b_1 c_2 a_3 \quad +c_1 a_2 b_3$$

3. Find the product of the three elements along the secondary diagonal, and the products of the three elements along the parallel diagonals to the right, as shown in the following. Subtracting these three products from the previous sum gives the value of the determinant.

$$+a_3 b_2 c_1 \quad +b_3 c_2 a_1 \quad +c_3 a_2 b_1$$

$$\begin{vmatrix} a_1 & b_1 & c_1 \\ a_2 & b_2 & c_2 \\ a_3 & b_3 & c_3 \end{vmatrix} \quad \begin{matrix} a_1 & b_1 \\ a_2 & b_2 \\ a_3 & b_3 \end{matrix}$$

───── **EXAMPLE 1–19** ─────

Find the value of the following third-order determinant:

$$\begin{vmatrix} 4 & -3 & 2 \\ 0 & -2 & 4 \\ 2 & 0 & -5 \end{vmatrix}$$

Solution.   Following the steps described above, we find

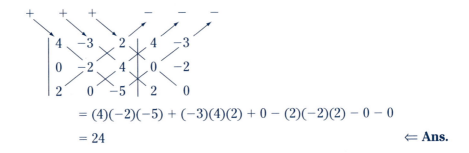

$$= (4)(-2)(-5) + (-3)(4)(2) + 0 - (2)(-2)(2) - 0 - 0$$

$$= 24 \qquad\qquad\qquad\qquad\qquad\qquad\qquad \Leftarrow \textbf{Ans.}$$

──── **EXAMPLE 1-20** ────────────────────────────────────

Solve the following equations for x, y, and z by using Cramer's rule.

$$3x \qquad\;\; + \;\; z = 0 \qquad\qquad\qquad\qquad\qquad (a)$$

$$2x - \;\; y \; + 4z = 8 \qquad\qquad\qquad\qquad (b)$$

$$4x - 3y \; + \;\; z = -7 \qquad\qquad\qquad\qquad (c)$$

Solution. First, we set up and evaluate determinant D using the coefficients of x, y, and z as elements:

$$D = \begin{vmatrix} 3 & 0 & 1 \\ 2 & -1 & 4 \\ 4 & -3 & 1 \end{vmatrix} \begin{matrix} 3 & 0 \\ 2 & -1 \\ 4 & -3 \end{matrix}$$

$$= (3)(-1)(1) + 0 + (1)(2)(-3) - (4)(-1)(1) - (-3)(4)(3) - 0$$

$$= 31$$

Replacing the elements of the first column of D (the coefficient of x) with the right-hand constants, we get

$$D_x = \begin{vmatrix} 0 & 0 & 1 \\ 8 & -1 & 4 \\ -7 & -3 & 1 \end{vmatrix} \begin{matrix} 0 & 0 \\ 8 & -1 \\ -7 & -3 \end{matrix}$$

$$= 0 + 0 + (1)(8)(-3) - (-7)(-1)(1) - 0 - 0$$

$$= -31$$

By Cramer's rule, we find

$$x = \frac{D_x}{D} = \frac{-31}{31} = -1 \qquad\qquad\qquad \Leftarrow \textbf{Ans.}$$

Substituting into Equation (a), we find

$$z = -3x = -3(-1) = 3 \qquad\qquad\qquad \Leftarrow \textbf{Ans.}$$

Substituting into Equation (b), we find

$$y = 2x + 4z - 8$$
$$= 2(-1) + 4(3) - 8 = 2 \qquad \Leftarrow \textbf{Ans.}$$

1–13
GENERAL PROCEDURE FOR PROBLEM SOLUTION

Extensive applications of statics and strength of materials are based on a few simple principles. The most effective way of learning this subject is to solve problems of different levels of complexity. The following general procedure is helpful:

1. Read the problem carefully. Identify the given data and the unknown quantities to be determined.
2. Make a neat sketch showing all the quantities involved. For some problems, it may be helpful to tabulate the given data and the computed results.
3. Apply the relevant principles and express the physical conditions in mathematical form. The solution must be based on the principles and theorems presented in the text and must be executed in a logical manner.
4. The equations obtained must be dimensionally homogeneous. Values in consistent units must be used for substitution. The answer obtained must be rounded off to the proper degree of accuracy or precision.
5. Use your common sense and judgment to determine if the answer obtained is reasonable. In some problems, there are conditions in which answers can be checked. If such conditions are available, always use them to check the answers.
6. The engineering profession requires work that meets high standards. Students preparing to enter an engineering career must present their work in a neat and organized fashion.

1–14
SUMMARY

Forces. *Mechanics* is a physical science that studies the effects of forces. Forces are vector quantities. *Vector quantities* are characterized by a magnitude, a point of application, and a direction.

Types of Forces. Forces can be applied on a body by *direct contact* or through *remote action*. Forces can be *concentrated* at a point or *distributed* along a length, over an area, or throughout the entire body. *External forces* are exerted on the body by another body. *Internal forces* are the resisting forces within a body.

Types of Force Systems. Force systems can be classified into the following three types, depending on whether they are coplanar or spatial, concurrent or nonconcurrent.

1. Concurrent-coplanar force system
2. Nonconcurrent-coplanar force system
3. Spatial force system

Newton's Three Laws. These three laws form the foundation for the study of *Newtonian mechanics.* The first law deals with conditions for equilibrium of a particle and thereby lays the foundation for the study of statics. The second law provides the basic formulation for the study of dynamics. The third law provides the basic understanding for the nature of action and reaction forces.

The Principle of Transmissibility. The point of application of a force may be placed anywhere along the line of action of the force without changing the external effects of the force. However, the line of action and the direction of a force must be well defined. For the internal effect or the deformation of a body, a force acting on the body must have a fixed point of application, and therefore the principle of superposition does not apply.

System of Units. Two systems of units are used in this book: the U.S. customary units and the SI units. The base units in the U.S. system are the foot, second, and pound. The base unit for force (or weight), the pound, is dependent on gravitational attraction; it is therefore a *gravitational system.* The base units in the SI system are the meter, second, and kilogram. The base unit for mass, the kilogram, is independent of gravitational attraction; it is therefore an *absolute system.*

Rules for Numerical Computations. Calculated results should always be rounded off according to the following rules:

Rule 1   When approximate numbers are *multiplied* or *divided,* the result is expressed to the *same accuracy* as the least accurate number.

Rule 2   When approximate numbers are *added* or *subtracted,* the result is expressed to the *same precision* as the least precise number.

Mathematics Used in Mechanics. Some fundamental mathematical skills are required of the student. For example, students must be able to perform elementary algebraic manipulations, solve a right triangle using the Pythagorean theorem and trigonometric functions, solve an oblique triangle using the law of sines and/or the law of cosines, and solve two or three simultaneous linear equations.

General Procedure for Problem Solution. Problems must be solved in a logical and orderly manner. Students must learn to analyze the problem carefully. Make necessary sketches and apply the relevant principles. Equations must be solved by using proper mathematical operations. Results should be checked against certain required conditions or judged to be reasonable using common sense. Work must be presented in a neat and organized fashion.

PROBLEMS

Section 1–1   Introduction to Mechanics

1–1   (*a*)   What are the characteristics of a rigid body?
      (*b*)   In the study of statics and dynamics, bodies are considered rigid. Why?
      (*c*)   In the study of strength of materials, why is it important to consider the bodies deformable?

1–2   Identify whether each of the following is a topic in statics, dynamics, or strength of materials.
      (*a*)   Determining the size of a beam
      (*b*)   Calculating the reactions on a ladder
      (*c*)   Studying the motion of a projectile
      (*d*)   Calculating the deflections of a beam
      (*e*)   Determining the forces in truss members
      (*f*)   Studying the motion of a pendulum

Section 1–3   Scalar and Vector Quantities

1–3   What are the characteristics of a vector quantity?

1–4   In each of the following, indicate whether it is a scalar or a vector quantity.
      (*a*)   60 minutes
      (*b*)   A displacement of 300 feet due east
      (*c*)   An upward force of 5 kN
      (*d*) $1000.00
      (*e*)   A downward gravitational acceleration of 9.81 m/s^2
      (*f*)   A 50-kg mass

Section 1–5   Types of Force Systems

1–5   Name the force system in which all the forces are on a single plane and passing through a common point.

1–6   Name the force system in which the spatial forces do not meet at a common point.

Section 1-6 Newton's Laws

1-7   What is the meaning and significance of the mass of a body?

1-8   What is the significance of Newton's third law?

Section 1-7 The Principle of Transmissibility

1-9   Do we have to define the point of application of a force all the time?

1-10   What are the conditions when the principle of transmissibility can or cannot be applied?

Section 1-8 Systems of Units

1-11   What are the differences between the gravitational system of units and the absolute system of units?

1-12   What is the weight of a 5-slug mass in pounds?

1-13   What is the mass in slugs of a body weighing 500 lb?

1-14   What is the weight in newtons of a 10-kg mass?

1-15   What is the mass in kilograms of a body weighing 1000 N?

1-16   What is the weight of a 10-Mg mass in kilonewtons (kN)?

1-17   What is the mass in kilograms of a body weighing 100 N?

1-18   An astronaut weighs 150 lb on the surface of the earth. Determine (*a*) the mass of the astronaut in slugs, and (*b*) his weight in pounds on the moon, where the gravitational acceleration is 5.30 ft/s^2. What is his mass on the moon?

1-19   Reduce the following SI units to the units indicated.
(*a*)   6.38 Gg to kg
(*b*)   900 km to m
(*c*) 3.76×10^7 g to Mg
(*d*)   70 mm to m
(*e*)   23 400 N to kN

Section 1-9 Unit Conversion

1-20   A car travels at 60 mph. What is the equivalent speed in ft/min?

1-21   The world record for the men's 100-m dash is 9.82 s. What is the equivalent speed in mph?

1–22   The specific weight (weight per unit volume) of concrete is 150 lb/ft³. What is its equivalent value in kN/m³?

1–23   The mean radius of the earth is 6371 km. Determine its equivalent value in miles.

1–24   Use the conversion factors listed in Table 1–2 to convert the following SI units into the U.S. customary units indicated.
(*a*)   9.81 m/s² to ft/s²
(*b*)   100 MN/m² to ksi (kips/in.²)
(*c*)   10 m/s to mph

1–25   Use the conversion factors listed in Table 1–2 to convert the following U.S. customary units into the SI units indicated.
(*a*)   200 lb-ft to N · m
(*b*)   600 mph to km/h
(*c*)   100 hp to kW

Section 1–10   Consistency of Units in an Equation

Section 1–11   Rules for Numerical Computations

In Problems 1–26 to 1–36, evaluate the given formula for the quantity indicated. Round off the results to a proper number of significant digits.

1–26   A circular area can be computed by the formula $A = \pi r^2$. Find the area of a circle having a radius of 3.25 ft.

1–27   Use the formula $A = \pi r^2$ for circular areas to find the area of a circular lot of 268 ft diameter in acres (1 acre = 43 560 ft²).

1–28   The vertical distance y traveled by a freely falling body can be computed from the formula

$$y = v_0 t + \frac{1}{2}gt^2$$

where v_0 is the initial velocity, g is the gravitational acceleration, and t is the time of falling. Find the distance traveled by a falling body with an initial downward velocity of 2.25 m/s for 30 s.

1–29   Use the equation in Problem 1–28 to find the distance traveled by a body falling with an initial downward velocity of 25.0 ft/s for 15.0 s.

1–30   Use the formula in Problem 1–28 to find the time it takes a body to fall a vertical distance 1250 ft starting from rest.

1–31   The force F in a linear spring is given by $F = kx$, where k is the spring constant (force per unit length of spring deflection) and x is the spring deflection. Find the force in a spring with a spring constant of 100 lb/ft and a deflection of 3.00 in.

1–32 The frequency f in Hz (cycles per second) of an oscillating body can be computed from

$$f = \frac{1}{2\pi} \sqrt{\frac{k}{m}}$$

where k is the spring constant and m is the mass of the body. Find the frequency of vibration of a 30.5-kg mass supported on a spring with a spring constant of 1.57 kN/m.

1–33 Using the formula in Example 1–5, find the elongation in a steel wire of 2.00-mm diameter and a length of 10.0 m that is subjected to an axial force of 400 N. For steel, $E = 270$ GN/m².

1–34 An object falling from rest through a height h reaches a velocity $v = \sqrt{2gh}$, where g is the gravitational acceleration. If a rock falls from a cliff 125 ft above the ground, what is its velocity when it hits the ground?

1–35 Using the formula in Problem 1–34, find the velocity of a rock after falling 40 m from a cliff.

1–36 The area of a triangle with three given sides a, b, and c can be computed from the formula

$$A = \sqrt{s(s - a)(s - b)(s - c)}$$

where $s = (a + b + c)/2$. Find the area of a triangle with sides 5.45 ft, 6.85 ft, and 7.39 ft.

Section 1–12 A Brief Review of Mathematics

1–37 The hypotenuse of a right triangle is 700 mm and one of the acute angles is 35°. Find the lengths of the other two sides.

1–38 The hypotenuse and one side of a right triangle are 15 in. and 10 in., respectively. Determine the angle between the hypotenuse and the shorter side.

1–39 Determine the distance between two points BC across the river shown in Fig. P1–39 if the angle at C is laid out at an angle of 90°, the distance CA is laid out 400 ft away, and angle A is measured to be 49.5°.

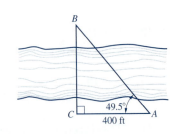

FIGURE P1–39

1–40   Find the angle between the wings of
the toggle bolt shown in Fig. P1.40.

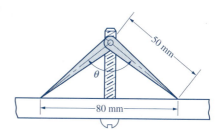

FIGURE P1–40

1–41   The flagpole in Fig. P1.41 has two
sections, *AB* and *BC*. The angles α
and β, measured at *D* at a distance
of 50 m from the pole, are 40° and
30°, respectively. Find the heights *a*
and *b* of the two sections of the
pole.

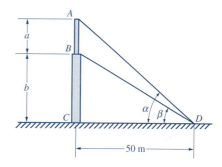

FIGURE P1–41

1–42   Determine the height *h* and the lengths *a* and *b* of the roof truss in Fig.
P1–42. (*Hint:* Draw *AD* perpendicular to *BC*; *AD* bisects both *BC* and *EF*.
Solve *AD* = *h* from the right triangle *ABD*, and *ED* = *b*/2 from the right tri-
angle *AED*.)

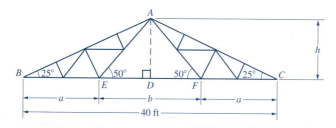

FIGURE P1–42

*In Problems 1–43 to 1–47, find the unknown elements of an oblique triangle if
three elements are given. See Fig. 1–7 for the notations used.*

1–43 $a = 100$ mm, $A = 35°$, $B = 65°$

1–44 $a = 3.5$ ft, $B = 32°$, $C = 105°$

1–45 $b = 12$ m, $c = 15$ m, $A = 45°$

1–46 $a = 9$ in., $b = 10$ in., $C = 120°$

1–47 $a = 2.3$ m, $b = 4.5$ m, $c = 5.4$ m

1–48   The reciprocal engine in Fig. P1–48 consists of a crankshaft *OA* 100 mm long and a connecting rod *AB* 250 mm long. In the crankshaft position shown, α = 40°. Determine the angle β and the distance *OB*.

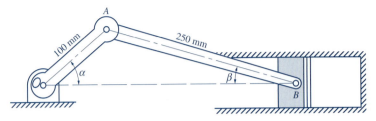

FIGURE P1–48

1–49   A ship sails 70 miles due north and then 90 miles in the N60°E direction, as shown in Fig. P1–49. How far is the ship from its starting point *O*?

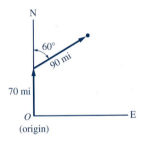

FIGURE P1–49

In Problems 1–50 to 1–54, solve the given system of linear equations by addition or subtraction.

1–50 $3x + 5y = -8$
$5x - 3y = 15$

1–51 $3.45x - 2.65y = 2.77$
$1.86x + 3.76y = 9.85$

1–52 $T\sin10° - P\sin40° = 0$
$T\cos10° - P\cos40° = 200$ lb

1–53 $-0.429P + 0.231Q \qquad\quad = 1920$ lb
$-0.857P - 0.923Q - 0.923R = 2880$ lb
$0.286P - 0.308Q - 0.385R = 2160$ lb

1–54 $-0.444x - 0.857y + 0.667z = 0$
$0.444x + 0.429y + 0.667z = 17$ kN
$0.778x - 0.286y - 0.333z = 0$

1–55   Solve the equations in Problem 1–50 by Cramer's rule.

1–56   Solve the equations in Problem 1–51 by Cramer's rule.

1–57   Solve the equations in Problem 1–52 by Cramer's rule.

1–58   Solve the equations in Problem 1–53 by Cramer's rule.

1–59   Solve the equations in Problem 1–54 by Cramer's rule.

Computer Program Assignments

For each of the following problems, write a computer program using an appropriate programming language with which you are most familiar. Make the program user friendly by incorporating plenty of comments and input prompts so that the user will understand the input data to be entered and the limitations of their values. The output should include the data entered and the computed results, and they must be well labeled to identify each quantity.

C1–1 Write a computer program that can be used to solve a system of two equations with two unknowns using Cramer's rule. The user input should be the coefficients of x and y, and the right-hand constant of each equation. The output must include the given equations, the value of the determinants, and the solution to the equations. If D, D_x, and D_y are all equal to zero, the two equations are dependent and there is no unique solution. If D is zero, but either one or both D_x and D_y are not zero, the two equations are inconsistent and there is no solution. Use this program to solve (*a*) Example 1–16, (*b*) Problem 1–51, and (*c*) Problem 1–52.

C1–2 Modify the program written for Problem C1–1 so that it can be used to solve a system of three equations with three unknowns using Cramer's rule. Use this program to solve (*a*) Example 1–20, (*b*) Problem 1–53, and (*c*) Problem 1–54.

CHAPTER 2

RESULTANT OF COPLANAR FORCE SYSTEMS

2–1
INTRODUCTION

Two systems of forces are said to be *equivalent* if they produce the same mechanical effect on a rigid body. A single force that is equivalent to a given force system is called the *resultant* of the force system.

We shall first introduce the *parallelogram law* and use it to find the resultant of concurrent coplanar forces. Then the rectangular components of forces are discussed and used to find the resultants.

As we shall see later, any system of nonconcurrent coplanar forces can be replaced by a *single resultant* that is equivalent to the given force system. The location of the line of action of the resultant is not immediately known. To determine the line of action of the resultant of a nonconcurrent coplanar force system, we will introduce the concepts of the *moment* of a force first. The resultant of some simple types of distributed line loads will be discussed in this chapter also.

2–2
VECTOR REPRESENTATION

Notations. In this book, vector quantities will be distinguished from scalar quantities through the use of boldface type, such as **P.** An italic type, such as *P*, will be used to denoted the magnitude of a vector. In long-hand writing, vectors may be represented by the notation $\vec{P}$.

Graphical Representation. A force **F** (or any vector quantity) is represented graphically by a line segment *AB* with an arrowhead at one end, as shown in Fig. 2–1. *A* is the point of application and *x* is a reference coordinate axis. The length of the line segment *AB* represents the magnitude of the force measured according to some convenient scale. The direction is indicated by the angle θ from the reference axis.

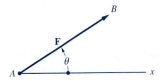

FIGURE 2–1

Equal Vectors.   Two vectors having the same magnitude and the same direction are said to be equal. Two equal vectors may or may not have the same line of action (Fig. 2–2). Equal vectors may be denoted by the same letter.

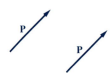

FIGURE 2–2

Negative Vector.   The negative vector of a given vector **P** is defined as a vector having the same magnitude as **P** and a direction opposite to that of **P** (Fig. 2–3). The negative vector of **P** is denoted by −**P**. According to Newton's law of action and reaction presented in Section 1–6, the forces of action and reaction must always be equal in magnitude and opposite in direction. Thus, the forces of action and reaction may be represented by **P** and −**P**.

FIGURE 2–3

2–3
RESULTANT OF CONCURRENT FORCES

Parallelogram Law.   As mentioned in Section 1–3, vectors are added according to the *parallelogram law*. Figure 2–4 shows two vectors that are added according to this law. The two vectors **P** and **Q** are placed at the same point *A* and a parallelogram is constructed using **P** and **Q** as its two adjacent sides. The diagonal of the parallelogram from *A* to the opposite corner represents the sum of **P** and **Q** and is denoted by **P** + **Q**. Note that, in general, the magnitude of the vector sum **P** + **Q** is not equal to the algebraic sum of the magnitudes *P* and *Q*.

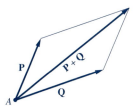

FIGURE 2–4

Triangle Rule. The sum of two vectors can also be determined by constructing one-half of the parallelogram, or a triangle. This method is called the *triangle rule*. To find the vector sum **P** + **Q**, we first lay out **P** at *A* (Fig. 2–5a), then lay out **Q** from the tip of **P** in a tip-to-tail fashion. The closing side of the triangle, drawn from *A* to the tip of **Q**, represents the sum of the two vectors. Figure 2–5b shows that the same result is obtained if the vector **Q** is laid out first. Hence, the vector sum is not affected by the order in which the vectors are added; that is, *vector addition is commutative:*

$$\mathbf{P} + \mathbf{Q} = \mathbf{Q} + \mathbf{P} \qquad (2\text{–}1)$$

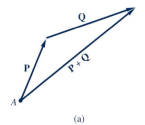

(a)

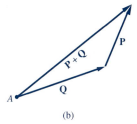

(b)

FIGURE 2–5

Polygon Rule. The sum of three or more concurrent coplanar vectors may be carried out by adding two vectors successively. For example, the sum of three coplanar concurrent vectors, **P**, **Q**, and **S** (Fig. 2–6a) can be obtained by first finding **P** + **Q**, then adding **S** to **P** + **Q** to find **P** + **Q** + **S**, as shown in Fig. 2–6b. Notice that the dotted line in the figure could be omitted, and the sum of the vectors can be obtained directly by laying out the given vectors in a tip-to-tail fashion to form the sides of a polygon. The closing side of the polygon, from the starting point to the final point, represents the sum of the vectors. This is known as the *polygon rule* for the addition of vectors. A polygon formed by forces is called a *force polygon.*

Since the vector sum is commutative, *the order in which the vectors are added is arbitrary.* In Fig. 2–6c the vectors are added in the order of **P**, **S**, and **Q**. We see that, although the shape changes, the resultant obtained remains the same.

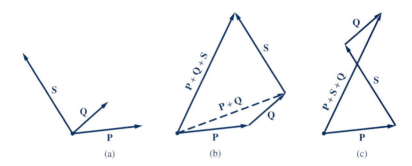

(a)   (b)   (c)

FIGURE 2–6

    Resultant.   A given system of concurrent coplanar forces acting on a rigid body may be replaced by a single force, called the *resultant,* equal to the vector sum of the given forces. The resultant will produce the same effect on the rigid body as the given force system.

─── **EXAMPLE 2–1** ───────────────────────────────────────

Determine the resultant of two forces **P** and **Q** acting on the hook in Fig. E2–1(1).

    Solution. Two methods are presented here.

    (a) The Graphical Method. A force triangle *ABC* is drawn as shown in Fig. E2–1(2). *AB* represents force **P** and *BC* represents force **Q**. The magnitude of each force is laid out by using a properly chosen linear scale. The direction of each force is measured by using a protractor. The closing side of the triangle, *AC,* is the resultant. The magnitude and direction of the resultant are measured to be

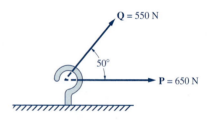

FIGURE E2–1(1)

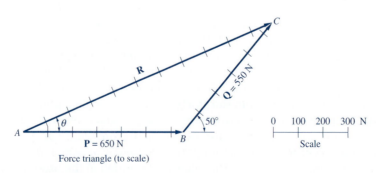

Force triangle (to scale)

FIGURE E2–1(2)

$$R = 1090 \text{ N} \qquad \theta = 23°$$

$$\mathbf{R} = 1090 \text{ N} \measuredangle \ 23° \qquad\qquad \Leftarrow \textbf{Ans.}$$

(b) *The Trigonometric Method.* First, the force triangle *ABC* is sketched. [See Fig. E2–1(3).] A freehand sketch is usually sufficient for this purpose. The triangle has two known sides and a known angle between the two sides. The magnitude of the resultant can be computed by applying the law of cosines.

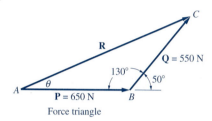

Force triangle

FIGURE E2–1(3)

$$R = \sqrt{P^2 + Q^2 - 2PQ \cos B}$$
$$= \sqrt{(650 \text{ N})^2 + (550 \text{ N})^2 - 2(650 \text{ N})(550 \text{ N}) \cos 130°}$$
$$= 1088 \text{ N}$$

To find the direction of the resultant, we compute angle *A(θ)* by applying the law of sines.

$$\frac{\sin A}{Q} = \frac{\sin B}{R}$$
$$A = \sin^{-1}\!\left(\frac{Q \sin B}{R}\right)$$
$$= \sin^{-1}\!\left(\frac{550 \sin 130°}{1088}\right)$$
$$= 22.8°$$

Thus, the resultant is

$$\mathbf{R} = 1088 \text{ N} \measuredangle \ 22.8° \qquad\qquad \Leftarrow \textbf{Ans.}$$

Comparison of the two methods indicates clearly that the trigonometric method gives a more accurate solution. The degree of accuracy of the graphical solution can be improved, however, if a larger scale is used and more care is exercised when making the drawing. Much greater accuracy may be obtained if computer-aided drafting is used.

——— **EXAMPLE 2–2** —————————————————————————

A 250-lb weight is lifted by pulling the two cords shown in Fig. E2–2(1). To lift the weight, the resultant of the two tensions T_1 and T_2 must be 250 lb acting vertically upward. Determine (a) the tension in each rope, knowing that $\theta = 40°$, and (b) the angle θ, for which the tension T_2 is a minimum.

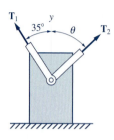

FIGURE E2–2(1)

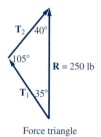

Force triangle

FIGURE E2–2(2)

Solution.

(*a*) *Tensions for* $\theta = 40°$.   A force triangle is drawn with $\mathbf{R} = 250$ lb vertically upward, and $\mathbf{T}_1$ and $\mathbf{T}_2$ in the directions shown in Fig. E2–2(2). By the law of sines, we write

$$\frac{T_1}{\sin 40°} = \frac{T_2}{\sin 35°} = \frac{250 \text{ lb}}{\sin 105°}$$

From which, we get

$$T_1 = 166 \text{ lb} \qquad T_2 = 148 \text{ lb} \qquad\qquad \Leftarrow \textbf{Ans.}$$

(*b*) *Values of* θ *for Minimum* $\mathbf{T}_2$.   Refer to Fig. E2–2(3). Using the triangular rule, we first draw line AB to represent the known resultant. Then we draw line $A1$ from A along the known direction of $\mathbf{T}_1$. Several possible directions of $\mathbf{T}_2$ are represented by the lines marked $B2$. Among these lines, the shortest one representing $(\mathbf{T}_2)_{\min}$ is perpendicular to $\mathbf{T}_1$. Thus,

$$\theta = 90° - 35°$$
$$= 55° \;\triangleright \qquad\qquad \Leftarrow \textbf{Ans.}$$

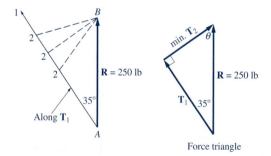

FIGURE E2–2(3)

───── **EXAMPLE 2–3** ─────

Determine the resultant of the five forces shown in Fig. E2–3(1) by the graphical method.

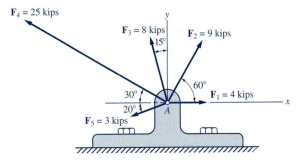

FIGURE E2–3(1)

Solution. Refer to Fig. E2–3(2). Starting from point O, draw Oa, ab, bc, cd, and de, representing forces $\mathbf{F}_1$, $\mathbf{F}_2$, $\mathbf{F}_3$, $\mathbf{F}_4$, and $\mathbf{F}_5$, respectively, in a head-to-tail fashion.

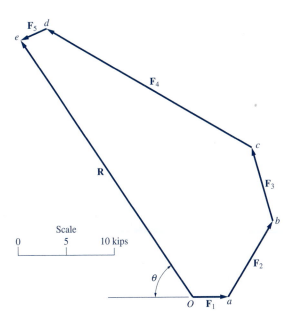

FIGURE E2–3(2)

The closing side of the polygon Oe is the desired resultant **R.** The magnitude and direction of **R** are measured to be

$$\mathbf{R} = 32.5 \text{ kips} \ \measuredangle \ 56° \qquad \Leftarrow \textbf{Ans.}$$

2–4
RECTANGULAR COMPONENTS

Any two or more forces whose resultant is equal to a force **F** are called the *components* of the force **F**. In Fig. 2–7a, **F**$_1$ and **F**$_2$ are the components of force **F** along the *O*1 and *O*2 directions. Two mutually perpendicular components are called the *rectangular components*. In Fig. 2–7b, **F**$_x$ and **F**$_y$ are the rectangular components of **F** in the *x* and *y* directions. We write

$$\mathbf{F} = \mathbf{F}_x + \mathbf{F}_y$$

The *x* and *y* axes may be chosen in any two perpendicular directions. Usually the axes are chosen along horizontal and vertical directions.

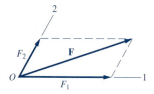

(a) Components of a force along two arbitrary directions 1 and 2

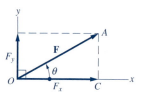

(b) Rectangular components of a force along two perpendicular directions

FIGURE 2–7

Rectangular Components. If the magnitude *F* and the direction angle θ of a force are known, then, from the right triangle *OAC* in Fig. 2–7b, the rectangular components are

$$F_x = F\cos\theta \quad \text{and} \quad F_y = F\sin\theta \tag{2–2}$$

In Equation 2–2, the direction angle θ must be measured in the *standard position*; that is, it is measured from the positive *x* axis to the force vector **F**. Counterclockwise measurement is regarded as positive; clockwise measurement is regarded as negative, as indicated in Fig. 2–8. The components along the positive coordinate axes are positive and those along the negative coordinate axes are negative. If the direction angle is in the standard position, Equation 2–2 will yield the correct sign for the components.

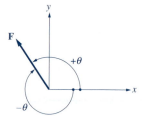

FIGURE 2–8

Magnitude and Direction. When the scalar components F_x and F_y of the force **F** are given, the magnitude of **F** may be determined from

$$F = \sqrt{F_x^2 + F_y^2} \qquad (2\text{–}3)$$

and the reference angle α (a positive acute angle between the positive or negative x axis and the force vector) is

$$\alpha = \tan^{-1} \left| \frac{F_y}{F_x} \right| \qquad (2\text{–}4)$$

Depending on which quadrant the force vector is in, the direction angle θ, in the standard position, is

$$\text{First quadrant: } \theta = \alpha \qquad (2\text{–}5\text{a})$$

$$\text{Second quadrant: } \theta = 180° - \alpha \qquad (2\text{–}5\text{b})$$

$$\text{Third quadrant: } \theta = 180° + \alpha \qquad (2\text{–}5\text{c})$$

$$\text{Fourth quadrant: } \theta = 360° - \alpha \quad \text{or} \quad \theta = -\alpha \qquad (2\text{–}5\text{d})$$

The quadrant that a force vector is in may be determined by using a sketch. For example, if both components are negative, a simple sketch will indicate that the force vector is in the third quadrant.

──────── **EXAMPLE 2–4** ────────

Resolve the 500-N force exerted on the hook in Fig. E2–4(1) into horizontal and vertical components.

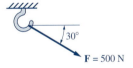

FIGURE E2–4(1)

Solution. The coordinate axes are chosen as shown in Fig. E2–4(2), where the positive x axis is horizontal to the right and the positive y axis is vertically upward. The reference angle is 30°. Since the force is in the fourth quadrant, angle θ in the standard position is

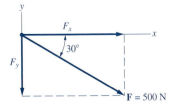

FIGURE E2–4(2)

$$\theta = 360° - \alpha$$
$$= 360° - 30°$$
$$= 330° \qquad (\text{or} -30°)$$

Using Equation 2–2, we get

$$F_x = F \cos \theta$$
$$= (500 \text{ N}) \cos 330°$$
$$= +433 \text{ N} \qquad \Leftarrow \textbf{Ans.}$$
$$F_y = F \sin \theta$$
$$= (500 \text{ N}) \sin 330°$$
$$= -250 \text{ N} \qquad \Leftarrow \textbf{Ans.}$$

─────── **EXAMPLE 2–5** ───────

Refer to Fig. E2–5(1). The tension in guy wire BC is $T = 3.5$ kips. Resolve this force into the x and y components.

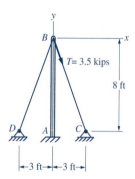

FIGURE E2–5(1)

Solution. The tension **T** is along wire BC, which has a slope of

$$\text{Horizontal : Vertical} = 3{:}8$$

as indicated in the slope triangle shown in Fig. E2–5(2). The hypotenuse in the slope triangle is

$$\sqrt{3^2 + 8^2} = \sqrt{73}$$

From the slope triangle, we get

$$\sin \alpha = \frac{8}{\sqrt{73}}$$

$$\cos \alpha = \frac{3}{\sqrt{73}}$$

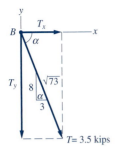

FIGURE E2–5(2)

Using Equation 2–2, we get

$$T_x = +T \cos \alpha$$
$$= +(3.5 \text{ kips}) \left(\frac{3}{\sqrt{73}} \right)$$
$$= +1.23 \text{ kips} \qquad \Leftarrow \textbf{Ans.}$$

$$T_y = -T \sin \alpha$$
$$= -(3.5 \text{ kips}) \left(\frac{8}{\sqrt{73}} \right)$$
$$= -3.28 \text{ kips} \qquad \Leftarrow \textbf{Ans.}$$

Since the angle α used is not in the standard position, we must assign a proper sign to each component by inspection.

EXAMPLE 2–6

Refer to Fig. E2–6(1). Resolve the 100-lb weight into components along the incline and normal to the incline.

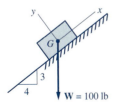

FIGURE E2–6(1)

Solution. The x and y axes are chosen along and normal to the incline, respectively. The components W_x and W_y are shown in Fig. E2–6(2). By inspection, we see that both components are negative. Thus,

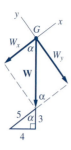

FIGURE E2-6(2)

$$W_x = -W \cos \alpha \qquad W_y = -W \sin \alpha$$

From the slope triangle (a 3-4-5 right triangle), we have

$$\cos \alpha = \frac{3}{5} \qquad \sin \alpha = \frac{4}{5}$$

Therefore,

$$W_x = -(100 \text{ lb}) \frac{3}{5}$$

$$= -60 \text{ lb} \qquad\qquad \Leftarrow \textbf{Ans.}$$

$$W_y = -(100 \text{ lb}) \frac{4}{5}$$

$$= -80 \text{ lb} \qquad\qquad \Leftarrow \textbf{Ans.}$$

EXAMPLE 2-7

Refer to Fig. E2-7. The rectangular components of a force **F** are given as $F_x = -30$ lb and $F_y = +40$ lb. Determine the magnitude and direction of the force.

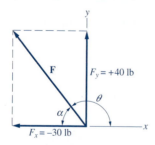

FIGURE E2-7

Solution. From Equation 2-3, the magnitude of the force is

$$F = \sqrt{F_x^2 + F_y^2}$$

$$= \sqrt{(-30 \text{ lb})^2 + (40 \text{ lb})^2}$$

$$= 50 \text{ lb}$$

From Equation 2–4, the reference angle α is

$$\alpha = \tan^{-1} \left| \frac{F_y}{F_x} \right|$$

$$= \tan^{-1} \left| \frac{40}{-30} \right|$$

$$= 53.1°$$

Since the force vector is in the second quadrant, from Equation 2–5b, the angle θ in the standard position is

$$\theta = 180° - \alpha$$

$$= 180° - 53.1°$$

$$= 126.9°$$

Thus,

$$\mathbf{F} = 50 \text{ lb } \angle 126.9° \qquad \Leftarrow \textbf{Ans.}$$

2–5
RESULTANTS BY RECTANGULAR COMPONENTS

The resultant of any number of concurrent coplanar forces can be determined by using their rectangular components. Consider three coplanar forces $\mathbf{F}_1$, $\mathbf{F}_2$, and $\mathbf{F}_3$ acting at point O, as shown in Fig. 2–9. The resultant $\mathbf{R}$ of the three forces is

$$\mathbf{R} = \mathbf{F}_1 + \mathbf{F}_2 + \mathbf{F}_3$$

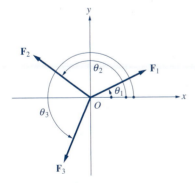

FIGURE 2–9

Each force is resolved into its rectangular components. All the x components are in the horizontal direction; hence, these components may be added algebraically. Similarly, all the y components may be added algebraically. In terms of the direction angle in the standard position, we write

$$R_x = (F_x)_1 + (F_x)_2 + (F_x)_3$$
$$= F_1 \cos \theta_1 + F_2 \cos \theta_2 + F_3 \cos \theta_3$$
$$R_y = (F_y)_1 + (F_y)_2 + (F_y)_3$$
$$= F_1 \sin \theta_1 + F_2 \sin \theta_2 + F_3 \sin \theta_3$$

The components of the resultant are the algebraic sums of the corresponding components of the forces. In general, for a system of coplanar forces, we write

$$R_x = \Sigma F_x = \Sigma F \cos \theta \qquad R_y = \Sigma F_y = \Sigma F \sin \theta \qquad (2\text{--}6)$$

where the symbol Σ (Greek capital letter sigma) stands for "summation." The direction angle θ must be in the standard position. Once the scalar components of the resultant are obtained, the magnitude and direction of the resultant can be obtained from Equations 2–3, 2–4, and 2–5.

───── **EXAMPLE 2–8** ──────────────────────────────────

Determine the resultant of the two forces $\mathbf{F}_1$ and $\mathbf{F}_2$ acting on the eye-bolt shown in Fig. E2–8(1).

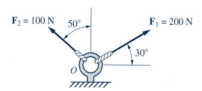

FIGURE E2–8(1)

Solution. The direction angles of the forces in the standard position are indicated in Fig. E2–8(2). Using Equation 2–6, we write

$$R_x = F_1 \cos \theta_1 + F_2 \cos \theta_2$$
$$= (200 \text{ N}) \cos 30° + (100 \text{ N}) \cos 140°$$
$$= +96.6 \text{ N}$$

$$R_y = F_1 \sin \theta_1 + F_2 \sin \theta_2$$
$$= (200 \text{ N}) \sin 30° + (100 \text{ N}) \sin 140°$$
$$= +164.3 \text{ N}$$

FIGURE E2–8(2)

Note that if the 50° angle from the y axis for $\mathbf{F}_2$ is used, we must pay close attention to the sign and the proper trigonometric functions to be used. We write

$$R_x = (200 \text{ N}) \; \cos 30° - (100 \text{ N}) \; \sin 50°$$
$$= +96.6 \text{ N}$$

$$R_y = (200 \text{ N}) \; \sin 30° + (100 \text{ N}) \; \cos 50°$$
$$= +164.3 \text{ N}$$

The magnitude of the resultant is

$$R = \sqrt{(96.6 \text{ N})^2 + (164.3 \text{ N})^2} = 190.6 \text{ N}$$

Since both components are positive, the resultant is in the first quadrant. Thus, the direction angle is

$$\theta = \alpha = \tan^{-1}\left(\frac{R_y}{R_x}\right)$$

$$= \tan^{-1}\left(\frac{164.3}{96.6}\right) = 59.5°$$

$$\mathbf{R} = 190.6 \text{ N} \; \angle \; 59.5° \qquad\qquad \Leftarrow \textbf{Ans.}$$

EXAMPLE 2–9

Determine the resultant of the five forces in Example 2–3 by using rectangular components. The diagram is reproduced here as Fig. E2–9.

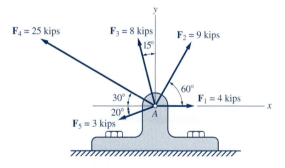

FIGURE E2–9

Solution. The solution is presented in two methods.

(a) By Equation. Using the given angles as they are, we find

$$R_x = \Sigma F_x = (F_x)_1 + (F_x)_2 + (F_x)_3 + (F_x)_4 + (F_x)_5$$
$$= \quad 4 + (9) \cos 60° - (8) \sin 15° - (25) \cos 30° - (3) \cos 20°$$
$$= -18.04 \text{ kips}$$

$$R_y = \Sigma F_x = (F_y)_1 + (F_y)_2 + (F_y)_3 + (F_y)_4 + (F_y)_5$$
$$= 0 + (9)\sin 60° - (8)\cos 15° - (25)\sin 30° - (3)\sin 20°$$
$$= +27.0 \text{ kips}$$

In the above equations, a plus or minus sign is assigned to each component by inspection. Components along positive directions of the axes are positive, and those along negative directions of the axes are negative. To decide which trigonometric function to use, just remember that to get a component opposite to the angle, the sine function of the angle must be used, and to get a component adjacent to the angle, the cosine function must be used. When making the calculation with a calculator, make sure that the calculator is in the degree mode.

With the x and y components determined, the magnitude of the resultant is

$$R = \sqrt{(18.04 \text{ kips})^2 + (27.0 \text{ kips})^2} = 32.5 \text{ kips}$$

With a negative x component and a positive y component, the resultant is in the second quadrant. Thus,

$$\alpha = \tan^{-1}\left|\frac{R_y}{R_x}\right| = \tan^{-1}\left|\frac{27.0}{-18.04}\right| = 56.3°$$

$$\theta = 180° - \alpha = 123.7°$$

$$\mathbf{R} = 32.5 \text{ kips} \quad \angle\!\!\!\diagdown \ 123.7° \qquad\qquad \Leftarrow \textbf{Ans.}$$

(b) By Tabulation. When the number of forces involved is greater than three, a solution in tabulated form is recommended. The x and y components of each force are listed in the table below. The angle θ of each force must be in the standard position.

Force	Magnitude F (kips)	Angle θ (deg)	x component F_x (kips) $= F\cos\theta$	y component F_y (kips) $= F\sin\theta$
F_1	4	0	4.00	0
F_2	9	60	4.50	7.79
F_3	8	105	−2.07	7.73
F_4	25	150	−21.65	12.50
F_5	3	200	−2.82	−1.03
Σ			−18.04	+27.0

Thus,

$$R_x = -18.04 \text{ kips} \qquad R_x = -18.04 \text{ kips}$$

These are the same results obtained before.

2-6
MOMENT OF A FORCE

Effects of a Force. A force tends to move a body along its line of action. It also tends to rotate a body about an axis. For example, a pull on a door knob (Fig. 2–10) causes the door to rotate about the axis through the hinges. The ability of a force to cause a body to rotate is measured by a quantity called the *moment* of the force.

FIGURE 2–10

Consider a wrench used to tighten a bolt, as shown in Fig. 2–11. The rotating moment (also called the torque) produced by the applied force **F** depends not only on the magnitude of the force, but also on the perpendicular distance *d* from the center *O* of the bolt to the line of action of the force. In fact, the turning effect of the force is measured by the product of *F* and *d*.

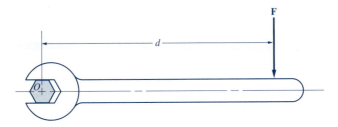

FIGURE 2–11

Definition of Moment. In the two-dimensional case, the moment M_O (Fig. 2–12) of a force F about a point O (called the *moment center*) is equal to the magnitude of the force **F** multiplied by the perpendicular distance d (called the *moment arm*) from O to the line of action of the force:

$$M_o = Fd \qquad\qquad (2\text{-}7)$$

The units for moment are lb · ft or lb · in. in the U.S. customary units, and N · m or kN · m in the SI units.

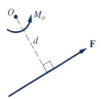

FIGURE 2–12

Direction of Moments. In Fig. 2–13, we see that the two forces **P** and **Q** cause the lever to rotate about the pivot point *O* in opposite directions. The force **P** causes a counterclockwise (c.c.w.) rotation; thus, the moment of **P** about *O* is c.c.w. The force **Q**, on the other hand, causes a clockwise (c.w.) rotation; thus, the moment of **Q** about *O* is c.w. It is important to distinguish whether a moment is c.w. or c.c.w.

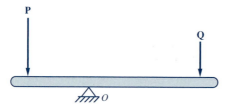

FIGURE 2–13

Summation of Moments. In the two-dimensional case, moments may be added algebraically. A proper sign must be assigned to the moment, depending on whether it is c.w. or c.c.w. In this book, unless stated otherwise, *a c.c.w. moment will be considered positive and a c.w. moment will be considered negative.*

───── **EXAMPLE 2–10** ─────

A 500-N force is applied to the end of a lever pivoted at point *O* [see Fig. E2–10(1)]. Determine the moment of the force about *O* if (*a*) $\theta = 30°$, (*b*) $\theta = 120°$, (*c*) $\theta = 90°$, and (*d*) $\theta = 50°$.

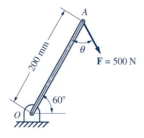

FIGURE E2–10(1)

Solution. The moment about O will be determined by definition as follows.

(a) $\theta = 30°$ for a Vertical Force.   Refer to Fig. E2–10(2). The moment arm d is

$$d = (0.2 \text{ m}) \cos 60°$$

$$= 0.1 \text{ m}$$

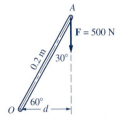

FIGURE E2–10(2)

The moment of the force about O is

$$M_O = Fd$$

$$= -(500 \text{ N})(0.1 \text{ m})$$

$$= -50 \text{ N} \cdot \text{m}$$

Since the force tends to rotate the lever clockwise about O, the moment is clockwise.

$$M_O = 50 \text{ N} \cdot \text{m} \ \circlearrowright \qquad\qquad \Leftarrow \textbf{Ans.}$$

(b) $\theta = 120°$ for a Horizontal Force.   Refer to Fig. E2–10(3). The moment arm and the moment are

$$d = (0.2 \text{ m}) \sin 60°$$

$$= 0.1732 \text{ m}$$

$$M_O = Fd$$

$$= -(500 \text{ N})(0.1732 \text{ m})$$

$$= -86.6 \text{ N} \cdot \text{m}$$

$$= 86.6 \text{ N} \cdot \text{m} \ \circlearrowright \qquad\qquad \Leftarrow \textbf{Ans.}$$

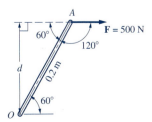

FIGURE E2–10(3)

(c) $\theta = 90°$ *for a Perpendicular Force.* See Fig. E2–10(4). In this case the entire length of the lever is the moment arm and the moment of the force about O has the maximum value.

$$M_O = Fd$$
$$= -(500 \text{ N})(0.2 \text{ m})$$
$$= -100 \text{ N} \cdot \text{m}$$
$$= 100 \text{ N} \cdot \text{m} \quad \circlearrowleft \qquad\qquad \Leftarrow \textbf{Ans.}$$

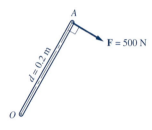

FIGURE E2–10(4)

(d) $\theta = 50°$ *for an Inclined Force.* Figure E2–10(5) represents a more general case. The moment arm can still be determined easily as:

$$d = (0.2 \text{ m}) \sin 50°$$
$$= 0.1532 \text{ m}$$

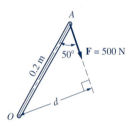

FIGURE E2–10(5)

The moment of the force about O is

$$M_O = Fd$$
$$= -(500 \text{ N})(0.1532 \text{ m})$$
$$= -76.6 \text{ N} \cdot \text{m}$$
$$= 76.6 \text{ N} \cdot \text{m} \quad \circlearrowleft \qquad\qquad \Leftarrow \textbf{Ans.}$$

2–7
VARIGNON'S THEOREM

Varignon's theorem states that *the moment of a force about any point is equal to the sum of the moments produced by the components of the forces about the same point.* This theorem was established by the French mathematician Varignon (1654–1722). A formal proof of this theorem will not be given here. Intuitively, we see that any force can be resolved into components without altering its effect, so the sum of the moments of the components must be the same as the moment of the force itself.

Since the moment arm of a force is often hard or impossible to determine, Varignon's theorem is very useful for finding the moment of a force. In Fig. 2–14, if the coordinates of the point of application A of the force are (x_A, y_A), then the moment of the force about the origin O in terms of its rectangular components is

$$M_O = F_y x_A - F_x y_A \qquad (2\text{–}8)$$

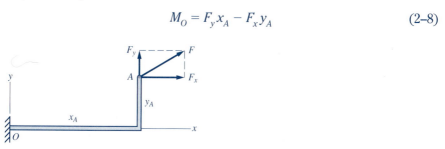

FIGURE 2–14

The *principle of transmissibility* is also helpful in calculating the moment of a force. Using this principle, the point of application of a force acting on a rigid body may be placed anywhere along its line of action. We see that the moment arm is clearly independent of the point of application of a force. Therefore, as long as the magnitude, the direction, and the line of action of a force are defined, the moment of a force about a given point may be determined by placing the force at *any* point along its line of action. For example, to find the moment of the force $\mathbf{F}$ (at A) in Fig. 2–15a about point O, we can resolve the force into rectangular components at B on the line of action of the force, as shown in Fig. 2–15b. Since the component F_x passes through the moment center O, it produces no moment about O. As a result, the moment of the force about O is simply

$$M_O = F_y x_B$$

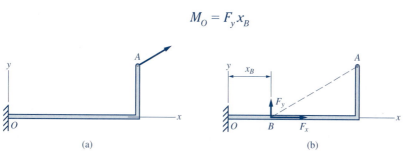

(a)   (b)

FIGURE 2–15

━━━━━ **EXAMPLE 2–11** ━━━━━━━━━━━━━━━━━━━━━━━━━━━━━━━━

Determine the moment of the 100-lb force about point B in Fig. E2–11(1).

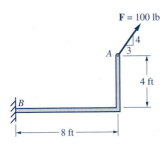

FIGURE E2–11(1)

Solution. The moment will be determined in three ways to illustrate different methods of solution.

(a) By Definition. From the geometry shown in Fig. E2–11(2), the moment arm d is

$$CD = AD/\tan \alpha$$
$$= (4 \text{ ft}) / (4/3)$$
$$= 3 \text{ ft}$$
$$BC = BD - CD$$
$$= 8 \text{ ft} - 3 \text{ ft}$$
$$= 5 \text{ ft}$$
$$d = BC \sin \alpha$$
$$= (5 \text{ ft})(4/5)$$
$$= 4 \text{ ft}$$

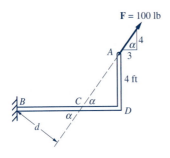

FIGURE E2–11(2)

By the definition of moment, we get

$$M_B = Fd$$
$$= +(100 \text{ lb})(4 \text{ ft})$$
$$= + 400 \text{ lb} \cdot \text{ft}$$

which is a counterclockwise moment. Thus,

$$M_B = 400 \text{ lb} \cdot \text{ft} \;\circlearrowleft \qquad\qquad \Leftarrow \textbf{Ans.}$$

(b) By Varignon's Theorem. Refer to Fig. E2–11(3). Resolve the force into horizontal and vertical components at A. The vertical component produces a counterclockwise moment and the horizontal component produces a clockwise moment. Thus,

$$M_B = (80 \text{ lb})(8 \text{ ft}) - (60 \text{ lb})(4 \text{ ft})$$
$$= +400 \text{ lb} \cdot \text{ft} \;\circlearrowleft$$

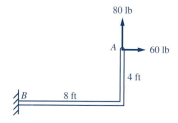

FIGURE E2–11(3)

(c) By the Principle of Transmissibility. Refer to Fig. E2–11(4). The force may be considered to act at C, where the force is resolved into horizontal and vertical components. The line of action of the horizontal component passes through point B and produces no moment about the point. Hence,

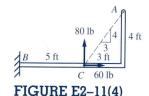

FIGURE E2–11(4)

$$M_B = +(80 \text{ lb})(5 \text{ ft})$$
$$= +400 \text{ lb} \cdot \text{ft} \;\circlearrowleft$$

EXAMPLE 2–12

Determine the maximum clockwise moment that can be produced by a 10-kN force exerted on the rectangular plate about the corner A in Fig. E2–12(1).

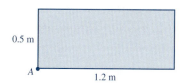

FIGURE E2–12(1)

Solution. To produce the maximum moment, the moment arm must be a maximum. See Fig. E2–12(2). This occurs when the point of application is at the opposite corner B and the line of action of the force is perpendicular to the diagonal AB. The moment arm is

$$d = AB$$
$$= \sqrt{(1.2 \text{ m})^2 + (0.5 \text{ m})^2}$$
$$= 1.3 \text{ m}$$

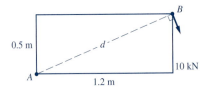

FIGURE E2–12(2)

The maximum moment is

$$M_A = Fd$$
$$= (10 \text{ kN})(1.3 \text{ m})$$
$$= 13 \text{ kN} \cdot \text{m} \; \circlearrowleft \qquad\qquad \Leftarrow \textbf{ Ans.}$$

2–8
COUPLE

Effect of a Couple. Two equal and opposite forces having parallel lines of action form a couple. Figure 2–16a shows a couple formed by two such forces. The sum of the two forces is zero. The sum of the moment of the two forces, however, is not zero. The effect of a couple acting on a rigid body, therefore, is to cause the rigid body to rotate about an axis perpendicular to the plane of the forces.

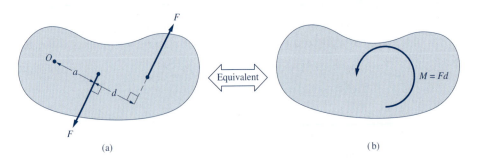

(a)   (b)

FIGURE 2–16

Moment of a Couple. Denoting the perpendicular distance between the two forces by d, the moment of a couple about an arbitrary point O is

$$M = F(a + d) - Fa = Fd \qquad (2\text{–}9)$$

Since O is an arbitrary point, *the moment of a couple about any point is equal to the magnitude of the forces times the perpendicular distance between the forces.* Fig. 2–16b shows an alternative representation of a couple. A couple can be placed anywhere in the plane of the forces.

Equivalent Couples. Two couples acting on the same plane or parallel planes are equivalent if they have the same moment acting in the same direction; that is, two couples are equivalent if both the magnitude and the direction of their moments are equal.

Addition of Couples. The addition of two or more couples in a plane or parallel planes is the algebraic sum of their moments. Unless specified otherwise, we will treat a c.c.w moment as positive and a c.w. moment as negative.

────── **EXAMPLE 2–13** ──────────────────

Determine the moment of the couple applied to the torque wrench shown in Fig. E2–13 if the magnitude of **F** is 40 lb.

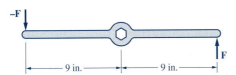

FIGURE E2–13

Solution. The perpendicular distance between the two forces is

$$d = 9 \text{ in.} + 9 \text{ in.}$$
$$= 18 \text{ in.} = 1.5 \text{ ft}$$

By definition, the moment of the couple is

$$M = Fd$$
$$= +(40 \text{ lb})(1.5 \text{ ft})$$
$$= +60 \text{ lb} \cdot \text{ft}$$

which produces a c.c.w. torque on the nut. Thus, the moment of the couple is

$$M = 60 \text{ lb} \cdot \text{ft} \ \circlearrowleft \qquad \qquad \Leftarrow \textbf{Ans.}$$

── EXAMPLE 2–14 ──────────────────────

Two couples act on the rectangular plate shown in Fig. E2–14(1).
(a) Determine the resultant moment of the two couples.
(b) Replace the couples by an equivalent couple formed by two smaller forces applied at the corners A and C.

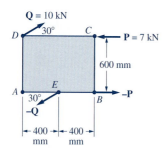

FIGURE E2–14(1)

Solution.

   (a) Resultant Moment. Because the perpendicular distance between Q and $-Q$ cannot be determined readily, the forces are resolved into horizontal and vertical components, as shown in Fig. E2–14(2). Now the two horizontal components form a c.w. couple and the two vertical components also form a c.w. couple. The resultant moment of the couples is the algebraic sum of the moment of each couple. Thus,

$$M = +(7 \text{ kN})(0.6 \text{ m}) - (8.66 \text{ kN})(0.6 \text{ m}) - (5 \text{ kN})(0.4 \text{ m})$$
$$= -3.00 \text{ kN} \cdot \text{m}$$
$$= 3.00 \text{ kN} \cdot \text{m} \; \circlearrowright \qquad\qquad\qquad\qquad \Leftarrow \textbf{Ans.}$$

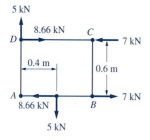

FIGURE E2–14(2)

   (b) Equivalent Couple. The equivalent couple is formed by two equal and opposite forces $\mathbf{F}$ and $-\mathbf{F}$ applied at the corners A and C shown in Fig. E2–14(3). If the forces are to be the smallest, they must be perpendicular to the diagonal AC. Thus,

$$d = \sqrt{(0.8 \text{ m})^2 + (0.6 \text{ m})^2}$$

$$= 1.0 \text{ m}$$

$$Fd = M$$

$$F = \frac{M}{d}$$

$$= \frac{3.00 \text{ kN} \cdot \text{m}}{1.0 \text{ m}}$$

$$= 3.00 \text{ kN}$$

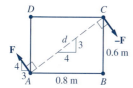

FIGURE E2–14(3)

The equivalent couple is formed by two forces

$$\mathbf{F} = 3.00 \text{ kN} \quad \text{at } A \text{ and } -\mathbf{F} \text{ at } C \qquad \Leftarrow \textbf{Ans.}$$

2–9
REPLACING A FORCE WITH A FORCE-COUPLE SYSTEM

Two systems of forces are said to be equivalent if they produce the same mechanical effect on a rigid body. The mechanical effect of any system of forces on a rigid body is characterized entirely by the resultant force and the resultant moment of the system. Hence, we define *equivalent force systems* as follows.

Equivalent Force Systems. Systems of forces are said to be *equivalent* if they have *the same resultant force and the same resultant moment about the same point.*

Force-Couple System. Consider a force **F** acting on a rigid body at point *A* (Fig. 2–17a). Suppose that it is necessary to move the force to another point *B*. From the principle of transmissibility, we know that we can move the force to any point along its line of action, but we cannot move the force to *any* point *B*. We may, however, add two equal and opposite forces **F** and −**F** at point *B* (Fig. 2–17b) without altering the mechanical effect of the original force. Now we have a force **F** at *B*, and a couple formed by force **F** at *A* and force −**F** at *B*. The moment of the couple is

$$M = Fd \qquad\qquad (2\text{–}10)$$

where d is the perpendicular distance from point B to the line of action of the given force at A. We see that the moment M is simply the moment of the original force at A about point B. Since the moment of the couple about any point is the same, it may be placed anywhere in the plane. For convenience, however, the force and the couple are usually shown to act at the same point B (Fig. 2–17c) and we refer to this combination as a *force-couple system*. Thus, *any given force may be moved to another point without changing its mechanical effect, provided that an appropriate couple is added. The couple has a moment produced by the given force about the point where the force is to be located.*

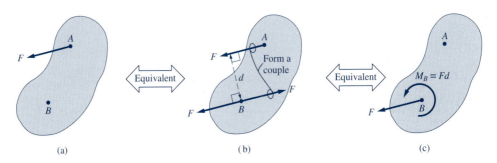

(a)   (b)   (c)

FIGURE 2–17

EXAMPLE 2–15

Replace the 2-kN force **F** shown in Fig. E2–15(1) with the force-couple system at point B.

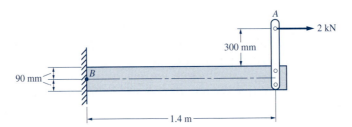

FIGURE E2–15(1)

Solution. Refer to Fig. E2–15(2). The force **F** can be moved from point A to point B, provided that a suitable couple is introduced. The moment of the couple is equal to the moment of the force at A about point B. We have

$$M_B = -\,(2\text{ kN})(0.3\text{ m} + 0.09\text{ m})$$
$$= -0.78\text{ kN} \cdot \text{m} \; \circlearrowleft \qquad\qquad \Leftarrow \textbf{Ans.}$$

Thus, the single force at A is now replaced by an equivalent force-couple system at point B, as shown in Fig. E2–15(2).

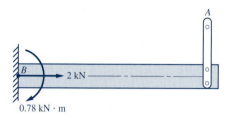

FIGURE E2-15(2)

2-10
RESULTANT OF A NONCONCURRENT COPLANAR FORCE SYSTEM

In a concurrent coplanar force system, the line of action of the resultant passes through the common point. In a nonconcurrent coplanar force system, there is no point of concurrency, so the location of the line of action of the resultant is not immediately known.

The magnitude and direction of the resultant can be calculated by using the rectangular components of the forces, similar to the method used for the concurrent coplanar force system. First, we choose convenient x and y coordinate axes and then resolve each force into rectangular components. *The components of the resultant are the algebraic sums of the corresponding components of all the forces in the system.* We write

$$R_x = \Sigma F_x \qquad R_y = \Sigma F_y \qquad (2\text{--}11)$$

From these components, the magnitude and direction of the resultant can be determined. Now the location of the line of action of the resultant can be determined by the requirement of the moments. If two force systems are equivalent, the resultant moments of the two systems about an arbitrary point must be equal. Consider the given force system $\mathbf{F}_1$, $\mathbf{F}_2$, $\mathbf{F}_3$, and $\mathbf{F}_4$ acting on the beam shown in Fig. 2-18. The resultant $\mathbf{R}$ of the given force system is assumed to act through point C at distance $\bar{x}$ to the right of point A. The moment of $\mathbf{R}$ about A must be the same as the sum of the moments of the given forces about A. Note that R_x passes through A, so it produces no moment about A. We write

$$R_y \bar{x} = \Sigma M_A$$

From which $\bar{x}$ can be solved.

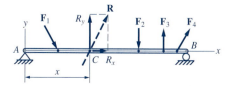

FIGURE 2-18

—— **EXAMPLE 2–16** ——————————————————————————————

Three forces are applied to the beam shown in Fig. E2–16(1). Find the resultant of the three forces and the location of the resultant.

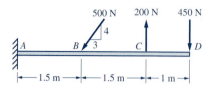

FIGURE E2–16(1)

Solution. The x and y coordinate axes are selected as shown in Fig. E2–16(2). The 500-N force is resolved into x and y components.

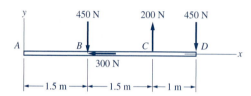

FIGURE E2–16(2)

Resultant.   Summing up the corresponding components, we get

$$R_x = \Sigma R_x$$
$$= -300 \text{ N} \leftarrow$$
$$R_y = \Sigma R_y$$
$$= -400 \text{ N} + 200 \text{ N} - 450 \text{ N}$$
$$= -650 \text{ N} \downarrow$$

From the components of the resultant, we see that it is a vector in the third quadrant. The magnitude and direction of the resultant are

$$R = \sqrt{(-300 \text{ N})^2 + (-650 \text{ N})^2} = 716 \text{ N}$$
$$\alpha = \tan^{-1} \left| \frac{650}{300} \right|$$
$$= 65.2°$$
$$\theta = 180° + 65.2°$$
$$= 245.2°$$
$$\mathbf{R} = 716 \text{ N} \ \angle\ 245.2° \qquad\qquad \Leftarrow \textbf{Ans.}$$

Location of Resultant.   To determine the location of the resultant, we need to find the sum of the moments of the forces about A.

$$\Sigma M_A = -(400\ \text{N})\ (1.5\ \text{m}) + (200\ \text{N})\ (3.0\ \text{m}) - (450\ \text{N})(4.0\ \text{m})$$
$$= -1800\ \text{N} \cdot \text{m} \ \circlearrowleft$$

Since R_y is downward, the resultant **R** must be to the right of point A to produce a clockwise moment about A. Equating the moment of **R** about point A to the moment found above, we write

$$(\circlearrowleft +)\colon (650\ \text{N})\ \bar{x} = 1800\ \text{N} \cdot \text{m}$$

From which we get

$$\bar{x} = +2.77\ \text{m} \qquad\qquad \Leftarrow \textbf{Ans.}$$

EXAMPLE 2–17

For the given force system acting on the bracket shown in Fig. E2–17(1), find the resultant and the point of intersection of the resultant along AD.

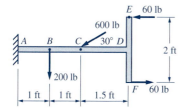

FIGURE E2–17(1)

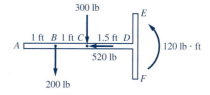

FIGURE E2–17(2)

Solution. Refer to Fig. E2–17(2). The two 60-lb forces acting in opposite directions form a couple. The moment of the couple is

$$M = (60\ \text{lb})(2\ \text{ft})$$
$$= 120\ \text{lb} \cdot \text{ft} \ \circlearrowleft$$

The couple is represented by its moment, and the 600-lb force is replaced by its rectangular components.

Resultant. The components of the resultant are

$$R_x = \Sigma F_x$$
$$= -520\ \text{lb} \ \longleftarrow$$
$$R_y = \Sigma F_y$$
$$= -200\ \text{lb} - 300\ \text{lb}$$
$$= -500\ \text{lb} \ \downarrow$$

The resultant is a vector in the third quadrant. The magnitude and direction of the resultant are

$$R = \sqrt{(-520 \text{ lb})^2 + (-500 \text{ lb})^2}$$

$$= 721 \text{ lb}$$

$$\alpha = \tan^{-1}\left|\frac{500}{520}\right| = 43.9°$$

$$\theta = 180° + 43.9°$$

$$= 223.9°$$

$$\mathbf{R} = 721 \text{ lb} \quad \angle \quad 223.9° \qquad \Leftarrow \textbf{Ans.}$$

Location of Resultant. To determine the location of the resultant, we need to sum up the moment of the forces about A.

$$\Sigma M_A = -(200 \text{ lb})(1.0 \text{ ft}) - (300 \text{ lb})(2.0 \text{ ft}) + 120 \text{ lb} \cdot \text{ft}$$

$$= -680 \text{ lb} \cdot \text{ft} \quad \circlearrowright$$

Note that the moment of a couple is independent of the moment center, so the moment of the couple about A remains 120 lb · ft. Since R_y is downward, the resultant **R** must be to the right of point A to produce a clockwise moment about A. Equating the moment of **R** about A to the moment found above, we write

$$(\circlearrowright +): (500 \text{ lb}) \bar{x} = 680 \text{ lb} \cdot \text{ft}$$

From which we get

$$\bar{x} = +1.36 \text{ ft} \qquad \Leftarrow \textbf{Ans.}$$

EXAMPLE 2–18

Determine the resultant of the three forces and a couple acting on the plate shown in Fig. E2–18(1). Find the point of intersection of the resultant along the x and y axes.

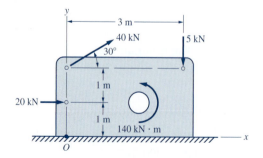

FIGURE E2–18(1)

Solution. The 40-kN force is replaced by its rectangular components and the couple is moved to point O, as shown in Fig. E2–18(2).

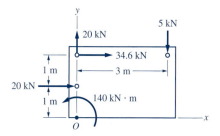

FIGURE E2–18(2)

Resultant.   The components of the resultant are

$$R_x = \Sigma F_x$$
$$= 20 \text{ kN} + 34.6 \text{ kN}$$
$$= 54.6 \text{ kN} \longrightarrow$$
$$R_y = \Sigma F_y$$
$$= 20 \text{ kN} - 5 \text{ kN}$$
$$= 15.0 \text{ kN} \uparrow$$

Thus, the resultant is a vector in the first quadrant. The magnitude and direction of the resultant are

$$R = \sqrt{(54.6 \text{ kN})^2 + (15 \text{ kN})^2}$$
$$= 56.6 \text{ kN}$$
$$\theta = \alpha = \tan^{-1} \frac{15}{54.6}$$
$$= 15.4°$$
$$\mathbf{R} = 56.6 \text{ kN} \measuredangle 15.4° \qquad\qquad \Leftarrow \textbf{Ans.}$$

Location of Resultant.   To determine the location of the resultant, we need to find the sum of the moments of the forces about O.

$$\Sigma M_O = -(20 \text{ kN})(1.0 \text{ m}) - (34.6 \text{ kN})(2.0 \text{ m}) - (5.0 \text{ kN})(3.0 \text{ m}) + 140 \text{ kN} \cdot \text{m}$$
$$= +35.7 \text{ kN} \cdot \text{m} \; \circlearrowleft$$

The resultant $\mathbf{R}$ must be to the right and below point O to produce a counterclockwise moment about O.

At the x intercept, R_x produces no moment about O. Equating the moment of R_y about O to the moment found above, we write

$$(\circlearrowleft +)\text{: } (15.0 \text{ kN}) \, \bar{x} = 35.7 \text{ kN} \cdot \text{m}$$

From which we get

$$\bar{x} = 2.38 \text{ m} \qquad\qquad \Leftarrow \textbf{Ans.}$$

At the y intercept, R_y produces no moment about O. Equating the moment of R_x about O to the moment found above, we write

$$(\circlearrowleft+): (54.6 \text{ kN}) \, |\bar{y}| = 35.7 \text{ kN} \cdot \text{m}$$

From which we get

$$|\bar{y}| = 0.654 \text{ m} \qquad \qquad \Leftarrow \textbf{Ans.}$$

The location of the resultant is indicated in Fig. E2–18(3).

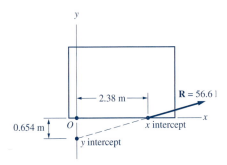

FIGURE E2–18(3)

2–11
RESULTANT OF DISTRIBUTED LINE LOADS

Distributed Load.   A *distributed load* occurs whenever the load applied to a body is not concentrated at a point. A distributed load could be exerted along a line, over an area, or throughout an entire solid body. This section deals only with distributed line loads. Examples of such loading include the weight of a beam or the load from the floor system that the beam supports, and loads caused by wind or liquid pressure.

Load Intensity.   A distributed load along a line is characterized by a load intensity expressed as force per unit length. For example, a load intensity of 1000 lb/ft, or 1 kip/ft, means that a load of 1000 lb, or 1 kip, is distributed over 1 ft length. In S.I. units, the load intensity is in N/m or kN/m.

Uniform Load.   A distributed load with a constant intensity is called a *uniform load.* A uniform load may be represented by a loading diagram in the shape of a rectangular block, as shown in Fig. 2–19a. The uniform intensity w is represented by the height of the block, and the length of distribution b is represented by the width of the block. The weight of a beam of uniform cross-section is an example of a uniform load.

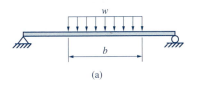 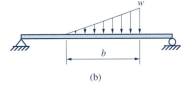

(a)   (b)

FIGURE 2–19

Triangular Load. A *triangular load* is a distributed load whose intensity varies linearly from zero to a maximum intensity w. A triangular load may be represented by a loading diagram in the shape of a triangle, as shown in Fig. 2–19b. Liquid pressure can be represented by a triangular load.

Equivalent Concentrated Force. For the purpose of determining the resultant of a force system, each distributed load may be replaced by its equivalent concentrated force. It will be established later in Section 7–5 that *a distributed line load may be replaced by an equivalent concentrated force having a magnitude equal to the area of the loading diagram and a line of action passing through the centroid of the loading diagram.* The equivalent concentrated forces of a uniform load and a triangular load are shown in Fig. 2–20a and b, respectively.

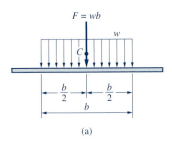 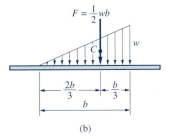

(a)   (b)

FIGURE 2–20

Trapezoidal Load. A load diagram in the shape of a trapezoid can be treated as a uniform load plus a triangular load, as shown in Fig. 2–21a. The general case of distributed load shown in Fig. 2–21b will be treated later in Section 7–5.

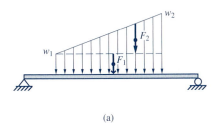 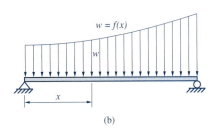

(a)   (b)

FIGURE 2–21

━━━━ **EXAMPLE 2–19** ━━

Determine the equivalent resultant force of the distributed loads on the beam in Fig. E2–19(1), and specify its location along the beam.

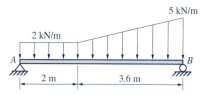

FIGURE E2–19(1)

Solution. The loading diagram is divided into a rectangle and a triangle, as shown in Fig. E2–19(2). The rectangle represents a uniformly distributed load of a constant intensity of 2 kN/m. The triangle represents a load with an intensity varying linearly from 0 to 3 kN/m.

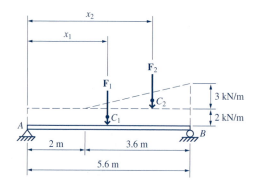

FIGURE E2–19(2)

Resultant.   The distributed loads may be replaced by their equivalent concentrated forces of magnitudes equal to their associated areas. We have

$$F_1 = (2 \text{ kN/m})(5.6 \text{ m})$$
$$= 11.2 \text{ kN}$$
$$F_2 = \frac{1}{2}(3 \text{ kN/m})(3.6 \text{ m})$$
$$= 5.4 \text{ kN}$$

The line of action of each equivalent concentrated force passes through the *centroid* of the associated area of its loading diagram. The distances from A to the lines of action of the forces are

$$x_1 = \frac{1}{2}(5.6 \text{ m})$$
$$= 2.8 \text{ m}$$
$$x_2 = 2 \text{ m} + \frac{2}{3}(3.6 \text{ m})$$
$$= 4.4 \text{ m}$$

The magnitude of the resultant is

$$(\downarrow +): R = F_1 + F_2$$
$$= 11.2 \text{ kN} + 5.4 \text{ kN}$$
$$R = 16.6 \text{ kN} \downarrow \qquad\qquad \Leftarrow \textbf{Ans.}$$

Location of Resultant. Refer to Fig. E2–19(3). The distance $\bar{x}$ from A to the line of action of **R** may be obtained by equating the moment of **R** about A to the sum of the moments of $\mathbf{F}_1$ and $\mathbf{F}_2$ about A. We write

$$(\circlearrowleft +): (16.6 \text{ kN})\bar{x} = (11.2 \text{ kN})(2.8 \text{ m}) + (5.4 \text{ kN})(4.4 \text{ m})$$

From which we get

$$\bar{x} = 3.32 \text{ m} \qquad\qquad \Leftarrow \textbf{Ans.}$$

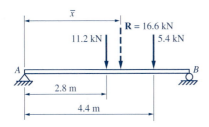

FIGURE E2–19(3)

EXAMPLE 2–20

Determine the equivalent resultant force of the loads acting on the beam shown in Fig. E2–20(1), and specify the location of the resultant along the beam.

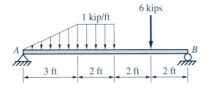

FIGURE E2–20(1)

Solution. Note that this combined loading consists of a concentrated load, a uniform load, and a triangular load.

Resultant. The two distributed loads may be replaced by their equivalent concentrated forces of magnitudes:

$$F_1 = \frac{1}{2}(1 \text{ kip/ft})(3 \text{ ft})$$
$$= 1.5 \text{ kips}$$
$$F_2 = (1 \text{ kip/ft})(2 \text{ ft})$$
$$= 2 \text{ kips}$$

Each of the equivalent concentrated forces acts vertically downward through the *centroid* of the associated area of its loading diagram, as shown in Fig. E2–20(2). The magnitude of the resultant is

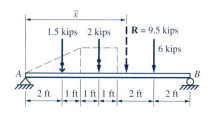

FIGURE E2–20(2)

$$(\downarrow +): R = \Sigma F_y = 1.5 \text{ kips} + 2 \text{ kips} + 6 \text{ kips}$$
$$= 9.5 \text{ kips}$$
$$\mathbf{R} = 9.5 \text{ kips} \downarrow \qquad \Leftarrow \textbf{Ans.}$$

Location of Resultant. The distance $\bar{x}$ from point A to the line of action of **R** may be obtained by equating the moment of **R** about A to the sum of the moment of the forces about A. We write

$$(\circlearrowleft +): (9.5 \text{ kips})\bar{x} = (1.5 \text{ kips})(2 \text{ ft}) + (2 \text{ kips})(4 \text{ ft}) + (6 \text{ kips})(7 \text{ ft})$$

From which we get

$$\bar{x} = 5.58 \text{ ft} \qquad \Leftarrow \textbf{Ans.}$$

─── **EXAMPLE 2–21** ───

The bracket ABC is subjected to a uniform load and a trapezoidal load, as shown in Fig. E2–21(1). Replace the loads with an equivalent resultant force and specify its location along BC measured from the fixed end C.

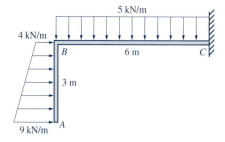

FIGURE E2–21(1)

Solution. The trapezoidal load is divided into a uniform load and a triangular load, as shown in Fig. E2–21(2).

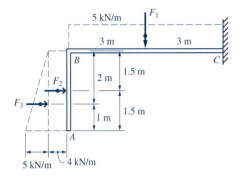

FIGURE E2–21(2)

 Resultant.   The equivalent concentrated forces of the distributed
loads are

$$F_1 = (5 \text{ kN/m}) \, (6 \text{ m}) \ = 30 \text{ kN}$$

$$F_2 = (4 \text{ kN/m}) \, (3 \text{ m}) \ = 12 \text{ kN}$$

$$F_3 = \frac{1}{3}(5 \text{ kN/m})(3 \text{ m}) = 5 \text{ kN}$$

Each equivalent concentrated force passes through the centroid of the
associated area of its loading diagram, as shown in Fig. E2–21(2). Refer to
Fig. E2–21(3). The components of the resultant force are

$$R_x = \Sigma F_x = 12 \text{ kN} + 5 \text{ kN} = 17 \text{ kN} \ \longrightarrow$$

$$R_y = \Sigma F_y = -30 \text{ kN} \ \downarrow$$

$$\mathbf{R} = 34.5 \text{ kN} \ \diagdown \ 60.5° \qquad\qquad \Leftarrow \textbf{Ans.}$$

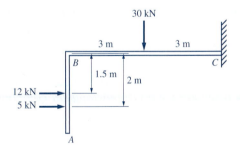

FIGURE E2–21(3)

 Location of Resultant.   Refer to Fig. E2–21(3). The resultant moment
of the forces about *C* is

$$(\circlearrowleft +): \Sigma M_C = (30 \text{ kN})(3\text{m}) + (12 \text{ kN})(1.5 \text{ m}) + (5 \text{ kN})(2 \text{ m})$$

$$= 118 \text{ kN} \cdot \text{m} \ \circlearrowleft$$

The single resultant force acting at point *D* along *BC* in Fig. E2–21(4) must
produce the same c.c.w. moment about *C* as calculated above. We write

$$(\circlearrowleft +)\colon (30 \text{ kN})\,\bar{x} = 118 \text{ kN} \cdot \text{m}$$

From which we get

$$\bar{x} = 3.93 \text{ m} \qquad\qquad \Leftarrow \textbf{Ans.}$$

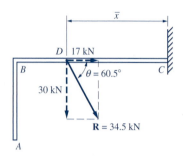

FIGURE E2–21(4)

2–12
SUMMARY

Resultant. A system of coplanar forces acting on a rigid body may be replaced by a single equivalent force called its *resultant*, which will produce the same mechanical effect to the rigid body as the given system. The determination of the resultant of a given coplanar force system is the major topic of this chapter.

Parallelogram Law. The resultant of two concurrent coplanar forces can be determined graphically by using the *parallelogram law* or the *triangular rule*. The resultant of three or more concurrent coplanar forces can be determined graphically by using the *polygon rule*.

Rectangular Components. Two mutually perpendicular components whose resultant is equal to a force are called the *rectangular components* of the force. If the magnitude and direction angle of a force are known, then the rectangular components are

$$F_x = F \cos\theta \qquad F_y = F \sin\theta \tag{2–2}$$

where the direction angle θ must be measured in the *standard position*. When the rectangular components of the force are given, the magnitude and direction of the force may be determined from

$$F = \sqrt{F_x^2 + F_y^2} \tag{2–3}$$

and the reference angle α is

$$\alpha = \tan^{-1} \left| \frac{F_y}{F_x} \right| \tag{2–4}$$

The direction angle θ in the standard position can then be determined.

Component Method. The resultant of any number of concurrent coplanar forces may be obtained by using the *component method.* First, each force is resolved into the x and y components. Then the components of the resultant can be determined from

$$R_x = \Sigma F_x \qquad R_y = \Sigma F_y \qquad\qquad (2\text{--}6)$$

From the two components R_x and R_y, the magnitude and direction of the resultant **R** can be determined. The line of action of the resultant **R** must pass through the common point of the given concurrent force system.

Moment. The *moment of a force* about a *moment center* is defined as the product of the magnitude of the force and the *moment arm,* which is the perpendicular distance from the *moment center* to the line of action of the force. Sometimes it is more convenient to determine the moment of a force by using its components. *Varignon's theorem* states that the moment of a force about a point is equal to the sum of the moments of the components of the force about the same point. By using the *principle of transmissibility,* it is more convenient to resolve the force into rectangular components at a point along its line of action, where only one component produces a moment about the given point.

Couple. A *couple* is produced by two equal, opposite, and noncollinear forces. The moment of a couple is equal to the magnitude of the force multiplied by the perpendicular distance between the two forces. A couple is characterized by its moment, which is independent of the moment center. Two couples on the same plane or parallel planes are equivalent if they have moments of the same magnitude and direction.

Force-Couple System. A force acting on a rigid body may be replaced by an equivalent *force-couple system* at an arbitrary point O consisting of the force applied at O and a couple having a moment equal to the moment about O of the given force at the original location.

Resultant of a Nonconcurrent Coplanar Force System. A nonconcurrent coplanar force system can be replaced by a single resultant force. The components of the resultant force may be determined the same way as those of the concurrent coplanar force system. From the two components, the magnitude and direction of the resultant can be determined. The location of the line of action of the resultant must be such that its moment about O will be equal to the sum of the moments of the given forces about O.

Resultant of Distributed Line Loads. For the purpose of determining the resultant of a force system, each distributed force is replaced by its equivalent concentrated force as follows:

For *uniform load:* $\mathbf{F} = wb$ acting through the midpoint of length b

For *triangular load:* $\mathbf{F} = \frac{1}{2}wb$ acting through the centroid of the triangle

PROBLEMS

Section 2–3   Resultant of Concurrent Forces

2–1   Determine graphically the magnitude and direction of the resultant of the two forces shown in Fig. P2–1 using (*a*) the parallelogram law and (*b*) the triangle rule.

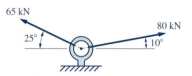

FIGURE P2–1

2–2   Solve Problem 2–1 by the trigonometric method.

2–3   Determine the magnitude and direction of the resultant of two forces acting on the eye hook shown in Fig. P2–3 by (*a*) the graphical method and (*b*) the trigonometric method.

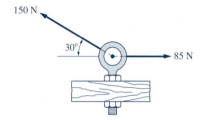

FIGURE P2–3

2–4   Determine the magnitude and direction of the resultant of the two forces acting on the bracket shown in Fig. P2–4.

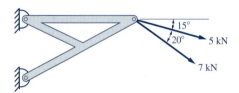

FIGURE P2–4

2–5   Determine the magnitude of the force **P** so that the resultant of the two forces acting on the block shown in Fig. P2–5 is vertical.

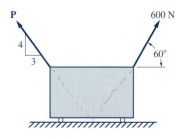

FIGURE P2–5

2–6 A trolley is acted on by two forces as shown in Fig. P2–6. If $\mathbf{P} = 3.5$ kN, find the value of angle α so that the resultant of the forces is in the vertical direction.

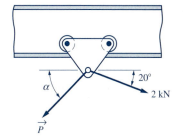

2–7 If $\alpha = 40°$ for the force $\mathbf{P}$ acting on the trolley shown in Fig. P2–6, find the magnitude of force $\mathbf{P}$ so that the resultant of the two forces is vertical.

FIGURE P2–6

2–8 If the resultant of the two forces $\mathbf{P}$ and $\mathbf{Q}$ acting on the ring in Fig. P2–8 is a vertical force equal to 45 kN, find the magnitude and direction of force $\mathbf{Q}$.

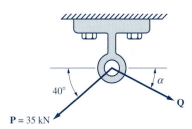

FIGURE P2–8

2–9 A barge is pulled by two tugboats as shown in Fig. P2–9. The tension in cable AC is 1000 lb. Determine the tension in cable AB if the resultant of the cable tensions is along the x axis.

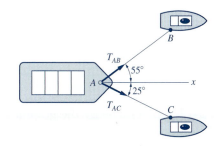

FIGURE P2–9

2–10 The resultant of cable tension $\mathbf{T}$ and a 5-kN weight must act along the axis of boom AB of the derrick shown in Fig. P2–10. Determine (*a*) the magnitude of tension $\mathbf{T}$ if $\theta = 30°$, and (*b*) the value of angle θ for which the tension $\mathbf{T}$ is a minimum.

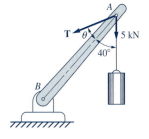

FIGURE P2–10

2–11   To have compressive soil pressure exist over the entire base of a gravity dam, the resultant of the forces acting on the dam above the base must pass through the middle third of the base. For the gravity dam shown in Fig. P2–11, the force **W** represents the weight of the dam for a one-foot section. The total water pressure acting on the one-foot section is represented by the horizontal force **P**. Determine the resultant of **W** and **P**. Is it within the middle third of the base?

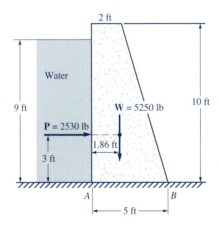

FIGURE P2–11

2–12   Determine by the graphical method the resultant of the three forces acting on the eye hook shown in Fig. P2–12.

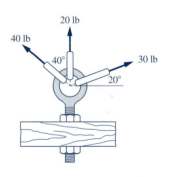

FIGURE P2–12

2–13   Determine by the graphical method the magnitude and direction of the resultant of the four forces shown in Fig. P2–13.

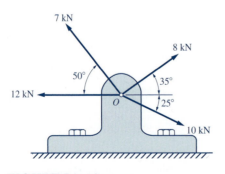

FIGURE P2–13

Section 2–4 Rectangular Components

2–14 to **2–17** Determine the x and y components of the forces shown in Figs. P2–14 to P2–17.

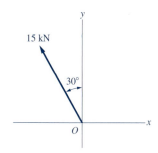

FIGURE P2–14

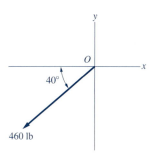

FIGURE P2–15

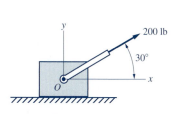

FIGURE P2–16

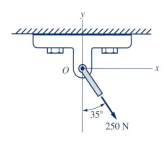

FIGURE P2–17

2–18 Refer to Fig. P2–18. Prove that the x and y components of a force acting in a direction indicated by the ratio $h{:}v$ are

$$F_x = \frac{h}{\sqrt{h^2 + v^2}}\, F \qquad F_y = \frac{v}{\sqrt{h^2 + v^2}}\, F$$

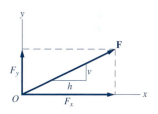

FIGURE P2–18

2–19 and **2–20** Find the x and y components of the forces shown in Figs. P2–19 and P2–20 by using the formulas in Problem 2–18.

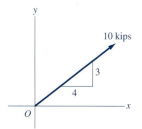

FIGURE P2–19

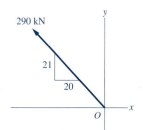

FIGURE P2–20

2–21 and **2–22**   Find the x and y components of the forces **P** and **Q** shown in Figs. P2–21 and P2–22.

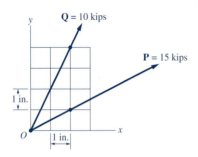

FIGURE P2–21

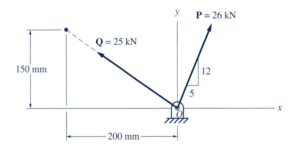

FIGURE P2–22

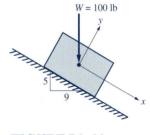

2–23   Find the x and y components of the weight of the block shown in Fig. P2–23.

FIGURE P2–23

Section 2–5   Resultants by Rectangular Components

2–24 to **2–27**   Determine the magnitude and direction of the resultant of the force systems shown in Figs. P2–24 to P2–27.

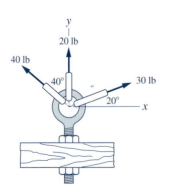

FIGURE P2–24

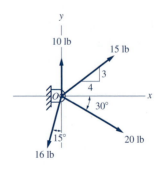

FIGURE P2–25

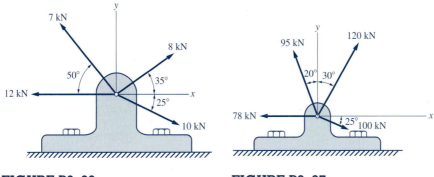

FIGURE P2–26   **FIGURE P2–27**

2–28   A collar that slides freely on a horizontal rod is subjected to the three forces shown in Fig. P2–28. Determine the angle θ for which the resultant of the three forces is vertical.

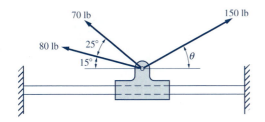

FIGURE P2–28

Section 2–6   Moment of a Force
Section 2–7   Varignon's Theorem

2–29   Refer to Fig. P2–29. Determine the moment of the 10-kN force about point O.

FIGURE P2–29

2–30   Refer to Fig. P2–30. Determine the moment of the 10-kip force shown about A by (a) using the definition directly and (b) resolving the force into horizontal and vertical components.

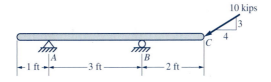

FIGURE P2–30

2–31 Refer to Fig. P2–31. Determine the moment of the 50-lb force about point *A* by resolving the force into horizontal and vertical components.

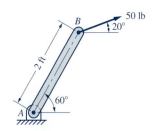

2–32 Rework Problem 2–31 by resolving the force into components along and perpendicular to *AB*.

FIGURE P2–31

2–33 Refer to Fig. P2–33. Determine the moment of the 2000-N force shown about *A* by (*a*) using the definition directly, (*b*) resolving the force into horizontal and vertical components at *C*, and (*c*) resolving the force into components at *D*.

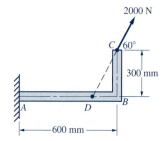

FIGURE P2–33

2–34 Refer to Fig. P2–34. Determine the moment of the 200-N force about point *B* if α is 60°.

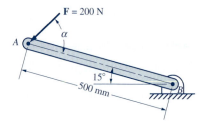

2–35 Refer to Fig. P2–34. Determine (*a*) angle α for which the moment of the 200-N force **F** about point *B* is a maximum and (*b*) the maximum moment.

FIGURE P2–34

2–36 Refer to Fig. P2–36. Determine the total moment of the two forces about point *O*.

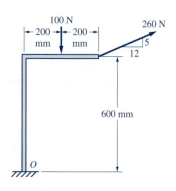

FIGURE P2–36

2–37 Determine the total moment of the three forces about point D in Fig. P2–37.

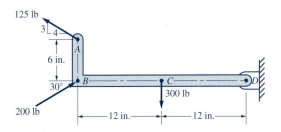

FIGURE P2–37

2–38 Refer to Fig. P2–38. Determine (*a*) the moment of the 400-lb force about point O, (*b*) the magnitude and direction of a vertical force applied at B that will produce the same moment about O, and (*c*) the smallest force applied at C that will produce the same moment about O.

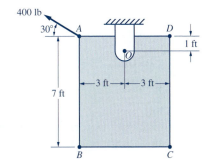

FIGURE P2–38

2–39 Refer to Fig. P2–39. Determine the moment of the 50-kN force about (*a*) the center O and (*b*) point B.

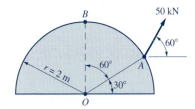

FIGURE P2–39

Section 2–8 Couple

2–40 to **2–42** Determine the moment of the couple acting on the bodies shown in Figs. P2–40 to P2–42.

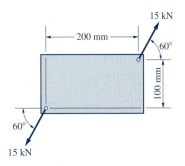

FIGURE P2–40

FIGURE P2–41

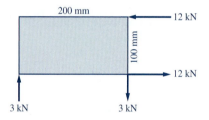

FIGURE P2–42

2–43 and **2–44**   Determine the resultant moment of the couples acting on the bodies shown in Figs. P2–43 and P2–44.

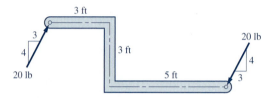

FIGURE P2–43

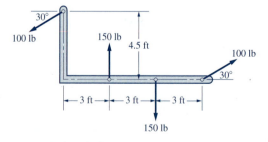

FIGURE P2–44

2–45   Two couples are applied to the 3-ft by 4-ft rectangular plate shown in Fig. P2–45. Prove that the total moment of the couples is zero by (*a*) adding the moment of the couples and (*b*) showing that the resultant of the two forces acting at the corner A is equal to, opposite to, and collinear with the resultant of the two forces acting at the corner B.

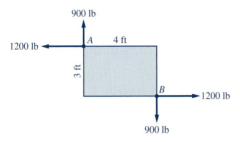

FIGURE P2-45

2-46   The angle bracket shown in Fig. P2–46 is subjected to the two 5-kN forces applied at points A and B. Replace these forces with an equivalent system consisting of the 7-kN force applied at point C and a second force applied at point D. Determine the magnitude and the direction of the second force at D and the distance CD.

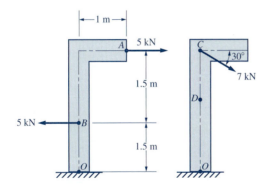

FIGURE P2-46

2-47   The plate in Fig. P2–47 is subjected to a system of forces that form three couples as shown. Determine the resultant moment of the couples.

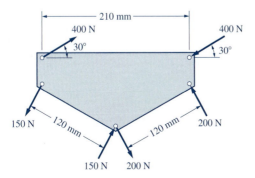

FIGURE P2-47

Section 2–9 Replacing a Force with a Force-Couple System

2–48 Replace the 5-kN horizontal force on the lever in Fig. P2–48 with an equivalent force-couple system at *O*.

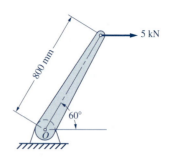

5 kN

800 mm

60°

O

FIGURE P2–48

2–49 Replace the 10-kip force acting on the post in Fig. P2–49 with an equivalent force-couple system at *C*.

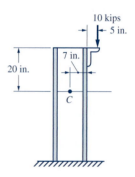

10 kips

5 in.

7 in.

20 in.

C

FIGURE P2–49

2–50 Replace the 2-kip force in Fig. P2–50 with an equivalent force-couple system at *B*.

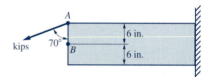

A

6 in.

kips   70°   *B*

6 in.

FIGURE P2–50

2–51 Replace the 600-lb force acting on the connection in Fig. P2–51 with an equivalent force-couple system at the center of rivet *B*.

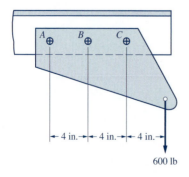

A   *B*   *C*

← 4 in. →← 4 in. →← 4 in. →

600 lb

FIGURE P2–51

2–52 Refer to Fig. P2–52. Replace the 500-N force applied to the bracket at *A* with an equivalent force-couple system at *B*.

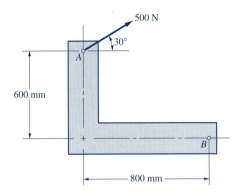

FIGURE P2–52

2–53 Replace the 20-lb force exerted on the wrench in Fig. P2–53 with an equivalent force-couple at *O*.

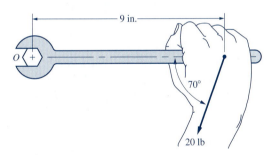

FIGURE P2–53

2–54 In designing the lift hook shown in Fig. P2–54, we must replace the 5-ton force with a force-couple at point *B* of section *a–a*. If the moment of the couple is 2500 lb-ft, determine the distance *d*.

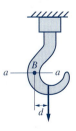

FIGURE P2–54

Section 2–10   Resultant of a Nonconcurrent Coplanar Force System

2–55   Replace the force and couple in Fig. P2–55 with a single force applied at a point on the diameter *AB*. Determine the distance from the center *O* to the point of application of the single force.

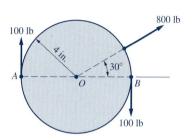

FIGURE P2–55

2–56   Replace the force and couple in Fig. P2–56 with a single force acting at a point along *AB*.

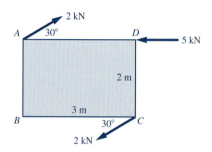

FIGURE P2–56

2–57   Replace the force and couple in Fig. P2–57 with a single force applied at a point along *AB*.

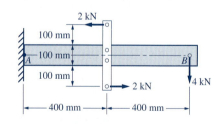

FIGURE P2–57

2–58   If the force-couple exerted on the beam in Fig. P2–58 can be replaced with an equivalent single force at *B*, find the magnitude of force **F**.

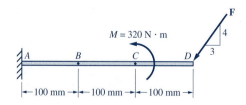

FIGURE P2–58

2–59 If the 350-N force and the couple M in Fig. P2–59 can be replaced with an equivalent single force at the corner C, determine the moment of the couple.

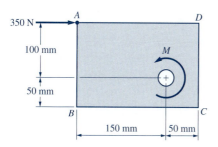

FIGURE P2–59

2–60 Refer to Fig. P2–60. Reduce the forces acting on the beam to a single resultant force and determine its point of application along AB.

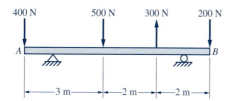

FIGURE P2–60

2–61 Refer to Fig. P2–61. Determine the height of the point above the base B through which the resultant of the three forces passes.

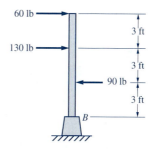

FIGURE P2–61

2–62 The trolley shown in Fig. P2–62 can be moved freely along the rail. Determine the location of the resultant of the two forces from point A when (a) $a = 2$ m and (b) $a = 3$ m.

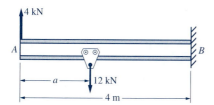

FIGURE P2–62

2–63 Explain why the resultant of the three forces acting on the beam in Fig. P2–63 always passes through point A for any value of force **F**.

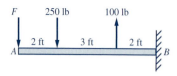

FIGURE P2–63

2–64 Find the magnitude, direction, and location of the resultant of the three forces acting on the beam in Fig. P2–64.

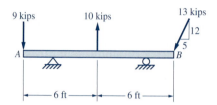

FIGURE P2–64

2–65 Refer to Fig. P2–65. Determine the magnitude of the vertical force **F** if the resultant of the three forces acting on the crank passes through the bearing O.

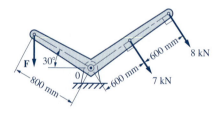

FIGURE P2–65

2–66 Determine the resultant of the four forces on the truss in Fig. P2–66 and its location along AB.

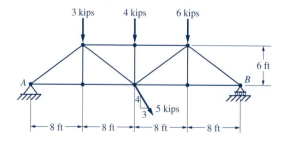

FIGURE P2–66

2–67 Replace the three forces acting on the frame in Fig. P2–67 with an equivalent force-couple system at *A*. Find the location of the single resultant above *A*.

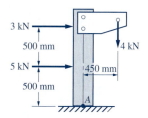

FIGURE P2–67

2–68 Determine the magnitude, direction, and location of the resultant of the two forces and a couple acting on the beam in Fig. P2–68.

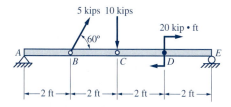

FIGURE P2–68

2–69 For the angle bracket in Fig. P2–69 subjected to the system of forces and couple shown, determine the resultant force and the points of intersection of its line of action with the *x* and *y* axes.

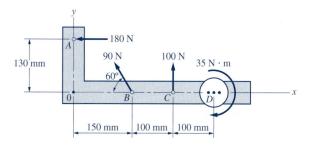

FIGURE P2–69

2–70 For the bracket in Fig. P2–70 subjected to the system of forces and couples shown, determine the single resultant and the point of intersection of its line of action with line *BC*.

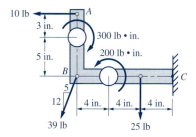

FIGURE P2–70

2–71   For the bracket in Fig. P2–71 subjected to the system of forces and couples shown, determine the point along line *AC* where the single resultant passes through.

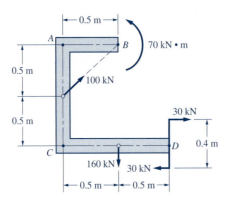

FIGURE P2–71

2–72   Reduce the force system acting on the bracket in Fig. P2–72 to the simplest form.

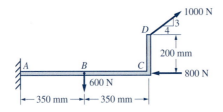

FIGURE P2–72

2–73   To have compressive soil pressure exerted over the entire base of a retaining wall, the resultant of the forces acting on the wall above the base must pass through the middle third of the base. For the retaining wall shown in Fig. P2–73, the vertical forces represent the weight of the concrete wall and footing, and the weight of the earth above the footing for a one-foot section of the wall. The horizontal force represents the total earth pressure acting on a one-foot section. Determine the location where the resultant passes through the base. Is it within the middle third of the base?

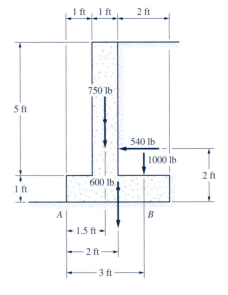

FIGURE P2–73

2-74 To have compressive soil pressure exerted over the entire base of a gravity dam, the resultant of the forces acting on the dam above the base must pass through the middle third of the base. For the gravity dam shown in Fig. P2–74, the weight of two parts of the dam for a one-meter section is shown. The total water pressure acting on the one-meter section is represented by the horizontal force. Determine the location where the resultant passes through the base. Is it within the middle third of the base?

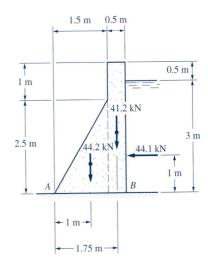

FIGURE P2–74

Section 2–11   Resultant of Distributed Line Loads

2-75 The loading on the bookshelf can be considered as three uniform loads, as shown in Fig. P2–75. Find the equivalent resultant force and specify its location along the shelf from point A.

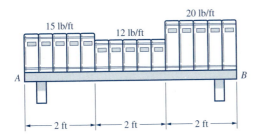

FIGURE P2–75

2-76 to 2-81 Replace the loading on the beams shown in Figs. P2–76 to Fig. P2–81 with an equivalent resultant force and specify their location along each beam measured from the left-hand end A.

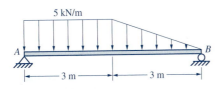

FIGURE P2–76

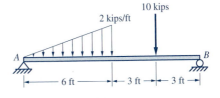

FIGURE P2–77

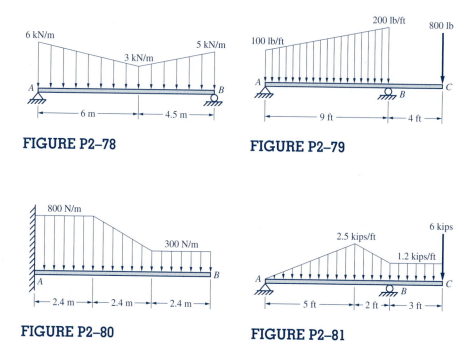

FIGURE P2–78   **FIGURE P2–79**

FIGURE P2–80   **FIGURE P2–81**

2–82   Replace the loading on the vertical post in Fig. P2–82 with an equivalent resultant force and specify its location along the post from the fixed end A.

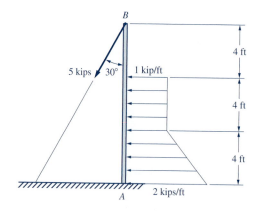

FIGURE P2–82

2–83 and **2–84**   Replace the loading on the brackets in Fig. P2–83 and Fig. P2–84 with an equivalent resultant force and specify its location along AB measured from a convenient point.

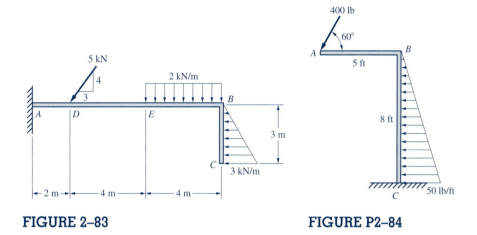

FIGURE 2–83   FIGURE P2–84

2–85   The beam in Fig. P2–85 is subjected to the distributed loading shown. Determine the distances a and b of the uniform load such that the resultant force and the resultant couple moment of the loading are zero.

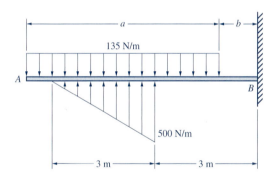

FIGURE P2–85

2–86   Determine the distances a and b of the triangular load in Fig. P2–86 so that the resultant force of the loading is a 200-lb force acting downward at the midpoint of the beam.

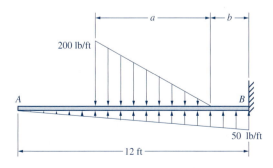

FIGURE P2–86

2–87   To have compressive soil pressure exerted over the entire base of a retaining wall, the resultant of the forces acting on the wall above the base must pass through the middle third of the base. The retaining wall in Fig. P2–87 is of concrete with a specific weight (weight per unit volume) of 150 lb/ft³. The specific weight of the earth is 100 lb/ft³. The lateral pressure exerted by the earth on a one-foot section of the wall is a triangular load as shown. Using a one-foot section of the wall, determine the location where the resultant passes through the base. Is it within the middle third of the base?

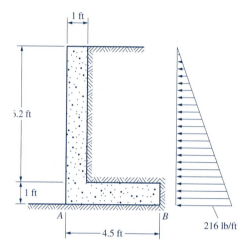

FIGURE P2–87

2–88   To have compressive soil pressure exerted over the entire base of a gravity dam, the resultant of the forces acting on the dam above the base must pass through the middle third of the base. The gravity dam in Fig. P2–88 is of masonry with a specific weight (weight per unit volume) of 23.6 kN/m³. The lateral pressure of water on a one-meter section of the dam is a triangular load as shown. Using a one-meter section of the dam, determine the location where the resultant passes through the base. Is it within the middle third of the base?

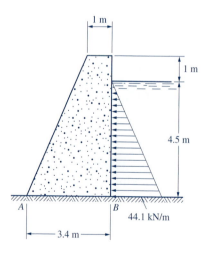

FIGURE P2–88

Computer Program Assignments

For each of the following problems, write a computer program using an appropriate programming language with which you are most familiar. Make the program user friendly by incorporating plenty of comments and input prompts so that the user will understand the input data to be entered and the limitations of their values. The output should include the data entered and the computed results, and they must be well labeled to identify each quantity. If a tabulated format is used, a proper heading must be included at the top of the table. Do not limit the program to any specific unit system. Indicate the consistent U.S. customary or SI units that can be used.

C2–1 Write a computer program that can be used to determine the resultant of a concurrent coplanar force system as shown in Fig. C2–1. The user input should include (1) the number of forces n, and (2) the magnitude and the direction angle in the standard position of each force F_i and θ_i. The output results must include the x and y components, and the magnitude and direction angle (in the standard position) of the resultant. Use this program to solve (a) Example 2–9, (b) Problem 2–25, and (c) Problem 2–26.

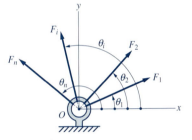

FIGURE C2–1

C2–2 Write a computer program that can be used to determine the resultant of a parallel coplanar force system as shown in Fig. C2–2. The user input should include (1) the number of forces n, and (2) the magnitude and location of each force F_i and d_i. Treat the downward force as positive and the upward force as negative. The output results must include the magnitude and direction of the resultant and its location along AB. Use this program to solve (a) Problem 2–60 and (b) Problem 2–61.

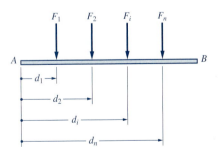

FIGURE C2–2

101

CHAPTER 3

EQUILIBRIUM OF COPLANAR FORCE SYSTEMS

3–1
INTRODUCTION

Statics deals primarily with the equilibrium of structures or machines. This chapter deals with the equilibrium of coplanar force systems. The equilibrium conditions are basic for the analysis of two-dimensional structures. A thorough understanding of the concept of equilibrium and its applications is of the utmost importance.

We shall make use of the concepts developed in Chapter 2, including forces, moments, and couples, as we apply the conditions of equilibrium to the two-dimensional problems. First, the equilibrium conditions are introduced. As we shall see later, the importance of free-body diagrams cannot be overemphasized, and they will be discussed in detail. They will be followed by the discussion of equilibrium problems involving concurrent-coplanar force systems. Equilibrium problems involving a two-force body and a three-force body are covered next. Finally, equilibrium problems involving a general nonconcurrent coplanar force system are discussed.

3–2
EQUILIBRIUM EQUATIONS

A body is in *equilibrium* when the resultant of all the external forces acting on the body is zero. This means that the forces and moments acting on the body are completely balanced. Under this condition, Newton's first law asserts that the body either remains at rest or continues to move along a straight line with constant velocity.

Equilibrium of a body requires that the resultant of all the forces acting on the body be equal to zero. Consequently, the sums of the corresponding rectangular components of all the external forces, ΣF_x and ΣF_y, must each be equal to zero. Since a body in equilibrium does not rotate, the algebraic sum of the moments, ΣM_A, of all the external forces about an arbitrary point A must be equal to zero also. Thus, the equilibrium equations for the equilibrium of the coplanar force system are

$$\Sigma F_x = 0 \qquad \Sigma F_y = 0 \qquad \Sigma M_A = 0 \qquad\qquad (3\text{--}1)$$

where the moment center A is an arbitrary point in the plane of the forces.

3–3
THE FREE-BODY DIAGRAM

When solving an equilibrium problem, it is important that the external forces acting on the body be described precisely. Any omission of forces acting on the body or the inclusion of extra forces will change the conditions of equilibrium and produce erroneous results. Therefore, a diagram indicating all the forces acting on the body must be drawn before the equilibrium equations are written. A *free-body diagram* (FBD) is a sketch of a body isolated from its supports or other connected bodies with all the *external forces* indicated clearly on the diagram. Such a diagram gives a detailed account of all the external forces acting on the body under consideration. *Constructing an appropriate free-body diagram is the single most important step for the solution of mechanics problems. Correct solution of a statics problem always depends on the successful completion of the free-body diagram.*

Construction of Free-Body Diagram. First, a particular body is selected for consideration as a free body. It could be the entire body or a portion of it. The free body selected is then isolated from its support or other part of the body, and a schematic diagram of the free body is drawn. On it, we must indicate clearly the magnitude, direction, and location of all the external forces, including weight, applied forces, reactions, and dimensions and angles.

Weight. The weight of the body represents the gravitational attraction of the earth. The weight of a body always acts vertically downward through the *center of gravity* of the body. (See Chapter 7 for details on the location of the center of gravity of a body.)

Applied Forces. The applied forces include the loads that are applied to the free body. When these forces are shown on the free-body diagram, their magnitudes, directions, and locations must be indicated clearly.

Reactions. The reactions represent the constraining forces exerted on the free body by the supports or by the connected bodies. Reactions from various types of supports will be discussed in detail in the next section. Since the reactions are usually unknown quantities, they are designated by symbols. Newton's law of action and reaction must be observed carefully when showing the directions of the reactions.

Dimensions and Angles. A free-body diagram must contain the significant dimensions and angles necessary for specifying the direction and location of the forces.

Note that *the internal forces within the body must never be drawn on the free-body diagram.* For example, the coupling force at a joint between two connected bodies should not be shown if two bodies are not separated at the joint.

3–4
TYPES OF SUPPORTS

A two-dimensional structure contains members that lie on the same plane. Supports for these structures also lie on the same plane. Basically, there are three types of supports for two-dimensional structures.

Roller Supports with One Unknown Reaction Element. Figure 3–1 shows examples of this type of support. The reaction from a roller support (Fig. 3–1a), rollers support (Fig. 3–1b), rocker support (Fig. 3–1c), or smooth surface (Fig. 3-1d) is a force perpendicular to the plane of support. The reaction from a cable support (Fig. 3-1e) or link support (Fig. 3-1f) is along the cable or the link. Hence, this type of support contains only one reaction element.

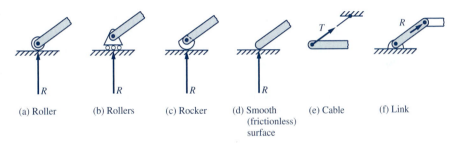

| (a) Roller | (b) Rollers | (c) Rocker | (d) Smooth (frictionless) surface | (e) Cable | (f) Link |

FIGURE 3–1

Hinge Support with Two Unknown Reaction Elements. Figure 3–2 shows examples of this type of support. These supports prevent the structure from moving along any direction at the point of support. In a hinge support (Fig. 3–2a), the hinge is assumed to be frictionless; hence, there is no moment component on the reaction. The knife-edge support (Fig. 3–2b) is a schematic drawing of a hinge support. The rough surface (Fig. 3–2c) produces friction force that prevents the structure from sliding. The reactions from this type of support contain two unknown components: either R and θ, or R_x and R_y.

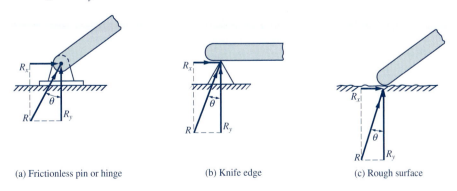

| (a) Frictionless pin or hinge | (b) Knife edge | (c) Rough surface |

FIGURE 3–2

Fixed Support with Three Unknown Reaction Elements. Figure 3–3 shows examples of this type of support. These supports offer complete constraint and resist linear and rotational motion of the body at the face of the support. Reactions from a built-in support (Fig. 3–3a) or a fixed support (Fig. 3–3b) contain three unknown elements: two unknown force components and an unknown moment.

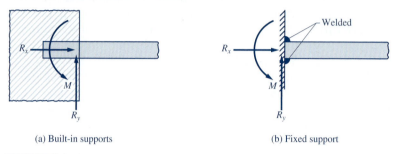

(a) Built-in supports   (b) Fixed support

FIGURE 3–3

In some cases, the direction of an unknown reaction component is not immediately apparent; however, it may be assumed arbitrarily. After a reaction component is solved, a positive result indicates that the assumed direction is correct; a negative result indicates that the assumed direction must be reversed.

Two additional types of supports, *pulleys* and *linear springs,* are frequently encountered in engineering applications.

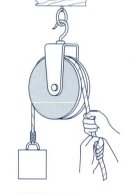

Pulley. The simple pulley shown in Fig. 3–4 is a grooved wheel over which a rope or belt is passed for the purpose of modifying the load application. The center of a pulley is usually supported on an axle or a bearing. For the pulley in Fig. 3-5a and its free-body diagram in Fig. 3–5b, the equilibrium equation, $\Sigma M_O = 0$, is

$$\Sigma M_O = T_1\,(r) - T_2\,(r) = 0$$

From which we get

$$T_1 = T_2 \qquad\qquad (3\text{-}2)$$

Therefore, *if the axle of a pulley is frictionless and the pulley is in equilibrium, the magnitudes of the tensile forces of the rope on the two sides of the pulley are the same.*

FIGURE 3–4

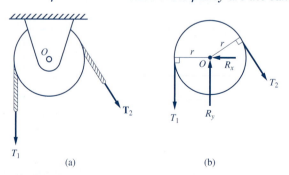

(a)   (b)

FIGURE 3–5

Linear Spring. Springs are elastic elements capable of developing elastic restoring forces when deformed. A *linear spring* develops a tensile or compressive force proportional to the amount of elongation or contraction of the spring. The *free* or *undeformed length* of a spring is the length of the spring in its undeformed state. The *deformation* (elongation or contraction) of a spring is the change in the length of the spring from its free length. For an elongation x from a free length L in the linear spring shown in Fig. 3–6, the tensile force applied to the spring is

$$F = kx \qquad\qquad (3\text{–}3)$$

where k is the *spring constant* or *stiffness,* which is the force required to cause a deformation of one unit length in the spring. The unit of spring constant is N/m in SI units, and lb/ft or lb/in. in U.S. customary units. For example, to produce a 2-m stretch from the free length in a spring with a spring constant of 3 kN/m, it will need a tensile force of

$$F = (3 \text{ kN/m})(2\text{m})$$
$$= 6 \text{ kN}$$

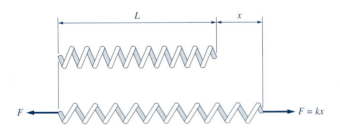

FIGURE 3–6

———— **EXAMPLE 3–1** ————

Draw the free-body diagram of a 30-lb wheel supported as shown in Fig. E3–1(1). The vertical surface is assumed to be frictionless.

Solution. Refer to Fig. E3–1(2). The wheel is taken as a free body and is isolated from its supports. To draw the free-body diagram, we need to show all the forces acting on the wheel and the necessary dimensions and angles to specify the direction and location of the forces. The 30-lb

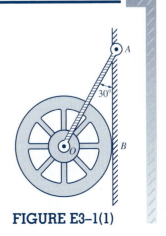

FIGURE E3–1(1)

weight is acting vertically downward through the center of the wheel. The tension T in the cable is drawn along the direction of the cable, which is 30° from the vertical or 60° from the horizontal. The reaction R from the wall is normal to the frictionless wall, so it is shown to act in the horizontal direction. Note that the detailed features of the wheel may be omitted. All the forces pass through point O, so the radius of the wheel is not needed to specify the location of the forces.

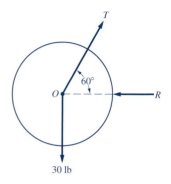

FIGURE E3–1(2)

EXAMPLE 3–2

Draw the free-body diagram of the beam shown in Fig. E3–2(1).

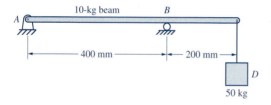

FIGURE E3–2(1)

Solution. A free-body diagram may be drawn for the entire system, or two free-body diagrams, one for beam AC and one for block D, may be drawn separately. For illustration purposes, both ways will be shown.

(a) The FBD of the Entire System. The beam and the block together will be considered as a free body. See Fig. E3–2(2). The weight of block D is $mg = 50 \times 9.81 = 490.5$ N, which is shown to act vertically downward through the center of the block. The weight of the beam AC is $10 \times 9.81 = 98.1$ N, which is shown to act through the center of gravity G (the midpoint) of the beam. The reaction $\mathbf{B}_y$ of the roller support at B acts vertically upward. For the hinge support at A, two unknown reaction components, $\mathbf{A}_x$ and $\mathbf{A}_y$, are shown. The tension in the cable is not shown in the free-body diagram because the block is not detached from the beam and the cable tension is an internal force. The dimensions are shown in meters to specify the locations of the forces.

(b) The FBDs of the Beam and the Block Drawn Separately. We cut the cable to separate the beam and the block. See Fig. E3–2(3). In this case, the tension in the cable becomes an external force and it must be shown on the free-body diagram. The tension acting on the block and that acting on

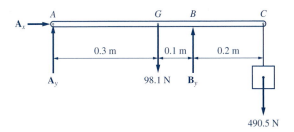

FIGURE E3–2(2)

the beam are action and reaction to each other, so they are shown with the same symbol **T**, but in opposite directions. The other forces are the same as in part (a).

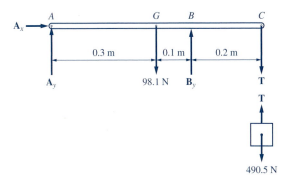

FIGURE E3–2(3)

EXAMPLE 3–3

The 100-lb bracket shown in Fig. E3–3(1) is supported by a hinge at A and a linear spring with a spring constant of 20 lb/in. at B. The cable attached to the bracket at C passes over a small pulley that is mounted on a frictionless axle. The other end of the cable is attached to a 50-lb weight. Draw the free-body diagram of the block.

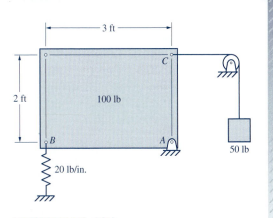

FIGURE E3–3(1)

Solution. The free-body diagram of the bracket is sketched as shown in Fig. E3–3(2). The 100-lb weight of the bracket acts vertically downward through the center of gravity G of the bracket. The tension of the cable at C is 50 lb due to the weight attached. Assuming that the spring deflection is x in. in compression, the compressive force in the spring is $20x$ lb. By the law of action and reaction, the spring in compression would push joint B with the same force of $20x$ lb. So the spring force is shown acting vertically upward on the bracket at B. The reactions at the hinge support are designated by $\mathbf{A}_x$ and $\mathbf{A}_y$.

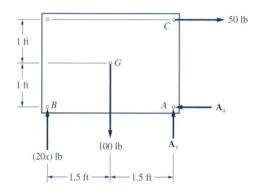

FIGURE E3–3(2)

3–5
EQUILIBRIUM OF A CONCURRENT COPLANAR FORCE SYSTEM

A body subjected to concurrent coplanar forces is in equilibrium if the resultant of the forces acting on the body is zero. The moment of the forces about the common point is always zero; hence, the equilibrium conditions of a concurrent coplanar force system consist of only two scalar equations:

$$\Sigma F_x = 0 \qquad \Sigma F_y = 0 \qquad\qquad (3\text{–}4)$$

In general, two unknowns may be solved from these equations.

Recall from Section 2–3 that the resultant of a concurrent coplanar force system may be determined by the polygon rule, in which the forces are connected in a tip-to-tail fashion with the resultant represented by the closing side of the force polygon. A balanced concurrent coplanar force system has zero resultant; therefore, the forces of the system must form a closed polygon. When three concurrent forces are in equilibrium, they must form a closed force triangle. In this case, the unknowns can be determined by solving the force triangle.

Two-Force Member. Problems involving concurrent coplanar force systems often involve members that are in equilibrium under the action of only two forces. Such members are called two-force members. *Equilibrium conditions require that the two forces be equal in magnitude, opposite in direction, and acting along the line joining the two points of application.* A two-force member can be either straight (Fig. 3–7a) or of any shape (Fig. 3–7b).

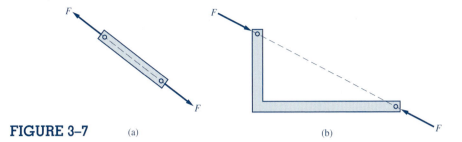

FIGURE 3–7   (a)   (b)

─────── **EXAMPLE 3–4** ───

A 50-kg mass is suspended by a boom hinged to the wall at one end and tied to the wall by a cable, as shown in Fig. E3–4(1). Determine the tension in cable *AB* and the axial force in boom *AC*.

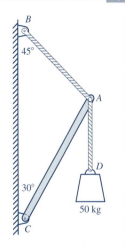

Solution. Cables *AB* and *AD* and boom *AC* are all two-force members; the forces in these members are all in the axial directions. These forces are concurrent at joint *A,* and thus the free-body diagram of joint *A* can be sketched as shown in Fig. E3–4(2). The tension in cable *AD* is equal to the weight of the mass: 50(9.81) = 491 N. The boom is assumed to be in compression, so the force $\mathbf{F}_{AC}$ is shown pushing joint *A*. The two unknown forces may be found by either the force triangle or the equilibrium equations. For illustrative purposes, both methods will be shown.

FIGURE E3–4(1)

(a) By Force Triangle. Since the three forces acting on joint *A* are in equilibrium, they must form a closed triangle, which is sketched in Fig. E3–4(3). The 491-N vertical downward weight is drawn first. Then the two sides of the triangle representing $\mathbf{T}_{AB}$ and $\mathbf{F}_{AC}$ are connected in a head-to-tail fashion. The forces are acting in the same directions as shown in the free-body diagram, indicating that the assumed directions are correct. Thus, the boom is in compression, as assumed. The magnitude of the forces may be determined by using the law of sines. We write

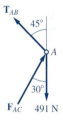

FIGURE E3–4(2)

$$\frac{T_{AB}}{\sin 30°} = \frac{F_{AC}}{\sin 45°} = \frac{491 \text{ N}}{\sin 105°}$$

From which we get

$$T_{AB} = \frac{(491 \text{ N}) \sin 30°}{\sin 105°}$$
$$= 254 \text{ N} \qquad \Leftarrow \textbf{Ans.}$$

$$F_{AC} = \frac{(491 \text{ N}) \sin 45°}{\sin 105°}$$
$$= 359 \text{ N} \qquad \Leftarrow \textbf{Ans.}$$

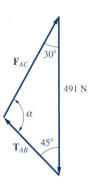

FIGURE E3–4(3)

(b) By the Equilibrium Equations. Refer to Fig. E3–4(4). The equilibrium equations for the free-body diagram along the x and y axes are

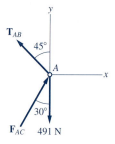

$$\Sigma F_x = -T_{AB} \sin 45° + F_{AC} \sin 30° = 0 \qquad \text{(a)}$$

$$\Sigma F_y = T_{AB} \cos 45° + F_{AC} \cos 30° - 491 \text{ lb} = 0 \qquad \text{(b)}$$

FIGURE E3–4(4)

From Equation (a),

$$F_{AC} = \frac{\sin 45°}{\sin 30°} T_{AB} \qquad \text{(c)}$$

$$= 1.414 T_{AB}$$

Substituting into Equation (b), we get

$$T_{AB} (\cos 45° + 1.414 \cos 30°) = 491 \text{ N}$$

From which we get

$$T_{AB} = 254 \text{ N} \qquad \Longleftarrow \textbf{Ans.}$$

Substituting into Equation (c), we get

$$F_{AC} = 359 \text{ N} \qquad \Longleftarrow \textbf{Ans.}$$

EXAMPLE 3–5

The 800-lb horizontal load and 1500-lb vertical load in Fig. E3–5(1) are supported by two bars, AB and AC, connected at A by a pin. The bars are hinged to the wall at B and C. Find the reactions at the supports at B and C.

Solution. Bars AB and AC are two-force members. The force in each member is along the axial direction; hence, the direction of reaction $\mathbf{R}_A$ must be along BA and the direction of reaction $\mathbf{R}_C$ must be along CA, as shown in the free-body diagram in Fig. E3–5(2).

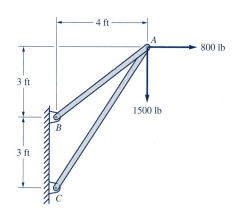

FIGURE E3–5(1)

Note that the directions (as indicated by the arrowhead) of the reactions are assumed arbitrarily. The correct direction will be determined by the sign of the computed results of the reactions: a positive sign indicates that the direction of the reaction is the same as assumed; a negative sign indicates that the direction of the reaction is opposite to that assumed. The reactions may be found by setting up the equilibrium equations along the x and y axes. We write

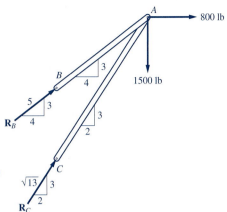

$$\Sigma F_x = \frac{4}{5}R_B + \frac{2}{\sqrt{13}}R_C + 800 = 0 \quad \text{(a)}$$

$$\Sigma F_y = \frac{3}{5}R_B + \frac{3}{\sqrt{13}}R_C - 1500 = 0 \quad \text{(b)}$$

To solve the two equations, we perform the following operations:

FIGURE E3–5(2)

$$3 \times \text{(a):} \quad \frac{12}{5}R_B + \frac{6}{\sqrt{13}}R_C + 2400 = 0 \tag{c}$$

$$2 \times \text{(b):} \quad \frac{6}{5}R_B + \frac{6}{\sqrt{13}}R_C - 3000 = 0 \tag{d}$$

$$\text{(c)} - \text{(d):} \quad \frac{6}{5}R_B + 5400 = 0$$

$$R_B = -4500 \text{ lb} \qquad \mathbf{R}_B = 4500 \text{ lb} \qquad \Leftarrow \textbf{Ans.}$$

Substituting into Equation (a), we get

$$R_C = +5050 \text{ lb} \qquad \mathbf{R}_C = 5050 \text{ lb} \qquad \Leftarrow \textbf{Ans.}$$

Note that the negative sign in R_B indicates that its assumed direction should be reversed.

EXAMPLE 3–6

A 200-kg crate B is lifted with the rope-and-pulley arrangement shown in Fig. E3–6(1). Determine the magnitude of the applied force $\mathbf{P}$ required to hold the crate in equilibrium. Assume that the pulleys are weightless and frictionless.

Solution. The free-body diagrams of the lower pulley A and the crate B are drawn in Fig. E3–6(2). The rope that passes through the frictionless pulleys must have the same tension P at every segment of the rope. The tension in the rope attached to crate B is equal to the weight of the crate, which is equal to $200(9.81) = 1962$ N. Applying the equilibrium condition to pulley A in the vertical direction, we get

$$\Sigma F_y = 3P - 1962 \text{ N} = 0$$

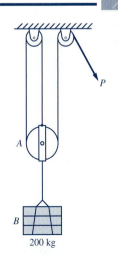

FIGURE E3–6(1)

From which we get

$$P = 654 \text{ N} \qquad \Leftarrow \textbf{Ans.}$$

We see that the pulleys not only make the force application more convenient, but they also reduce the effort greatly.

FIGURE E3–6(2)

EXAMPLE 3–7

The two blocks A and B are held in the position shown in Fig. E3–7(1) by a spring with a stiffness of 20 lb/in. and a free length of 8 in. What are the weights of the blocks?

Solution. The free-body diagram of joint C is shown in Fig. E3–7(2). For the position shown, the spring has a stretched length of 10 in., so its elongation is

$$x = 10 \text{ in.} - 8 \text{ in.} = 2 \text{ in.}$$

The tension in the spring is

$$F_s = kx = (20 \text{ lb/in.})(2 \text{ lb/in.}) = 40 \text{ lb}$$

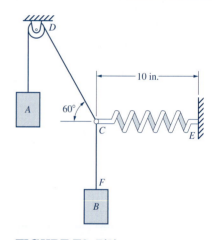

FIGURE E3–7(1)

The tension in cable CD is equal to W_A, and the tension in cable CF is equal to W_B. The three forces acting on joint C must form a closed triangle, as shown in Fig. E3–7(3). From the right triangle we obtain

$$W_A = 40 \text{ lb} / \cos 60° = 80 \text{ lb} \qquad \Leftarrow \textbf{Ans.}$$
$$W_B = 40 \text{ lb} \tan 60° = 69.3 \text{ lb} \qquad \Leftarrow \textbf{Ans.}$$

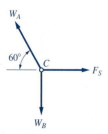

FIGURE E3–7(2)

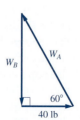

FIGURE E3–7(3)

3–6
EQUILIBRIUM OF A NONCONCURRENT COPLANAR FORCE SYSTEM

Equilibrium Equations.   It was stated in Section 3-2 that the equilibrium equations for a coplanar force system are

$$\Sigma F_x = 0 \qquad \Sigma F_y = 0 \qquad \Sigma M_A = 0 \qquad\qquad (3\text{--}5)$$

The first two equations require that there be no resultant force. The third equation requires that there be no resultant moment about an arbitrary point A. The equilibrium conditions may be expressed in two other forms as described below.

Alternative Equations Containing Two Moment Equations.   The alternative equilibrium equations for this situation are as follows:

$$\Sigma F_x = 0 \qquad \Sigma M_A = 0 \qquad \Sigma M_B = 0 \qquad\qquad (3\text{--}6)$$

where the x axis may be chosen arbitrarily, and A and B are arbitrary points, except that line AB must not be perpendicular to the x axis. To satisfy the first two equations, $\Sigma F_x = 0$ and $\Sigma M_A = 0$, the resultant, if any, can be reduced to a single force along the line perpendicular to the x axis passing through A. Since point B is not on this line, the resultant must be zero to satisfy the third equation $\Sigma M_B = 0$. The body is therefore in equilibrium if all three equations are satisfied.

Alternative Equations Containing Three Moment Equations.   The three alternative equilibrium equations for this situation are as follows:

$$\Sigma M_A = 0 \qquad \Sigma M_B = 0 \qquad \Sigma M_C = 0 \qquad\qquad (3\text{--}7)$$

where A, B, and C are arbitrary but noncollinear points. To satisfy the first two equations, $\Sigma M_A = 0$ and $\Sigma M_B = 0$, the resultant, if any, must be along line AB. Since point C is not along this line, the resultant must be zero to satisfy the third equation, $\Sigma M_C = 0$.

Independent Equations.   Any one of the three sets of equilibrium equations listed above will produce three *independent equations* that may be used to solve for three unknowns. Any set of three equilibrium equations may be used, as long as they are independent. If the three equilibrium equations are not independent, we will not have enough equations to solve for three unknowns. For example, three moment equations written about three points along a straight line are not independent and are not sufficient for equilibrium because a force system with a nonzero resultant force passing through the three points would satisfy the three equations.

Choice of Equations.   By choosing the equations properly, it is possible to set up an equilibrium equation containing only one unknown that can be solved immediately without the need for solving the simultaneous equations. *Equations containing one unknown may be obtained by writing the moment equation about the point of intersection of the other two unknowns.*

Or, if two of the unknown forces are parallel, the third unknown may be determined by summing the force components in a direction perpendicular to the two parallel unknown forces. For example, consider the beam shown in Fig. 3–8a and its free-body diagram shown in Fig. 3–8b. Two of the three unknown reactions are in the y direction. Hence, the equilibrium equation $\Sigma F_x = 0$ contains only the unknown A_x. The two unknowns A_x and B_y intersect at point B. The moment equation $\Sigma M_B = 0$ contains only the unknown A_y. Similarly, the moment equation $\Sigma M_A = 0$ contains only the unknown B_y. Thus, each equation contains only one unknown and the equations can be solved independently of each other.

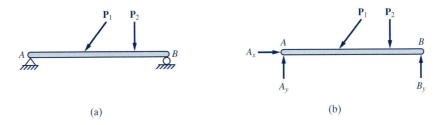

(a)   (b)

FIGURE 3–8

Three-Force Body.   A special case occurs when a body in equilibrium is subjected to three forces. Such a body is called a *three-force body.* Figure 3–9a and b shows examples of three-force bodies. *Equilibrium conditions require that the three forces be coplanar and concurrent.* Suppose that the forces were not concurrent. One of the forces would then produce an unbalanced moment about the point of intersection of the other two forces. Hence, the three forces must be concurrent. The only exception occurs when the three forces are parallel. In this case, we may consider the point of concurrency to be at infinity.

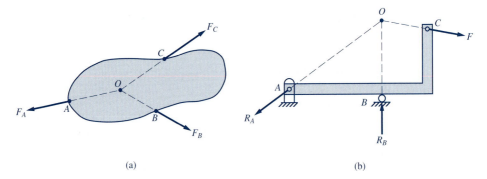

(a)   (b)

FIGURE 3–9

The equilibrium conditions of a three-force body are very useful in solving some equilibrium problems. In Fig. 3–9b, if the applied force F and the reaction R_B of the roller support are known to intersect at O, then the reaction R_A at the hinge support must also pass through O. The three forces form a force triangle that can be used to solve for the unknown forces.

──────── **EXAMPLE 3-8** ────────

Refer to Fig. E3–8(1). A 600-lb roller with a 12-in. radius is to be pulled over a 5-in. curb by a force F applied at the center O of the roller. Find the magnitude and direction of the minimum force F required to start to roll the roller over the curb.

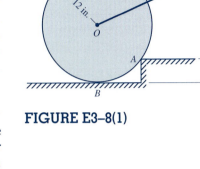

FIGURE E3-8(1)

Solution. When the roller starts to roll over the curb, there is no contact between the roller and the floor, and all the reaction is concentrated at A. Hence, the roller is subjected to only three forces, as shown in the free-body diagram in Fig. E3–8(2). Since the applied force F and the weight pass through the center O, the reaction R_A must also pass through O. In the right triangle AOC, we have

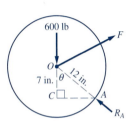

FIGURE E3-8(2)

$$OA = 12 \text{ in.}$$

$$OC = 12 \text{ in.} - 5 \text{ in.} = 7 \text{ in.}$$

Thus

$$\theta = \cos^{-1}\left(\frac{7}{12}\right) = 54.3°$$

Since the body is in equilibrium, the three forces must form a closed triangle. [See Fig. E3–8(3)]. To draw the force triangle, we lay out a vertical line DE starting from point D to represent the 600-lb weight. From E we draw line EG' along the direction of reaction R_A. The force F can be represented by any line from a point on line EG' to D. The minimum force F_{min} must be perpendicular to EG', as shown in the force triangle. From the force triangle, we get

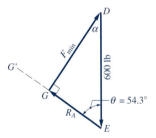

FIGURE E3-8(3)

$$F_{min} = (600 \text{ lb}) \sin 54.3° = 487 \text{ lb}$$

$$= 487 \text{ lb} \angle 54.3° \qquad \Leftarrow \textbf{Ans.}$$

—— **EXAMPLE 3–9** ——

Refer to Fig. E3–9(1). The lever is supported by a hinge at A and a short link BD. Find the axial force in the link and the reaction at the hinge support due to the 70-lb force applied to the handle.

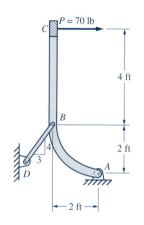

FIGURE E3–9(1)

Solution. Link BD is a two-force member. The axial force F_{BD}, assumed to be in tension, is along BD. Lever ABC is a three-force body. The equilibrium conditions of the lever are used for solving the reactions. Two methods are presented.

(a) By Force Triangle. In the free-body diagram in Fig. E3–9(2), we see that two of the three forces, P and F_{BD}, intersect at point O. The third force R_A must also pass through O. Angles α and β are

$$\alpha = \tan^{-1}\left(\frac{4\text{ ft}}{3\text{ ft}}\right) = 53.1°$$

$$\beta = \tan^{-1}\left(\frac{6\text{ ft}}{1\text{ ft}}\right) = 80.5°$$

The three forces must form a closed triangle, which is sketched in Fig. E3–9(3). Applying the law of sines to the force triangle, we get

$$\frac{F_{BD}}{\sin(180° - \beta)} = \frac{R_A}{\sin \alpha} = \frac{70\text{ lb}}{\sin(\beta - \alpha)}$$

$$\frac{F_{BD}}{\sin 99.5°} = \frac{R_A}{\sin 53.1°} = \frac{70\text{ lb}}{\sin 27.4°}$$

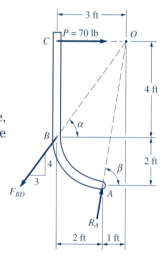

FIGURE E3–9(2)

From which we get

$$F_{BD} = 150\text{ lb (T)} \qquad \Leftarrow \textbf{Ans.}$$

$$R_A = 121.6\text{ lb} \ \angle\ 80.5° \qquad \Leftarrow \textbf{Ans.}$$

(b) By Equilibrium Equations. We will treat the problem as a general type of coplanar force system. In the free-body diagram of the lever shown in Fig. E3–9(4), the axial force F_{BD} in the link is along the direction of the link. The reaction at the hinge support is represented by the horizontal and vertical components. The link is assumed to be in tension, and the directions of A_x and A_y are assumed arbitrarily. Theoretically, we can write any three independent equations and solve the equations

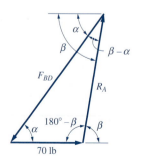

FIGURE E3–9(3)

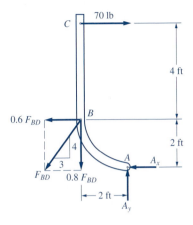

FIGURE E3–9(4)

simultaneously for the three unknowns. However, for ease of solution, we will write three equilibrium equations, each of which contains only one unknown so that the equations can be solved independently of each other.

$$\Sigma M_A = (0.6F_{BD})(2 \text{ ft}) + (0.8F_{BD})(2 \text{ ft}) - (70 \text{ lb})(6 \text{ ft}) = 0$$
$$F_{BD} = +150 \text{ lb (T)} \qquad \Leftarrow \textbf{Ans.}$$

$$\Sigma F_x = 70 \text{ lb} - 0.6(150 \text{ lb}) - A_x = 0$$
$$A_x = -20 \text{ lb} \longrightarrow \qquad \Leftarrow \textbf{Ans.}$$

$$\Sigma F_y = A_y - 0.8(150 \text{ lb}) = 0$$
$$A_y = 120 \text{ lb} \uparrow \qquad \Leftarrow \textbf{Ans.}$$

Note that the negative sign for A_x indicates that its assumed direction (horizontally to the left) must be reversed and is actually acting horizontally to the right.

EXAMPLE 3–10

The beam is subjected to a uniform load and a concentrated load as shown in Fig. E3–10(1). Determine the reactions at the supports.

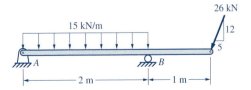

FIGURE E3–10(1)

Solution. We will first draw the FBD and then write equilibrium equations.

Free-Body Diagram. In the free-body diagram of the beam in Fig. E3–10(2), the concentrated load is resolved into its horizontal and vertical components, and the uniform load is replaced by its equivalent concentrated force. The reaction at the roller support is represented by B_y, and the reaction at the hinge support is represented by the components A_x and A_y.

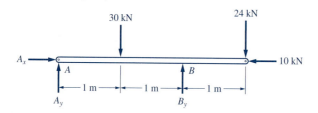

FIGURE E3–10(2)

Equilibrium Equations. Note that the following equations are set up so that they can be solved independently of each other.

$$\Sigma F_x = A_x - 10 \text{ kN} = 0$$
$$A_x = 10 \text{ kN} \longrightarrow \qquad \Leftarrow \textbf{Ans.}$$

$$\Sigma M_B = -A_y\,(2 \text{ m}) + (30 \text{ kN})(1 \text{ m}) - (24 \text{ kN})(1 \text{ m}) = 0$$
$$A_y = 3 \text{ kN} \uparrow \qquad \Leftarrow \textbf{Ans.}$$

$$\Sigma M_A = B_y\,(2 \text{ m}) - (30 \text{ kN})(1 \text{ m}) - (24 \text{ kN})(3 \text{ m}) = 0$$
$$B_y = 51 \text{ kN} \uparrow \qquad \Leftarrow \textbf{Ans.}$$

Check. The results can be checked by

$$\Sigma F_y = 3 + 51 - 30 - 24 = 0 \qquad \text{(Checks)}$$

EXAMPLE 3–11

The streetlight shown in Fig. E3–11(1) is supported by a fixed support at A and a guy wire BC, which has a tension of 200 lb. The total weight of the structure is 500 lb acting through point G. The 40-lb horizontal force represents the effect of the wind. Determine the reactions at the fixed support.

Solution.

Free-Body Diagram. In the free-body diagram of the frame in Fig. E3–11(2), the 200-lb tension at B is resolved into horizontal and vertical components. The reaction at the fixed support is represented by three elements, A_x, A_y, and M_A.

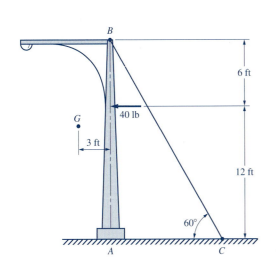

FIGURE E3–11(1)

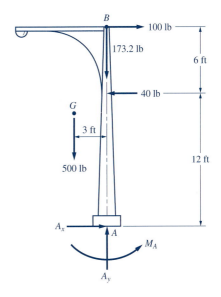

FIGURE E3–11(2)

Equilibrium Equations.

$$\Sigma F_x = A_x + 100 \text{ lb} - 40 \text{ lb} = 0$$
$$A_x = -60 \text{ lb} \longleftarrow \qquad \Leftarrow \textbf{Ans.}$$

$$\Sigma F_y = A_y - 500 \text{ lb} - 173.2 \text{ lb} = 0$$
$$A_y = 673 \text{ lb} \uparrow \qquad \Leftarrow \textbf{Ans.}$$

$$\Sigma M_A = M_A + (500 \text{ lb})(3 \text{ ft}) - (100 \text{ lb})(18 \text{ ft}) + (40 \text{ lb})(12 \text{ ft}) = 0$$
$$M_A = -180 \text{ lb} \cdot \text{ft} \quad \circlearrowleft \qquad \Leftarrow \textbf{Ans.}$$

EXAMPLE 3–12

A bracket ABC is supported by a hinge at A and a roller at C. Determine the reactions at the supports due to the loading shown in Fig. E3–12(1).

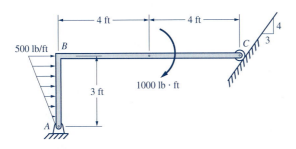

FIGURE E3–12(1)

Solution.

Free-Body Diagram. In the free-body diagram of the bracket in Fig. E3–12(2), the triangular load is replaced by its equivalent concentrated force, and the reaction R_C at the roller support is perpendicular to the supporting plane. The reaction at hinge A is represented by its horizontal and vertical components A_x and A_y. The line of action of R_C is extended both ways to intersect A_y at D and A_x at E.

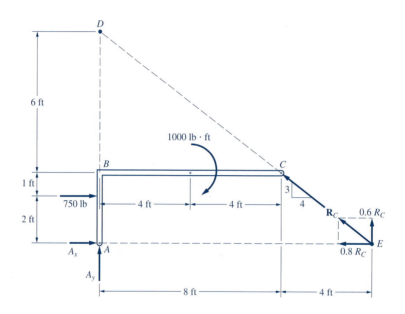

FIGURE E3–12(2)

Equilibrium Equations. To find the reaction R_C, we write the moment equation about A to avoid the unknowns A_x and A_y. To simplify the solution, the reaction R_C is resolved into rectangular components at E so that its horizontal component produces no moment about A. Note that, because a couple is independent of the moment center, the moment of a couple about any point is the same.

$$\Sigma M_A = (0.6R_C)(12 \text{ ft}) - (750 \text{ lb})(2 \text{ ft}) - 1000 \text{ lb} \cdot \text{ft} = 0$$

$$R_C = +347 \text{ lb} \quad \overset{3}{\underset{4}{\diagdown}} \qquad \Leftarrow \textbf{Ans.}$$

Now, by summing moments about D, we can avoid the unknowns R_C and A_y since they both pass through point D.

$$\Sigma M_D = A_x(9 \text{ ft}) + (750 \text{ lb})(7 \text{ ft}) - 1000 \text{ lb} \cdot \text{ft} = 0$$

$$A_x = -472 \text{ lb} \quad \longleftarrow \qquad \Leftarrow \textbf{Ans.}$$

Similarly, by summing moments about E, we can avoid the unknowns R_C and A_x.

$$\Sigma M_E = -A_y(12 \text{ ft}) - (750 \text{ lb})(2 \text{ ft}) - 1000 \text{ lb} \cdot \text{ft} = 0$$
$$A_y = -208 \text{ lb} \downarrow \qquad\qquad \Leftarrow \textbf{Ans.}$$

Check. The solution can be checked by

$$\Sigma F_x = -472 + 750 - 0.8(347) = 0.4$$
$$\Sigma F_y = -208 + 0.6(347) = 0.2$$

where the small errors are due to truncation, so the results check out all right.

Remark. In this solution, the unknowns have been solved independently of each other. If the result of one of the unknowns is in error, it will not affect the results of the other two unknowns. As an alternative solution, we could set up three equations $\Sigma M_A = 0$, $\Sigma F_x = 0$, $\Sigma F_y = 0$. However, in this case, the value of A_x and A_y will be dependent on the value of R_C. An erroneous answer to the value of R_C will spoil the solutions to the other two unknowns.

3–7
SUMMARY

Equilibrium of a Coplanar Force System. A rigid body subjected to a coplanar force system is said to be in equilibrium if, and only if, the resultant of the system is zero and the resultant moment about an arbitrary point is zero.

Free-Body Diagram. When solving an equilibrium problem, it is essential to include all the external forces acting on the body. Therefore, the first step in the solution of the equilibrium problem is to draw a *free-body diagram* showing all the external forces acting on the body under consideration.

Equilibrium of a Concurrent Coplanar Force System. When concurrent forces are in equilibrium, the resultant of the forces must be zero; that is,

$$\Sigma F_x = 0 \qquad \Sigma F_y = 0 \qquad\qquad\qquad (3\text{–}4)$$

Two-Force Member. A member in equilibrium subjected to only two forces is call a *two-force member.* Equilibrium conditions require that *the forces be equal and opposite and acting along the line joining the two points of application.*

Equilibrium of a Nonconcurrent Coplanar Force System. The equilibrium equations may be any one of the following three sets:

$$\Sigma F_x = 0 \qquad \Sigma F_y = 0 \qquad \Sigma M_A = 0 \qquad (3\text{–}2)$$

where A is an arbitrary point in the plane.

$$\Sigma F_x = 0 \qquad \Sigma M_A = 0 \qquad \Sigma M_B = 0 \qquad\qquad (3\text{–}6)$$

where line AB must not be perpendicular to the x axis.

$$\Sigma M_A = 0 \qquad \Sigma M_B = 0 \qquad \Sigma M_C = 0 \qquad\qquad (3\text{–}7)$$

where the points A, B, and C must not be on the same straight line.

Any one set of the above equations gives rise to three independent equations that may be solved for three unknowns. A judicious choice of equations may simplify the solution greatly. An equation containing one unknown may be obtained by writing the moment equation about the point of intersection of the other two unknowns. If two of the unknowns are parallel, the third one can be solved from the equation obtained by summing the force components in the perpendicular direction.

Three-Force Body. A body in equilibrium subjected to three forces is called a *three-force body.* Equilibrium conditions require that *the three forces be coplanar and concurrent, and that the three forces form a closed force triangle.* The force triangle can be used as an alternative approach to the solution by equilibrium equations.

PROBLEMS

Section 3–3　The Free-Body Diagram

Section 3–4　Types of Supports

3–1 to 3–5　In each of these problems on this page and the next, an incomplete free-body diagram (FBD) of the isolated body is shown. Add the forces necessary to complete the free-body diagram. For simplicity, dimensions are not shown.

Problem	Description	Body	Incomplete FBD
3–1	Uniform beam of 20-lb weight supported by roller at A and hinge at B.		
3–2	Uniform pole of 10-kg mass being hoisted into position by a winch.		

Problem	Description	Body	Incomplete FBD
3–3	A 20-lb ladder supported by a smooth wall and a rough floor.	200 lb, B, C, 20 lb, G, A	200 lb, R_B, B, C, G, A
3–4	Bell crank of negligible mass holding 500-kg mass supported by a pin at A and a cable at B.	B, 500 kg, C, A, 60°	B, 60°, C, A, T
3–5	Traffic-signal pole weighing 900 lb with a 100-lb traffic signal supported by a fixed support at A.	B, G, A	B, 100 lb, G, 1000lb, 900 lb, x, A, M

3–6 Draw the free-body diagram of the 50-kg homogeneous cylinder that is supported on two
smooth surfaces, as shown in Fig. P3–6.

3–7 Draw the free-body diagram of the beam in Fig.
P3–7.

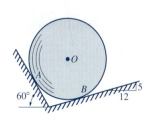

FIGURE P3–6

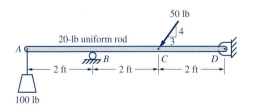

FIGURE P3–7

3–8   Draw separate free-body diagrams of the 50-kg homoge-
neous cylinder and the rod *AB*
in Fig. P3–8. Neglect the weights
of the rod and the cable and
assume that all contact surfaces are smooth.

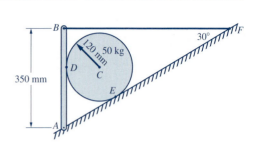

FIGURE P3–8

3–9   Draw separate free-body diagrams of beams *AB* and *CD*
in Fig. P3–9. Assume that the
weights of the beams are
negligible.

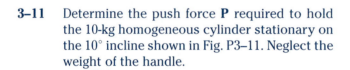

Section 3–5   Equilibrium of a Concurrent Coplanar Force System

FIGURE P3–9

3–10   Find the horizontal force **F**
required to hold the 10-lb
weight in the position shown
in Fig. P3–10. Neglect the
weight of the wire.

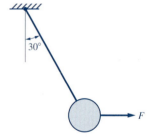

3–11   Determine the push force **P** required to hold
the 10-kg homogeneous cylinder stationary on
the 10° incline shown in Fig. P3–11. Neglect the
weight of the handle.

3–12   Determine the force **P** required to suspend the
200-lb crate in the position shown in Fig.
P3–12. Solve the problem by using (*a*) the
force triangle and (*b*) the equilibrium equations along the *x* and *y* axes.

FIGURE P3–10

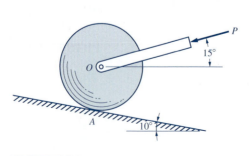

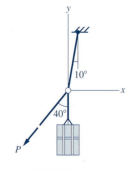

FIGURE P3–11   **FIGURE P3–12**

3–13 Determine the reactions at A and B on the smooth inclines supporting the 75-kg homogeneous cylinder shown in Fig. P3–13.

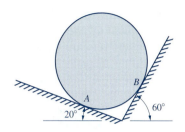

FIGURE P3–13

3–14 to 3–17 Two cables, AB and AC, are fastened together at A as shown in Figs. P3–14 to P3–17. Determine the tensions in each cable.

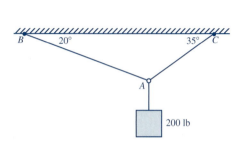

FIGURE P3–14

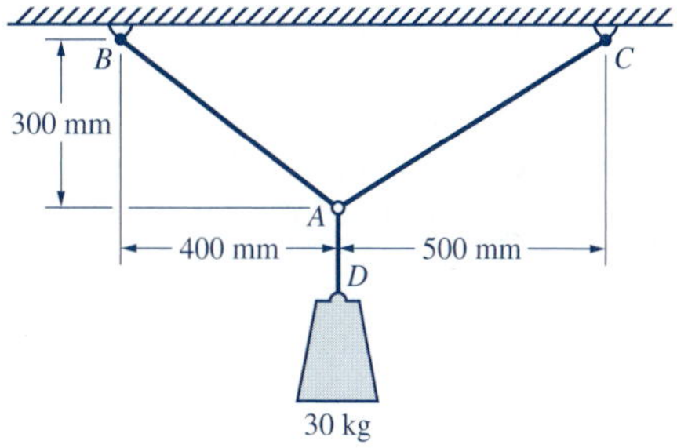

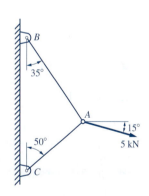

FIGURE P3–15

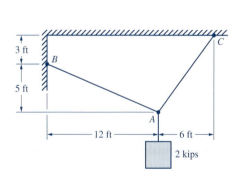

FIGURE P3–16

FIGURE P3–17

3–18 A 90-lb block is supported by two cables, AB and AC, as shown in Fig. P3–18. Determine (*a*) the tension in each cable if $\theta = 45°$ and (*b*) the value of angle θ for which the tension in cable AC is a minimum.

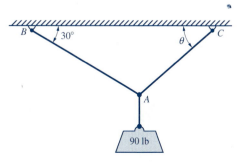

FIGURE P3–18

3–19   Refer to Fig. P3–19. A 100-kg mass is suspended by a boom hinged to the wall at one end and tied to the wall at the other end by a cable. Determine the tension in cable *AB* and the axial force in boom *AC.*

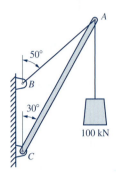

FIGURE P3–19

3–20   Refer to Fig. P3–20. Determine the reaction at the hinge support at *A* due to a 500-lb load on the derrick.

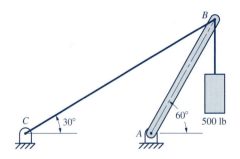

FIGURE P3–20

3–21   Refer to Fig. P3–21. Two identical linear springs have a spring constant of 100 lb/ft and are undeformed when the springs are in the vertical position along *AC.* Determine the horizontal force **F** necessary to hold the springs in the position shown.

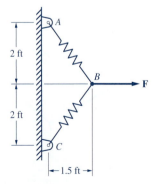

FIGURE P3–21

3–22 See Fig. P3–22. Determine the weight of block C and the value of angle θ for the equilibrium of the system.

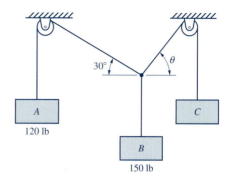

FIGURE P3–22

3–23 A 200-kg crate is supported by the rope and pulley arrangement shown in Fig. P3–23. Determine the tension **T** necessary to maintain equilibrium.

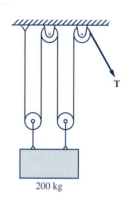

FIGURE P3–23

3–24 A 100-lb block is suspended by cables AB and AC as shown in Fig. P3–24. Determine the tension in each cable.

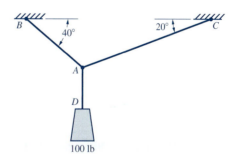

FIGURE P3–24

3–25 Refer to Fig. P3–25. Three identical spheres, each of 10-in. diameter and weighing 20 lb, fit inside the smooth 26-in. wide box shown. Find the reactions at A and B.

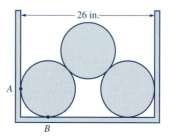

FIGURE P3–25

3-26 Compute the tensions in cables AB and AC attached to the 200-lb crate shown in Fig. P3–26.

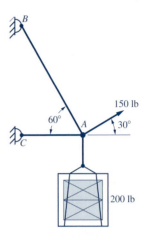

FIGURE P3–26

3-27 Refer to Fig. P3–27. Three blocks with weights expressed in terms of W are in equilibrium suspended by the pulley and cable system shown. Determine the angles α and β for equilibrium.

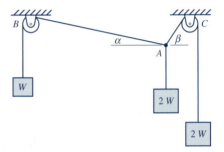

FIGURE P3–27

3-28 The 50-lb sliding block in Fig. P3–28 may slide freely on the frictionless slot. The spring is undeformed when it is in the horizontal position. Determine the spring constant k if the slider is in equilibrium in the position shown.

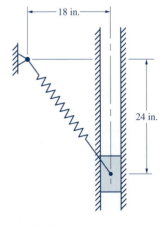

FIGURE P3–28

3–29 Determine the amount of stretch in each spring caused by the 100-N force shown in Fig. P3–29.

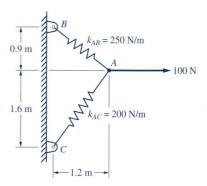

FIGURE P3–29

3–30 In the bridge-truss joint shown in Fig. P3–30, the forces acting on the members form a concurrent force system. Knowing that the joint is in equilibrium, determine the forces C and T.

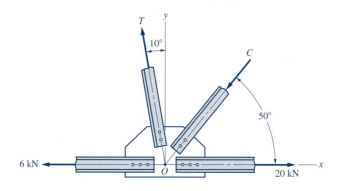

FIGURE P3–30

Section 3–6 Equilibrium of a Nonconcurrent Coplanar Force System

3–31 A 30-lb, 16-ft ladder leans against a smooth wall with its lower end resting on a rough ground. See Fig. P3–31. The angle between the ladder and the wall is 20°. Knowing that the ladder will not slip on its lower end, determine the reactions at both ends of the ladder by (*a*) the force triangle and (*b*) equilibrium equations.

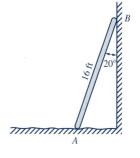

FIGURE P3–31

3–32 Refer to Fig. P3–32. A bell crank of negligible mass is holding a 100-kg mass and is supported by a link AD and a pin at B. Determine the reactions at A and B by (a) the force triangle and (b) equilibrium equations.

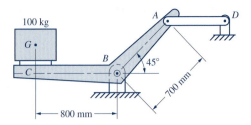

FIGURE P3–32

3–33 Determine the reactions at supports A and B of the bracket in Fig. P3–33 by (a) the force triangle and (b) equilibrium equations.

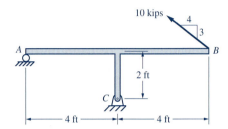

FIGURE P3–33

3–34 Refer to Fig. P3–34. Determine the horizontal force $\mathbf{F}$ required to start the 400-lb roller of 2-ft radius rolling over a 6-in. curb by (a) the force triangle and (b) equilibrium equations.

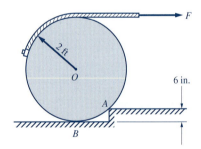

FIGURE P3–34

3–35 Refer to Fig. P3–35. A 50-lb pole of length 10 ft is raised and held in position by the force $\mathbf{P}$ applied to the rope. Find the force $\mathbf{P}$ and the reaction at A by (a) the force triangle and (b) equilibrium equations.

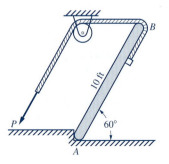

FIGURE P3–35

3–36   Refer to Fig. P3–36. Determine the reactions at supports C and D due to the 100-N load shown by (a) the force triangle and (b) equilibrium equations.

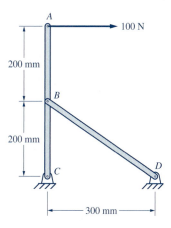

FIGURE P3–36

3–37   Refer to Fig. P3–37. Determine the reactions at the supports of the beam due to the 500-N load shown by (a) the force triangle and (b) equilibrium equations.

FIGURE P3–37

3–38   Refer to Fig. P3–38. Determine the reactions of the supports at A and B due to the 3-kN load shown by (a) the force triangle and (b) equilibrium equations.

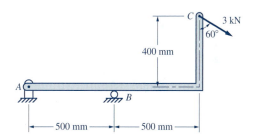

FIGURE P3–38

3–39   See Fig. P3–39. Determine the reactions of the supports at A and D due to the 400-lb load applied to the frame shown by (a) the force triangle and (b) equilibrium equations.

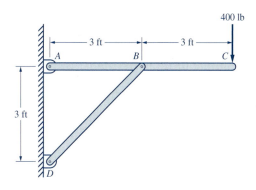

FIGURE P3–39

3–40   The hook wrench shown in Fig. P3–40 is used to turn a shaft. A pin fits in a hole at A; the peg at B is smooth. Determine the reactions at A and B due to a force $\mathbf{P} = 50$ lb by (*a*) the force triangle and (*b*) equilibrium equations.

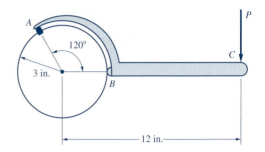

FIGURE P3–40

3–41   Refer to Fig. P3–41. Determine the reactions at A and D due to the 2-kN force shown by (*a*) the force triangle and (*b*) equilibrium equations.

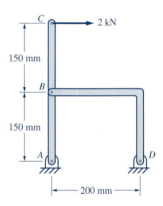

FIGURE P3–41

3–42   Refer to Fig. P3–42. Determine the reaction components at the roller support at B and the hinge support at C due to the applied load shown.

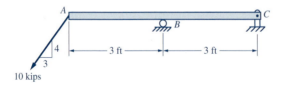

FIGURE P3–42

3–43   Refer to Fig. P3–43. Determine the reaction components at supports A and D due to the loads shown.

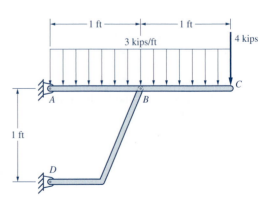

FIGURE P3–43

3–44 See Fig. P3–44. Determine the reaction components at the supports of the beam subjected to a uniform load as shown.

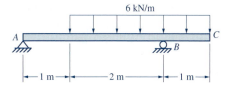

FIGURE P3–44

3–45 Refer to Fig. P3–45. Determine the tension in cable AC and the reaction components at the hinge support at B of the crane due to the 20-kN crate and the 4-kN weight of member AB if $d = 3$ m.

3–46 For the crane supporting the movable load shown in Fig. P3–45, determine the greatest distance d from support B to the 20-kN crate so that the tension in cable AC will not exceed 34 kN. The weight of member AB is 4 kN.

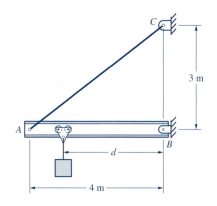

FIGURE P3–45

3–47 The truss shown in Fig. P3–47 is supported by a hinge at A and a link BC. Determine the components of the reactions at A and B.

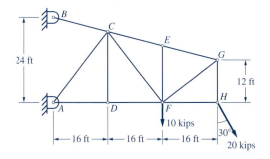

FIGURE P3–47

3–48 Refer to Fig. P3–48. Determine the reaction components at the fixed support of the cantilever beam due to the loads shown.

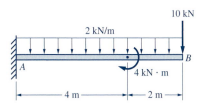

FIGURE P3–48

3–49   See Fig. P3–49. Determine the
reaction components at the
supports of the beam due to
the distributed loads shown.

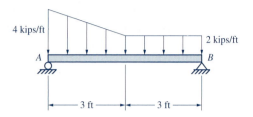

FIGURE P3–49

3–50   Refer to Fig. P3–50. Determine the reaction components at supports A and
B due to the loads applied to the bracket shown.

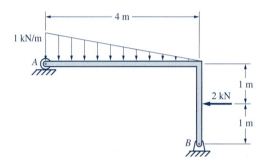

FIGURE P3–50

3–51   A fixed crane having a mass of 1000 kg is used to lift a crate of 3000 kg mass.
The center of gravity G of the crane is located as shown in Fig. P3–51.
Determine the reaction components at the rocker support at A and the
hinge support at B.

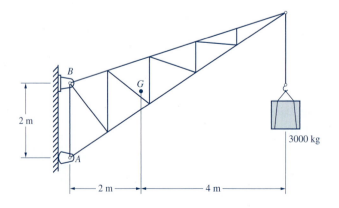

FIGURE P3–51

3–52 Refer to Fig. P3–52. Determine the reaction components at the hinge support at A and the roller support at B due to the four forces acting on the truss shown.

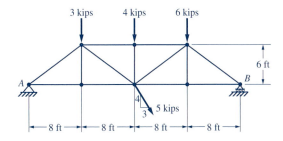

FIGURE P3–52

3–53 See Fig. P3–53. Determine the reaction components at the supports of the beam due to the loads shown.

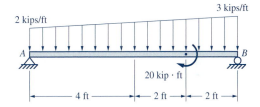

FIGURE P3–53

3–54 See Fig. P3–54. Determine the reaction components at the supports of the beam due to the loads shown.

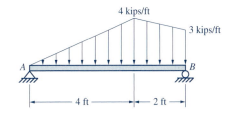

FIGURE P3–54

3–55 Refer to Fig. P3–55. Determine the reactions at the roller support at B and the hinge support at D due to given loads on the beam shown.

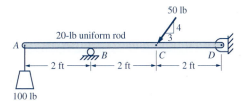

FIGURE P3–55

3–56 See Fig. P3–56. Determine the reactions at supports A and B due to the applied loads shown.

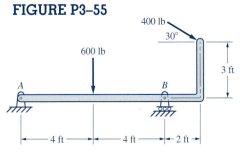

FIGURE P3–56

3–57   See Fig. P3–57. Determine the reactions at the supports of the beam due to the applied loads shown.

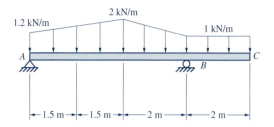

FIGURE P3–57

3–58   Refer to Fig. P3–58. Member *ABC* is supported by a pin at *A* and a cable passing over a pulley at *D*. Find the tension in the cable and the reaction components at support *A* due to the 10-kN load.

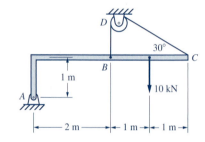

FIGURE P3–58

3–59   See Fig. P3–59. Determine the reactions at *A* and *D* due to the masses that the frame supports. Neglect the weights of members *ABC* and *BD*.

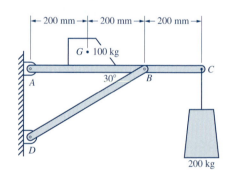

FIGURE P3–59

3–60   Refer to Fig. P3–60. A frame *ABC* is supported by a hinge at *C* and a smooth collar at *B*. Determine the reaction components at these supports due to a 100-lb load and the 50-lb weight of the frame acting through its center of gravity *G*.

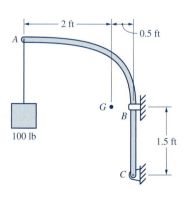

FIGURE P3–60

3–61 The two pulleys in Fig. P3–61 are fastened together to form an integral unit. The bearing at O is frictionless. A force $\mathbf{P} = 2$ kN exerted on the cable attached to the bigger pulley is balanced by the force exerted by the linear spring attached to the smaller pulley. If the spring constant is 20 kN/m, find the deflection in the spring.

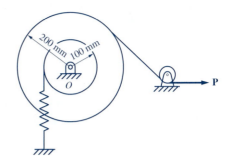

FIGURE P3–61

3–62 Refer to Fig. P3–62. Determine the reactions at the hinge support at A and the fixed support at D of the beams shown.

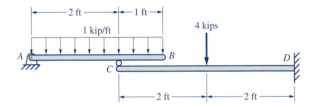

FIGURE P3–62

3–63 A 40-kg boy standing on a 20-kg platform is pulling the cable to lift the platform to the equilibrium position shown in Fig. P3–63. Determine the tension in the cable and the reaction at A.

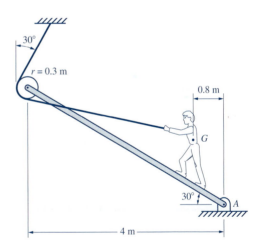

FIGURE P3–63

3–64 See Fig. P3–64. Determine the reactions at the supports of the bracket subjected to the distributed loads shown.

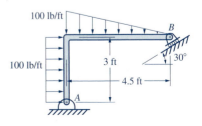

FIGURE P3–64

Computer Program Assignments

For each of the following problems, write a computer program using an appropriate programming language with which you are most familiar. Make the program user friendly by incorporating plenty of comments and input prompts so that the user will understand the input data to be entered and the limitations of their values. The output should include the data entered and the computed results, and they must be well labeled to identify each quantity. If a tabulated format is used, a proper heading must be included at the top of the table. Do not limit the program to any specific unit system. Indicate the consistent U.S. customary or SI units that can be used.

C3–1 Refer to Fig. C3–1. Write a computer program that can be used to determine two unknown forces P and Q of the truss joint loaded as shown. The user input should include (1) the direction angles θ_P and θ_Q of the two unknown forces P and Q, and (2) the magnitude F_i and the direction angle θ_i of each of the known forces. The direction angles must be in the standard position. Indicate in the output results whether a member is in tension or compression. Use this program to solve (*a*) Example 3–4, (*b*) Problem 3–17, and (*c*) Problem 3–30.

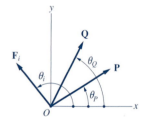

FIGURE C3–1

C3–2 Write a computer program that can be used to determine the reactions of an overhanging beam (Fig. C3–2a) or a cantilever beam (Fig. C3–2b) subjected to the loads shown. The user input should include (1) the span length L and the length of the overhang a for an overhanging beam; (2) the magnitude and location of the concentrated force P and b; (3) the intensity and location of the uniform loads w, c_1, and c_2; and (4) the magnitude of the couple moment M. Treat the loads shown in the same direction in the figure as positive. The output results are to be the magnitude and direction of the reactions R_A and R_B for the overhanging beam, and V_A and M_A for the cantilever beam. Use this program to solve (a) Problem 3–44 and (b) Problem 3–48.

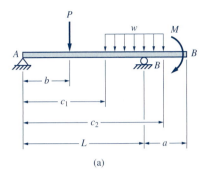

(a)

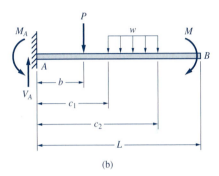

(b)

FIGURE C3–2

CHAPTER 4

ANALYSIS OF STRUCTURES

4-1
INTRODUCTION

A *structure* is a connected system of members designed to support loads safely or to transfer forces effectively. In this chapter, we shall apply the equilibrium conditions to the analysis of structures formed by pin-connected members, including trusses, frames, and machines. The analysis of structures involves analyzing the equilibrium of the entire structure, part of the structure, or an individual member. In these types of problems, we shall determine the external reactions as well as the internal forces in the members. The *internal forces* are the forces within the members responsible for holding the structure together.

We shall limit our analysis only to *statically determinate structures*, which can be solved by the equilibrium equations alone. These problems require careful observation of Newton's *law of action and reaction*, which states that *the forces of action and reaction between interactive bodies must have the same magnitude, the same line of action, and opposite directions.*

4-2
EXTERNAL AND INTERNAL FORCES IN A STRUCTURE

Forces acting on a structure are classified into two types: external and internal.

External Forces. The *external forces* include the weight, the externally applied forces on the structure, and the reactions from the supports. These forces are responsible for the equilibrium of the entire structure.

Internal Forces. The *internal forces* are the forces inside the structural members and are responsible for holding the structure together.

Limitation of the Principle of Transmissibility. When dealing with external forces on a structure, the structure can be considered as a rigid body and the *principle of transmissibility* can be applied freely, which means that a force may be applied at any point along its line of action without altering its external effects on the rigid body. When the concern is the internal forces or the deformation of a structure, the principle of transmissibility is not applicable. For example, consider the truss subjected to the load **P** shown in Fig. 4–1. The external reactions at supports *A* and *C* are not affected, whether the force is acting at *B* (Fig. 4–1a) or at *D* (Fig. 4–1b). As we will see in later analysis, however, the internal forces in the truss members will be different for the two cases shown in Fig. 4–1a and b. Therefore, as far as the internal effects are concerned, the principle of transmissibility is not applicable.

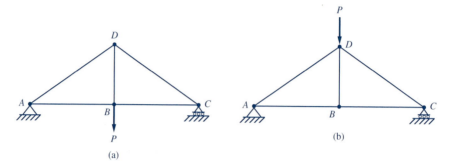

FIGURE 4–1

4–3
TRUSSES

Definition. Trusses are frameworks composed of slender members joined together at the ends of the members. The members are straight and commonly consist of wooden struts, metal bars, angles, channels, and wide flange or built-up sections. For steel trusses, the joint connections are usually formed by bolting, riveting, or welding the ends of the members to a connection plate, called a *gusset plate*, as shown in Fig. 4–2a. For timber trusses, the members are connected by a nailing brace, as shown in Fig. 4–2b.

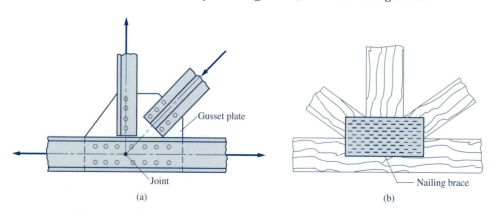

FIGURE 4–2

When the members of a truss lie essentially in a single plane, the truss is known as a *plane truss*. Bridge trusses are usually plane trusses placed side by side, one on each side of the roadway, as shown in Fig. 4–3. The trusses are connected together by lateral bracing. A floor system consisting of a *deck*, *stringers*, and *floor beams* provides the means for transmitting the load. The deck is supported on the stringers, which transmit the deck loads to the cross beams. The cross beams, connected to truss joints, transmit the loads to the joints.

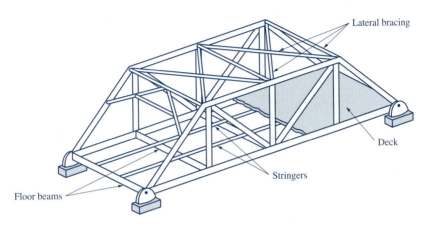

FIGURE 4–3

In the case of a roof truss, the roof loads are transmitted to the truss joints by means of *purlins*, which are attached to the truss at the joints.

Assumptions. Two assumptions are essential for truss analysis:

1. *All loadings are applied at the joints.* For the bridge and roof trusses, this assumption is true by design. The weight of the member, small compared to the joint load, can usually be neglected. If the weights of the members are to be included in the analysis, half the weight of each member is applied as a vertical load on each joint that the member connects.
2. *The members are joined together by smooth pins.* When the members are connected as shown in Fig. 4–2a or b, the assumption of a pin-jointed connection is usually satisfactory if the centerline of the members are concurrent at the joint.

Based on these assumptions, *each truss member is a two-force member.* The entire truss can be considered an assembly of two-force members connected at their ends by frictionless pins. Thus, a truss member is subjected to axial forces only, either in tension or in compression, as shown in Fig. 4–4. If the axial forces tend to pull the member apart, the member is in tension (Fig. 4–4a). On the other hand, if the forces tend to compress the member, the member is in compression (Fig. 4–4b).

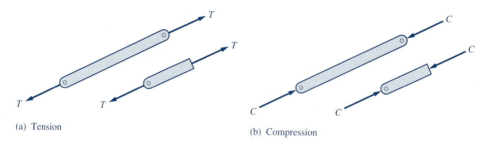

(a) Tension   (b) Compression

FIGURE 4–4

Simple Trusses. A truss must be *rigid*. The term *rigid* is used here to indicate that the truss will be able to maintain its initial shape and will not collapse under ordinary loading conditions. Three bars pinned together at their ends in the form of a triangle (Fig. 4–5a) constitute a rigid framework. Since loads applied to the triangular truss will produce only a small change in the lengths of the bars, the shape of the truss will remain unchanged.

The triangular truss may be expanded by adding more members and more joints on the same plane. Each time two more members are introduced, a new joint is added, as shown in Fig. 4–5b, c, and d. These trusses are all rigid. Any truss that can be formed in this manner is called a *simple truss*.

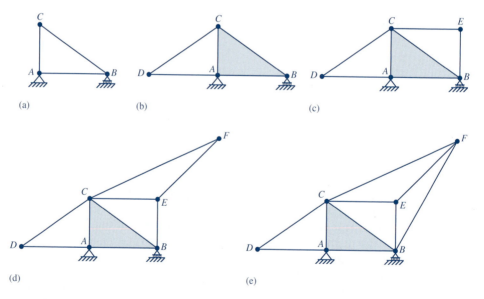

FIGURE 4–5

Statical Determinacy. In a simple truss, the basic triangle is statically determinate. For each new joint added, a new pair of equilibrium equations is available for calculating the forces in the two new members. Thus, simple trusses are statically determinate. Should a new member be introduced without adding a new joint, such as member *BF* in Fig. 4–5e, the truss would become statically indeterminate because no new static equilibrium equations are available to determine the unknown force in the new member.

A general method for determining the statical determinacy of a truss is by comparing the total number of independent equations available to the total number of unknowns. If we let j be the number of joints in a truss, m the number of members, and r the number of reaction components, then the number of equilibrium equations available is $2j$ and the number of unknowns is $m + r$. For a statically determinate truss, the number of equations must be equal to the number of unknowns, that is

$$2j = m + r \qquad\qquad (4\text{--}1)$$

On the other hand, if $2j < m + r$, then the truss is statically indeterminate, and if $2j > m + r$, the truss is unstable (not rigid). For example, in Fig. 4–5d, $j = 6$, $m = 9$, and $r = 3$. We have $2j = 12$ equations for $m + r = 12$ unknowns, so the truss is statically determinate. In Fig. 4–5e, $j = 6$, $m = 10$, and $r = 3$. We have $2j = 12$ equations, which is not enough to solve for $m + r = 13$ unknowns, so the truss is statically indeterminate.

Several typical roof trusses are shown in Fig. 4–6, and several typical bridge trusses are shown in Fig. 4–7.

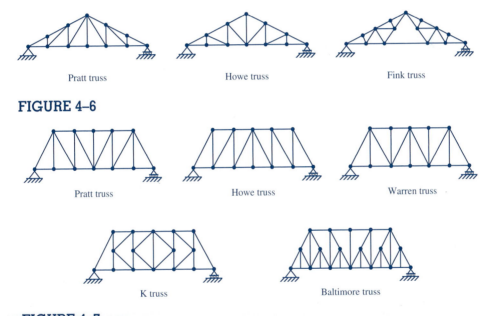

Pratt truss   Howe truss   Fink truss

FIGURE 4–6

Pratt truss   Howe truss   Warren truss

K truss   Baltimore truss

FIGURE 4–7

4–4
METHOD OF JOINTS

Truss Analysis. Analyzing a truss means determining the internal axial force in each member of a truss. These determinations are necessary prerequisites to the design of a truss because the force of truss members must be known before their sizes and connections can be designed properly. The analytic approach of truss analysis uses either the *method of joints* or the *method of sections*. Both are based on the principle of equilibrium: if a truss is in equilibrium, then each part of the truss, either a joint or part of the truss separated by sections, must also be in equilibrium. The method of joints is discussed in this section. The method of sections will be covered in the next section.

The Method of Joints. Each joint is isolated from the rest of the truss, and the equilibrium conditions of each joint are considered. Since the loads are applied on the joints only and truss members are all two-force bodies subjected to axial forces only, the forces acting on each joint form a balanced concurrent force system. For a plane truss, the equilibrium equations for each joint are

$$\Sigma F_x = 0 \qquad \Sigma F_y = 0$$

Free-Body Diagrams. In the simple truss shown in Fig. 4–8a, a free-body diagram can be drawn for each joint and each member, as shown in Fig. 4–8b. Each member is acted on by two equal and opposite forces along the axial direction of the member. Newton's third law requires that the action and reaction forces between a member and the connected joint be equal and opposite. Therefore, the forces exerted by a member on two joints at its ends must be equal and opposite.

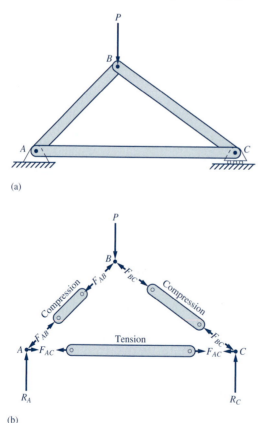

(a)

(b)

FIGURE 4–8

Arrow Sign Convention. When a member is in tension, it is being pulled and it, in turn, pulls the joint. Hence, a tension member exerts a force pointing away from the joint. Thus, *a tension member is represented by arrows on each end of the member pointing away from the connected joints.*

On the other hand, *a compression member is represented by arrows on each end of the member pointing toward the connected joints.*

Sometimes it is difficult to tell whether a member is in tension or in compression before the joint is analyzed. In this case, we may arbitrarily assume that the member is in tension. A positive result will indicate that the member is indeed in tension, while a negative result will indicate that the member is actually in compression. Since the design criteria for tensile and compressive members are different, it is important to label tension (T) or compression (C) in each member.

—————— **EXAMPLE 4–1** ——————————————————————

Refer to Fig. E4–1(1). Determine the force in each member of the truss subjected to the loading shown.

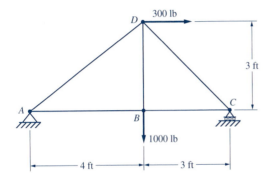

FIGURE E4–1(1)

Solution.   The reactions are determined first by considering the equilibrium of the entire truss. Then the equilibrium of each joint is analyzed to determine the internal forces in the members.

Equilibrium of the Entire Truss.   A free-body diagram of the entire truss [see Fig. E4–1(2)] is drawn showing the external forces and the unknown reaction components. The equilibrium equations give

$$\Sigma M_C = -A_y(7 \text{ ft}) + (1000 \text{ lb})(3 \text{ ft}) - (300 \text{ lb})(3 \text{ ft}) = 0$$
$$A_y = +300 \text{ lb} \uparrow$$

$$\Sigma M_A = C_y(7 \text{ ft}) - (1000 \text{ lb})(4 \text{ ft}) - (300 \text{ lb})(3 \text{ ft}) = 0$$
$$C_y = +700 \text{ lb} \uparrow$$

$$\Sigma F_x = A_x + 300 \text{ lb} = 0$$
$$A_x = -300 \text{ lb} \longleftarrow$$

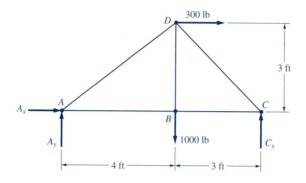

FIGURE E4–1(2)

Joint A.   The free-body diagram of this joint shows two unknown member forces: F_{AB} and F_{AD} [see Fig. E4–1(3)]. Both are assumed to be in tension. The force F_{AD} is resolved into horizontal and vertical components, H_{AD} and V_{AD}, respectively.

$$\Sigma F_y = V_{AD} + 300 \text{ lb} = 0$$

$$V_{AD} = -300 \text{ lb}$$

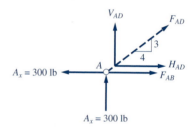

FIGURE E4–1(3)

The minus sign indicates that the member is in compression. Since F_{AD} is along the axial direction of the member, its components must be in the same proportion as the slope of the member. Thus,

$$\frac{H_{AD}}{V_{AD}} = \frac{4}{3} \qquad H_{AD} = \frac{4}{3}(-300 \text{ lb}) = -400 \text{ lb}$$

$$|F_{AD}| = \sqrt{(300 \text{ lb})^2 + (400 \text{ lb})^2} = 500 \text{ lb}$$

$$F_{AD} = 500 \text{ lb (C)} \qquad\qquad\qquad \Leftarrow\textbf{Ans.}$$

$$\Sigma F_x = F_{AB} + H_{AD} - 300 \text{ lb} = 0$$

$$F_{AB} = 300 \text{ lb} - (-400 \text{ lb}) = +700 \text{ lb}$$

$$F_{AB} = 700 \text{ lb (T)} \qquad\qquad\qquad \Leftarrow\textbf{Ans.}$$

Joint B.   The force in member AB is 700 lb in tension. The force exerted by member AB, shown in the free–body diagram of Fig. E4–1(4), is a 700-lb pulling force, pointing away from the joint. The two unknowns, F_{BC} and F_{BD}, can be solved from the equilibrium conditions of this joint. We arbitrarily assume that these members are both in tension. The equilibrium equations give

$$\Sigma F_x = F_{BC} - 700 \text{ lb} = 0$$

$$F_{BC} = +700 \text{ lb (T)} \qquad \Leftarrow \textbf{Ans.}$$

$$\Sigma F_y = F_{BD} - 1000 \text{ lb} = 0$$

$$F_{BD} = +1000 \text{ lb (T)} \qquad \Leftarrow \textbf{Ans.}$$

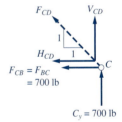

FIGURE E4–1(4)

Joint C. Now we have only one unknown member, force F_{CD}, left unsolved. The unknown force is again assumed to be in tension in the free–body diagram of joint C [see Fig. E4–1(5)]. The equilibrium equation along the y direction gives

$$\Sigma F_y = V_{CD} + 700 \text{ lb} = 0$$

$$V_{CD} = -700 \text{ lb}$$

From the slope 1:1 of member CD, we have

$$H_{CD} = V_{CD} = -700 \text{ lb}$$

$$|F_{CD}| = \sqrt{(700 \text{ lb})^2 + (700 \text{ lb})^2} = 990 \text{ lb}$$

$$F_{CD} = 990 \text{ lb (C)} \qquad \Leftarrow\textbf{Ans.}$$

FIGURE E4–1(5)

Equilibrium condition along the x direction gives the first check on the computations.

$$\Sigma F_x = 700 + H_{CD} = 700 + (-700) = 0 \qquad \text{(Checks)}$$

Joint D. Now that all the member forces have been solved, the equilibrium conditions of joint D [see the free–body diagram of the joint in Fig. E4–1(6)] provide two more checks.

$$\Sigma F_x = 400 + 300 - 700 = 0 \qquad \text{(Checks)}$$

$$\Sigma F_y = 300 + 700 - 1000 = 0 \qquad \text{(Checks)}$$

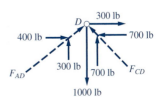

FIGURE E4–1(6)

The above results can be summarized as shown in Fig. E4–1(7). Note that the arrow sign convention is used to indicate whether a member is in tension or in compression. The horizontal and vertical components are shown on the sides of the triangle attached to the inclined members. Figure E4–1(7) shows the equilibrium of each joint. For example, at joint A, the 300-lb upward reaction is balanced by $V_{AD} = 300$ lb, which is acting downward. In the horizontal direction, the 300-lb horizontal reaction, plus $H_{AD} = 400$ lb acting to the left, is balanced by $H_{AB} = 700$ lb, which is acting to the right.

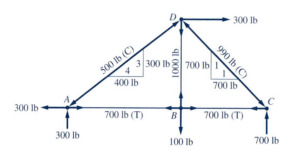

FIGURE E4–1(7)

EXAMPLE 4–2

Refer to Fig. E4–2(1). Determine the forces in all the members of the truss subjected to the loading shown.

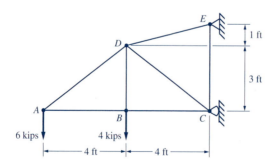

FIGURE E4–2(1)

Solution.   The reactions are determined first by considering the equilibrium of the entire truss. Then the equilibrium of each joint is analyzed to determine the internal forces in the members.

Equilibrium of the Entire Truss.   The equilibrium equations written for the free-body diagram of the entire truss [Fig. E4–2(2)] give

$$\Sigma M_E = -C_x(4 \text{ ft}) + (6 \text{ kips})(8 \text{ ft}) + (4 \text{ kips})(4 \text{ ft}) = 0$$

$$C_x = +16 \text{ kips} \longleftarrow$$

$$\Sigma M_C = -E_x(4\text{ ft}) + (6\text{ kips})(8\text{ ft}) + (4\text{ kips})(4\text{ ft}) = 0$$

$$E_x = +16\text{ kips} \longrightarrow$$

$$\Sigma F_y = E_y - 6\text{ kips} - 4\text{ kips} = 0$$

$$E_y = +10\text{ kips} \uparrow$$

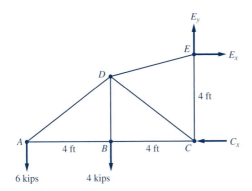

FIGURE E4–2(2)

Joint A.   Refer to Fig. E4–2(3). There are two unknown member forces: F_{AD} and F_{AB}. Since member AB is in the horizontal direction, the unbalanced vertical force of 6 kips (due to the applied downward load) must be balanced by $V_{AD} = 6$ kips in tension. From the slope of member AD (written inside the triangle attached to the member), we find $H_{AD} = 8$ kips. Now the unbalanced horizontal force of 8 kips (due to H_{AD} acting to the right) can be balanced by $F_{AB} = 8$ kips in compression.

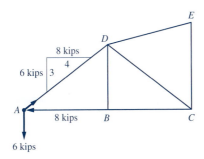

FIGURE E4–2(3)

Joint B.   Refer to Fig. E4–2(4). The unbalanced vertical force of 4 kips (due to the applied downward load) must be balanced by $F_{BD} = 4$ kips in tension. The unbalanced horizontal force of 8 kips (due to F_{AB} acting to the right) can be balanced by $F_{BC} = 8$ kips in compression. At this point, we see that any one of the three remaining joints, C, D, and E, can be considered next because each joint involves only two unknowns. But the two unknown members in joint D are both inclined, which requires setting up two simultaneous equilibrium equations. So we choose to solve joint C next.

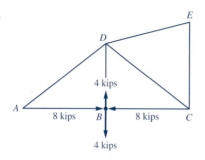

FIGURE E4–2(4)

Joint C.   Refer to Fig. E4–2(5). There are two unknown member forces: F_{CE} and F_{CD}. Since CE is in the vertical direction, the unbalanced horizontal force of 8 kips (16 kips − 8 kips, acting to the left) can be balanced by H_{CD} = 8 kips in compression. From the slope of the member, we find V_{CD} = 6 kips. Now the unbalanced vertical force of 6 kips (due to V_{CD} acting downward) can be balanced by F_{CE} = 6 kips in tension.

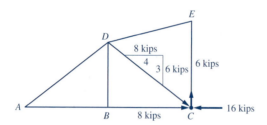

FIGURE E4–2(5)

Joint E.   Refer to Fig. E4–2(6). The unbalanced force in the horizontal direction is 16 kips (due to the horizontal reaction acting to the right); it can be balanced by H_{DE} = 16 kips in tension. From the slope of member DE, we find V_{DE} = 4 kips. Since F_{CE} is already solved in joint C, the equilibrium condition in the vertical direction provides the first check to the solution:

$$\Sigma F_y = 10 - 4 - 6 = 0 \qquad \text{(Checks)}$$

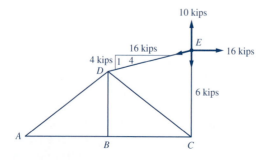

FIGURE E4–2(6)

Joint D. See Fig. E4–2(7). Now that the forces in all members have been solved, the equilibrium conditions of this joint will provide two more useful checks to the solution:

$$\Sigma F_x = 16 - 8 - 8 = 0 \qquad \text{(Checks)}$$

$$\Sigma F_y = 6 + 4 - 6 - 4 = 0 \qquad \text{(Checks)}$$

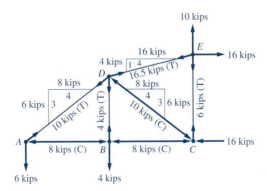

FIGURE E4–2(7)

The force in an inclined member can be determined from the equation

$$F = \sqrt{H^2 + V^2}$$

The final results are shown in the truss diagram in Fig. E4–2(8). Note that the truss diagram has been repeated several times for explanation purposes. Actually only one truss diagram is needed for the complete solution.

FIGURE E4–2(8)

4–5
ZERO-FORCE MEMBERS

Under certain loading conditions, some truss members carry no loads. Such members are called *zero-force members*. When loading conditions change, however, such members may no longer be zero-force members. Under a given loading, most of the zero-force members can be spotted by inspection. Consider a joint that is formed by three members, two of which are collinear (Fig. 4–9a). If the joint is not subjected to any load or support reaction, then the equilibrium condition along the x direction requires that F_3 be zero; thus, member 3 must be a zero-force member.

Another case where the zero-force members can be spotted immediately is shown in Fig. 4–9b, where two noncollinear members meet at a joint. If the joint is not subjected to any load or support reaction, the equilibrium condition along the y direction requires that F_2 be zero, and the equilibrium condition along the x direction requires that F_1 be zero. Therefore, members 1 and 2 are both zero-force members.

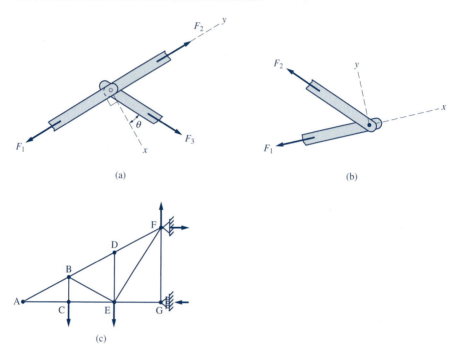

(a)   (b)

(c)

FIGURE 4–9

From the above discussions, we have the following two cases where zero-force members can be identified readily:

1. *If a nonloaded truss joint is formed by three members, two of which are collinear, the third member must be a zero-force member.*
2. *If a nonloaded truss joint is formed by two noncollinear members, then both of them must be zero-force members.*

As an example, for the truss shown in Fig. 4–9c, joint A is nonloaded. The two members, AB and AC, meeting at the joint must both be zero-force members (case 2). At the nonloaded joint D, two of the three members meeting at the joint are collinear. The third member DE must be a zero-force member (case 1). There is no way we can conclude that DE is a zero-force member at joint E. However, since it has already been identified as a zero-force member, it must exert no force to joint E. Thus, we see that a member may not appear to be a zero-force member at one end (such as member DE at joint E), but it may be obvious that it is a zero-force member at the other end (such as member DE at joint D). Therefore, to identify zero-force members, *all joints must be examined.* At joint G, the reaction and member EG are in the horizontal direction, and there is no force exerted in the vertical direction to produce any force in member FG, so it must be a zero-force member.

──────── **EXAMPLE 4–3** ────────────────────────────────────

Refer to Fig. E4–3(1). Identify the zero-force members in the truss due to the load shown.

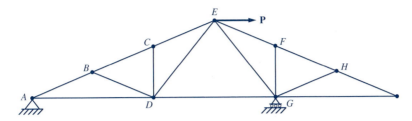

FIGURE E4–3(1)

Solution. The free-body diagram of the entire truss is sketched in Fig. E4–3(2). The values of force P and reactions need not be known for the purpose of identifying the zero-force members. The zero-force members are identified by considering the following joints.

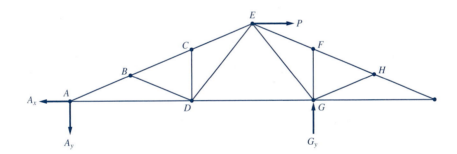

FIGURE E4–3(2)

Joint B. Two of the three members meeting at this nonloaded joint are collinear. The third member BD must be a zero-force member (case 1).

Joint C. This joint is similar to joint B (case 1). We conclude that CD is a zero-force member.

Joint D. Since BD and CD are identified as zero-force members, they do not exert any force on joint D. Two of the three remaining members meeting at this joint are collinear; thus, we conclude that DE is a zero-force member (case 1).

Joint I. At this nonloaded joint, both members HI and GI must be zero-force members (case 2).

Joint H.   Since *HI* is a zero-force member, it exerts no force on joint *H*. The remaining two members, *FH* and *GH*, meeting at this nonloaded joint must both be zero-force members (case 2).

Joint F.   Since *FH* is a zero-force member, it exerts no force on joint *F*. The remaining two members, *EF* and *FG*, meeting at this nonloaded joint must both be zero-force members (case 2).

The zero-force members are marked with a circle on the members, as shown in Fig. E4–3(3).

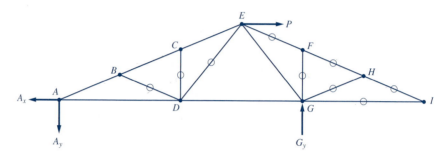

FIGURE E4–3(3)

4–6
METHOD OF SECTIONS

The *method of sections* consists of passing a section through the members in question. The truss is separated into two parts by the section. Either part of the truss may be used as a free body. The part involving fewer forces usually yields a simpler solution. Since there are three independent equations available, we can solve up to three unknowns. The main advantage of the method of sections is that the force in a member can be determined directly by passing a section through the member.

The reactions must be determined before applying the method of sections. Suppose we want to determine the forces in members *BD*, *CD*, and *CE* of the truss shown in Fig. 4–10a. A vertical section *m–m* cutting through three members is used. This section separates the truss into two parts. The part to the left of the section involves fewer forces, so the free-body diagram of this part is drawn, as shown in Fig. 4–10b. When a member has been cut, its internal forces are exposed and they become external forces, so they must be shown on the free-body diagram. The three unknown forces, F_{BD}, F_{CD}, and F_{CE}, are all assumed to be in tension (shown acting away from the section). A positive result indicates that the member is indeed in tension, while a negative result indicates that the member is actually in compression.

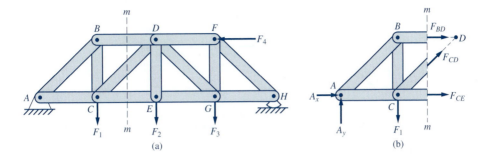

FIGURE 4–10

The three unknown forces can be determined independently by using the following equilibrium equations:

$$\Sigma F_y = 0 \ \text{ for } F_{CD}$$

$$\Sigma M_D = 0 \ \text{ for } F_{CE}$$

$$\Sigma M_C = 0 \ \text{ for } F_{BD}$$

EXAMPLE 4–4

Determine the internal forces of members DF, EF, and EG of the truss shown in Fig. E4–4(1).

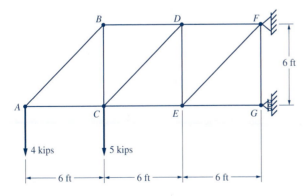

FIGURE E4–4(1)

Solution. The internal forces in members DF, EF, and EG can be determined by passing section m–m through the three members. The support reactions are not needed if the part of the truss to the left of the section is isolated as a free body. The free-body diagram is shown in Fig. E4–4(2). Three unknowns are involved. The unknown forces are assumed to be in tension. The three unknowns can be solved independently with the following equilibrium equations:

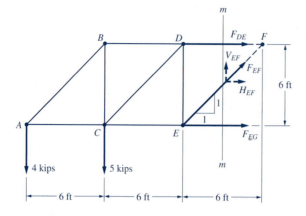

FIGURE E4–4(2)

$$\Sigma M_E = -F_{DF}(6 \text{ ft}) + (4 \text{ kips})(12 \text{ ft}) + (5 \text{ kips})(6 \text{ ft}) = 0$$
$$F_{DF} = +13 \text{ kips (T)} \qquad \Leftarrow \textbf{Ans.}$$

$$\Sigma F_y = V_{EF} - 4 \text{ kips} - 5 \text{ kips} = 0$$
$$V_{EF} = +9 \text{ kips}$$

The slope of member *EF* is 1:1, hence

$$H_{EF} = +9 \text{ kips}$$
$$F_{EF} = 12.73 \text{ kips (T)} \qquad \Leftarrow \textbf{Ans.}$$

$$\Sigma M_F = +F_{EG}(6 \text{ ft}) + (4 \text{ kips})(18 \text{ ft}) + (5 \text{ kips})(12 \text{ ft}) = 0$$
$$F_{EG} = -22 \text{ kips (C)} \qquad \Leftarrow \textbf{Ans.}$$

Check. The results can be checked by

$$\Sigma F_x = F_{DF} + H_{EF} + F_{FG}$$
$$= +13 + 9 - 22 = 0 \qquad \text{(Checks)}$$

EXAMPLE 4–5

Determine the forces in members *FH*, *GH*, and *GI* of the Howe roof truss shown in Fig. E4–5(1).

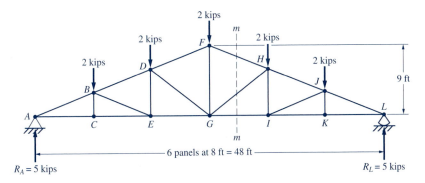

FIGURE E4–5(1)

Solution. From the condition of symmetry, the vertical reactions at A and L must be equal to half the total load. The internal forces in members FH, GH, and GI can be determined by passing a section m–m through the three members. The free-body diagram of the part of the truss to the right of the section is shown in Fig. E4–5(2). The three unknown forces are assumed to be in tension. Note that F_{FH} is resolved into horizontal and vertical components, H_{FH} and V_{FH}, respectively, at joint F. The force F_{GH} is resolved into horizontal and vertical components, H_{GH} and V_{GH}, respectively, at joint G.

The three unknown member forces can be computed independently by using the following equilibrium equations:

$$\Sigma M_G = +H_{FH}(9 \text{ ft}) - (2 \text{ kips})(8 \text{ ft}) - (2 \text{ kips})(16 \text{ ft}) + (5 \text{ kips})(24 \text{ ft}) = 0$$

$$H_{FH} = -8 \text{ kips}$$

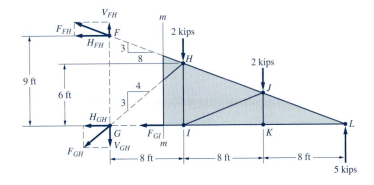

FIGURE E4–5(2)

Since the slope of member FH is 8:3, we get

$$V_{FH} = \frac{3}{8}(-8 \text{ kips}) = -3 \text{ kips}$$

$$F_{FH} = \sqrt{(-8 \text{ kips})^2 + (-3 \text{ kips})^2} = 8.54 \text{ kips}$$

$$F_{FH} = 8.54 \text{ kips (C)} \qquad \Leftarrow \textbf{Ans.}$$

$$\Sigma M_L = +V_{GH}(24 \text{ ft}) + (2 \text{ kips})(16 \text{ ft}) + (2 \text{ kips})(8 \text{ ft}) = 0$$

$$V_{GH} = -2 \text{ kips}$$

Since the slope of member GH is 4:3, we get

$$H_{GH} = \frac{4}{3}(-2 \text{ kips}) = -2.67 \text{ kips}$$

$$|F_{GH}| = \sqrt{(-2 \text{ kips})^2 + (-2.67 \text{ kips})^2} = 3.33 \text{ kips}$$

$$F_{GH} = 3.33 \text{ kips (C)} \qquad \qquad \Leftarrow \textbf{Ans.}$$

$$\Sigma M_H = -F_{GI}(6 \text{ ft}) - (2 \text{ kips})(8 \text{ ft}) + (5 \text{ kips})(16 \text{ ft}) = 0$$

$$F_{GI} = +10.67 \text{ kips (T)} \qquad \qquad \Leftarrow \textbf{Ans.}$$

Check. The results can be checked by

$$\Sigma F_x = -H_{FH} - H_{GH} - F_{GI}$$
$$= -(-8) - (-2.67) - (+10.67) = 0 \qquad \text{(Checks)}$$

$$\Sigma F_y = V_{FH} - V_{GH} - 2 - 2 + 5$$
$$= (-3) - (-2) - 2 - 2 + 5 = 0 \qquad \text{(Checks)}$$

EXAMPLE 4–6

See Fig. E4–6(1). Find the forces in all the members of the truss due to the loading shown.

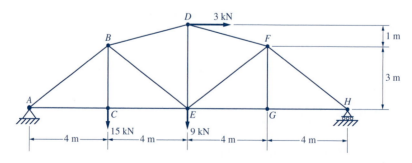

FIGURE E4–6(1)

Solution. Refer to Fig. E4–6(2). The reactions at the supports can be computed by considering the equilibrium of the entire truss.

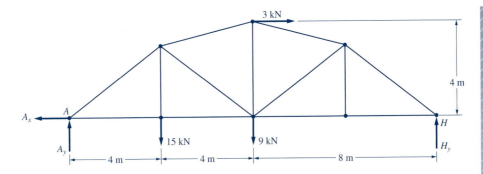

FIGURE E4–6(2)

$$\Sigma F_x = -A_x + 3 \text{ kN} = 0$$
$$A_x = +3 \text{ kN} \quad \longleftarrow$$

$$\Sigma M_H = -A_y(16 \text{ m}) + (15 \text{ kN})(12 \text{ m}) + (9 \text{ kN})(8 \text{ m}) - (3 \text{ kN})(4 \text{ m}) = 0$$
$$A_y = +15 \text{ kN} \quad \uparrow$$

$$\Sigma M_A = +H_y(16 \text{ m}) - (15 \text{ kN})(4 \text{ m}) - (9 \text{ kN})(8 \text{ m}) - (3 \text{ kN})(4 \text{ m}) = 0$$
$$H_y = +9 \text{ kN} \quad \uparrow$$

The results can be checked by:

$$\Sigma F_y = +15 + 9 - 15 - 9 = 0 \qquad \text{(Checks)}$$

Method of Joints. To find the forces in all the members, we will start by using the method of joints. Joint A is analyzed first; then Joint C can be analyzed as shown in Fig. E4–6(3).

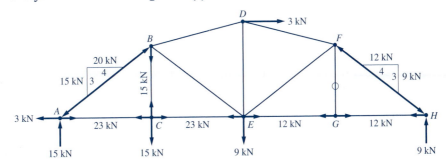

FIGURE E4–6(3)

At this point, we see that it is impossible to analyze joint E because this joint involves more than two unknowns. Although it is possible to analyze joint B, the two unsolved members are both inclined, which will require setting up two simultaneous equilibrium equations. However, we could analyze joints H and G. The results are shown in Fig. E4–6(3). At joint F, we run into the same problem as joint B. To avoid the need for solving simultaneous equations, we will apply the method of section.

Method of Section.   See Fig. E4–6(4). Pass a vertical section through member *BD* and sketch the free-body diagram of the part of the truss to the left of the section, as shown. The force in member *BD* is assumed to be in tension and is resolved into horizontal and vertical components at *D*. We write

$$\Sigma M_E = -H_{BD}(4 \text{ m}) - (15 \text{ kN})(8 \text{ m}) + (15 \text{ kN})(4 \text{ m}) = 0$$

$$H_{BD} = -15 \text{ kN}$$

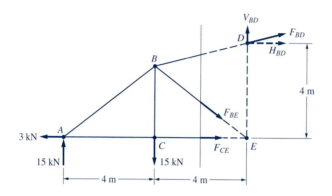

FIGURE E4–6(4)

Using this value, joint *B* can be analyzed. We then proceed to joints *D* and *F*. The equilibrium conditions of joint *E* provide useful checks. The final results are shown in the truss diagram in Fig. E4–6(5).

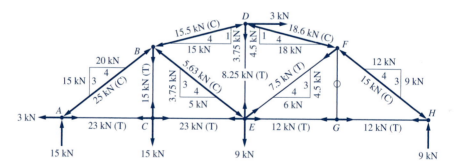

FIGURE E4–6(5)

4–7
FRAMES

Pin-connected frames are structures containing one or more *multiforce members*. A multiforce member is acted upon by three or more forces that are not all in the axial direction of the member. Figure 4–11 shows several examples of framed structures. Limitations imposed on the truss members no longer apply. Members in a frame may not always be straight and they may have loads or connections at some intermediate points.

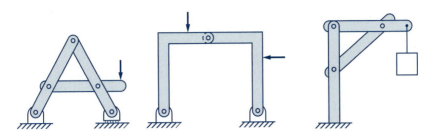

FIGURE 4-11

Analysis of a Frame. The external reactions at the supports of most frames can be determined by considering the free-body diagram of the entire frame. The internal forces of the members at the joints are determined by considering the free-body diagrams of the members separately. Close attention must be paid to the following when drawing the free-body diagrams of the members.

1. Identify all the two-force members in the frame. Note that a two-force member may or may not be straight. The two forces in a two-force member are equal and opposite and directed along the line joining the two points of application, regardless of the shape of the member. For example, member *DE* in Fig. 4-12a is a two-force member. Since the member is straight, the force in this member is along the axial direction of the member, as shown in Fig. 4-12b.

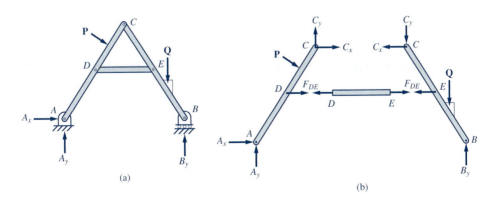

FIGURE 4-12

2. The force at a pin-connected joint between two multiforce members is usually represented by the horizontal and vertical components, as in joint *C* shown in Fig. 4-12b.

3. Newton's law of action and reaction must be observed when forces are drawn on the free-body diagrams of two connected members. Hence, the horizontal and vertical components at joint C between members ADC and BEC must be equal in magnitude (represented by the same notation) and opposite in direction, as shown in Fig. 4–12b.

4. Generally three equilibrium equations can be written for the free-body diagram of a multiforce member.

5. The directions of the unknown force components can be assumed arbitrarily as long as the law of action and reaction is observed. A negative value in the result indicates that the assumed direction of the force must be reversed.

6. Sometimes not all the reaction components or joint forces can be solved from the free-body diagram of the entire frame or a member. In such cases, equilibrium equations can be set up in terms of two unknowns, which can be used later when one of the two unknowns is determined from the equilibrium conditions of another member.

EXAMPLE 4–7

Refer to Fig. E4–7(1). Determine the forces in each member of the A-frame due to the loads shown.

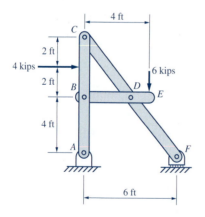

FIGURE E4–7(1)

Solution. The support reactions at A and F can be determined first by considering the equilibrium of the entire frame.

Entire Frame. The three reaction elements can be computed independently by setting up the following equilibrium equations for the free-body diagram of the entire frame shown in Fig. E4–7(2).

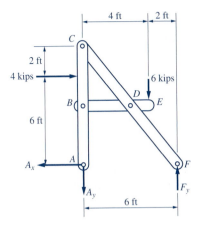

FIGURE E4–7(2)

$$\Sigma F_x = -A_x + 4 \text{ kips} = 0$$
$$A_x = +4 \text{ kips} \quad \longleftarrow$$

$$\Sigma M_F = +A_y(6 \text{ ft}) - (4 \text{ kips})(6 \text{ ft}) + (6 \text{ kips})(2 \text{ ft}) = 0$$
$$A_y = +2 \text{ kips} \quad \downarrow$$

$$\Sigma M_A = +F_y(6 \text{ ft}) - (4 \text{ kips})(6 \text{ ft}) - (6 \text{ kips})(4 \text{ ft}) = 0$$
$$F_y = +8 \text{ kips} \quad \uparrow$$

$$\Sigma F_y = -2 - 6 + 8 = 0 \qquad\qquad \text{(Checks)}$$

To determine the internal forces in the members, consider the equilibrium of each member.

Member BDE. Refer to Fig. E4–7(3). The equilibrium equations written for the free-body diagram of the member give

$$\Sigma M_D = +B_y(3 \text{ ft}) - (6 \text{ kips})(1 \text{ ft}) = 0$$
$$B_y = +2 \text{ kips} \qquad\qquad \Leftarrow\textbf{Ans.}$$

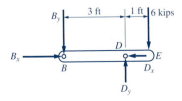

FIGURE E4–7(3)

$$\Sigma M_B = +D_y(3 \text{ ft}) - (6 \text{ kips})(4 \text{ ft}) = 0$$

$$D_y = +8 \text{ kips} \qquad \qquad \Leftarrow\textbf{Ans.}$$

$$\Sigma F_y = -2 + 8 - 6 = 0 \qquad \qquad \text{(Checks)}$$

$$\Sigma F_x = B_x - D_x = 0$$

$$D_x = B_x \qquad \qquad \text{(a)}$$

Note that the positive signs for B_y and D_y indicate that their assumed directions are correct. The components B_x and D_x cannot yet be determined.

Member ABC.   The equilibrium equations written for the free-body diagram of member ABC in Fig. E4–7(4) give

$$\Sigma M_C = -B_x(4 \text{ ft}) - (4 \text{ kips})(8 \text{ ft}) + (4 \text{ kips})(2 \text{ ft}) = 0$$

$$B_x = -6 \text{ kips} \qquad \qquad \Leftarrow\textbf{Ans.}$$

$$\Sigma M_B = -C_x(4 \text{ ft}) - (4 \text{ kips})(4 \text{ ft}) - (4 \text{ kips})(2 \text{ ft}) = 0$$

$$C_x = -6 \text{ kips} \qquad \qquad \Leftarrow\textbf{Ans.}$$

$$\Sigma F_y = C_y + 2 \text{ kips} - 2 \text{ kips} = 0$$

$$C_y = 0 \qquad \qquad \Leftarrow\textbf{Ans.}$$

$$\Sigma F_x = -4 + 6 + 4 - 6 = 0 \qquad \qquad \text{(Checks)}$$

From Equation (a), we get

$$D_x = B_x = -6 \text{ kips} \qquad \qquad \Leftarrow\textbf{Ans.}$$

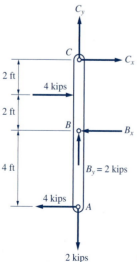

FIGURE E4–7(4)

Note that the negative signs for B_x, C_x, and D_x indicate that their assumed directions must be reversed. Now that all the forces have been determined, the results are shown in Fig. E4–7(5).

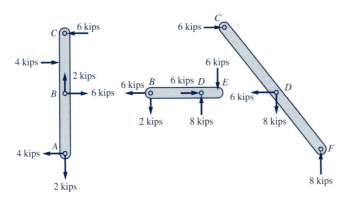

FIGURE E4–7(5)

Check. The equilibrium conditions of member *CDF* provide useful checks of the results:

$$\Sigma F_x = +6 - 6 = 0 \qquad \text{(Checks)}$$
$$\Sigma F_y = +8 - 8 = 0 \qquad \text{(Checks)}$$

EXAMPLE 4–8

See Fig. E4–8(1). Determine the forces in each member of the pin-connected frame due to the 1200-N load applied at joint *C*.

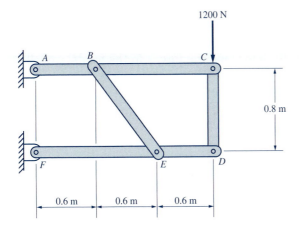

FIGURE E4–8(1)

Solution. First, we will consider the equilibrium of the entire frame.

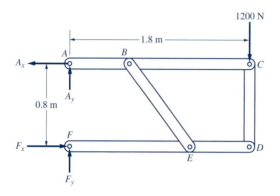

FIGURE E4–8(2)

Entire Frame. Since the free-body diagram in Fig. E4–8(2) involves four unknowns, not all the unknown reaction elements can be determined immediately. However, three equilibrium equations can be set up.

$$\Sigma M_F = +A_x(0.8 \text{ m}) - (1200 \text{ N})(1.8 \text{ m}) = 0$$
$$A_x = +2700 \text{ N} \; \longleftarrow$$

$$\Sigma F_x = +F_x - 2700 \text{ N} = 0$$
$$F_x = +2700 \text{ N} \; \longrightarrow$$

$$\Sigma F_y = A_y + F_y - 1200 \text{ N} = 0$$
$$F_y = -A_y + 1200 \text{ N} \qquad \qquad \text{(a)}$$

Note that A_y and F_y cannot be determined yet.

Members *BE* and *CD* are recognized as two-force bodies. These members are subjected to axial forces. We will arbitrarily assume that the two members are in tension. To determine the forces in the joints, we will consider the equilibrium of each member.

Member ABC. The equilibrium equations written for the free-body diagram of the member in Fig. E4–8(3) give

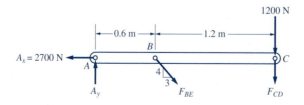

FIGURE E4–8(3)

$$\Sigma F_X = \frac{3}{5}F_{BE} - 2700 \text{ N} = 0$$

$$F_{BE} = +4500 \text{ N (T)} \qquad\qquad \Leftarrow\textbf{Ans.}$$

$$\Sigma M_A = -F_{CD}(1.8 \text{ m}) - (1200 \text{ N})(1.8 \text{ m}) - \left(\frac{4}{5} \times 4500 \text{ N}\right)(0.6 \text{ m}) = 0$$

$$F_{CD} = -2400 \text{ N (C)} \qquad\qquad \Leftarrow\textbf{Ans.}$$

$$\Sigma F_y = A_y - \frac{4}{5}(4500 \text{ N}) - 1200 \text{ N} + 2400 \text{ N} = 0$$

$$A_y = +2400 \text{ N} \uparrow \qquad\qquad \Leftarrow\textbf{Ans.}$$

Substituting in Equation (a) gives

$$F_y = -2400 \text{ N} + 1200 \text{ N} = -1200 \text{ N}$$

$$F_y = 1200 \text{ N} \downarrow \qquad\qquad \Leftarrow\textbf{Ans.}$$

Now that all the forces have been determined, the results are shown in Fig. E4–8(4).

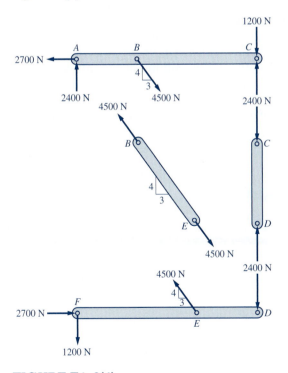

FIGURE E4–8(4)

Check. The equilibrium conditions of member *FED* provide useful checks of the results:

$$\Sigma F_y = 2700 - \frac{3}{5}(4500) = 0 \qquad\qquad \text{(Checks)}$$

$$\Sigma F_y = -1200 + \frac{4}{5}(4500) - 2400 = 0 \qquad \text{(Checks)}$$

4–8
MACHINES

Definition.   *Machines* are structures that consist of one or more movable parts, and at least one of the members or part of the machine is a multiforce member. Machines are designed to transmit forces or to alter the effect of forces.

Analysis of Machines.   The analysis of a machine is similar to that of a frame, except a machine may not be fully constrained because of the movable parts. The analysis usually involves the equilibrium of part of the machine or a member isolated from the other part of the machine. Consider, for example, a pair of pliers used to exert two equal and opposite forces on a nut (Fig. 4–13a). If we assume that the hand-gripping force is concentrated at A and B, the *input forces* (P) on the handles are transmitted by the pliers to produce the *output forces* (Q) acting on the nut, as shown in Fig. 4–13b. To relate the input force P to the output force Q, the free-body diagrams of the component parts are drawn (Fig. 4–14a and b). From either one of the free-body diagrams, the relation $Pa = Qb$ or $Q = (a/b)P$ can be established by writing the moment equation $\Sigma M_C = 0$. For a plier, a is much greater than b; consequently, the magnitude of the output force Q will be much greater than the magnitude of the input force P. The *mechanical advantage* is defined as the ratio of the output force to the input force. The mechanical advantage of a machine is usually greater than 1.

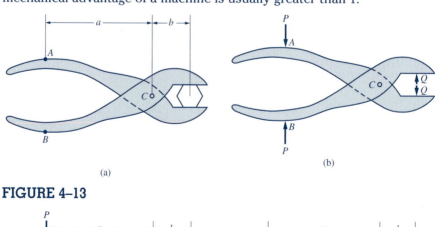

(a)

FIGURE 4–13

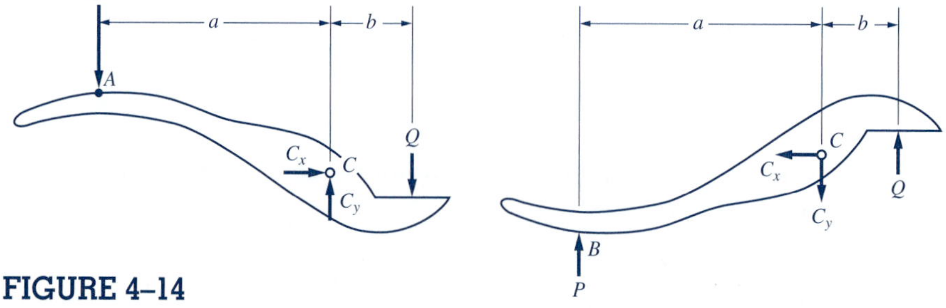

FIGURE 4–14

———— **EXAMPLE 4–9** ————————————————————————————————————

A 50-lb ice block is picked up by a pair of ice pickers as shown in Fig. E4–9(1). Determine the forces acting on tong *CDE*.

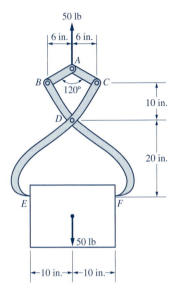

FIGURE E4–9(1)

Solution. Note that members *AB* and *AC* are two-force bodies. The axial forces of the members can be determined by the equilibrium conditions of joint *A*.

Joint A. The equilibrium equations for the free-body diagram of the joint in Fig. E4–9(2) give

$$\Sigma F_x = +F_{AC} \cos 30° - F_{AB} \cos 30° = 0$$

$$F_{AC} = F_{AB}$$

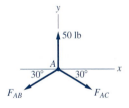

FIGURE E4–9(2)

$$\Sigma F_y = -F_{AC} \sin 30° - F_{AB} \sin 30° + 50 \text{ lb} = 0$$

$$(F_{AC} + F_{AB}) \sin 30° = 50 \text{ lb}$$

$$F_{AC} = F_{AB} = 50 \text{ lb (T)} \qquad \Leftarrow \textbf{Ans.}$$

Ice Block. The free-body diagram of the ice block in Fig. E4–9(3) is symmetrical with respect to the vertical axis through the center of gravity. By symmetry, we see that $E_y = F_y$. The equilibrium equation in the vertical direction gives

$$\Sigma F_y = E_y + F_y - 50 \text{ lb} = 0$$

$$E_y = F_y = +25 \text{ lb} \qquad \Leftarrow \textbf{Ans.}$$

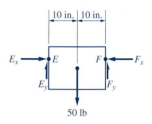

FIGURE E4–9(3)

Tong CDE. The equilibrium equations for the free-body diagram of the tong in Fig. E4–9(4) give

$$\Sigma M_D = -E_x(20 \text{ in.}) + (25 \text{ lb})(10 \text{ in.}) + (50 \text{ lb})(\sin 30°)(6 \text{ in.})$$
$$+ (50 \text{ lb})(\cos 30°)(10 \text{ in.}) = 0$$

$$E_x = +41.65 \text{ lb} \qquad \Leftarrow \textbf{Ans.}$$

$$\Sigma F_x = D_x - 41.65 \text{ lb} - (50 \text{ lb}) \cos 30° = 0$$

$$D_x = +84.95 \text{ lb} \qquad \Leftarrow \textbf{Ans.}$$

$$\Sigma F_y = D_y - 25 \text{ lb} + (50 \text{ lb}) \sin 30° = 0$$

$$D_y = 0 \qquad \Leftarrow \textbf{Ans.}$$

The results are sketched in Fig. E4–9(5).

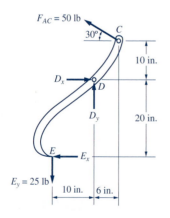

FIGURE E4–9(4)

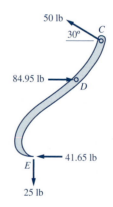

FIGURE E4–9(5)

EXAMPLE 4–10

In the press shown in Fig. E4–10(1), the resultant pressure acting on the piston at A is 500 N. Determine the compressive forces acting on the workpiece at E.

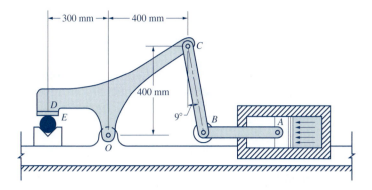

FIGURE E4–10(1)

Solution. The resultant pressure on the piston causes a compressive force of 500 N in member AB, which is a two-force member. The axial force in member BC, another two-force member, can be determined by the equilibrium of joint B.

Joint B. The free-body diagram of the joint and the corresponding force triangle are shown in Fig. E4–10(2). From the force triangle, we get

$$F_{BC} = \frac{500 \text{ N}}{\sin 9°} = 3196 \text{ N (C)}$$

FIGURE E4–10(2)

Member COD. The moment equation about the pivot point O applied to the free-body diagram of the member in Fig. E4–10(3) gives

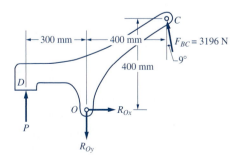

FIGURE E4–10(3)

$$\Sigma M_O = -P(0.3 \text{ m}) + (3196 \text{ N})(\cos 9°)(0.4 \text{ m}) + (3196 \text{ N}) (\sin 9°)(0.4 \text{ m}) = 0$$

$$P = +4880 \text{ N} \qquad\qquad \Leftarrow\textbf{Ans.}$$

Thus, an output force of 4880 N on the workpiece is a result of an input force of 500 N acting on the piston. Therefore, the mechanical advantage of the press is

$$\frac{4990 \text{ N}}{500 \text{ N}} = 9.98$$

which is almost tenfold.

4–9
SUMMARY

Structure. A *structure* is a connected system of members designed to support loads safely or to transfer forces effectively.

Truss. A *truss* is a structure consisting of straight members connected at the ends of the members. All the loads on a truss are applied at the joints. All the truss members are two-force bodies subjected to axial tension or compression.

Simple Truss. A rigid, triangular framework may be expanded by adding two new members connected at a new joint. A truss formed in this manner is called a *simple truss.*

Statical Determinacy. A truss is *statically determinate* if its total number of joints j and the total number of members m are such that

$$2j = m + r \qquad\qquad (4\text{–}1)$$

where r is the number of reaction components. A truss is *statically indeterminate* if $2j < m + r$.

Method of Joints. The equilibrium of each joint is considered. The forces exerted by each truss member on the connected joints must act in the axial direction of the member. A tension member exerts a force away from the joint; a compression member exerts a force toward the joint. Only two equilibrium equations can be written for each joint, so we can solve only two unknowns from a joint. If one of the two unknown forces is either horizontal or vertical, then the two components of the unknown forces can be determined by inspection.

Method of Sections. The truss is divided into two parts by passing an imaginary section through the members of interest. Either part of the truss separated by the section may be considered a free body. Three independent equilibrium equations can be written for the free body, and they may be solved for three unknown member forces.

Frames and Machines. *Frames* and *machines* are structures containing *multiforce members*, i.e., members acted on by three or more forces. Members in a frame or a machine may not always be straight and they may have loads or connections at some intermediate points. Frames are designed to support loads and are usually stationary and fully constrained. Machines are designed to transmit or modify forces, and they consist of one or more movable parts.

Forces on the joints of a frame or a machine between connected members are determined by considering the equilibrium of the isolated members. Newton's law of action and reaction must be strictly followed when drawing the forces between two connected members in the free-body diagram of each member.

PROBLEMS

Section 4–4 Method of Joints

4–1 to **4–10** Refer to Figs. P4–1 to P4–10. Determine the forces in all members of the trusses shown using the method of joints. Indicate the results on the truss diagram using the arrow sign convention.

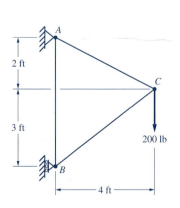

FIGURE P4–1

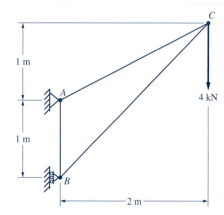

FIGURE P4–2

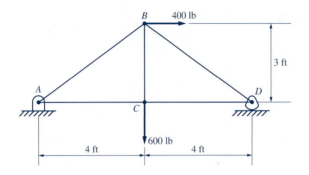

FIGURE P4–3

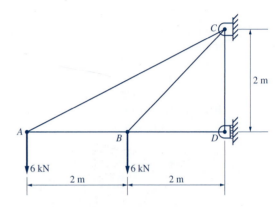

FIGURE P4–4

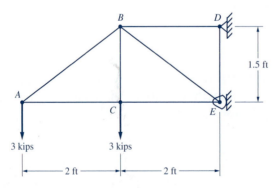

FIGURE P4–5

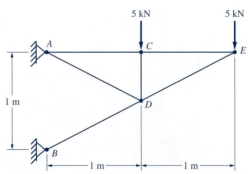

FIGURE P4–6

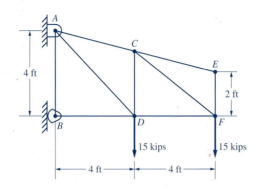

FIGURE P4-7

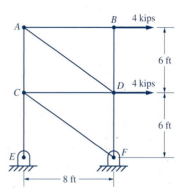

FIGURE P4-8

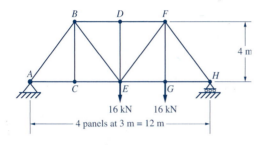

FIGURE P4-9

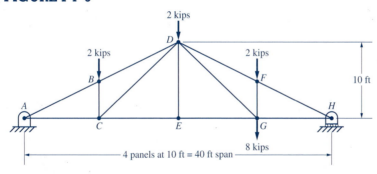

FIGURE P4-10

Section 4-5   Zero-Force Members

4-11 to **4-15** See Figs. P4-11 to P4-15. Identify the zero-force members in the trusses for the loading shown.

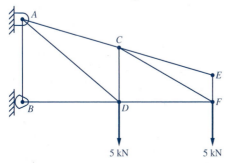

FIGURE P4-11

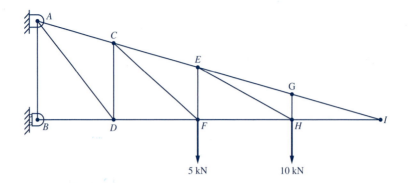

FIGURE P4–12

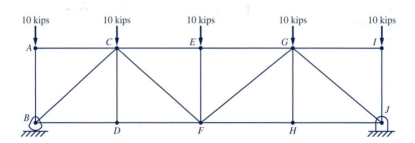

FIGURE P4–13

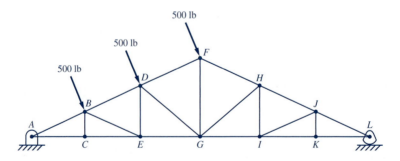

FIGURE P4–14

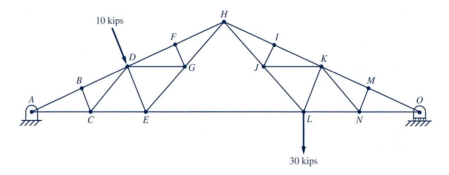

FIGURE P4–15

Section 4–6 Method of Sections

4–16 Determine the forces in members *BD*, *BE*, and *CE* of the truss and loading shown in Fig. P4–16 by the method of sections.

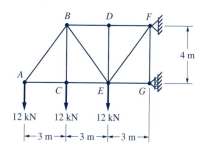

FIGURE P4–16

4–17 See Fig. P4–17. Determine the forces in members *FH*, *FI*, and *GI* of the roof truss subjected to the loading and reactions shown using the method of sections.

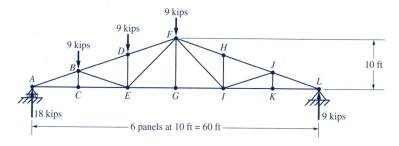

FIGURE P4–17

4–18 Determine the forces in members *DF*, *EF*, and *EG* of the roof truss shown in Fig. P4–17.

4–19 Determine the forces in members *BD*, *CD*, and *CE* of the truss due to the loading shown in Fig. P4–19.

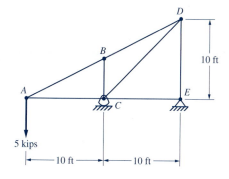

FIGURE P4–19

4–20 Refer to Fig. P4–20. Determine the forces in members *AC*, *CD*, and *DF* of the truss due to the loading shown.

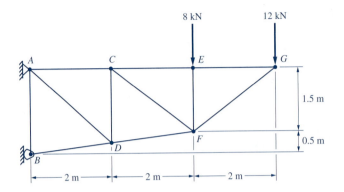

FIGURE P4–20

4–21   See Fig. P4–21. Determine the forces in members *DF*, *DG*, and *EG* of the Parker truss subjected to the loading and reactions shown.

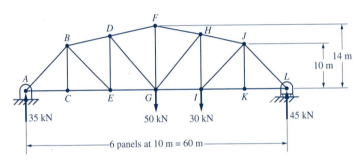

FIGURE P4–21

4–22   Determine the forces in members *FH*, *GH*, and *GI* of the Parker truss due to the loading shown in Fig. P4–21.

4–23   Refer to Fig. P4–23. Determine the forces in members *DF*, *DG*, and *EG* of the bridge truss subjected to the loading and reactions shown.

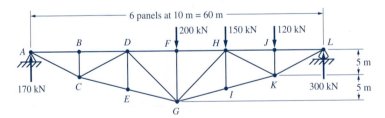

FIGURE P4–23

4–24 to 4–27 See Figs. P4–24 to P4–27. Determine the forces in all the members of the trusses by combined use of the method of sections and the method of joints so that the solution of simultaneous equations can be avoided.

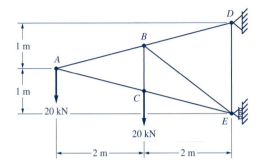

FIGURE P4–24

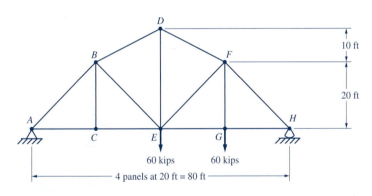

FIGURE P4–25

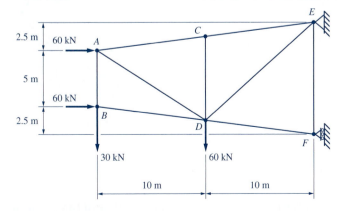

FIGURE P4–26

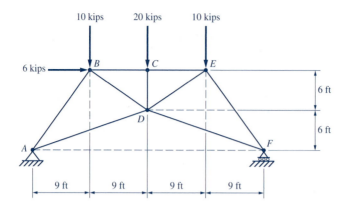

FIGURE P4–27

Section 4–7 Frames

4–28 See Fig. P4–28. Determine the reactions at *A* and *B* of the beam due to the loads shown.

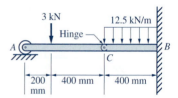

FIGURE P4–28

4–29 Refer to Fig. P4–29. Determine the forces in each member of the frame subjected to the load shown.

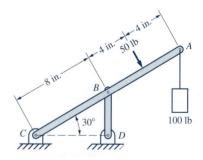

FIGURE P4–29

4–30 See Fig. P4–30. Determine the forces in each member of the frame subjected to the loads shown.

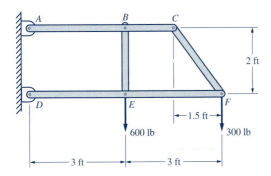

FIGURE P4–30

4–31 Refer to Fig. P4–31. Determine the forces acting on vertical member *ABC* due to the 2-kN load and the uniformly distributed wind load shown.

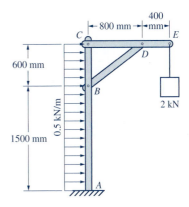

FIGURE P4–31

4–32 The three-hinged arch *ACB* in Fig. P4–32 is subjected to the loads shown. Determine the reactions at supports *A* and *B*.

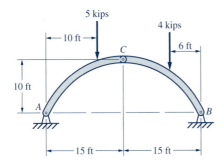

FIGURE P4–32

4–33 The three-hinged frame in Fig. P4–33, p. 186, is subjected to the loads shown. Find the reactions at supports *A* and *B*.

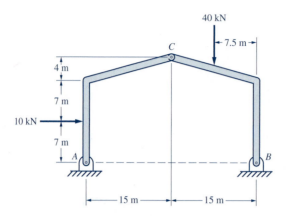

FIGURE P4–33

4–34 Refer to Fig. P4–34. Determine the forces acting on each member of the frame due to the 4-kN load shown. Neglect the weights of all members.

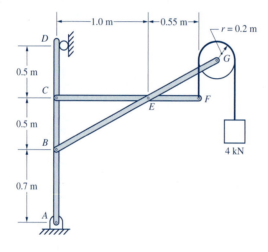

FIGURE P4–34

4–35 Determine the forces acting on the horizontal member *ACB* in Fig. P4–35 due to the load shown. The weight of the pulley is 10 lb and the weight of the beam is 15 lb.

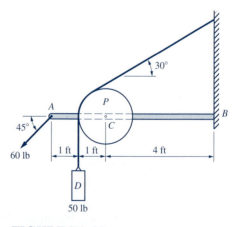

FIGURE P4–35

4–36 Determine the forces acting on each member of the frame in Fig. P4–36.

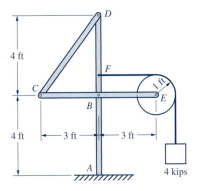

FIGURE P4–36

4–37 Determine the forces acting on each member of the frame in Fig. P4–37 due to the 1-kip load shown.

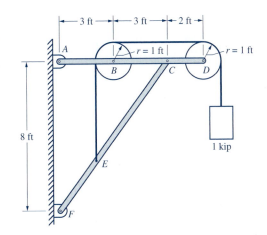

FIGURE P4–37

4–38 Two sawhorses support a log weighing 160 lb. (Only one sawhorse is shown in Fig. P4–38.) Each sawhorse carries one-half the weight of the log. If the radius of the log is 5 in. and the floor is smooth, determine the forces acting on each member.

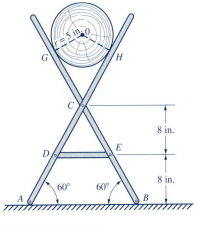

FIGURE P4–38

Section 4–8 Machines

4–39 See Fig. P4–39, p. 188. Determine the magnitude of the vertical force F on the pry bar required to lift the 2000-kg crate.

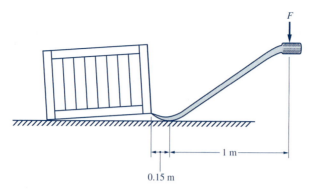

FIGURE P4–39

4–40   A horizontal force F of 40 N is applied to the claw hammer in Fig. P4–40. Determine the force exerted on the nail by the claw hammer.

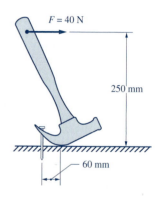

FIGURE P4–40

4–41   The pipe in Fig. P4–41 is held by a joint plier with a clamping force of 60 lb. Determine (a) the force P applied to the handles and (b) the force exerted by the pin E on portion AB of the plier.

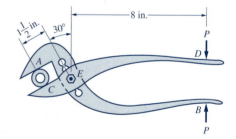

FIGURE P4–41

4–42   Refer to Fig. P4–42. A 100-lb force is applied to the handle of the toggle press at C. Determine the compressive force exerted by the press on bar E.

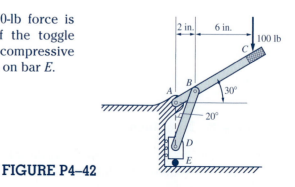

FIGURE P4–42

4–43 The resultant force of the pressure acting on the piston of the engine system in Fig. P4–43 is 2 kN. Determine the couple M required to hold the system in equilibrium.

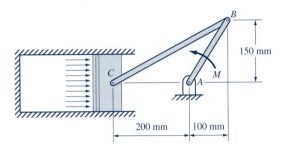

FIGURE P4–43

4–44 A 200-kg mass is supported by a pair of tongs, as shown in Fig. P4–44. Determine the forces acting on tong *CDE*.

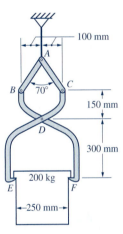

FIGURE P4–44

4–45 A 900-lb crate is lifted by a pair of tongs, as shown in Fig. P4–45. The tongs cross without touching at *H*. Determine the force in member *DE*.

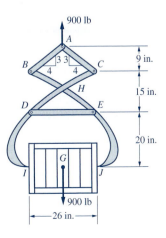

FIGURE P4–45

4–46 The hydraulic press in Fig. P4–46 consists of symmetrical links. The press transmits and magnifies the 8-kN resultant pressure force acting on the piston. Determine the compressive force acting on block G. Neglect friction and the weights of all components.

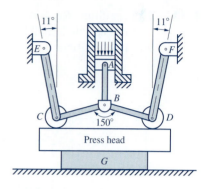

FIGURE P4–46

Computer
Program
Assignments

For each of the following problems, write a computer program using an appropriate programming language with which you are most familiar. Make the program user friendly by incorporating plenty of comments and input prompts so that the user will understand the input data to be entered and the limitations of their values. The output should include the data entered and the computed results, and they must be well labeled to identify each quantity. If a tabulated format is used, a proper heading must be included at the top of the table. Do not limit the program to any specific unit system. Indicate the consistent U.S. customary or SI units that can be used.

C4–1 Refer to Fig. C4–1. Write a computer program that can be used to analyze the truss subjected to the load shown with specified magnitude and direction. Run the program for (*a*) $F = 1000$ lb, $\theta = 50°$ and for (*b*) $F = 5000$ N, $\theta = 60°$.

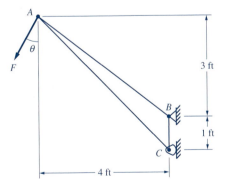

FIGURE C4–1

C4–2 Modify the program in Problem C4–1 so that it will produce a table listing the force in each member for a specified load F and with angle θ varying from 0° to 90°, at 5° steps. Run the program with (*a*) $F = 1000$ lb and (*b*) $F = 5000$ N.

CHAPTER 5

FRICTION

5–1
INTRODUCTION

When two surfaces are in contact, forces tangent to the contact surfaces, known as *friction forces,* always exist if a surface has a tendency to slide on another surface or if the two surfaces are actually in sliding contact. Two main types of friction frequently encountered in engineering practice are dry friction and fluid friction. *Dry friction,* also referred to as *Coulomb friction,* occurs when the unlubricated surfaces of two solid bodies are in sliding contact or have a tendency to slide. *Fluid friction* occurs in fluid flow or problems involving lubricated mechanisms or power transmission devices. Such problems are treated in fluid mechanics and will not be studied in this text.

Friction is an ever-present phenomenon in our daily activities and in many engineering systems. Without friction, it would be impossible for us to walk or to pick up anything. Friction forces are also needed in the moving and stopping of cars and in the transmission of power with belt drives. In these cases, the design consideration is to maximize the friction forces. In other cases, where friction causes energy loss or wear of the component parts such as a journal bearing, the primary concern is to minimize friction forces.

In this chapter, we will study first the laws governing dry friction and their applications to some general equilibrium problems involving friction forces. A few specific applications where dry friction plays a major role, such as wedges, square-threaded screws, belt friction, and rolling resistance, will also be presented.

5–2
LAWS OF DRY FRICTION

Variation of Friction Force.   The laws of dry friction can be understood by considering an experiment that involves the application of a horizontal force P to a block of weight W resting on a horizontal plane, as shown in Fig. 5–1a. A spring scale is used to measure the magnitude of P.

Figure 5–1b shows the free-body diagram of the block, where N and F represent the rectangular components of the reaction. The component N is normal to the contact surface and is called the *normal force.* The component F is tangent to the contact surface and is called the *friction force.* The friction force always acts in the direction opposite to the direction of motion or the tendency of the motion of the block.

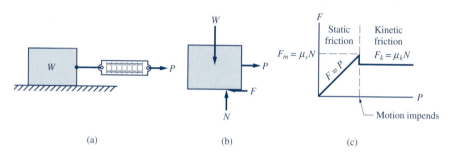

(a) (b) (c)

FIGURE 5–1

Let the force P be varied gradually from zero to a value large enough to cause the block to slide along the horizontal plane. As P increases, F increases by the same amount but in the opposite direction, as required by the equilibrium condition. The relationship between F and P is plotted in Fig. 5–1c. When the force P reaches a value large enough to overcome the static friction force, the block will start to slide. Once the block is in motion, the friction force drops abruptly to a lower value and remains essentially constant at this value.

Static Friction Force. In Fig. 5–1c, in the interval between $P = 0$ and the point where motion is impending, the body is in equilibrium, so the friction force is called the *static friction force.* The maximum static friction force occurs at the point of *impending motion* before sliding actually occurs. Experimental evidence indicates that *the maximum static friction force, denoted by F_m, is proportional to the normal force and is independent of the size of the contact area.* We write

$$F_m = \mu_s N \tag{5–1}$$

where μ_s is a constant called the *coefficient of static friction.*

Kinetic Friction Force. After sliding occurs, the friction force is reduced to a smaller value called the *kinetic friction force,* denoted by F_k. The value of F_k is also proportional to the normal force and is independent of the size of the contact area and the velocity of motion. We write

$$F_k = \mu_k N \tag{5–2}$$

where μ_k is a constant called the *coefficient of kinetic friction.*

Coefficients of Static and Kinetic Friction. The friction coefficients μ_s and μ_k are independent of the size and shape of the contact area. The values of both coefficients are dependent, however, primarily on the materials and the conditions of the contact surfaces. The ranges of values of these coefficients for common contact surfaces are listed in Table 5–1. Note that the values of μ_k are generally 20% to 25% lower than the values of μ_s for the same contact surfaces. This is why it is harder to make an object to start to move than to keep it moving.

TABLE 5–1 Range of Values of Coefficients of Static and Kinetic Friction for Common Contact Materials

Contact Materials	μ_s	μ_k
Steel on ice	0.03–0.04	0.02–0.03
Cast iron on brake lining	0.40–0.50	0.30–0.40
Leather on wood	0.40–0.50	0.30–0.40
Wood on wood	0.30–0.60	0.25–0.50
Rubber tire on concrete	0.80–0.90	0.65–0.70

Angles of Friction. Instead of using the normal and the tangential components N and F of the reaction, it is sometimes convenient to use the reaction force **R** itself, as shown in Fig. 5–2. The angle between **R** and the normal component N is denoted by ϕ in the figure. By the definition of tangent function, we write

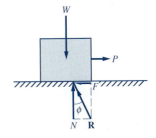

FIGURE 5–2

$$\tan \phi = \frac{F}{N}$$

When the friction force reaches the maximum static force F_m, the corresponding angle ϕ reaches a maximum value ϕ_s. Thus,

$$\tan \phi_s = \frac{F_m}{N} = \frac{\mu_s N}{N} = \mu_s$$

which gives

$$\phi_s = \tan^{-1} \mu_s \qquad (5\text{–}3)$$

The angle ϕ_s is known as the *angle of static friction*. Thus, when a body is on the verge of impending motion, the reaction **R** at the contact surface makes an angle ϕ_s with the normal force N, where $\phi_s = \tan^{-1}\mu_s$.

If sliding occurs between the contact surfaces, the friction force becomes $F_k = \mu_k N$. The corresponding angle ϕ_k is

$$\tan \phi_k = \frac{F_k}{N} = \frac{\mu_k N}{N} = \mu_k$$

which gives

$$\phi_k = \tan^{-1} \mu_k \qquad (5\text{–}4)$$

The angle ϕ_k is known as the *angle of kinetic friction*. Thus, when sliding occurs between two contact surfaces, the reaction at the contact surface makes an angle ϕ_k with the normal force N, where $\phi_k = \tan^{-1}\mu_k$.

5–3
PROBLEMS INVOLVING DRY FRICTION

Many engineering problems are related to dry friction. This section deals with some general equilibrium problems involving friction forces. From what has been discussed on dry friction in the preceding section, we conclude:

1. If a body is not in motion, the static friction force acting on the body is just enough to maintain the body in equilibrium. The maximum static friction force that can be developed in the contact surface is $F_m = \mu_s N$ (which occurs when the body is on the verge of slipping), where N is the normal force between the contact surfaces and μ_s is the static friction coefficient. The friction force always acts in the direction opposite to the impending motion. If there are only three forces acting on a free body, replace $\mathbf{N}$ and $\mathbf{F}$ with $\mathbf{R}$, which acts in the direction at an angle ϕ_s from the normal direction. The unknowns can be solved by solving the force triangle formed by the three forces.

2. If a body is sliding, the kinetic friction force acting on the body is $F_k = \mu_k N$ in the direction *opposite* to that of the sliding motion of the body.

Direction of Friction Forces. When two bodies are in contact, the forces between them are equal and opposite, as required by Newton's third law. Therefore, friction forces between contacting surfaces must be equal and opposite. Since friction force is resistant in nature, *it must act in the direction opposite to the direction of motion (or tendency of motion).* Just remember, friction forces always tend to slow down or stop motion; they never act to generate motion.

Tipping About a Point. The block in Fig. 5–3a has a width b, height h, and weight W. It rests on a horizontal plane where the static friction coefficient is μ_s. We need to find the smallest force P that will cause the block to move. There are two distinct possibilities of motion—slipping and tipping, as shown in Fig. 5–3b and c, respectively. Figure 5–3b shows the situation where the block is on the verge of slipping. In this case, the friction force is $F_m = \mu_s N$. The equilibrium condition in the horizontal direction requires

$$P_1 = F_m = \mu_s N$$

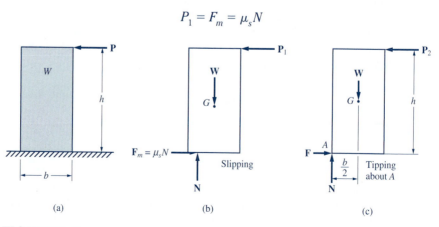

(a)   (b)   (c)

FIGURE 5–3

Figure 5–3c shows the situation where the block is on the verge of tipping about point A. In this case, the normal and friction forces are concentrated at A and the resultant moment of the forces on the block about point A must be zero. We write

$$\Sigma M_A = P_2(h) - W\left(\frac{b}{2}\right) = 0$$

$$P_2 = \frac{b}{2h}W$$

The smallest force P that will cause the block to move is equal to the smaller one of P_1 and P_2. If P_1 is smaller, the block will slip first; if P_2 is smaller, the block will tip about corner A first.

———— **EXAMPLE 5–1** ————

A 50-kg block rests on a 20° inclined plane, as shown in Fig. E5–1(1). The coefficient of static friction between the block and the plane is $\mu_s = 0.40$. Determine the maximum horizontal force P that can be applied to the block without causing it to slide.

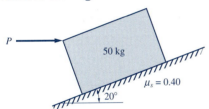

P

50 kg

$\mu_s = 0.40$

20°

FIGURE E5–1(1)

Solution. For illustration purposes, two solutions will be presented in the following.

(a) By Equilibrium Equations.   The maximum force P_{max} is required when the block is on the verge of sliding, at which point the maximum static friction force $F_m = \mu_s N$ develops between the contact surfaces. The force must act to the left to oppose the impending motion of the block. The free-body diagram of the block is sketched in Fig. E5–1(2). The equilibrium equations of the block are

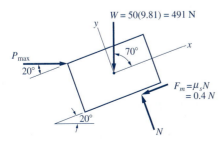

$W = 50(9.81) = 491$ N

y

P_{max}

20°

70°

x

$F_m = \mu_s N$
$= 0.4 N$

20°

N

FIGURE E5–1(2)

$$\Sigma F_x = P_{max} \cos 20° - (491 \text{ N}) \cos 70° - 0.4N = 0$$

$$0.940 P_{max} - 0.4N = 168 \text{ N} \qquad \text{(a)}$$

$$\Sigma F_y = -P_{max} \sin 20° - (491 \text{ N}) \sin 70° + N = 0$$

$$-0.342 P_{max} + N = 461 \text{ N} \qquad \text{(b)}$$

Solving Equations (a) and (b) simultaneously, we find

$$P_{max} = 439 \text{ N} \qquad \Leftarrow \textbf{Ans.}$$

(b) By Force Triangle. Refer to the free-body diagram of the block in Fig. E5–1(3). The normal and friction forces are replaced by their resultant **R** acting in the direction at the angle ϕ_s from the normal to the incline. We have

$$\phi_s = \tan^{-1}\mu_s = \tan^{-1} 0.4 = 21.8°$$

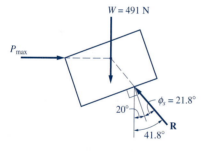

FIGURE E5–1(3)

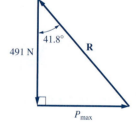

FIGURE E5–1(4)

Now the block is subjected to three forces. The three forces must form a closed triangle as shown in Fig. E5–1(4). Since the force triangle is a right triangle, by the definition of the tangent function, we write

$$P_{max} = (491 \text{ N}) \tan 41.8° = 439 \text{ N} \qquad \Leftarrow \textbf{Ans.}$$

EXAMPLE 5–2

The 60-lb block in Fig. E5–2(1) is kept from sliding down the 40° incline by the counterweight W. Knowing that the coefficient of static friction between the block and the incline is 0.25, determine the smallest counterweight W_{min} that will keep the block from sliding.

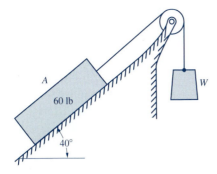

FIGURE E5–2(1)

Solution. Assume that the axle of the pulley is frictionless. Then the tension in the cable attached to the block is equal to the counterweight. For illustration purposes, two solutions will be presented in the following.

(a) By Equilibrium Equations. When the counterweight is a minimum, the block is on the verge of sliding and the friction force reaches a maximum value:

$$F_m = \mu_s N = 0.25N$$

as shown in the free-body diagram in Fig. E5–2(2). The equilibrium equations of the block are

$$\Sigma F_y = -(60 \text{ lb}) \sin 50° + N = 0$$

$$N = 46.0 \text{ lb}$$

$$\Sigma F_x = W_{min} - (60 \text{ lb}) \cos 50° + 0.25N = 0$$

$$W_{min} = (60 \text{ lb}) \cos 50° - 0.25(46.0 \text{ lb})$$

$$= 27.1 \text{ lb} \qquad \Leftarrow \textbf{Ans.}$$

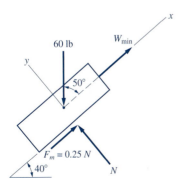

FIGURE E5–2(2)

(b) By Force Triangle. The normal and friction forces can be replaced by their resultant **R,** which makes an angle ϕ_s from the normal to the incline. We have

$$\phi_s = \tan^{-1}\mu_s = \tan^{-1} 0.25 = 14.0°$$

Now the block is subjected to three forces, as shown in the free-body diagram in Fig. E5–2(3). The three forces must form a closed triangle [sketched in Fig. E5–2(4)]. Two of the interior angles are indicated. The third angle α must be

$$\alpha = 180° - 50° - 26° = 104°$$

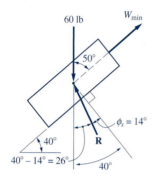

FIGURE E5–2(3)

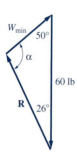

FIGURE E5–2(4)

Applying the law of sines to the triangle, we write

$$\frac{W_{min}}{\sin 26°} = \frac{60 \text{ lb}}{\sin 104°}$$

$$W_{min} = 27.1 \text{ lb} \qquad\qquad \Leftarrow \textbf{Ans.}$$

EXAMPLE 5–3

The movable bracket shown in Fig. E5–3(1) is a self-locking device. When the load P is placed on the bracket at a point far enough from the axis of the pipe, the bracket will not slide downward. Find the minimum distance d at which the force $P = 160$ lb can be applied without causing the bracket to slip if the static friction coefficient in all contact points is 0.25. The weight of the bracket is 20 lb, concentrated at its center of gravity G.

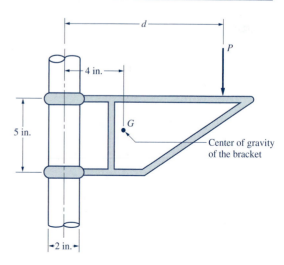

FIGURE E5–3(1)

Solution.   When the distance d is less than the minimum value, the bracket will slide along the pipe. When the distance is equal to the minimum value, the bracket is on the verge of sliding and the friction forces at the contact points A and B reach the maximum static friction, as shown in the free-body diagram in Fig. E5–3(2). The equilibrium equations give

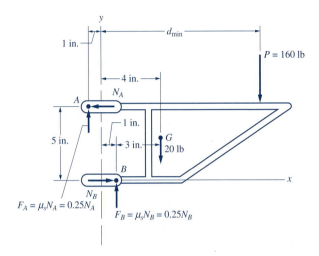

FIGURE E5–3(2)

$$\Sigma F_x = N_B - N_A = 0$$

$$N_A = N_B$$

$$\Sigma F_y = 0.25N_A + 0.25N_B - 20 \text{ lb} - 160 \text{ lb} = 0$$

$$N_A = N_B = 360 \text{ lb}$$

$$\Sigma M_B = -(160 \text{ lb})(d_{min} - 1 \text{ in.}) - (20 \text{ lb})(3 \text{ in.}) + (360 \text{ lb})(5 \text{ in.})$$

$$- 0.25(360 \text{ lb})(2 \text{ in.}) = 0$$

From which we get

$$d_{min} = 10.75 \text{ in.} \qquad\qquad \Leftarrow \textbf{Ans.}$$

──────── **EXAMPLE 5–4** ────────

Refer to Fig. E5–4(1). Determine the largest angle θ for which the adjustable incline may be raised before the block of weight W, width $b = 40$ mm, and height $h = 100$ mm begins to move. The coefficient of static friction between the block and the inclined surface is 0.45.

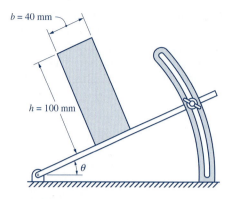

b = 40 mm

h = 100 mm

θ

FIGURE E5–4(1)

Solution. At the largest angle θ, the block may either slip or tip over. Which motion occurs first depends on which one takes place at a smaller angle.

*(a) **When Slipping Occurs.*** When the block is on the verge of slipping, the friction force reaches the maximum value of

$$F_m = \mu_s N = 0.45N$$

The free-body diagram of the block is sketched in Fig. E5–4(2). The equilibrium conditions require

$$\Sigma F_y = N - W\cos\theta = 0$$

$$N = W\cos\theta$$

$$\Sigma F_x = 0.45N - W\sin\theta = 0$$

$$0.45W\cos\theta = W\sin\theta$$

From which we get

$$\sin\theta/\cos\theta = \tan\theta = 0.45$$

$$\theta = 24.2°$$

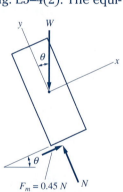

FIGURE E5–4(2)

*(b) **When Tipping Occurs.*** Refer to Fig. E5–4(3). When the block is on the verge of tipping, the reaction must be acting at corner A and the resultant moment about A must be zero. Thus,

$$\Sigma M_A = W\sin\theta\left(\frac{h}{2}\right) - W\cos\theta\left(\frac{b}{2}\right) = 0$$

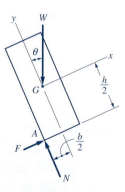

FIGURE E5–4(3)

From which we get

$$\frac{\sin \theta}{\cos \theta} = \tan \theta = \frac{b}{h} = \frac{40 \text{ mm}}{100 \text{ mm}} = 0.40$$

$$\theta = 21.8° \qquad\qquad \Leftarrow \textbf{Ans.}$$

This angle is smaller than the angle when sliding occurs. Thus, tipping of the block occurs first when the angle θ reaches $21.8°$.

5–4
WEDGES

Wedges are small blocks with small angles between two of their faces. They can be used to apply large forces, raise heavy loads, or make small adjustments in positioning heavy machines.

Free-Body Diagrams. Figure 5–4a shows a wedge used to raise a heavy block. The free-body diagrams of the block and the wedge are shown in Fig. 5–4b and c, respectively. We need to determine the smallest force P required to cause the block to move upward. Since the static friction must be overcome before the motion can be initiated, the friction forces on all contact surfaces must be the maximum static friction forces. The direction of friction forces must be opposite to the tendency of motion. The friction force F_1 on the block must act downward because the block tends to move upward. The friction forces F_2 and F_3 on the wedge must act to the left because the wedge tends to move to the right. Note that, in the contact surfaces between the block and the wedge, the forces of action and reaction must be equal and opposite. Note also that the weight of the wedge is usually neglected.

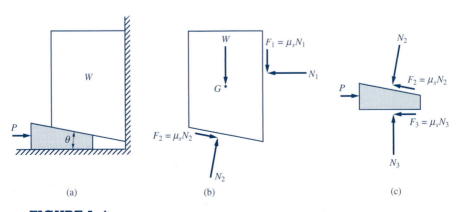

(a)   (b)   (c)

FIGURE 5–4

Self-Locking. A wedge is *self-locking* if it will remain in place when the force *P* is removed. For a self-locking wedge, a pulling force is needed to withdraw the wedge. Wedges with small wedge angles θ are usually self-locking.

Mechanical Advantage. The *mechanical advantage* (M.A.) for a machine is defined in general as the ratio of the output force to the input force. For wedges, it is defined as

$$\text{M.A.} = \frac{\text{Direct force required without wedge}}{\text{Force required with wedge}} \tag{5-5}$$

For the wedge illustrated in Fig. 5–4, a relatively smaller force *P* can be used to raise a heavy load. The wedge also modifies the method of load application. Force *P* can be applied simply by a tap on the wedge, which is more convenient than raising the load directly. Furthermore, if the wedge is self-locking, the block can be held in a raised position by the wedge.

─── **EXAMPLE 5–5** ───

A 6° wedge of negligible weight is used to adjust the position of a 2000-lb machine block *B*, as shown in Fig. E5–5(1). The coefficient of static friction is 0.35 between all contact surfaces. Determine the minimum force *P* required to move the block a little to the right and determine the mechanical advantage of the system.

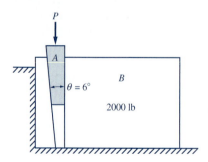

FIGURE E5–5(1)

Solution. To cause motion to be impending, the maximum static friction forces must be overcome. Thus, the static friction angle

$$\phi_s = \tan^{-1} \mu_s = \tan^{-1} 0.35 = 19.3°$$

must be used to define the direction of the reactions from the normal direction.

Equilibrium of the Block. The free-body diagram of the block is sketched in Fig. E5–5(2). As the wedge is driven downward, it causes a downward friction force on the block. Thus, the reaction R_1 is shown on the upper side of the normal so that its friction component is downward. The block tends to slide to the right; the friction force from the floor must act to the left. Thus, the reaction R_2 is shown on the right side of the normal so that its friction component is to the left. The three forces acting on the block form a closed triangle, sketched in Fig. E5–5(3). From the force triangle, we get

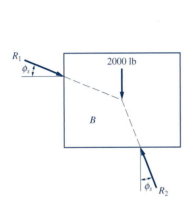

FIGURE E5–5(2)

FIGURE E5–5(3)

$$\frac{R_1}{\sin \phi_s} = \frac{2000 \text{ lb}}{\sin(90° - 2\phi_s)}$$

$$\frac{R_1}{\sin 19.3°} = \frac{2000 \text{ lb}}{\sin 51.4}$$

$$R_1 = 846 \text{ lb}$$

Equilibrium of the Wedge. The free-body diagram of the wedge is sketched in Fig. E5–5(4). Since the wedge is driven downward, the reaction R_1 on the wedge is drawn below the normal so that its friction component is upward. Note that the reaction R_1 on the wedge is opposite to that on the block. Similarly, the reaction R_3 is shown below the normal so that its friction component is upward to resist the downward motion of the wedge. The three forces acting on the wedge form a closed triangle, sketched in Fig. E5–5(5). From the force triangle, we get

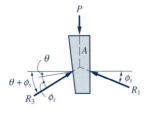

FIGURE E5–5(4)

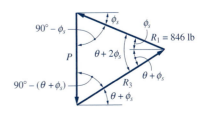

FIGURE E5–5(5)

$$\frac{P}{\sin(\theta + 2\phi_s)} = \frac{846 \text{ lb}}{\sin[90° - (\theta + \phi_s)]}$$

$$\frac{P}{\sin 44.6°} = \frac{846 \text{ lb}}{\sin 64.7°}$$

$$P = 657 \text{ lb} \downarrow \qquad\qquad \Leftarrow \textbf{Ans.}$$

By definition, the mechanical advantage of the wedge is

$$\text{M.A.} = \frac{W}{P} = \frac{2000 \text{ lb}}{657 \text{ lb}} = 3.04 \qquad\qquad \Leftarrow \textbf{Ans.}$$

━━ EXAMPLE 5–6 ━━

The position of the concrete block A in Fig. E5–6(1) is adjusted by driving in the wedge B. The load on the block (including its own weight) is 8 kN. The wedge angle θ is 6° and the weight of the wedge is negligible. Knowing that the coefficient of static friction is 0.30 between all contact surfaces, determine (*a*) the force P required to raise the block, (*b*) the force P' required to lower the block, and (*c*) whether the system is self-locking.

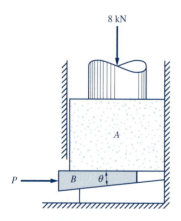

FIGURE E5–6(1)

Solution. To cause impending motion, the maximum static friction forces must be overcome. Thus, the static friction angle

$$\phi_s = \tan^{-1} \mu_s = \tan^{-1} 0.30 = 16.7°$$

must be used to define the direction of the reactions from the normal direction.

(*a*) ***To Raise the Block.*** The free-body diagram of the block and the corresponding force triangle are sketched in Fig. E5–6(2). From the force triangle, we get

$$\frac{R_2}{\sin(90° + \phi_s)} = \frac{8 \text{ kN}}{\sin(90° - 2\phi_s)}$$

$$\frac{R_2}{\sin 106.7°} = \frac{8 \text{ kN}}{\sin 56.6°}$$

$$R_2 = 9.18 \text{ kN}$$

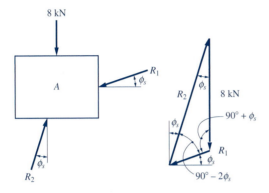

FIGURE E5–6(2)

The free-body diagram of the wedge and the corresponding force triangle are sketched in Fig. E5–6(3). From the force triangle, we get

$$\frac{P}{\sin (\theta + 2\phi_s)} = \frac{R_2}{\sin[90° - (\theta + \phi_s)]}$$

$$\frac{P}{\sin 39.4°} = \frac{9.18 \text{ kN}}{\sin 67.3°}$$

$$P = 6.32 \text{ kN} \longrightarrow \qquad \Leftarrow \textbf{Ans.}$$

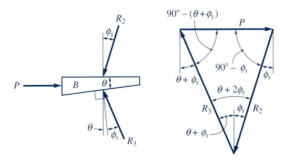

FIGURE E5–6(3)

(b) To Lower the Block. The free-body diagram of the block and the corresponding force triangle are sketched in Fig. E5–6(4). The force triangle is a right triangle, from which we get

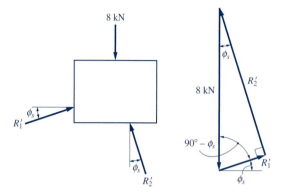

FIGURE E5–6(4)

$$R_2' = (8 \text{ kN}) \cos\phi_s = (8 \text{ kN}) \cos 16.7° = 7.66 \text{ kN}$$

The free-body diagram of the wedge and the corresponding force triangle are sketched in Fig. E5–6(5). From the force triangle, we get

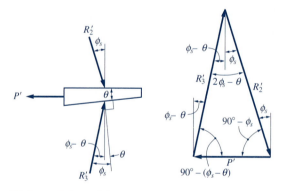

FIGURE E5–6(5)

$$\frac{P'}{\sin(2\phi_s - \theta)} = \frac{R_2'}{\sin[90° - (\phi_s - \theta)]}$$

$$\frac{P'}{\sin 27.4°} = \frac{7.66 \text{ kN}}{\sin 79.3°}$$

$$P' = 3.58 \text{ kN} \quad \longleftarrow \qquad\qquad \Leftarrow \textbf{Ans.}$$

(c) Self-Locking. Since a pulling force is needed to lower the block, the system is self-locking. (The block will remain in place if there is no external force acting on the wedge.)

5–5
SQUARE-THREADED SCREWS

Screws are used for fastening and for transforming couples into axial forces. Square-threaded screws are frequently used in jacks, presses, and other mechanisms. Consider the square-threaded jack in Fig. 5–5. The pair of equal and opposite forces (F) applied to a handle of length a produces a couple having a moment of $M = Fa$. The couple causes the screw to turn and move upward to raise the load W.

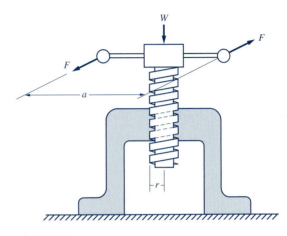

FIGURE 5–5

Lead Angle. The square thread can be regarded as an incline wrapped around a cylinder (Fig. 5–6a). When one turn of the thread is unwrapped, it develops into an incline, as shown in Fig. 5–6b. The horizontal distance of the incline is equal to $2\pi r$, where r is the *mean radius* of the thread, and the vertical rise is equal to the lead L of the screw. The *lead* is defined as the axial distance the screw advances for one revolution. A screw can be either single or multiple threaded. A single threaded screw has only one thread wrapping around the axis of the screw. A double or triple threaded screw has two or three parallel sets of threads wrapping around the axis of the screw. In general, we have

$$L = np \tag{5–6}$$

where n is the number of threads and p is the *pitch,* or the distance between two adjacent threads measured along the axial direction. The angle θ between the incline and the horizontal is called the *lead angle.* From Fig. 5–6b, we write

$$\tan \theta = \frac{L}{2\pi r} = \frac{np}{2\pi r}$$

From which we get

$$\theta = \tan^{-1}\left(\frac{np}{2\pi r}\right) \tag{5–7}$$

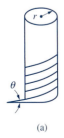

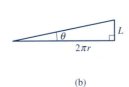

(a)   (b)

FIGURE 5–6

Moment Required for Loading. Since the friction force is independent of the contact area, the contact between the threads may be simulated by the contact between a block and an incline. Thus, the analysis of a square-threaded screw is treated as the problem of a block on an incline. Figure 5–7a shows the free-body diagram of a block of weight W on an incline at a lead angle θ from the horizontal. The forces on the block include its weight W, a horizontal force P, and a reaction R. The force P is an equivalent force caused by the couple applied at the handle. Since P must produce the same moment about the axis of the screw as the moment of the couple, we must have $Pr = M$, where r is the mean radius of the screw. To raise the load, the static friction force must be overcome. Therefore, the angle between the reaction R and the normal to the incline is the static friction angle ϕ_s. For the force triangle of the block shown in Fig. 5–7b, we write

$$P = W\tan(\phi_s + \theta)$$

Thus,

$$M = Pr = Wr\tan(\phi_s + \theta) \tag{5–8}$$

which gives the moment required to raise the load W.

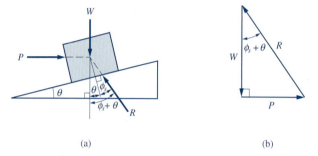

(a)   (b)

FIGURE 5–7

Self-Locking. The screw is said to be *self-locking* if the load W remains in place when the loading couple is released. That means the friction force alone is enough to support the load.

Releasing Moment. To lower the load for a self-locking screw, a moment M' in the opposite direction must be applied. This application causes an equivalent force P' acting in the direction opposite to the direction of P, as shown in Fig. 5–8a. Since the impending motion is downward, the friction force must act upward. The force triangle is shown in Fig. 5–8b, from which we get

$$P' = W\tan(\phi_s - \theta)$$

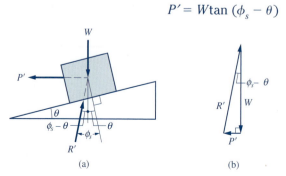

(a)   (b)   **FIGURE 5–8**

Thus,

$$M' = P'r = Wr \tan (\phi_s - \theta) \qquad (5\text{–}9)$$

which gives the couple required to lower the load. When $\theta > \phi_s$ the value of M' is negative, which means that a moment in the same direction as M is needed to hold the load and that the screw is not self-locking. Thus, a screw is self-locking provided $\theta \le \phi_s$.

Mechanical Advantage. The mechanical advantage (M.A.) of the square-threaded jack shown in Fig. 5–5 is defined as

$$\text{M.A.} = \frac{W}{F}$$

Since $M = Fa$, we have

$$\text{M.A.} = \frac{Wa}{M}$$

Substituting the expression for M from Equation 5–8, we get

$$\text{M.A.} = \frac{a}{r \tan (\phi_s + \theta)} \qquad (5\text{–}10)$$

where r is the mean radius of the thread, and a is the moment arm of the applied force. The mechanical advantage of square-threaded devices is usually very large; that is, a small effort yields a large output, which can be used to raise a very heavy load or cause very large compression.

───── **EXAMPLE 5–7** ─────

The press shown in Fig. E5–7 has a double-square thread of mean radius equal to 1 in., and 5 pitches per in. The coefficient of static friction in the threads is 0.10. To compress block A with a 2000-lb force, find (a) the torque required to compress the block and the corresponding mechanical advantage (the torque is applied by two equal and opposite forces on the rim of wheel B, which has a diameter of 12 in.) and (b) the torque required to loosen the press.

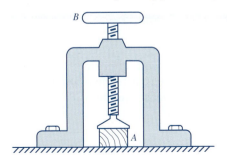

FIGURE E5–7

Solution. With five pitches per in., the pitch is

$$p = \frac{1}{5} \text{ in.} = 0.2 \text{ in.}$$

From Equation 5–7, the lead angle for the double-thread screw is

$$\theta = \tan^{-1}\left[\frac{np}{2\pi r}\right] = \tan^{-1}\left[\frac{2(0.2 \text{ in.})}{2\pi(1 \text{ in.})}\right] = 3.64°$$

The static friction angle is

$$\phi_s = \tan^{-1}\mu_s = \tan^{-1}0.10 = 5.71°$$

(a) Loading Torque. The torque required to produce the desired compressive load can be computed from Equation 5–8.

$$M = Wr \tan(\phi_s + \theta)$$

$$= (2000 \text{ lb})(1 \text{ in.}) \tan(5.71° + 3.64°)$$

$$= 329 \text{ lb} \cdot \text{in.} \qquad\qquad \Leftarrow \textbf{Ans.}$$

By definition, the mechanical advantage of the press is

$$\text{M.A.} = \frac{W}{F} = \frac{W}{M/a} = \frac{2000 \text{ lb}}{329 \text{ lb} \cdot \text{in.}/12 \text{ in.}}$$

$$= 72.9 \qquad\qquad \Leftarrow \textbf{Ans.}$$

Or from Equation 5–10 we get

$$\text{M.A.} = \frac{a}{r\tan(\phi_s + \theta)}$$

$$= \frac{12 \text{ in.}}{(1 \text{ in.}) \tan(5.71° + 3.64°)}$$

$$= 72.9$$

(b) Releasing Torque. Since the lead angle $\theta = 3.64°$ is less than $\phi_s = 5.71°$, the press is self-locking. Hence, a release moment opposite to the direction of M must be applied to release the load. This moment can be computed from Equation 5–9.

$$M' = Wr \tan(\phi_s - \theta)$$

$$= (2000 \text{ lb})(1 \text{ in.}) \tan(5.71° - 3.64°)$$

$$= 72.3 \text{ lb} \cdot \text{in.} \qquad\qquad \Leftarrow \textbf{Ans.}$$

5–6
BELT FRICTION

When a flexible belt is wrapped around a cylindrical drum as shown in Fig. 5–9a, the tensions on the two sides of the belt may differ substantially without causing the belt to slip on the drum. We will establish the *maximum* ratio between the two tensions before slipping occurs. This ratio is important in the design of belt drives, band brakes, hoisting rigs, and many other machines involving a flexible belt, rope, or band wrapping around a circular pulley or a cylindrical drum.

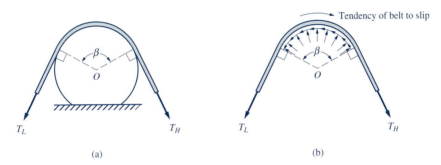

(a)   (b)

FIGURE 5–9

Maximum Ratio of Belt Tensions.   Figure 5–9b shows the free-body diagram of the belt in contact with the stationary drum. The forces acting on the belt include belt tensions T_H and T_L on the two sides of the belt. (The subscripts H and L indicate high and low belt tensions, respectively.) The distributed normal and friction forces on the curved part of the belt are in contact with the drum. Since the high tension occurs on the right side of the belt, the belt tends to slip to the right. Thus, the friction force must act to the left. The ratio of the two tensions T_H/T_L is maximum when the belt is on the verge of slipping. The maximum ratio can be established based on the law of dry friction and the equilibrium conditions through the use of integral calculus, which gives the following result:

$$\frac{T_H}{T_L} = e^{\mu\beta} \tag{5–11}$$

where T_H = the high belt tension

T_L = the low belt tension

μ = the coefficient of friction between the belt and the drum. If the belt is on the verge of slipping, the static friction coefficient μ_s must be used. If the belt slips relative to the drum, such as in the band brakes, the kinetic friction coefficient μ_k must be used.

β = the *angle of contact* between the belt and the drum in radians. (Recall that 1 radian = $180°/\pi$.)

e = the base of natural logarithms approximately equal to 2.718. The term $e^{\mu\beta}$ represents an exponential power of the number 2.718.

Equation 5–11 may be expressed in logarithmic form as

$$\ln\left(\frac{T_H}{T_L}\right) = \mu\beta \tag{5–12}$$

where $\ln(T_H/T_L)$ is the natural logarithm with base $e \approx 2.718$.

─── **EXAMPLE 5–8** ───────────────────────────

The 100-lb weight in Fig. E5–8(1) is suspended from a rope wrapped around a fixed shaft for more than one complete turn. A force T is applied to the free end of the rope at an angle of 30° from the horizontal. Knowing that the coefficient of static friction between the rope and the shaft is 0.20, determine the range of force T for which the block will neither be raised nor lowered.

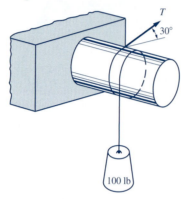

FIGURE E5–8(1)

Solution. The contact angle β shown in Fig. E5–8(2) is

$$\beta = 360° + 60°$$

$$= 420°$$

$$= 7.33 \text{ rad}$$

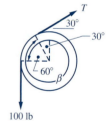

FIGURE E5–8(2)

From Equation 5–11, we find that the greatest ratio of the two tension forces is

$$\frac{T_H}{T_L} = e^{\mu_s\beta} = e^{(0.20)(7.33)} = 4.33$$

Maximum T. For the maximum value of T, the tension tends to pull the weight upward and T is greater than the 100-lb weight. Hence,

$$\frac{T_{\text{max}}}{100 \text{ lb}} = 4.33$$

$$T_{\text{max}} = 433 \text{ lb}$$

Minimum T. The minimum value of T is just enough to keep the weight from moving downward. Hence,

$$\frac{100 \text{ lb}}{T_{\min}} = 4.33$$

$$T_{\min} = 23.1 \text{ lb}$$

Thus, the range of T is

$$23.1 \text{ lb} \le T \le 433 \text{ lb} \qquad \qquad \Leftarrow \textbf{Ans.}$$

EXAMPLE 5–9

A boat is secured to a pier by wrapping a rope around the capstan in Fig. E5–9. One end of the rope is tied to the boat that exerts a force of 6000 N, while the other end is held by a dockworker who can exert a maximum force of 180 N. Knowing that the coefficient of static friction between the rope and the capstan is 0.30, determine the minimum number of loops that the rope must be wrapped around the capstan so that it will not slip.

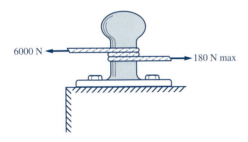

6000 N ← → 180 N max

FIGURE E5–9

Solution. We can find the minimum value of contact angle β based on the condition of impending slipping using μ_s in Equation 5–12.

$$\beta = \frac{1}{\mu_s} \ln\left(\frac{T_H}{T_L}\right) = \frac{1}{0.30} \ln\left(\frac{6000 \text{ N}}{180 \text{ N}}\right) = 11.69 \text{ rad}$$

$$= (11.69 \text{ rad}) \left(\frac{1 \text{ loop}}{2\pi \text{ rad}}\right) = 1.86 \text{ loop}$$

$$\text{No. of loops} = 2 \qquad \qquad \Leftarrow \textbf{Ans.}$$

EXAMPLE 5–10

Band brakes use belt friction to regulate the speed of rotating drums. For the band brake in Fig. E5–10(1), a force $P = 30$ lb is applied to the lever. Knowing that the coefficients of friction between the belt and the drum are $\mu_s = 0.40$ and $\mu_k = 0.35$, determine the torque produced by the belt tensions if the drum is rotating in the counterclockwise (c.c.w.) direction.

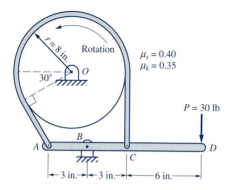

FIGURE E5–10(1)

Solution. Since the drum is rotating and the band is stationary, the band slips on the drum. Hence, μ_k must be used. The contact angle in radians is

$$\beta = (210°)\left(\frac{\pi\,\text{rad}}{180°}\right) = 3.67\text{ rad}$$

The ratio of belt tensions is

$$\frac{T_H}{T_L} = e^{\mu_k \beta} = e^{(0.35)(3.67)} = 3.61$$

$$T_H = 3.61T_L \tag{a}$$

Since the drum rotates in the c.c.w. direction, the friction force that the drum exerts on the band will make the tension on the right-hand side of the belt high tension. The free-body diagram of the lever is sketched in Fig. E5–10(2). The moment equation about the pivot point B is

$$\Sigma M_B = T_H\,(3\text{ in.}) - T_L \sin 60°(3\text{ in.}) - (30\text{ lb})(9\text{ in.}) = 0$$

$$T_H - 0.866T_L = 90\text{ lb} \tag{b}$$

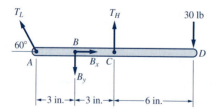

FIGURE E5–10(2)

Substituting T_H from Equation (a) into Equation (b) gives

$$(3.61 - 0.866)T_L = 90\text{ lb}$$

$$T_L = 32.8\text{ lb}$$

Substituting into Equation (a) gives

$$T_H = 118.4 \text{ lb}$$

The torque produced by belt tensions is

$$M = (T_H - T_L)r$$

$$= (118.4 \text{ lb} - 32.8 \text{ lb})(8 \text{ in.})$$

$$= 685 \text{ lb} \cdot \text{in.}$$

$$= 57.1 \text{ lb} \cdot \text{ft} \quad \circlearrowleft \qquad\qquad \Leftarrow \textbf{Ans.}$$

This torque will slow down the counterclockwise rotation of the drum.

EXAMPLE 5–11

Belt drives are used to transmit power between parallel shafts. The belt drive in Fig. E5–11(1) consists of a flat belt that passes over two pulleys rotating at constant angular velocities. Knowing that the coefficients of friction between the belt and the drum are $\mu_s = 0.40$, and $\mu_k = 0.30$, determine (a) the largest torque M_A that the motor may apply to pulley A without exceeding the maximum allowable belt tension of 3500 N, and (b) the corresponding torque M_B delivered to pulley B by the belt.

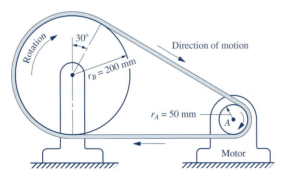

FIGURE E5–11(1)

Solution. As shown in Fig. E5–11(2), the contact angle in the smaller pulley is less than that in the bigger pulley. At the largest torque, the belt is on the verge of slipping on the smaller pulley. Thus, β_A and μ_s are used in the following computations.

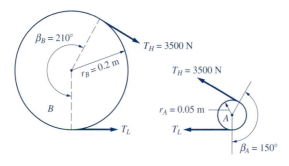

FIGURE E5–11(2)

$$\beta_A = (150°)\left(\frac{\pi \,\text{rad}}{180°}\right) = 2.62 \text{ rad}$$

$$\frac{T_H}{T_L} = \frac{3500 \text{ N}}{T_L} = e^{\mu_s \beta_A} = e^{(0.30)(2.62)} = 2.19$$

From which we get

$$T_L = 1598 \text{ N}$$

(a) The Largest Torque M_A.   Since the pulleys are rotating at constant rates, the moment of the belt tensions about A must be balanced by the torque M_A that the motor applies to the pulley. Thus,

$$M_A = (T_H - T_L)r_A$$
$$= (3500 \text{ N} - 1598 \text{ N})(0.05 \text{ m})$$
$$= 95.1 \text{ N} \cdot \text{m} \quad \circlearrowleft \qquad \Leftarrow \textbf{Ans.}$$

(b) Torque M_B Delivered to Pulley B by the Belt.   The torque M_B delivered to pulley B by the belt is equal to the moment of the belt tensions about the center of pulley B. Thus,

$$M_B = (T_H - T_L)r_B$$
$$= (3500 \text{ N} - 1598 \text{ N})(0.20 \text{ m})$$
$$= 380 \text{ N} \cdot \text{m} \quad \circlearrowleft \qquad \Leftarrow \textbf{Ans.}$$

5–7
ROLLING RESISTANCE

Rolling resistance refers to the resistance from the supporting surface to a rolling wheel that will gradually cause the wheel to slow down. Rolling resistance is not due to tangential friction forces; therefore, it is an entirely different phenomenon from that of dry friction.

Consider a wheel under a load W rolling on a horizontal surface, as shown in Fig. 5–10. Due to the load and the rolling action of the wheel, the wheel and the supporting surface undergo some deformation, which is shown in the figure greatly exaggerated. As a result, the reaction is distributed over an area, rather than at a single contact point. The resultant reaction R on the contact surface will act at some point A, and it must pass through point O to satisfy the equilibrium conditions. The resultant reaction R makes a small angle θ with the vertical. Since R has a horizontal component, a horizontal force P is required to maintain the rolling of the wheel at a constant speed.

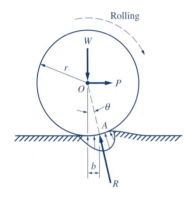

FIGURE 5–10

Coefficient of Rolling Resistance.   Since the wheel is rolling at constant speed, the forces acting on the wheel are in equilibrium. Thus, the sum of the moments of all forces about A must be zero. We write

$$\Sigma M_A = -P(r \cos \theta) + Wb = 0$$

Since angle θ is small, $\cos\theta \approx 1$. We get

$$P = \frac{b}{r}W \qquad\qquad (5\text{--}13)$$

where r = the radius of the wheel

$\qquad b$ = the horizontal distance between O and A

The distance b is commonly called the *coefficient of rolling resistance.* Note that b is not a dimensionless coefficient, as is μ_s or μ_k, but has a unit of length and is usually expressed in inches or millimeters. The value of b depends on the properties and conditions of the contact materials. Table 5–2 presents a list of the rolling coefficients of several contact materials. Note that the values listed in the table could vary by 50% or more, so no great accuracy can be expected from this table. The analysis presented here, however, can be used to explain the following phenomena:

1. The rolling resistance is proportional to the weight. Thus, a wheel of heavier weight would require a greater force to keep it rolling at a constant speed.

2. In general, rolling resistance is much smaller than sliding resistance. This is the reason for using wheels on vehicles and ball bearings in machines.

3. Tests indicate that the coefficient of rolling resistance varies only slightly with the wheel radius r. Neglecting this variation, we see that the rolling resistance varies inversely with the radius r of the wheel. This explains why the rolling resistance on a bigger wheel is less.

TABLE 5–2 Coefficients of Rolling Resistance

Contact Materials	b (in.)	b (mm)
Steel on steel	0.015	0.38
Steel on wood	0.08	2
Steel on soft ground	5	130
Pneumatic tire on pavement	0.025	0.6
Pneumatic tire on dirt road	0.05	1.3

EXAMPLE 5–12

A railroad car of mass 15 Mg carries a load of 100 kN. The car is supported on four wheels of 500-mm diameter. Using the coefficient of rolling resistance listed in Table 5–2 for steel wheels on steel rails, calculate the force necessary to move the car at a constant speed.

Solution. The total weight of the car and the freight is

$$W = (15\ 000\ \text{kg})(9.81\ \text{m/s}^2) + 100\ 000\ \text{N}$$

$$= 2.47 \times 10^5\ \text{N} = 247\ \text{kN}$$

Assuming that the weight is distributed equally on the four wheels, the load per wheel is 247 kN/4 = 61.8 kN. From Table 5–2, the coefficient of rolling resistance for steel on steel is $b = 0.38$ mm. With this value of b and $r = 250$ mm substituted in Equation 5–13, we get

$$P = \frac{bW}{r} = \frac{(0.38\ \text{mm})(61.8\ \text{kN})}{250\ \text{mm}} = 0.094\ \text{kN} = 94\ \text{N}$$

which is the rolling resistance per wheel. The total rolling resistance on the car is

$$4(94\ \text{N}) = 376\ \text{N} \qquad\qquad \Leftarrow \textbf{Ans.}$$

Remark. Rolling resistance is usually much less compared to sliding resistance. Even if we assume a smooth contact of steel on ice, with $\mu_k = 0.02$, the friction resistance is

$$\mu_k N = 0.02(247\ \text{kN}) = 4.94\ \text{kN} = 4940\ \text{N}$$

which is about thirteen times the rolling resistance for steel wheels on steel rails.

5–8
SUMMARY

Dry Friction. The friction forces between the contact surfaces of two bodies represent the resistance to the motion or the tendency of motion of the two bodies relative to each other. The friction force acting on a body is always tangent to the contact surface in the direction opposing the motion or the tendency of motion of the body.

Static Friction. When a body is stationary, the static friction force acting on the body is just enough to maintain the body in equilibrium. The maximum static friction force that can be developed in the contact surface is $F_m = \mu_s N$, which occurs when the body is on the verge of *impending motion* but before slipping actually occurs. N is the normal force between contact surfaces and μ_s is the *static friction coefficient,* the value of which depends on the materials and conditions of the contact surfaces. Static friction force always acts in the direction *opposite* to the tendency of motion of the body.

Kinetic Friction. When a body is sliding, the kinetic friction force acting on the body is $F_k = \mu_k N$, where μ_k is the *kinetic friction coefficient.* The kinetic friction force acts in the direction *opposite* to the direction of the sliding motion of the body.

Friction Angles. Sometimes it is more convenient to replace the normal force N and the friction force F with their resultant R. When the body is on the verge of impending motion, the friction force reaches a maximum. The angle between R and the normal to the contact surface also reaches a maximum value ϕ_s, called the *angle of static friction,* which is equal to $\phi_s = \tan^{-1}\mu_s$. Using the resultant R, a body may be subjected to only three forces. In this case, the force triangle of the three forces can be sketched and used to solve for the unknowns.

Wedges. *Wedges* are small blocks with small angles between two of their faces. Wedges are used to raise heavy loads, or make small adjustments in positioning heavy machines. Wedge problems are solved by considering the equilibrium of the heavy block and the wedge. The wedge and the block are considered to have impending motion, so the static friction force reaches a maximum value $F_m = \mu_s N$, or the reaction R makes a maximum angle ϕ_s with the normal to the contact surface.

Square-Threaded Screws. A *square-threaded screw* is an application of the inclined plane and can be used to move heavy loads or to transmit power, as in jacks, presses, and other mechanisms. The analysis of square-threaded screws can be reduced to the analysis of a block sliding on an incline. When one turn of the thread is unwrapped, it becomes an incline. The horizontal distance of the incline is equal to $2\pi r$, where r is the *mean radius* of the thread, and the vertical rise is equal to the lead L of the screw. The *lead* is defined as the axial distance the screw advances for one revolution. A screw can either be *single threaded* or *multiple threaded*. The lead *L* is the number of thread *n* times the pitch *p* between two adjacent threads, i.e., $L = np$. The *lead angle* θ between the incline and the horizontal is:

$$\theta = \tan^{-1}\left(\frac{np}{2\pi r}\right) \tag{5-7}$$

The moment required to raise the load *W* is:

$$M = Wr \tan (\phi_s + \theta) \tag{5-8}$$

The moment required to lower the load *W* is:

$$M' = Wr \tan (\phi_s - \theta) \tag{5-9}$$

When $\theta \le \phi_s$, the screw is said to be *self-locking*, which means the load will remain in place when the loading couple is released.

Belt Friction.   When a flexible belt is wrapped around a cylindrical drum, the maximum ratio T_H/T_L between the high belt tension T_H and the low belt tension T_L before slipping occurs is given by the formula

$$\frac{T_H}{T_L} = e^{\mu\beta} \tag{5-11}$$

where μ is the coefficient of friction between the belt and the drum. If the belt is on the verge of slipping, the static friction coefficient μ_s must be used; if the belt slips on the drum, the kinetic friction coefficient μ_k must be used. The term β is the *angle of contact* between the belt and the drum in radians, and *e* is the base of natural logarithms approximately equal to 2.718. Equation 5–11 may be expressed in logarithmic form as

$$\ln \left(\frac{T_H}{T_L}\right) = \mu\beta \tag{5-12}$$

Rolling Resistance.   *Rolling resistance* is the resistance from a supporting surface to a rolling wheel that tends to slow down the rolling of the wheel. Rolling resistance *P* can be computed from

$$P = \frac{b}{r}W \tag{5-13}$$

where the distance *b* is referred to as the *coefficient of rolling resistance*. This equation leads to the following conclusions:

1.   Heavier wheels encounter more rolling resistance.
2.   Rolling resistance is much smaller than sliding resistance.
3.   Rolling resistance varies inversely with the diameter of the wheel. Therefore, a bigger wheel encounters less rolling resistance.

PROBLEMS

Section 5–3   Problems Involving Dry Friction

5–1   The coefficients of friction between the 50-kg block and the plane in Fig. P5–1 are $\mu_s =$ 0.4 and $\mu_k = 0.3$. Determine the magnitude of the friction force acting on the block if $P = 150$ N.

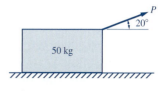

FIGURE P5–1

5–2   In Problem 5–1, determine the magnitude of the friction force acting on the block if the applied force P is increased to 250 N.

5–3   In Problem 5–1, determine the maximum force P that can be applied to the block without causing it to slide.

5–4   Determine the maximum force P that can be applied to the block in Fig. P5–4 without causing it to slide upward along the incline.

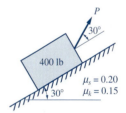

FIGURE P5–4

5–5   In Problem 5–4, determine the minimum force P required to prevent the block from sliding down the incline.

5–6   The 100-lb crate in Fig. P5–6 is resting on an adjustable incline. The coefficients of friction between the contact surfaces are $\mu_s = 0.3$ and $\mu_k = 0.25$. Determine whether the block is at rest or in motion if α is equal to (a) 10°, (b) 30°.

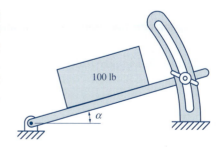

FIGURE P5–6

5–7   In Problem 5–6, determine the value of angle α for which the crate will start to slide down the incline if the angle is gradually increased from 0°.

5–8   The 20-kg block in Fig. P5–8 rests on a 30° incline. Knowing that $\mu_s = 0.25$, determine the magnitude and direction of the smallest force P required to cause the block to start sliding up the incline.

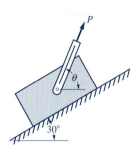

5–9   In Problem 5–8, determine the magnitude and direction of the smallest force P required to prevent the block from sliding down the incline.

FIGURE P5–8

5–10   Determine the minimum force P required to set either one or both of the two blocks in motion. The weight of each block and the coefficients of static friction between contact surfaces are indicated in Fig. P5–10.

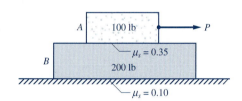

FIGURE P5–10

5–11   Block A in Fig. P5–11 has a 50-kg mass, and the coefficient of static friction in all contact surfaces is 0.30. Determine the maximum value of the mass of block B for which the blocks will not move. Assume that the pulley is frictionless and of negligible mass.

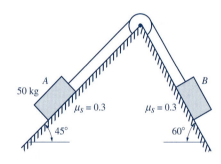

FIGURE P5–11

5–12   In Problem 5–11, determine the minimum value of the mass of block B for which the blocks will not move.

5–13 and **5–14**   Two blocks are arranged as shown in Figs. P5–13 and P5–14. The static friction coefficients are indicated on the contact surfaces. Determine the minimum force P required to cause block B to slide.

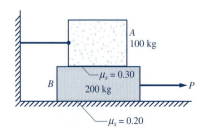

FIGURE P5–13

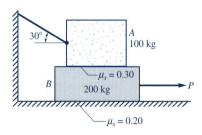

FIGURE P5–14

5–15 Refer to Fig. P5–15. A 160-lb man climbs a 25-ft ladder that weighs 20 lb. The coefficient of static friction between the ladder and the wall is 0.2, and that between the ladder and the floor is 0.6. Determine the highest point, as indicated by the maximum distance d, that the man can climb without causing the ladder to slip.

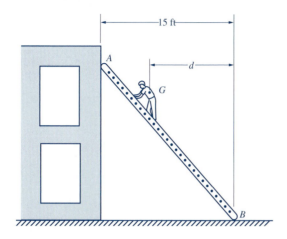

FIGURE P5–15

5–16 See Fig. P5–16. A force $P = 300$ lb is applied to a light bracket supported on a shaft of 6-in. diameter. Assume that the contact between the bracket and the shaft is made at points A and B. Find the minimum μ_s for the bracket to stay in place.

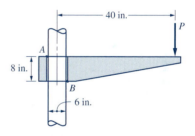

FIGURE P5–16

5–17 The file cabinet in Fig. P5–17 has a weight of 60 lb and a center of gravity G, indicated as shown. Knowing that the coefficient of static friction between the cabinet and the floor is $\mu_s = 0.30$, determine the minimum horizontal force P required to cause it to slide or tip over.

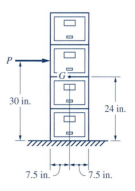

FIGURE P5–17

5–18   The 50-lb crate in Fig. P5–18 is pulled
as shown. The coefficient of static friction between the crate and the floor is
0.40. Determine whether the crate will
slide or tip, and determine the magnitude of *P* that will cause the motion.

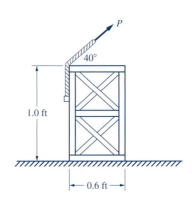

FIGURE P5–18

5–19   The 80-kg refrigerator in Fig. P5–19 is
pushed by a horizontal force *P.*
Knowing that the coefficient of static
friction between the bottom of the
refrigerator and the floor is 0.45, determine (*a*) the magnitude of the force *P*
required to move the refrigerator to
the left, and (*b*) the maximum *h* for
which the refrigerator will not tip over.

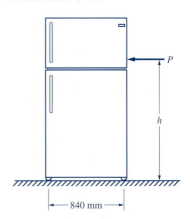

FIGURE P5–19

Section 5–4   Wedges

5–20   The wedge shown in Fig. P5–20 is used
to raise a 2000-lb block. Determine the
force *P* required to cause impending
motion, knowing that the coefficient
of static friction is 0.25 at all surfaces
of contact.

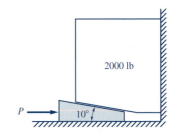

FIGURE P5–20

5–21   The two 6° wedges in Fig. P5–21
are used to adjust the horizontal
position of the 200-kg block.
Determine the minimum force *P*
required to move the block. The
static friction angle at all contact surfaces is 15°.

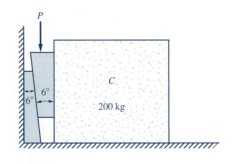

FIGURE P5–21

5–22 Determine the smallest weight W required to cause the wedge in Fig. P5–22 to be withdrawn from the block. The static friction angle at all contact surfaces is 15°. Neglect the friction at the rollers and at the pulley.

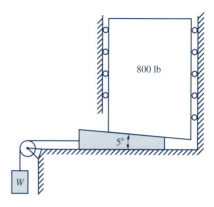

FIGURE P5–22

5–23 Determine the force P required to cause the wedge in Fig. P5–23 to start to move downward, knowing that $\mu_s = 0.30$ at all contact surfaces.

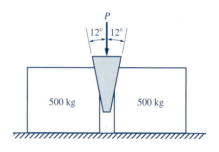

FIGURE P5–23

5–24 Two 10° wedges, A and B, are arranged as shown in Fig. P5–24. A downward force P applied at wedge A will result in the readjustment of the vertical position of the 1500-lb weight. The coefficient of static friction between all sliding surfaces is 0.25. Determine the minimum force P required to raise the load.

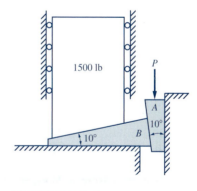

FIGURE P5–24

5–25 The 10° wedge in Fig. P5–25 is used to split a log. The coefficient of static friction between the wedge and the wood is 0.30. If a force $P = 500$ lb is needed to drive the wedge into the wood, determine the force exerted on the log by the wedge after it has been driven into the log.

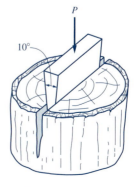

FIGURE P5–25

5–26   The horizontal position of the beam in Fig. P5–26 is adjusted by a 6° wedge. The coefficient of static friction between the beam and the wedge is 0.15, and that between the wedge and the support is 0.30. Determine the force P required to raise the beam.

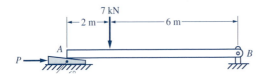

FIGURE P5–26

5–27   Refer to Fig. P5–27. The bracket subjected to the loading shown is leveled by a 5° wedge. The coefficient of static friction between all contact surfaces is 0.30. Determine the magnitude of the horizontal force P required to raise the bracket.

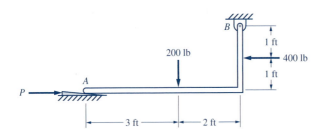

FIGURE P5–27

Section 5–5   Square-Threaded Screws

5–28   A jack of single square-threaded screw has a mean radius of 25 mm and a pitch of 10 mm. Knowing that the static friction coefficient is 0.10 and the mass of the load is 2000 kg, determine the torque required (a) to raise the load and (b) to lower the load.

5–29   A C-clamp is used to clamp a piece of wood to a workbench, as shown in Fig. P5–29. Its double square threads have a mean diameter of $\frac{3}{8}$ in. and a pitch of $\frac{1}{10}$ in. The static friction coefficient between the threads is 0.30. If a torque of 30 lb · in. is applied in tightening the clamp, determine (a) the compressive force on the wood and (b) the torque required to loosen the clamp.

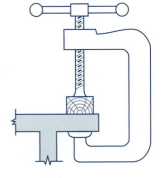

FIGURE P5–29

5–30   The turnbuckle in Fig. P5–30 is used for tightening two rods, A and B, that are subjected to a tensile force of 10 kips. Screws in each rod have a mean diameter of 1 in. and a single square thread with a pitch of $\frac{1}{4}$ in. Rod A has a right-handed thread and rod B has a left-handed thread. The static friction angle between the rods and the threaded sleeve is 8°. Determine the minimum torque M required to tighten the rods.

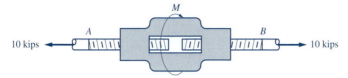

FIGURE P5–30

5–31   In Problem 5–30, determine the minimum torque required to loosen the rods.

5–32   The device shown in Fig. P5–32 is a scissors-type axle jack. The screw consists of two sections. The section on the left has right-handed threads, and the section on the right has left-handed threads. Each screw has double square threads with a mean diameter of 1.25 in. and a pitch of $\frac{1}{5}$ in. For $\alpha = 30°$ and $\mu_s = 0.15$ in the threads, determine the torque M that must be applied to the screw to raise a load of 2 kips.

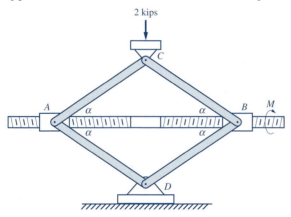

FIGURE P5–32

5–33   In Problem 5–32, determine the torque M' that must be applied to the screw to lower the 2-kip load.

5–34   The device shown in Fig. P5–34 consists of two members connected by a link and a double threaded screw. The mean radius of the screw is $\frac{3}{8}$ in., its pitch is $\frac{1}{8}$ in., and $\mu_s = 0.20$. The lower member is threaded at A. Determine the torque needed to tighten the screw and apply a compression of 150 lb to the wood blocks.

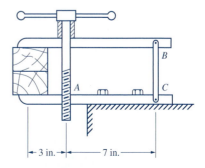

FIGURE P5–34

Section 5–6   Belt Friction

5–35   Find the smallest force P that will prevent the weight in Fig. P5–35 from moving downward.

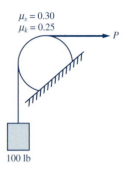

$\mu_s = 0.30$
$\mu_k = 0.25$

P

100 lb

FIGURE P5–35

5–36   See Fig. P5–36. Determine the smallest force P required to resist the 10-kN force if the static friction coefficient between the rope and the shaft is 0.25.

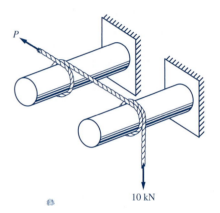

P

10 kN

FIGURE P5–36

5–37   In Problem 5–36, if a force $P = 400$ N is required to resist the 10-kN force, determine the coefficient of static friction between the rope and the shafts.

5–38   Determine the largest weight W that will not cause the 50-lb block in Fig. P5–38 to slide up the incline.

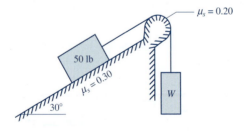

$\mu_s = 0.20$

50 lb

$\mu_s = 0.30$

W

30°

FIGURE P5–38

5–39   In Problem 5–38, determine the smallest weight W that will keep the 50-lb block from sliding down the incline.

5–40 See Fig. P5–40. Determine the largest mass m_A of block A that will not cause the system to start moving.

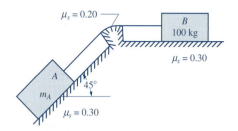

FIGURE P5–40

5–41 The band brake in Fig. P5–41 has a drum of 6-in. radius and a kinetic friction coefficient of 0.25. A force P of 20 lb is applied to the arm at D. Determine the braking torque acting on the drum if the drum rotates in the clockwise direction.

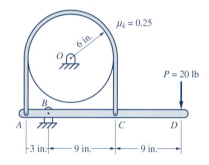

FIGURE P5–41

5–42 In Problem 5–41, determine the braking torque acting on the drum if the drum rotates in the counterclockwise direction.

5–43 The band brake shown in Fig. P5–43 is used to control the motion of the flywheel. A force of 100 N is applied to the lever at D. Determine the braking torque on the flywheel if it rotates in the clockwise direction.

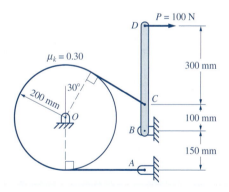

FIGURE P5–43

5–44 In Problem 5–43, determine the braking torque on the flywheel if it rotates in the counterclockwise direction.

5–45 The belt drive shown in Fig. P5–45 is used to transmit torque from pulley A to pulley B. The radius of each pulley is 5 in. and the coefficient of static friction is 0.30. Determine the largest torque that can be transmitted if the allowable belt tension is 1000 lb.

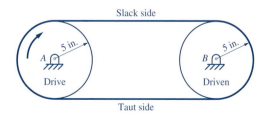

FIGURE P5–45

5–46   The motor shown in Fig. P5–46 drives pulley A and the torque is transmitted to pulley B by a flat belt. Pulley A is rotating at a constant angular velocity. Knowing that the allowable belt tension is 4000 N and that the coefficient of static friction is 0.40, determine the largest torque that the motor may apply to pulley A without causing the belt to slip on pulley A.

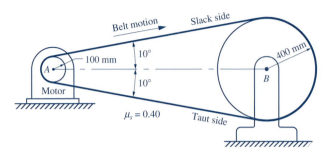

FIGURE P5–46

Section 5–7   Rolling Resistance

5–47   Do you expect to get better gas mileage driving a car with fully inflated rather than partially inflated tires? Why?

5–48   Why is the gas mileage of an automobile traveling on a paved road better than that traveling on a dirt road?

5–49   See Fig. P5–49. Compare the forces required to keep each form of transportation moving at the same constant speed if each is of the same weight.

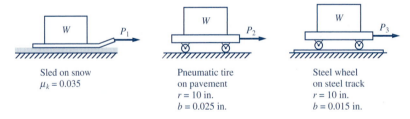

FIGURE P5–49

5–50   A railroad car and its cargo have a total weight of 50 kips. The car is supported on eight wheels, each of which has a diameter of 25 in. If the coefficient of rolling resistance is 0.015 in. between the wheels and the tracks, and the wheels share the load equally, determine the magnitude of the horizontal force required to overcome the rolling resistance of the wheels.

5–51   A baggage carrier piled with suitcases has a total mass of 300 kg. Each of the four wheels has a diameter of 50 mm, and the coefficient of rolling resistance is 4 mm between the wheels and the floor. Assuming that the four wheels carry equal loads, determine the horizontal force that must be applied to the carrier to move it at a constant speed.

5–52   A disabled car weighing 2200 lb must be pushed to the shoulder of a dirt road. Each of the tires has a diameter of 2.15 ft, and the coefficient of rolling resistance is 0.05 in. between the tires and the road. Determine the horizontal force that must be applied to the car to move it.

Computer Program Assignments

For each of the following problems, write a computer program using an appropriate programming language with which you are most familiar. Make the program user friendly by incorporating plenty of comments and input prompts so that the user will understand the input data to be entered and the limitations of their values. The output should include the data entered and the computed results, and they must be well labeled to identify each quantity. If a tabulated format is used, a proper heading must be included at the top of the table. Do not limit the program to any specific unit system. Indicate the consistent U.S. customary or SI units that can be used.

C5–1 Write a computer program that can be used to determine the torques required to load and to release the square-threaded screw shown in Fig. 5–5. Use Equations 5–8 and 5–9. The user input should include (1) the mean radius r, the number of threads n, and the pitch of the screw p; (2) the static friction coefficient μ_s; and (3) the load W and the arm a. The output results must include (1) the loading torque M and the forces F needed to produce this torque, (2) the mechanical advantage of the screw, and (3) the releasing torque M' and the corresponding forces F' if the screw is self-locking. Use this program to solve (a) Example 5–7; (b) Problem 5–28, assuming an arm length $a = 800$ mm; and (c) Problem 5–30, assuming an arm length $a = 4$ in.

C5–2 A weight W is suspended from a cable that wraps around a horizontal shaft and is held by a force P. Write a computer program that can be used to generate a table listing the values of W_{min} and W_{max} corresponding to the angle of contact β, varying from 1 wrap (or 2π radians) to 3 wraps, in quarter-wrap steps. Run the program using the data (a) $\mu_s = 0.20$ and $P = 25$ lb, and (b) $\mu_s = 0.30$ and $P = 180$ N.

CHAPTER 6

CONCURRENT SPATIAL FORCE SYSTEMS

6–1
INTRODUCTION

So far only coplanar force systems have been discussed. It is true that a majority of problems in mechanics involves forces in a plane. Some problems, however, involve force systems in three-dimensional space. Such a force system, called a *spatial force system*, requires three dimensions for its description. In this chapter, we will deal only with a *concurrent spatial force system*, in which the lines of action of all the forces in the system act through a common point.

6–2
RECTANGULAR COMPONENTS OF A SPATIAL FORCE

Consider a force **F** acting at the origin O of a rectangular coordinate system x, y, and z, as shown in Fig. 6–1a. The force **F** may be resolved into a vertical component F_y and a horizontal component F_h by projecting the force onto the y axis and the horizontal plane. The horizontal component F_h can be resolved further into components F_x and F_z along the x and z axes, respectively. The given force **F** has thus been resolved into three rectangular components F_x, F_y, and F_z.

Applying the Pythagorean theorem to the right triangles OBD and OAB in Fig. 6–1a, we obtain the following relationship between the magnitude F of the force and its three components:

$$F_h^{\,2} = F_x^{\,2} + F_z^{\,2}$$

$$F^2 = F_h^{\,2} + F_y^{\,2} = F_x^{\,2} + F_z^{\,2} + F_y^{\,2}$$

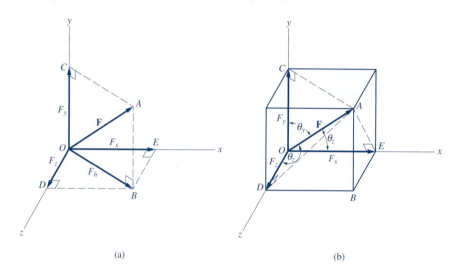

(a)   (b)

FIGURE 6–1

Thus,

$$F = \sqrt{F_x^2 + F_y^2 + F_z^2} \tag{6-1}$$

The relationship between the force **F** and the components of the force can be visualized more clearly if a box having the components as edges is drawn as shown in Fig. 6–1b. The force **F** is represented by the diagonal OA of the box. The angles θ_x, θ_y, and θ_z represent the angles that **F** forms with the x, y, and z axes, respectively. From the right triangles OAE, OAC, and OAD, we have the following relationships:

$$F_x = F\cos\theta_x \qquad F_y = F\cos\theta_y \qquad F_z = F\cos\theta_z \tag{6-2}$$

The three angles θ_x, θ_y, and θ_z are called the *direction angles* of the force **F** and $\cos\theta_x$, $\cos\theta_y$, and $\cos\theta_z$ are known as the *direction cosines* of the force **F**.

If the rectangular components F_x, F_y, and F_z are known, then the direction cosine can be computed from

$$\cos\theta_x = \frac{F_x}{F} \qquad \cos\theta_y = \frac{F_y}{F} \qquad \cos\theta_z = \frac{F_z}{F} \tag{6-3}$$

Substitution of Equation 6–2 into Equation 6–1 gives

$$F^2 = F^2\left(\cos^2\theta_x + \cos^2\theta_y + \cos^2\theta_z\right)$$

Thus,

$$\cos^2\theta_x + \cos^2\theta_y + \cos^2\theta_z = 1 \tag{6-4}$$

which shows that the direction cosines of a force are related. If any two of the three angles are known, the third one can be determined from this equation.

─────── **EXAMPLE 6–1** ───────────────────────────────────

Refer to Fig. E6–1. Express force **F** in terms of its rectangular components.

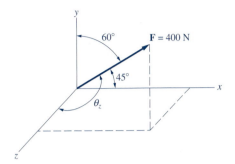

FIGURE E6–1

   Solution. From the figure, we identify that

$$\theta_x = 45° \qquad \theta_y = 60°$$

The third angle θ_z may be solved from Equation 6–4. Thus,

$$\cos^2\theta_z = 1 - \cos^2\theta_x - \cos^2\theta_y$$
$$= 1 - \cos^2 45° - \cos^2 60°$$
$$= 0.25$$

Taking the square root of both sides, we get

$$\cos\theta_z = \pm 0.5$$

Hence,

$$\theta_z = \cos^{-1}(0.5) = 60° \qquad \text{or} \qquad \theta_z = \cos^{-1}(-0.5) = 120°$$

From Fig. E6–1, we see that all the components of **F** are positive, so θ_z must be an acute angle. Thus, $\theta_z = 60°$. Using Equation 6–2, we get

$$F_x = F\cos\theta_x = (400 \text{ N}) \cos 45° = 283 \text{ N} \qquad \Leftarrow \textbf{Ans.}$$

$$F_y = F\cos\theta_y = (400 \text{ N}) \cos 60° = 200 \text{ N} \qquad \Leftarrow \textbf{Ans.}$$

$$F_z = F\cos\theta_z = (400 \text{ N}) \cos 60° = 200 \text{ N} \qquad \Leftarrow \textbf{Ans.}$$

EXAMPLE 6–2

Find the magnitude and direction angles of a force **F** if its rectangular components are 20.5 lb, 56.7 lb, and −34.8 lb along the x, y, and z axes, respectively.

Solution. The magnitude of the force can be computed from Equation 6–1 as

$$F = \sqrt{F_x^2 + F_y^2 + F_z^2}$$
$$= \sqrt{(20.5 \text{ lb})^2 + (56.7 \text{ lb})^2 + (-34.8 \text{ lb})^2}$$
$$= 69.6 \text{ lb} \qquad\qquad \Leftarrow \textbf{Ans.}$$

The direction angles are computed from Equation 6–3 as

$$\theta_x = \cos^{-1}\frac{F_x}{F} = \cos^{-1}\frac{20.5 \text{ lb}}{69.6 \text{ lb}} = 72.9° \qquad \Leftarrow \textbf{Ans.}$$

$$\theta_x = \cos^{-1}\frac{F_y}{F} = \cos^{-1}\frac{56.7 \text{ lb}}{69.6 \text{ lb}} = 35.4° \qquad \Leftarrow \textbf{Ans.}$$

$$\theta_x = \cos^{-1}\frac{F_z}{F} = \cos^{-1}\frac{-34.8 \text{ lb}}{69.6 \text{ lb}} = 119.8° \qquad \Leftarrow \textbf{Ans.}$$

The results can be checked by using Equation 6–4:

$$\cos^2 72.9° + \cos^2 35.4° + \cos^2 119.8° = 0.998 \qquad \text{(Checks)}$$

6–3
FORCE ACTING THROUGH TWO POINTS

In many applications of three-dimensional problems, the direction of a force is defined by two points of known coordinates along the line of action of the force. Consider a force **F** of known magnitude, directed from point A to point B of given coordinates, as shown in Fig. 6–2. The line vector **AB** has the three components d_x, d_y, and d_z, where

$$d_x = x_B - x_A \qquad d_y = y_B - y_A \qquad d_z = z_B - z_A \qquad (6\text{--}5)$$

Thus, the length of AB is

$$AB = d = \sqrt{d_x^2 + d_y^2 + d_z^2} \qquad (6\text{--}6)$$

and the direction cosines of the vector **AB** are

$$\cos \theta_x = \frac{d_x}{d} \qquad \cos \theta_y = \frac{d_y}{d} \qquad \cos \theta_z = \frac{d_z}{d} \qquad (6\text{--}7)$$

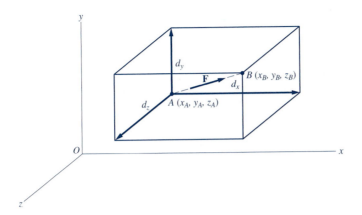

FIGURE 6–2

Since the force **F** acts along **AB**, it must have the same direction cosines as those of **AB**. Substituting Equation 6–7 into Equation 6–2, we obtain

$$F_x = \frac{d_x}{d}F \qquad F_y = \frac{d_y}{d}F \qquad F_z = \frac{d_z}{d}F \qquad (6\text{–}8)$$

─── **EXAMPLE 6–3** ───

In the bracket in Fig. E6–3, the turnbuckle is tightened until the tension in cable AB reaches 210 lb. Find the rectangular components of the tension **T** acting on the lever at A.

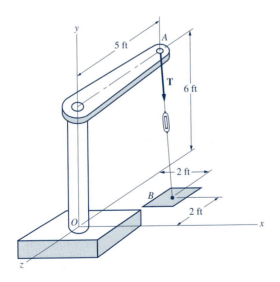

FIGURE E6–3

Solution. The tension **T** is directed from A to B. From Fig. E6–3, the coordinates of B and A are

$$B(2, 0, -2) \text{ ft} \qquad A(0, 6, -5) \text{ ft}$$

From Equation 6–5, the components of the line vector **AB** are

$$d_x = x_B - x_A = 2 - 0 = +2 \text{ ft}$$

$$d_y = y_B - y_A = 0 - 6 = -6 \text{ ft}$$

$$d_z = z_B - z_A = -2 - (-5) = +3 \text{ ft}$$

From Equation 6–6, the length AB is

$$d = \sqrt{d_x + d_y + d_z} = \sqrt{(+2)^2 + (-6)^2 + (+3)^2} = 7 \text{ ft}$$

From Equation 6–8, the rectangular components of **T** are

$$T_x = \frac{d_x}{d}T = \frac{+2}{7}(210 \text{ lb}) = +60 \text{ lb} \qquad \Leftarrow \textbf{Ans.}$$

$$T_y = \frac{d_y}{d}T = \frac{-6}{7}(210 \text{ lb}) = -180 \text{ lb} \qquad \Leftarrow \textbf{Ans.}$$

$$T_z = \frac{d_z}{d}T = \frac{+3}{7}(210 \text{ lb}) = +90 \text{ lb} \qquad \Leftarrow \textbf{Ans.}$$

6–4
RESULTANT OF A CONCURRENT SPATIAL FORCE SYSTEM

The resultant of two or more concurrent spatial forces can be determined by summing the rectangular components of the forces. Graphical or trigonometric methods are not practical in the case of spatial forces.

Each force is resolved into rectangular components, and the corresponding components are summed algebraically to get the components of the resultant:

$$R_x = \Sigma F_x \qquad R_y = \Sigma F_y \qquad R_z = \Sigma F_z \qquad (6\text{–}9)$$

Now the magnitude and direction cosines of the resultant may be computed from

$$R = \sqrt{R_x^2 + R_y^2 + R_z^2} \qquad (6\text{–}10)$$

$$\cos\theta_x = \frac{R_x}{R} \qquad \cos\theta_y = \frac{R_y}{R} \qquad \cos\theta_z = \frac{R_z}{R} \qquad (6\text{–}11)$$

———— **EXAMPLE 6–4** ————

Determine the resultant of the two forces **P** and **Q** applied at point A, as shown in Fig. E6–4.

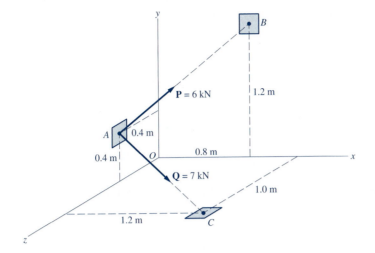

FIGURE E6–4

Solution. Force **P** acts from A to B; thus,

$$
\begin{array}{lccc}
B\,(\;\;0.8, & 1.2, & 0\;\;) \\
-)\;A\,(\;\;0, & 0.4, & 0.4\;) \\
\hline
d_x = 0.8 & d_y = 0.8 & d_z = -0.4
\end{array}
$$

$$
d_{AB} = \sqrt{(0.8)^2 + (0.8)^2 + (0.4)^2} = 1.2 \text{ m}
$$

$$
P_x = \frac{+0.8}{1.2}(6 \text{ kN}) = +4 \text{ kN}
$$

$$
P_y = \frac{+0.8}{1.2}(6 \text{ kN}) = +4 \text{ kN}
$$

$$
P_z = \frac{-0.4}{1.2}(6 \text{ kN}) = -2 \text{ kN}
$$

Force **Q** acts from A to C; thus,

$$
\begin{array}{lccc}
C\,(\;\;1.2, & 0, & 1.0\;) \\
-)\;A\,(\;\;0, & 0.4, & 0.4\;) \\
\hline
d_x = 1.2 & d_y = -0.4 & d_z = 0.6
\end{array}
$$

$$
d_{AC} = \sqrt{(1.2)^2 + (0.4)^2 + (0.6)^2} = 1.4 \text{ m}
$$

$$
Q_x = \frac{+1.2}{1.4}(7 \text{ kN}) = +6 \text{ kN}
$$

$$
Q_y = \frac{-0.4}{1.4}(7 \text{ kN}) = -2 \text{ kN}
$$

$$Q_z = \frac{+0.6}{1.4}(7 \text{ kN}) = +3 \text{ kN}$$

From Equation 6–9, the components of the resultant of the two forces are

$$R_x = P_x + Q_x = 4 + 6 = +10 \text{ kN}$$

$$R_y = P_y + Q_y = 4 - 2 = +2 \text{ kN}$$

$$R_z = P_z + Q_z = -2 + 3 = +1 \text{ kN}$$

From Equation 6–10, the magnitude of the resultant is

$$R = \sqrt{(10)^2 + (2)^2 + (1)^2} \text{ kN}$$
$$= 10.25 \text{ kN} \qquad\qquad \Leftarrow \textbf{Ans.}$$

From Equation 6–11, the direction angles of the resultant are

$$\theta_x = \cos^{-1}\frac{+10}{10.25} = 12.7° \qquad\qquad \Leftarrow \textbf{Ans.}$$

$$\theta_y = \cos^{-1}\frac{+2}{10.25} = 78.7° \qquad\qquad \Leftarrow \textbf{Ans.}$$

$$\theta_z = \cos^{-1}\frac{+1}{10.25} = 84.4° \qquad\qquad \Leftarrow \textbf{Ans.}$$

Check. The above results can be checked by

$$\cos^2 \theta_x + \cos^2 \theta_y + \cos^2 \theta_z = \cos^2 12.7° + \cos^2 78.7° + \cos^2 84.4°$$
$$= 1.000 \qquad\qquad \text{(Checks)}$$

6–5
EQUILIBRIUM OF A CONCURRENT SPATIAL FORCE SYSTEM

A body is in equilibrium if the resultant of all concurrent forces in space acting on it is zero. This requires that the sum of each of the corresponding force components be equal to zero. Thus,

$$\Sigma F_x = 0 \qquad \Sigma F_y = 0 \qquad \Sigma F_z = 0 \qquad\qquad (6\text{–}12)$$

These three equilibrium equations can be used to solve for three unknowns.

––––––– **EXAMPLE 6–5** –––––––

A mast *AB* of negligible weight is supported by a ball-and-socket joint at *A* and guy wires *BC* and *BD*, as shown in Fig. E6–5(1). A vertical load of 14 kips is applied at *B*. Determine the axial forces in the mast and the tensions in guy wires *BC* and *BD*.

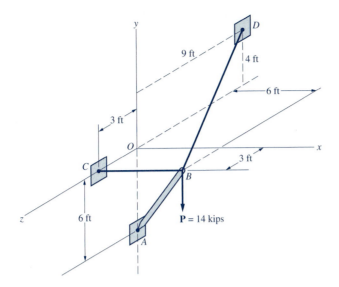

FIGURE E6–5(1)

Solution. Since mast AB is a two-force member, it is subjected to two equal and opposite forces along its axial direction. The free-body diagram of joint B is sketched in Fig. E6–5(2). Note that the mast is assumed to be in compression. The four forces exerted on the joint are concurrent. Three of them are of unknown magnitude. Each force, however, can be expressed in rectangular components, as shown in the following.

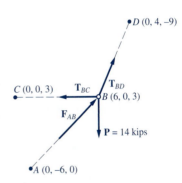

FIGURE E6–5(2)

Components of F_{AB}. This force is along **AB**; thus,

$$
\begin{array}{cccc}
B (& 6, & 0, & 3) \\
-) \; A (& 0, & -6, & 0) \\
\hline
d_x = 6 & d_y = 6 & d_z = 3 &
\end{array}
$$

$$d_{AB} = \sqrt{(6)^2 + (6)^2 + (3)^2} = 9 \text{ ft}$$

$$(F_{AB})_x = \frac{6}{9} \, (F_{AB}) = 0.667 F_{AB}$$

$$(F_{AB})_y = \frac{6}{9} \, (F_{AB}) = 0.667 F_{AB}$$

$$(F_{AB})_z = \frac{3}{9} \, (F_{AB}) = 0.333 F_{AB}$$

Components of T_{BD}.   This force is along **BD**; thus,

$$
\begin{array}{llll}
D(& 9, & 4, & -9\ ) \\
-)\ B(& 6, & 0, & 3\ ) \\
\hline
& d_x = -6 & d_y = 4 & d_z = -12
\end{array}
$$

$$d_{BD} = \sqrt{(6)^2 + (4)^2 + (12)^2} = 14 \text{ ft}$$

$$(T_{BD})_x = \frac{-6}{14}(T_{BD}) = -0.429 T_{BD}$$

$$(T_{BD})_y = \frac{4}{14}(T_{BD}) = 0.286 T_{BD}$$

$$(T_{BD})_z = \frac{-12}{14}(T_{BD}) = -0.857 T_{BD}$$

Components of T_{BC}.   This force is along the negative x axis; thus,

$$(T_{BC})_x = -T_{BC} \qquad (T_{BC})_y = 0 \qquad (T_{BC})_z = 0$$

Components of P.   The applied force $P = 14$ kips and acts along the negative y axis; thus,

$$P_x = 0 \qquad P_y = -14 \text{ kips} \qquad P_z = 0$$

Equilibrium Equations.   Equilibrium conditions require that the sum of the force components along the x, y, and z directions must each be equated to zero. Thus,

$$\Sigma F_x = 0.667 F_{AB} - 0.429 T_{BD} - T_{BC} = 0 \tag{a}$$

$$\Sigma F_y = 0.667 F_{AB} + 0.286 T_{BD} - 14 = 0 \tag{b}$$

$$\Sigma F_z = 0.333 F_{AB} - 0.857 T_{BD} = 0 \tag{c}$$

Solution of Equations.   Since Equations (b) and (c) contain only two unknowns, the two unknowns can be solved from these two equations. To eliminate T_{BD}, we add Equation (b) divided by 0.286 and Equation (c) divided by 0.857 to get

$$\left(\frac{0.667}{0.286} + \frac{0.333}{0.857}\right) F_{AB} - \frac{14}{0.286} = 0$$

From which we get

$$F_{AB} = 18.0 \text{ kips (C)} \qquad \Leftarrow \textbf{Ans.}$$

Recall that when the free-body diagram was drawn, member *AB* was assumed to be in compression; a positive result means that the member *is* in compression.

Substituting in Equation (c) gives

$$T_{BD} = 7.00 \text{ kips (T)} \qquad \Leftarrow \textbf{Ans.}$$

Substituting in Equation (a) gives

$$T_{BC} = 9.00 \text{ kips (T)} \qquad \Leftarrow \textbf{Ans.}$$

Remark. The method used above for solving the three linear equations is the method of addition or subtraction. There are many other methods that can be used to determine the solution to three linear equations with three unknowns. The following are three of them:

1. Use Cramer's rule in terms of third-order determinants (see Section 1–12).
2. Use matrix inversion with a graphing calculator. Refer to the user's manual of your calculator.
3. Use the computer program written for Program Assignment C1–2 or other computer software for linear equations.

6–6
SUMMARY

Rectangular Components. A force in three-dimensional space is called a *spatial force*. For a spatial force, if the magnitude *F* and the *direction angles* θ_x, θ_y, and θ_z between the force vector and the *x*, *y*, and *z* axes, respectively, are defined, the three *rectangular components* can be computed from

$$F_x = F \cos \theta_x \qquad F_y = F \cos \theta_y \qquad F_z = F \cos \theta_z \qquad (6\text{–}2)$$

On the other hand, if the three rectangular components F_x, F_y, and F_z of a spatial force are known, the magnitude and *direction cosines* can be computed from

$$F = \sqrt{F_x^2 + F_y^2 + F_z^2} \qquad (6\text{–}1)$$

$$\cos \theta_x = \frac{F_x}{F} \qquad \cos \theta_y = \frac{F_y}{F} \qquad \cos \theta_z = \frac{F_z}{F} \qquad (6\text{–}3)$$

The three direction cosines are related by the following equation:

$$\cos^2\theta_x + \cos^2\theta_y + \cos^2\theta_z = 1 \qquad (6\text{--}4)$$

Force Acting Through Two Points. If the direction of a force is defined by two points, A and B, of known coordinates A (x_A, y_A, z_A) and B (x_B, y_B, z_B) along the line of action of the force, the three rectangular components and the length of the line vector **AB** (directed from A to B) are

$$d_x = x_B - x_A \qquad d_y = y_B - y_A \qquad d_z = z_B - z_A \qquad (6\text{--}5)$$

$$d = \sqrt{d_x^2 + d_y^2 + d_z^2} \qquad (6\text{--}6)$$

The rectangular components of the force **F** along AB are

$$F_x = \frac{d_x}{d}F \qquad F_y = \frac{d_y}{d}F \qquad F_z = \frac{d_z}{d}F \qquad (6\text{--}8)$$

Resultant. The resultant of a *concurrent spatial force system* can be determined by summing the rectangular components of the forces:

$$R_x = \Sigma F_x \qquad R_y = \Sigma F_y \qquad R_z = \Sigma F_z \qquad (6\text{--}9)$$

Equilibrium. A body subjected to a *concurrent spatial force system* is in equilibrium if the resultant of all the forces in the system is zero. This requires that the sum of the force components along the x, y, and z directions must each be equal to zero:

$$\Sigma F_x = 0 \qquad \Sigma F_y = 0 \qquad \Sigma F_z = 0 \qquad (6\text{--}12)$$

PROBLEMS

Section 6–2　Rectangular Components of a Spatial Force

6–1　A 400-lb force **F** acts in the direction defined by the direction angles $\theta_x = 70°$ and $\theta_y = 50°$. The z component of the force is negative. Determine the components of the force.

6–2　A 150-N force acts in the direction defined by the direction angles $\theta_x = 35°$ and $\theta_y = 120°$. Knowing that the z component of the force is positive, determine the three components of the force.

6–3 and **6–4** Find the rectangular components of the forces shown in Figs. P6–3 and P6–4.

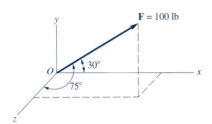

FIGURE P6–3

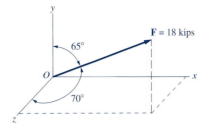

FIGURE P6–4

6–5 The three components of a force **F** are 600 N along the positive x axis, 800 N along the positive y axis, and 2400 N along the negative z axis. Determine the magnitude and direction angles of the force.

6–6 Determine the magnitude and the direction angles of a force **F** with the following components: 100 lb along the positive x axis, 105 lb along the positive y axis, and 145 lb along the negative z axis.

6–7 Determine the magnitude and the direction angles of a force **F** with components $F_x = -160$ N, $F_y = 300$ N, and $F_z = -340$ N.

6–8 A force acts in a direction defined by the angles $\theta_y = 40°$ and $\theta_z = 75°$. Knowing that the x component of the force is -260 lb, determine the magnitude of the force.

Section 6–3 Force Acting Through Two Points

6–9 Refer to Fig. P6–9. Given that the tension $\mathbf{T}_{AC}$ in cable AC is 900 N, determine the components of $\mathbf{T}_{AC}$ exerted on point C.

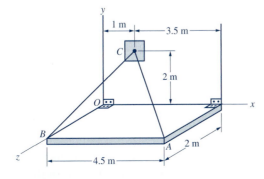

FIGURE P6–9

6–10 If the tension $\mathbf{T}_{BC}$ in cable BC of Fig. P6–9 is 600 N, determine the components of $\mathbf{T}_{BC}$ exerted on point C.

6–11 The ventilator door in Fig. P6–11 is held in the horizontal position by cable AB. Find the rectangular components of tension $\mathbf{T}$ in cable AB acting at A.

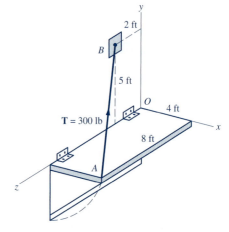

FIGURE P6–11

6–12 The weight in Fig. P6–12 is supported by the boom shown. Knowing that the magnitude of tension $\mathbf{T}_{BD}$ is 5 kN, find the rectangular components $\mathbf{T}_{BD}$ exerted at B.

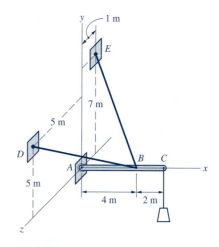

FIGURE P6–12

6–13 The access door in Fig. P6–13 is held in the position shown by cable AB. If the magnitude of tension $\mathbf{T}$ in the cable is 50 lb, find the rectangular components of $\mathbf{T}$ acting at A.

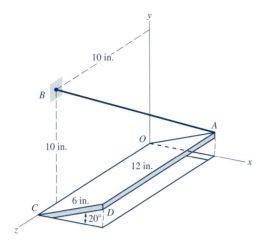

FIGURE P6-13

Section 6–4   Resultant of a Concurrent Spatial Force System

6–14 to **6–17** Refer to Figs. P6–14 to P6–17. Determine the components of the resultant of the two forces shown in each figure.

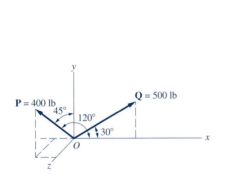

FIGURE P6–14

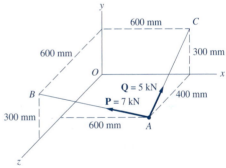

FIGURE P6–15

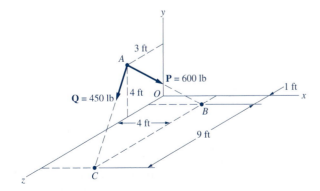

FIGURE P6–16

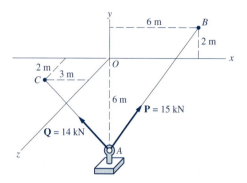

FIGURE P6–17

6–18   Refer to Fig. P6–18. Determine the magnitude and the direction angles of the resultant of the three forces $\mathbf{P}$, $\mathbf{T}_{AB}$, and $\mathbf{T}_{AC}$ exerted on the bracket at point A.

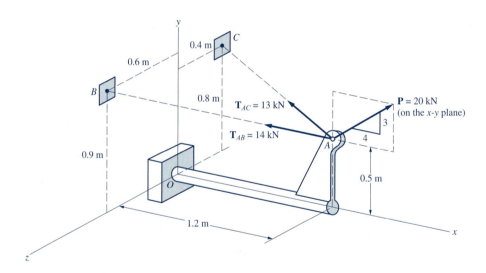

FIGURE P6–18

Section 6–5   Equilibrium of a Concurrent Spatial Force System

6–19   The boom AB in Fig. P6–19 supports a load $W = 720$ lb. The boom is supported by a ball-and-socket joint at A and by two cables, BC and BD. Determine the tension in each cable and the axial force in the boom. Neglect the weight of the boom.

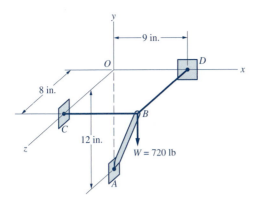

FIGURE P6–19

6–20 Determine the tension in each cable and the axial force in the boom, assuming that point D is 6 in. vertically above the position shown in Fig. P6–19 and the other dimensions are unchanged.

6–21 The 6-m pole shown in Fig. P6–21 supports a load $P = 20$ kN. The pole is supported by a ball-and-socket joint at A and by two cables, BC and BD. Determine the tension in each cable and the axial force in pole AB.

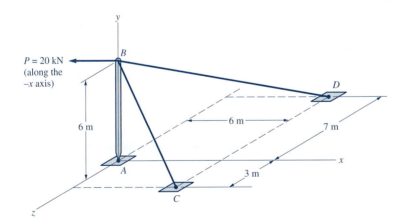

FIGURE P6–21

6–22 Each of the three members, AB, AC, and AD, of the tripod in Fig. P6–22 is secured by ball-and-socket connections and each is capable of exerting tension or compression. The turnbuckle in wire AE is tightened so that the wire tension is 4080 lb. Determine the axial force in each member. Neglect the weights of the members.

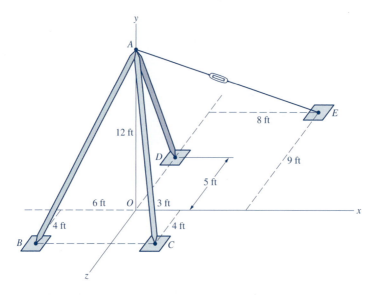

FIGURE P6–22

6–23 The boom AB in Fig. P6-23 supports a weight of 17 kN. The boom is supported by a ball-and-socket joint at A and two cables, BC and BD. Determine the tension in each cable and the axial force in boom AB. Neglect the weight of the boom.

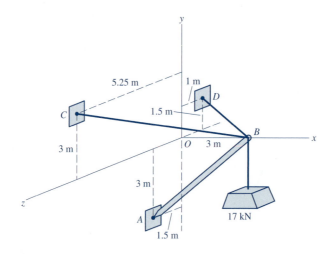

FIGURE P6–23

6–24 The 4-m × 8-m rectangular steel plate in Fig. P6–24 weighs 20 kN and is lifted by three cables joined at point A directly above the center of gravity G of the plate. Determine the tension in each cable. (*Hint*: The tension in the vertical cable AE is equal to the weight of the plate.)

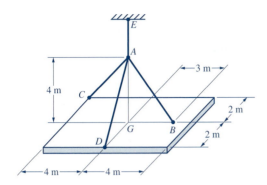

FIGURE P6–24

6–25 The steel ring in Fig. P6–25 is 6 ft in radius, weighs 800 lb, and is lifted by three cables joining at point A, which is 8 ft vertically above the center of gravity G of the ring. Determine the tension in each cable. (*Hint*: The tension in the vertical cable AE is equal to the weight of the plate.)

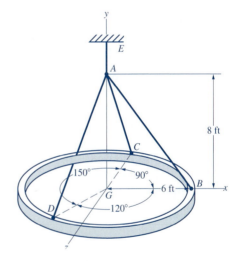

FIGURE P6–25

6–26 A 500-lb cylinder is held in place by the cables shown in Fig. P6–26. Determine the tension in each cable.

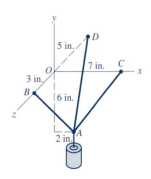

FIGURE P6–26

6–27   The 40-lb lamp in Fig. P6–27 is supported by a pole *AB* and wires *AC* and *AD*. The lower end of the pole is connected to a ball-and-socket joint. Neglect the weights of the pole and the wires. Determine the forces in the pole and the wires.

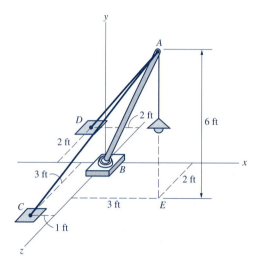

FIGURE P6–27

6–28   The derrick in Fig. P6–28 supports a 16-kN weight. The derrick is supported by a ball-and-socket joint at *A* and two cables *CD* and *CE*. Determine the tensions in the two cables, the tension in cable *BC*, and the axial forces in the two booms. Neglect the weights of the booms. (*Hint:* Consider first the equilibrium of joint *B*, then the equilibrium of joint *C*.)

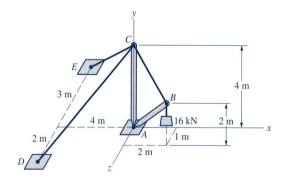

FIGURE P6–28

Computer Program Assignments

For each of the following problems, write a computer program using an appropriate programming language with which you are most familiar. Make the program user friendly by using plenty of comments and input prompts so that the user will understand the input data to be entered and the limitations of their values. The output should include the data entered and the computed results, and they must be well labeled to identify each quantity. If a tabulated format is used, a proper heading must be included at the top of the table. Do not limit the program to any specific unit system. Indicate the consistent U.S. customary or SI units that can be used.

C6–1 Refer to Fig. C6–1. Write a computer program that can be used to determine the rectangular components F_x, F_y, and F_z of a spatial force passing through two known points A and B. The user input should include (1) the magnitude F of the force; (2) the coordinates x_A, y_A, and z_A of point A; and (3) the coordinates x_B, y_B, and z_B of point B. Use this program to solve (*a*) Example 6–3, (*b*) Problem 6–11, and (*c*) Problem 6–12.

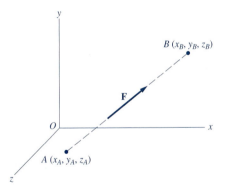

FIGURE C6–1

C6–2 Refer to Fig. C6–2. Write a computer program that can be used to determine the magnitude and the direction of the resultant of a concurrent spatial force system consisting of n forces. The program written for Problem C6–1 can be used as a subroutine. The user input should include (1) the total number of the forces n in the system; (2) the coordinates x_A, y_A, and z_A of the common point A; and (3) the magnitude and coordinates of a point on the line of action of each force F_i, $(x_B)_i$, $(y_B)_i$, and $(z_B)_i$. The output results must include (1) the rectangular components R_x, R_y, and R_z of the resultant; (2) the magnitude R of the resultant; and (3) the direction angles θ_x, θ_y, and θ_z of the resultant. Use this program to solve (a) Example 6–4, (b) Problem 6–16, and (c) Problem 6–18.

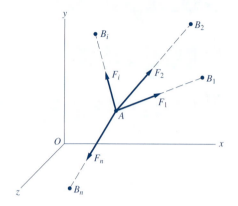

FIGURE C6–2

CHAPTER 7

CENTER OF GRAVITY AND CENTROIDS

7–1
INTRODUCTION

A solid body may be considered as a system of particles, and each particle has a weight due to the gravitational attraction of the earth. The total weight of all the particles represents the weight of the body. The resultant weight always act through a definite point, regardless of the orientation of the body. This point is called the *center of gravity*. As will be demonstrated in this chapter, the center of gravity can be determined by both the analytical and experimental methods. The same analytical method can be used to determine the *centroid* (geometric center) of a volume or an area. The method can also be applied to the determination of the resultant of a general type of distributed line load. Other topics related to distributed loads, such as liquid pressure and flexible cables, will also be covered in this chapter.

7–2
CENTER OF GRAVITY AND CENTROID OF A BODY

Center of Gravity. The gravitational attraction of the earth on a body is exerted on each particle of the body. The weights of the particles, being downward and pointing toward the center of the earth, can be considered parallel to each other for all practical purposes. The total weight of a body is thus the sum of the weights of all the particles.

Figure 7–1a shows a rigid body for which the center of gravity is to be determined. The body consists of a large number of small particles. A typical particle with weight ΔW is shown acting vertically downward. The weight of the body is the sum of the weight of all the particles; we write

$$W = \Sigma \Delta W$$

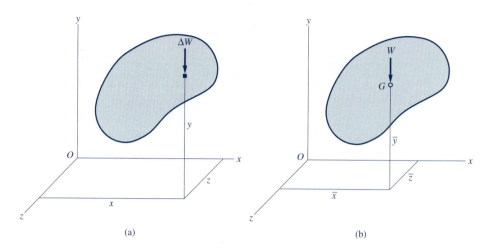

(a)   (b)

FIGURE 7–1

The total weight W must act through the center of gravity G of the body, as shown in Fig. 7–1b. The weight W through G must be equivalent to the distributed weights throughout the body. Hence, the moment of W about O must be equal to the sum of the moments of the weight of all the particles about O. Denoting the coordinates of G by $(\bar{x}, \bar{y}, \bar{z})$ and equating the moment about the z axis, we write

$$\bar{x} W = \Sigma x \Delta W$$

Equating the moment about the x axis, we write

$$\bar{z} W = \Sigma z \Delta W$$

Similarly we can also write

$$\bar{y} W = \Sigma y \Delta W$$

Thus, we obtain the following equations for locating the center of gravity of a solid body:

$$\bar{x} = \frac{\Sigma x \Delta W}{\Sigma \Delta W} \qquad \bar{y} = \frac{\Sigma y \Delta W}{\Sigma \Delta W} \qquad \bar{z} = \frac{\Sigma z \Delta W}{\Sigma \Delta W} \qquad (7\text{-}1)$$

Centroid of a Volume. The *centroid* of a volume is the geometric center of the volume. For a homogeneous body, the unit weight (weight per unit volume), γ (the Greek letter gamma), of the body is a constant. Thus, the weight of the typical particle becomes $\Delta W = \gamma \Delta V$ and the weight of the entire body becomes $W = \gamma V$. Substituting these in Equation 7–1 and dividing throughout by γ, we obtain

$$\bar{x} = \frac{\Sigma x \Delta V}{\Sigma \Delta V} \qquad \bar{y} = \frac{\Sigma y \Delta V}{\Sigma \Delta V} \qquad \bar{z} = \frac{\Sigma z \Delta V}{\Sigma \Delta V} \qquad (7\text{-}2)$$

These equations define the location of the *centroid* of a volume. The centroid of a homogeneous body is the same as its center of gravity. For a body of non-homogeneous material (such as a composite body of two materials), the centroid and the center of gravity of the body may be located at different points.

Centroid of Volumes of Common Shapes. Formulas for computing the volumes and the locations of centroids of common geometric shapes, such as cylinders, spheres, hemispheres, circular cones, and so on, have been determined mathematically and are listed in Table 7–1.

Centroid of Composite Volumes. A volume that can be divided into several component volumes of shapes such as those listed in Table 7–1 is called a *composite volume*. The coordinates of the centroid $C(\bar{x}, \bar{y}, \bar{z})$ of a composite volume may be computed from the following equations:

TABLE 7–1 Centroids of Volumes of Common Shapes

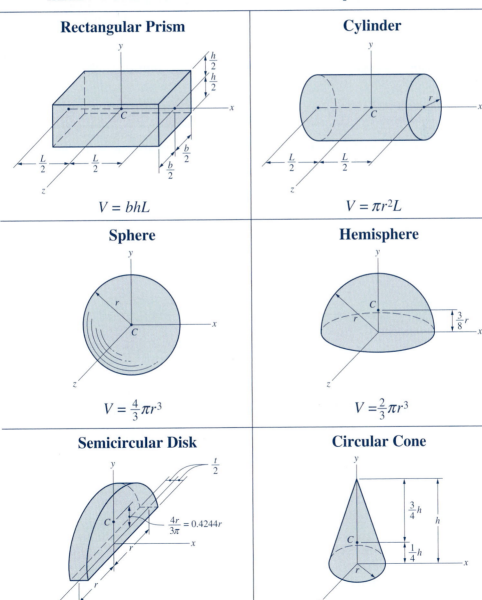

Rectangular Prism	Cylinder
$V = bhL$	$V = \pi r^2 L$
Sphere	**Hemisphere**
$V = \frac{4}{3}\pi r^3$	$V = \frac{2}{3}\pi r^3$
Semicircular Disk	**Circular Cone**
$V = \frac{1}{2}\pi r^2 t$	$V = \frac{1}{2}\pi r^2 h$

Within the Hemisphere diagram: $\frac{3}{8}r$

Within the Semicircular Disk diagram: $\frac{4r}{3\pi} = 0.4244r$

Within the Circular Cone diagram: $\frac{3}{4}h$, $\frac{1}{4}h$

$$\bar{x} = \frac{\Sigma Vx}{\Sigma V} = \frac{V_1 x_1 + V_2 x_2 + V_3 x_3 + \cdots}{V_1 + V_2 + V_3 + \cdots}$$

$$\bar{y} = \frac{\Sigma Vy}{\Sigma V} = \frac{V_1 y_1 + V_2 y_2 + V_3 y_3 + \cdots}{V_1 + V_2 + V_3 + \cdots}$$

$$\bar{z} = \frac{\Sigma Vz}{\Sigma V} = \frac{V_1 z_1 + V_2 z_2 + V_3 z_3 + \cdots}{V_1 + V_2 + V_3 + \cdots} \qquad (7\text{-}3)$$

where V_1, V_2, V_3, etc., are the component volumes and (x_1, y_1), (x_2, y_2), (x_3, y_3), etc., are the coordinates of the centroids of the component volumes.

When a body has a plane of symmetry, the centroid must be located on this plane. When a body has two mutually perpendicular planes of symmetry, the centroid must be located on the line of intersection of the two planes. *Recognizing the plane or the planes of symmetry will greatly simplify the necessary computations.*

—— **EXAMPLE 7–1** ——————————————————————

Locate the centroid of a body formed by a circular cone and a hemisphere, as shown in Fig. E7–1.

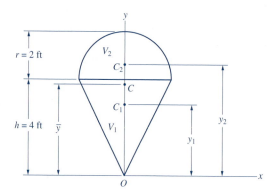

FIGURE E7–1

Solution. Note that the body is symmetrical with respect to the *xy* plane and the *yz* plane. The two planes intersect along the *y* axis. Thus, the centroid *C* must be along the *y* axis. The body consists of two components: a circular cone V_1 and a hemisphere V_2. The volumes can be computed from the formulas in Table 7–1. Thus,

$$V_1 = \frac{1}{3}\pi r^2 h = \frac{1}{3}\pi (2 \text{ ft})^2 (4 \text{ ft}) = 16.76 \text{ ft}^3$$

$$V_2 = \frac{2}{3}\pi r^3 = \frac{2}{3}\pi (2 \text{ ft})^3 = 16.76 \text{ ft}^3$$

The *y* coordinates of each volume are

$$y_1 = \frac{3}{4}h = \frac{3}{4}(4 \text{ ft}) = 3 \text{ ft}$$

$$y_2 = h + \frac{3}{8}r = 4 \text{ ft} + \frac{3}{8}(2 \text{ ft}) = 4.75 \text{ ft}$$

From Equation 7–3, we get

$$\bar{y} = \frac{V_1 y_1 + V_2 y_2}{V_1 + V_2}$$

$$= \frac{(16.76)(3) + (16.76)(4.75)}{16.76 + 16.76}$$

$$= 3.88 \text{ ft} \qquad \qquad \Leftarrow \textbf{Ans.}$$

EXAMPLE 7–2

Refer to Fig. E7–2(1). Locate the center of gravity of the plate of uniform thickness $t = 0.1$ m. The rectangular plate is made of steel, which has a unit weight $\gamma_{st} = 77$ kN/m³, and it has a circular cavity located as shown. The semicircular disk is made of copper, which has a unit weight of $\gamma_{cu} = 87.3$ kN/m³.

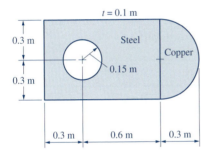

FIGURE E7–2(1)

Solution. The plate consists of three component weights: the rectangular plate W_1, the semicircular disk W_2, and the circular cavity W_3. [See Fig. E7–2(2).] The values of the weights are

$$W_1 = V_1 \gamma_{st} = (0.9 \text{ m})(0.6 \text{ m})(0.1 \text{ m})(77 \text{ kN/m}^3)$$
$$= 4.158 \text{ kN}$$

$$W_2 = V_2 \gamma_{cu} = \frac{1}{2}\pi(0.3 \text{ m})^2(0.1 \text{ m})(87.3 \text{ kN/m}^3)$$
$$= 1.234 \text{ kN}$$

$$W_3 = -V_3 \gamma_{st} = -\pi(0.15 \text{ m})^2(0.1 \text{ m})(77 \text{ kN/m}^3)$$
$$= -0.544 \text{ kN}$$

$$W = \Sigma W \quad = 4.848 \text{ kN}$$

Note that the weight of the cavity is considered negative.

The plate is symmetrical with respect to the xy plane and xz plane, which intersect at the x axis. Thus, the center of gravity must be along the x axis. The x coordinates of the center of gravities of the component weights are

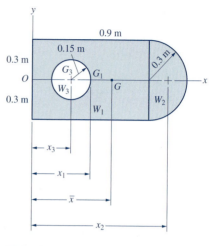

FIGURE E7–2(2)

$$x_1 = \frac{1}{2}(0.9 \text{ m}) = 0.45 \text{ m}$$

$$x_2 = 0.9 \text{ m} + 0.4244(0.3 \text{ m}) = 1.027 \text{ m}$$

$$x_3 = 0.3 \text{ m}$$

The center of gravity of the plate can be computed by

$$\bar{x} = \frac{W_1 x_1 + W_2 x_2 + W_3 x_3}{W_1 + W_2 + W_3}$$

$$= \frac{(4.158)(0.45) + (1.234)(1.027) + (-0.544)(0.3)}{4.848}$$

$$= 0.614 \text{ m} = 614 \text{ mm} \qquad\qquad \Leftarrow \textbf{Ans.}$$

The preceding calculations can be tabulated as follows:

Part	W (kN)	x (m)	Wx (kN · m)
1	4.158	0.45	1.871
2	1.234	1.027	1.267
3	−0.544	0.3	−0.163
Σ	4.848		2.975

Thus,

$$\bar{x} = \frac{\Sigma Wx}{\Sigma W} = \frac{2.975}{4.848} = 0.614 \text{ m} \qquad\qquad \Leftarrow \textbf{Ans.}$$

7–3
EXPERIMENTAL DETERMINATION OF THE CENTER OF GRAVITY

The shape of a body frequently does not possess well-defined boundaries, so an analytical solution is either too tedious or impossible. In such cases, the center of gravity of the body can be determined experimentally.

Consider a plate of irregular shape, as shown in Fig. 7–2. To determine its center of gravity experimentally, suspend the plate from any point, such as A. It is in equilibrium under the action of the tension T of the cable and its weight W acting through its center of gravity G. The two forces, T and W, must be collinear for equilibrium; the center of gravity G

must be on the vertical line AA' drawn through A. Similarly, if the plate is suspended from another point B, the center of gravity G must be on the vertical line BB' drawn through B. The point of intersection of lines AA' and BB' locates the center of gravity of the plate.

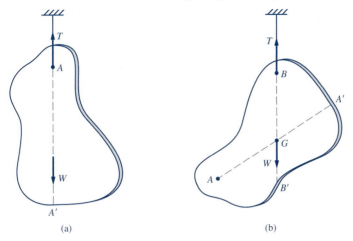

(a)   (b)

FIGURE 7–2

The method discussed above becomes impractical for some complicated shapes or heavy objects. In these instances, the weighing method may be employed. For example, the location of the center of gravity of an automobile can be determined by weighing its axle loads in two different inclinations.

For the car shown in Fig. 7–3a and b, A and B are the rear and front axles, respectively. The distance AB, called the wheel base, is denoted by b. The total weight of the car is W. The coordinates of the center of gravity G are denoted by $\bar{x}$ and $\bar{y}$. The front axle weight W_B is obtained first with line AB in the horizontal direction, as shown in Fig. 7–3a. The equilibrium of the car in this position requires

$$\Sigma M_A = -W\bar{x} + W_B b = 0$$

From which we get

$$\bar{x} = \frac{W_B}{W}b \qquad (7\text{–}4)$$

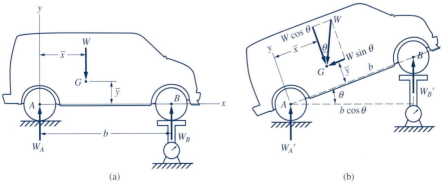

(a)   (b)

FIGURE 7–3

The front axle is then raised until AB is at some convenient angle θ from the horizontal, as shown in Fig. 7–3b. The front axle weight W_B' is obtained at this position. The equilibrium of the car in this position requires

$$\Sigma M_A = (W \sin\theta)\bar{y} - (W \cos\theta)\bar{x} + W_B'b \cos\theta = 0$$

Substituting $\bar{x}$ from Equation 7–4 and solving for $\bar{y}$, we get

$$\bar{y} = \frac{W_B b \cos\theta - W_B'b \cos\theta}{W\sin\theta}$$

or

$$\bar{y} = \frac{b}{W \tan\theta}(W_B - W_B') \qquad (7–5)$$

EXAMPLE 7–3

Assume that the minivan in Fig. 7–3 weighs 2000 lb and has a wheel base of 7 ft. The front axle weight is 1050 lb when the car is in the horizontal position. The front wheel is raised so that line AB is at 25° from the horizontal, and the front axle weight in this position is 850 lb. Locate the center of gravity of the car.

Solution. From Equation 7–4, we get

$$\bar{x} = \frac{W_B}{W}b = \frac{1050 \text{ lb}}{2000 \text{ lb}}(7 \text{ ft}) = 3.68 \text{ ft} \qquad \Leftarrow \textbf{Ans.}$$

From Equation 7–5, we get

$$\bar{y} = \frac{b}{W\tan\theta}(W_B - W_B')$$

$$= \frac{7 \text{ ft}}{(2000 \text{ lb})\tan 25°}(1050 \text{ lb} - 850 \text{ lb})$$

$$= 1.50 \text{ ft} \qquad \Leftarrow \textbf{Ans.}$$

7–4
CENTROID OF AN AREA

The *centroid* of a volume has been discussed in Section 7–2. The volume of a plate of uniform thickness t equals its surface area A times its thickness t, i.e., $V = At$. For a volume element, we have $\Delta V = t\Delta A$. Substituting these into Equation 7–2, we obtain

$$\bar{x} = \frac{\Sigma xt\Delta A}{\Sigma t\Delta A} \qquad \bar{y} = \frac{\Sigma yt\Delta A}{\Sigma t\Delta A}$$

Since t is the common factor in each term, it is canceled if both the numerator and the denominator are divided by t. Thus, we have

$$\bar{x} = \frac{\Sigma x\Delta A}{\Sigma \Delta A} \qquad \bar{y} = \frac{\Sigma y\Delta A}{\Sigma \Delta A} \qquad (7–6)$$

The equation defines the coordinates $\bar{x}$ and $\bar{y}$ of the *centroid* of the surface area. Physically, the centroid of an area is its geometric center. The quantities $\Sigma x \Delta A$ and $\Sigma y \Delta A$ are the *first moments* of the area about the x and y axes, respectively.

Axis of Symmetry. The area in Fig. 7–4a is *symmetrical* with respect to axis aa' because, for every point P in the area, there is a point P' that is the mirror image of P with respect to the axis. Line aa' is called the *axis of symmetry* of the area. If aa' is chosen to be the y axis, then for every area element ΔA at distance $+x$, there is a corresponding element at a distance $-x$. Hence, the terms $+x \Delta A$ and $-x \Delta A$ cancel each other and therefore $\bar{x} = 0$; that is, *the centroid of an area must be located on its axis of symmetry*. It follows that, if an area possesses two axes of symmetry, as shown in Fig. 7–4b, *the centroid of the area is located at the point of intersection of the two axes of symmetry.*

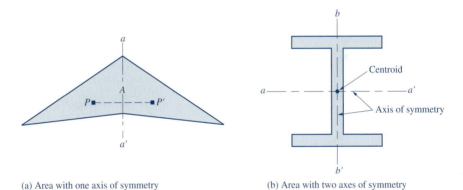

(a) Area with one axis of symmetry   (b) Area with two axes of symmetry

FIGURE 7–4

Centroid of Simple Areas. Formulas for the centroids of several simple geometric shapes, such as the triangle, semicircle, circular sector, parabolic area, and so on, have been determined mathematically and are listed in Table 7–2.

Centroid of Composite Areas. An area that can be divided into several simple geometric shapes, such as those listed in Table 7–2, is called a *composite area.* The area in Fig. 7–5 is divided into a rectangle, two triangles, and a semicircle. These areas are denoted by A_1, A_2, A_3, and A_4. The coordinates of the centroid of each area are denoted by x's and y's, with the subscripts corresponding to the area they represent. From Equation 7–6, the coordinates $\bar{x}$ and $\bar{y}$ of the centroid C of the composite area can be determined from the following equations.

$$\bar{x} = \frac{\Sigma Ax}{\Sigma A} = \frac{A_1 x_1 + A_2 x_2 + A_3 x_3 + A_4 x_4}{A_1 + A_2 + A_3 + A_4}$$

$$\bar{y} = \frac{\Sigma Ay}{\Sigma A} = \frac{A_1 y_1 + A_2 y_2 + A_3 y_3 + A_4 y_4}{A_1 + A_2 + A_3 + A_4}$$

(7–7)

TABLE 7–2 Centroids of Areas of Common Shapes

Rectangle

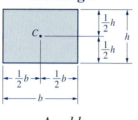

$$A = bh$$

Triangle

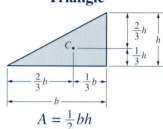

$$A = \frac{1}{2}bh$$

Circle

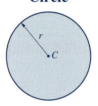

$$A = \pi r^2$$

Semicircle

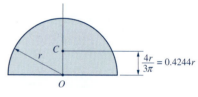

$$A = \frac{1}{2}\pi r^2$$

Quarter-Circle

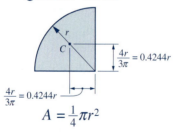

$$A = \frac{1}{4}\pi r^2$$

Sectors

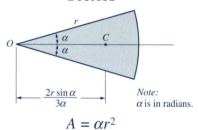

Note:
α is in radians.

$$A = \alpha r^2$$

Semiparabolic Area

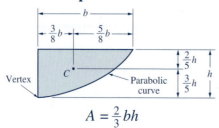

$$A = \frac{2}{3}bh$$

Parabolic Spandrel

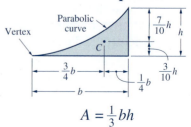

$$A = \frac{1}{3}bh$$

where A_1, A_2, A_3, and A_4 are the component areas, and (x_1, y_1), (x_2, y_2), (x_3, y_3), and (x_4, y_4) are the coordinates of the centroids of the component areas. The x and y axes are conveniently chosen so that the entire area is in the first quadrant and the coordinates of the centroids of all the areas have positive values. If there is an axis of symmetry, the centroid is located along this axis. The areas of holes, notches, and so on, should be treated as negative values because the areas in these parts are absent.

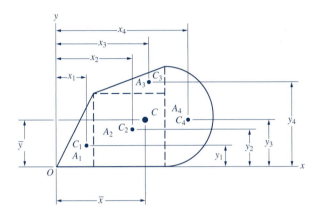

FIGURE 7–5

──────── **EXAMPLE 7–4** ────────

Locate the centroid of the shaded area shown in Fig. E7–4(1).

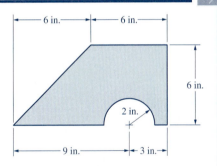

FIGURE E7–4(1)

Solution. The shaded area consists of three components: the triangle A_1, the square A_2, and the semicircular area A_3, as shown in Fig. E7–4(2). The values of the areas are

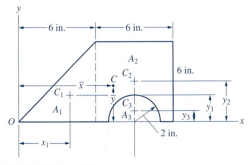

FIGURE 7–4(2)

$$A_1 = \frac{1}{2}(6 \text{ in.})(6 \text{ in.}) = 18 \text{ in.}^2$$

$$A_2 = (6 \text{ in.})(6 \text{ in.}) = 36 \text{ in.}^2$$

$$A_3 = -\frac{1}{2}\pi(2 \text{ in.})^2 = -6.3 \text{ in.}^2$$

$$A = \Sigma A = 47.7 \text{ in.}^2$$

Note that the area of the notch is considered negative. The coordinates of the centroid of each area referring to the x–y axes are

$$x_1 = \frac{2}{3}(6) = 4 \text{ in.} \qquad y_1 = \frac{1}{3}(6) = 2 \text{ in.}$$

$$x_2 = 9 \text{ in.} \qquad y_2 = 3 \text{ in.}$$

$$x_3 = 9 \text{ in.} \qquad y_3 = \frac{4r}{3\pi} = \frac{4(2)}{3\pi} = 0.849 \text{ in.}$$

The coordinates of the centroid of the entire area are given by

$$\bar{x} = \frac{\Sigma Ax}{\Sigma A} = \frac{A_1x_1 + A_2x_2 + A_3x_3}{A_1 + A_2 + A_3}$$

$$= \frac{(18)(4) + (36)(9) + (-6.3)(9)}{47.7}$$

$$= 7.11 \text{ in.} \qquad\qquad\qquad \Leftarrow \textbf{Ans.}$$

$$\bar{y} = \frac{\Sigma Ay}{\Sigma A} = \frac{A_1y_1 + A_2y_2 + A_3y_3}{A_1 + A_2 + A_3}$$

$$= \frac{(18)(2) + (36)(3) + (-6.3)(0.849)}{47.7}$$

$$= 2.91 \text{ in.} \qquad\qquad\qquad \Leftarrow \textbf{Ans.}$$

The preceding calculations can be tabulated as follows:

Part	A (in.^{2})	x (in.)	y (in.)	Ax (in.^{3})	Ay (in.^{3})
1	18.0	4.00	2.00	72.0	36.0
2	36.0	9.00	3.00	324.0	108.0
3	−6.3	9.00	0.849	−56.7	−5.3
Σ	47.7			339.3	138.7

Thus,

$$\bar{x} = \frac{\Sigma Ax}{\Sigma A} = \frac{339.3}{47.7} = 7.11 \text{ in.} \qquad \Leftarrow \textbf{Ans.}$$

$$\bar{y} = \frac{\Sigma Ay}{\Sigma A} = \frac{138.7}{47.7} = 2.91 \text{ in.} \qquad \Leftarrow \textbf{Ans.}$$

—————— **EXAMPLE 7–5** ——————

A homogeneous T-shaped plate of uniform thickness is attached to a cable at A, as shown in Fig. E7–5(1). Determine the angle θ that defines the equilibrium position of the plate.

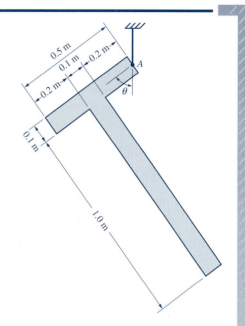

Solution. The center of gravity of the plate is determined first. Since the homogeneous plate is of uniform thickness, the center of gravity coincides with the centroid of the surface area. The coordinate axes are chosen as shown in Fig. E7–5(2). The area is symmetrical with respect to the x axis, so the centroid C is located along the x axis. Its location along the x axis can be computed from

FIGURE E7–5(1)

$$\bar{x} = \frac{A_1 x_1 + A_2 x_2}{A_1 + A_2}$$

$$= \frac{(0.1 \times 0.5)(0) + (0.1 \times 1.0)(0.55)}{0.1 \times 0.5 + 0.1 \times 1.0}$$

$$= 0.367 \text{ m}$$

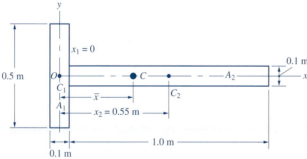

FIGURE E7–5(2)

The plate is subjected to two forces: the total weight W passing through C, and the tensile force T in the cable. The plate is thus a two-force member, and the two forces must act in the vertical direction through A, as shown in Fig. E7–5(3). From the right triangle ABC, we write

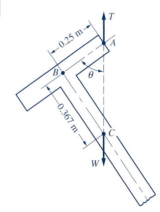

FIGURE E7–5(3)

$$\tan \theta = \frac{0.367}{0.25} = 1.467$$

From which we get

$$\theta = 55.7° \qquad\qquad \Leftarrow \textbf{Ans.}$$

7–5
GENERAL TYPES OF DISTRIBUTED LINE LOADS

A *distributive load* occurs whenever the load applied to a body is not con-centrated at a point. Two cases of distributed load, namely, the uniform load and the triangular load, have been discussed in Section 2–11. This sec-tion deals with the general case where the load intensity q is a function of x, a distance measured along the length of the beam, as represented in the loading diagram in Fig. 7–6a. The load over an incremental length Δx is $q\Delta x$, where q is the load intensity. The total load on the beam is the sum of loads over the entire length of the beam. Thus,

$$P = \Sigma q \Delta x$$

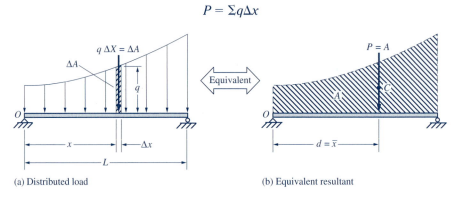

(a) Distributed load   (b) Equivalent resultant

FIGURE 7–6

Note that in Fig. 7–6a, the product $q\Delta x$ is the resultant of the distributed load over a short length Δx. It is physically equivalent to the area ΔA of the narrow strip. The total resultant over the entire length of the beam is

$$P = \Sigma q \Delta x = \Sigma \Delta A = A$$

Thus, *the resultant of a distributed line load is equal to the area of the load-ing diagram*, as shown in Fig. 7–6b.

The location of the resultant P along the beam may be determined by requiring that the moment of P about O be equal to the sum of the moments of the loads $q\Delta x$ over the entire beam about O. Let the distance from O to the load P be d. We write

$$Pd = \Sigma(q\Delta x)x = \Sigma(\Delta A)x = A\bar{x} = P\bar{x}$$

$$d = \bar{x}$$

Thus, a distributed line load can be replaced by a resultant having a magnitude equal to the area of the loading diagram and a line of action passing through the centroid of that area.

Note that a distributed load can be replaced by its resultant only for the determination of the reactions or the external effects of the load. For the determination of internal forces inside a member or the deflection of the member, the distributed load should *not* be replaced by its resultant.

───────── **EXAMPLE 7–6** ─────────

Refer to Fig. E7–6(1). Determine the reactions at the fixed support of the cantilever awning due to the distributed load produced by drifted snow as shown.

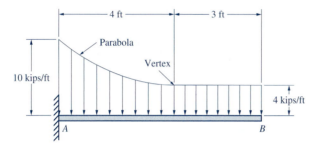

FIGURE E7–6(1)

Solution. The loading diagram is divided into a rectangular area and a parabolic spandrel, as shown in Fig. E7–6(2). The distributed loads are replaced by their resultants. Each resultant is equal to its associated areas and passes through the *centroid* of the associated area of the loading diagram. We have

$$F_1 = (4 \text{ kips/ft})(7 \text{ ft}) = 28 \text{ kips}$$

$$F_2 = \frac{1}{3}(6 \text{ kips/ft})(4 \text{ ft}) = 8 \text{ kips}$$

$$x_1 = \frac{1}{2}(7 \text{ ft}) = 3.5 \text{ ft}$$

$$x_2 = \frac{1}{4}(4 \text{ ft}) = 1 \text{ m}$$

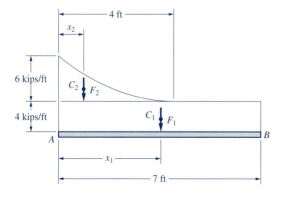

FIGURE E7–6(2)

In the free-body diagram of Fig. E7–6(3), the distributed loads are replaced by F_1 and F_2. The equilibrium equations give

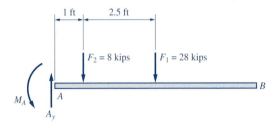

FIGURE E7–6(3)

$$\Sigma F_y = A_y - 28 \text{ kips} - 8 \text{ kips} = 0$$

$$A_y = +36 \text{ kips} \uparrow \qquad \Leftarrow \textbf{Ans.}$$

$$\Sigma M_A = M_A - (28 \text{ kips})(3.5 \text{ ft}) - (8 \text{ kips})(1 \text{ ft}) = 0$$

$$M_A = +106 \text{ kip} \cdot \text{ft} \circlearrowright \qquad \Leftarrow \textbf{Ans.}$$

7–6
LIQUID PRESSURE

The compressive force per unit area over which the force acts is called *pressure.* The liquid pressure on a solid surface is always perpendicular to the surface at every point. The pressure measured relative to the atmospheric pressure is called the *gage pressure.* A liquid is essentially incompressible and always possesses a free surface. For most practical purposes, the unit weight γ (weight per unit volume) can be considered constant throughout a liquid. At the free surface of a liquid exposed to the atmosphere, the gage pressure is zero (the same as the atmospheric pressure). Inside the liquid, the gage pressure at a point is proportional to the depth (the vertical distance) from the free surface. The gage pressure p at a point having a depth h in a liquid with unit weight γ is

$$p = \gamma h \qquad (7\text{–}8)$$

The units of pressure are lb/ft^2 (psf), lb/in.^2 (psi), or N/m^2 (pascal or Pa).
    Consider the rectangular plate AB shown in Fig. 7–7a. The pressures at A and B are

$$p_A = \gamma h_A \qquad p_B = \gamma h_B \qquad (7\text{–}9)$$

where h_A and h_B are the depths of points A and B, respectively. If the width of the plate is unity, then the pressure p becomes the same as the load intensity q in Section 7–5. Thus, the method used in Section 7–5 applies here also. Therefore, the resultant force P of the liquid pressure on the side of the plate is equal to the area under the pressure curve, and the resultant

P passes through the centroid of that area. Since the centroid of a trapezoid is not immediately known, it is convenient to divide the pressure diagram into two parts, as shown in Fig. 7–7b. Here, P_1 is the resultant of the uniform pressure γh_A over the length L; P_2 is the resultant of the linearly varying pressure from zero at A to $\gamma(h_B - h_A)$ at B.

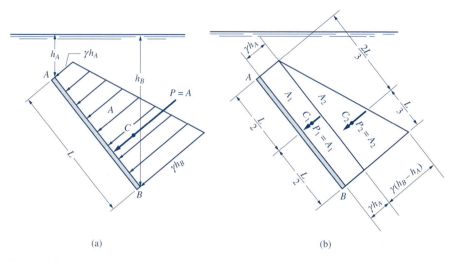

(a)   (b)

FIGURE 7–7

The method presented in this section may be used to determine the resultant force of the liquid pressure exerted on rectangular gates, vanes, and walls of containers, and on the plane surface of dams. Note that the method does not apply to planes of variable width or curved surfaces.

──────── **EXAMPLE 7–7** ──

The cross-section of a gravity dam is shown in Fig. E7–7(1). Consider a section of the dam 1 ft in length ($t = 1$ ft), and let the specific weight of concrete γ_c be 150 lb/ft³ and that of water γ_w be 62.4 lb/ft³. Determine (a) the resultant pressure force exerted by the water on side BC of the dam and (b) the resultant of the forces above the base AB.

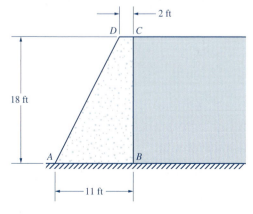

FIGURE E7–7(1)

Solution. The quantities calculated in the following are all based on a 1 ft length of the dam used as a free body.

(a) The Resultant Water Pressure.　　The side BC is vertical, so the water pressure acting on it is horizontal. Point B is 18 ft vertically below the free surface of water, where the gage pressure is

$$p_B = \gamma_w h_B = (62.4 \text{ lb/ft}^3)(18 \text{ ft}) = 1123 \text{ lb/ft}^2$$

The pressure at C is zero; hence, the loading diagram of water pressure is a triangle, as shown in Fig. E7–7(2). The resultant pressure force is

$$P = \frac{1}{2} p_B h_B t = \frac{1}{2} (1123 \text{ lb/ft}^2)(18 \text{ ft})(1 \text{ ft})$$

$$= 10\ 110 \text{ lb}$$

$$= 10.11 \text{ kips} \qquad\qquad \Leftarrow \textbf{Ans.}$$

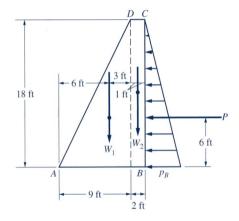

FIGURE E7–7(2)

which acts through the centroid of the triangular pressure diagram at

$$\frac{1}{3}(18 \text{ ft}) = 6 \text{ ft above } B \qquad\qquad \Leftarrow \textbf{Ans.}$$

(b) Resultant Force on the Base.　　The trapezoidal cross-section of the dam is divided into a triangle (W_1) and rectangle (W_2). The weights of the two parts are

$$W_1 = V_1 \gamma_c = \frac{1}{2}(9 \text{ ft})(18 \text{ ft})(1 \text{ ft})(150 \text{ lb/ft}^3)$$

$$= 12\ 150 \text{ lb} = 12.15 \text{ kips}$$

$$W_2 = V_2 \gamma_c = (2 \text{ ft})(18 \text{ ft})(1 \text{ ft})(150 \text{ lb/ft}^3)$$

$$= 5400 \text{ lb} = 5.40 \text{ kips}$$

The components of the resultant force are

$$R_x = \Sigma F_x = -10.11 \text{ kips}$$

$$R_y = \Sigma F_y = -12.15 \text{ kips} - 5.40 \text{ kips}$$

$$= -17.55 \text{ kips}$$

From these components, we find

$$\mathbf{R} = 20.3 \text{ kips } \quad \bigvee \quad 60.1° \qquad \qquad \Leftarrow \textbf{Ans.}$$

The resultant moment of the forces about A is

$$\Sigma M_A = (10.11 \text{ kips})(6 \text{ ft}) - (12.15 \text{ kips})(6 \text{ ft}) - (5.14 \text{ kips})(10 \text{ ft})$$
$$= -66.2 \text{ kip} \cdot \text{ft } \quad \circlearrowright$$

The single resultant force acting at point E in Fig. E7–7(3) along AB must produce the same moment about A. We write

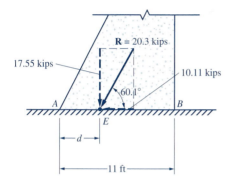

FIGURE E7–7(3)

$$(M_A \ \circlearrowright \ +): (17.55 \text{ kips}) \ d = 66.2 \text{ kip} \cdot \text{ft}$$

From which we get

$$d = 3.77 \text{ ft} \qquad \qquad \Leftarrow \textbf{Ans.}$$

For the gravity dam to be safe, the line of action of the resultant of all the forces acting on the dam above the base must pass through the middle third of the base. We see that this criterion is satisfied because

$$\frac{1}{3}(11 \text{ ft}) < 3.77 \text{ ft} < \frac{2}{3} (11 \text{ ft})$$

7–7
FLEXIBLE CABLES

Flexible cables are used in many engineering applications, such as suspension bridges, power transmission lines, guy wires for towers, telephone lines, etc. When analyzing such structural elements, cables are assumed to be *perfectly flexible* and *inextensible*. Because of its flexibility, a cable offers no resistance to bending. Therefore, the tensile force in a cable is either tangent to the cable or along the cable. The length of an inextensible cable remains constant before and after the loads are applied. Consequently, the geometry of the cable is fixed once the loads are applied, and the cable or segment of it can be treated as a rigid body, and equilibrium equations can be written as usual. A flexible cable may carry a series of vertical concentrated loads, or it may support loads uniformly distributed over its horizontal span. In these two cases, the weight of the cable is usually neglected. Sometimes a cable is subjected only to its own weight, which is uniformly distributed over its entire length.

This book deals only with cables subjected to a load uniformly distributed along the horizontal length of the cable, as shown in Fig. 7–8a. Cables of suspension bridges are approximately loaded this way. The uniform load per unit of horizontal length is denoted by q (in N/m or lb/ft) and the origin of the coordinate axes is located at the lowest point C of the cable.

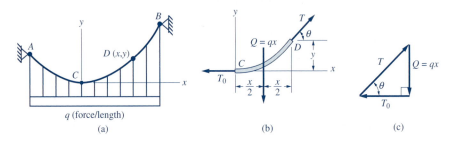

(a)   (b)   (c)

FIGURE 7–8

All flexible cables deflect vertically under load. *Sag* is defined as the difference in elevation between the lowest point of the cable and a support. When the supports are at different elevations, the sag measured from one support will be different from the sag measured from the other support. The *span* of a cable is the horizontal distance between the supports.

Derivation of Equations. The free-body diagram of a portion of the cable between the lowest point C and an arbitrary point D of the cable is shown in Fig. 7–8b. The forces acting on the free body are the horizontal tension T_0 at C; the tension T tangent to the cable at D; and the resultant of the uniform load $Q = qx$, which acts through the midpoint of the horizontal length between CD. The corresponding force triangle of the three forces is shown in Fig. 7–8c. From the force triangle, we obtain the following relations:

$$T = \sqrt{T_0^2 + q^2 x^2} \tag{7–10}$$

$$\tan \theta = \frac{qx}{T_0} \tag{7–11}$$

From Equation 7–10, we see that the term $q^2 x^2$ is always positive; hence, the minimum tension along the cable is T_0 at the lowest point C and the maximum tension occurs at the highest support having the greatest value of x.

The moment equation $\Sigma M_D = 0$ applied to the free body in Fig. 7–8b gives

$$\Sigma M_D = qx \left(\frac{x}{2} \right) - T_0 y = 0$$

From which we get

$$y = \frac{qx^2}{2T_0} \tag{7–12}$$

where $q/(2T_0)$ is a constant. This is the equation of a parabola with its vertex located at the lowest point C. Hence, the cable under a uniform horizontal load assumes the shape of a parabola and is called a *parabolic cable*.

Substituting the coordinates of support A or B in Equation 7–12 and solving for T_0, we obtain

$$T_0 = \frac{qx_A^2}{2y_A} = \frac{qx_B^2}{2y_B} \tag{7–13}$$

Supports at the Same Elevation. When the supports are at the same elevation, as shown in Fig. 7–9a, the sags from the two supports are the same. The expression for minimum tension T_0 in terms of span L, sag h, and the load q may be obtained by substituting the coordinates of support B $(L/2, h)$ into Equation 7–13:

$$T_0 = \frac{qL^2}{8h} \tag{7–14}$$

The magnitude and direction of the cable tension at any point may be computed from Equations 7–10 and 7–11, and the parabolic equation of the cable is defined by Equation 7–12.

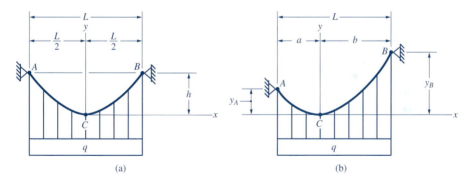

(a)   (b)

FIGURE 7–9

Supports at Different Elevations. Figure 7–9b shows the situation where the supports are at different elevations. In this case, the two sags are denoted by y_A and y_B from the two supports, A and B, respectively. Solving x from Equation 7–12, we get

$$x = \pm\sqrt{\frac{2T_0 y}{q}}$$

The absolute values of a and b in Fig. 7–9b are

$$a = \sqrt{\frac{2T_0 y_A}{q}} \qquad b = \sqrt{\frac{2T_0 y_B}{q}} \tag{7–15}$$

Since $a + b = L$, we must have

$$\sqrt{\frac{2T_0}{q}} \left(\sqrt{y_A} + \sqrt{y_B} \right) = L$$

Solving for T_0, we get

$$T_0 = \frac{qL^2}{2 \left(\sqrt{y_A} + \sqrt{y_B} \right)^2} \tag{7-16}$$

From which the minimum tension T_0 may be calculated. Then a and b can be computed from Equation 7–15 and the cable tension at A and B may be computed from Equations 7–10 and 7–11.

Cable Length. The length s of the cable from the lowest point C to any point $D(x, y)$ on the cable can be determined from the following formula:

$$s = x \left[1 + \frac{2}{3} \left(\frac{y}{x} \right)^2 - \frac{2}{5} \left(\frac{y}{x} \right)^4 \right] \tag{7-17}$$

── **EXAMPLE 7–8** ──────────────────────────

The light cable in Fig. E7–8(1) supports a load of 100 lb/ft, uniformly distributed along the horizontal length, and is suspended from two fixed points, A and B. Determine (a) the minimum tension, (b) the maximum tension, and (c) the total length of the cable.

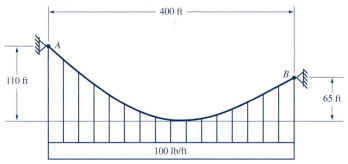

FIGURE E7–8(1)

Solution. Choose the x and y coordinates with the origin located at the lowest point, as shown in Fig. E7–8(2). The load is $q = 100$ lb/ft $= 0.1$ kip/ft.

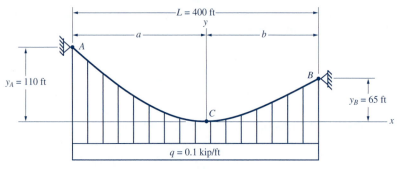

FIGURE E7–8(2)

(a) The Minimum Tension.   The minimum tension T_0 can be computed from Equation 7–16, from which we get

$$T_0 = \frac{qL^2}{2\left(\sqrt{y_A} + \sqrt{y_B}\right)^2}$$

$$= \frac{(0.1 \text{ kip/ft})(400 \text{ ft})^2}{2(\sqrt{110 \text{ ft}} + \sqrt{65 \text{ ft}})^2} = 23.25 \text{ kips} \qquad \Leftarrow \textbf{Ans.}$$

(b) The Maximum Tension.   The maximum tension in the cable occurs at the highest support A. The absolute value of a and b can be computed from Equation 7–15 as

$$a = \sqrt{\frac{2T_0 y_A}{q}} = \sqrt{\frac{2(23.25)(110)}{0.1}} = 226 \text{ ft}$$

$$b = \sqrt{\frac{2T_0 y_B}{q}} = \sqrt{\frac{2(23.25)(65)}{0.1}} = 174 \text{ ft}$$

Check.

$$a + b = 400 \text{ ft} \qquad \qquad \text{(Checks)}$$

Substituting $x_A = -a = -226$ ft into Equation 7–10, we get

$$T_{max} = \sqrt{T_0^{\,2} + q^2 x_A^2}$$
$$= \sqrt{(23.3)^2 + (0.1)^2 \, (-226)^2}$$
$$= 32.4 \text{ kips} \qquad \Leftarrow \textbf{Ans.}$$

(c) Cable Length.   The lengths of parts AC and BC of the cable must be computed separately. From Equation 7–17, we get

$$S_{AC} = x_A \left[1 + \frac{2}{3}\left(\frac{y_A}{x_A}\right)^2 - \frac{2}{5}\left(\frac{y_A}{x_A}\right)^4 \right]$$

$$= (226 \text{ ft})\left[1 + \frac{2}{3}\left(\frac{110}{226}\right)^2 - \frac{2}{5}\left(\frac{110}{226}\right)^4 \right]$$

$$= (226 \text{ ft})[1 + 0.158 - 0.022]$$

$$= 257 \text{ ft}$$

$$S_{BC} = x_B \left[1 + \frac{2}{3}\left(\frac{y_B}{x_B}\right)^2 - \frac{2}{5}\left(\frac{y_B}{x_B}\right)^4 \right]$$

$$= (174 \text{ ft})\left[1 + \frac{2}{3}\left(\frac{65 \text{ ft}}{174 \text{ ft}}\right)^2 - \frac{2}{5}\left(\frac{65 \text{ ft}}{174 \text{ ft}}\right)^4 \right]$$

$$= (174 \text{ ft})[1 + 0.093 - 0.008]$$

$$= 189 \text{ ft}$$

Thus, the total length of the cable is

$$S_{AB} = S_{AC} + S_{BC} = 446 \text{ ft} \qquad \Leftarrow \textbf{Ans.}$$

7–8
SUMMARY

Center of Gravity of a Body. The *center of gravity G* of a body is the point of application of the total weight of the body. The coordinates $(\bar{x}, \bar{y}, \bar{z})$ of G may be computed from the following equations:

$$\bar{x} = \frac{\Sigma x \Delta W}{\Sigma \Delta W} \qquad \bar{y} = \frac{\Sigma y \Delta W}{\Sigma \Delta W} \qquad \bar{z} = \frac{\Sigma z \Delta W}{\Sigma \Delta W} \qquad (7-1)$$

Centroid of a Volume. The *centroid C* of a volume is the geometric center of the volume. The coordinates $(\bar{x}, \bar{y}, \bar{z})$ of C may be computed from the following equations:

$$\bar{x} = \frac{\Sigma x \Delta V}{\Sigma \Delta V} \qquad \bar{y} = \frac{\Sigma y \Delta V}{\Sigma \Delta V} \qquad \bar{z} = \frac{\Sigma z \Delta V}{\Sigma \Delta V} \qquad (7-2)$$

The center of gravity of a body of homogeneous material is the same as the centroid of the body.

Centroid of Composite Volumes. A volume that can be divided into several components of known properties is called a *composite volume*. The coordinates $(\bar{x}, \bar{y}, \bar{z})$ of the centroid of a composite volume can be computed from the following equations:

$$\bar{x} = \frac{\Sigma V x}{\Sigma V} = \frac{V_1 x_1 + V_2 x_2 + V_3 x_3 + \cdots}{V_1 + V_2 + V_3 + \cdots}$$

$$\bar{y} = \frac{\Sigma V y}{\Sigma V} = \frac{V_1 y_1 + V_2 y_2 + V_3 y_3 + \cdots}{V_1 + V_2 + V_3 + \cdots}$$

$$\bar{z} = \frac{\Sigma V z}{\Sigma V} = \frac{V_1 z_1 + V_2 z_2 + V_3 z_3 + \cdots}{V_1 + V_2 + V_3 + \cdots}$$

where V_1, V_2, V_3, etc., are the component volumes, and (x_1, y_1), (x_2, y_2), (x_3, y_3), etc., are the coordinates of the centroids of the component volumes.

Experimental Determination of the Center of Gravity. When a body does not have well-defined boundaries, its center of gravity can be determined experimentally. For example, the location of the center of gravity of an automobile can be determined by weighing its axle loads in two different inclinations. The coordinates $(\bar{x}, \bar{y})$ of the center of gravity of the car can be computed from the following equations:

$$\bar{x} = \frac{W_B}{W} b \qquad (7-4)$$

$$\bar{y} = \frac{b}{W \tan \theta} (W_B - W_B') \qquad (7-5)$$

where b = the wheel base

W = the total weight of the car

W_B = the front axle weight measured when the wheel base is horizontal

W_B' = the front axle weight measured when the front wheel is raised so that the wheel base is at an angle θ from the horizontal

Centroid of an Area. The coordinates $(\bar{x}, \bar{y}, \bar{z})$ of the centroid of an area may be computed from the following equations:

$$\bar{x} = \frac{\Sigma x \Delta A}{\Sigma \Delta A} \qquad \bar{y} = \frac{\Sigma y \Delta A}{\Sigma \Delta A} \tag{7–6}$$

Centroid of Composite Areas. An area that can be divided into several components of known properties is called a *composite area*. The coordinates $(\bar{x}, \bar{y})$ of the centroid of a composite area may be computed from the following equations:

$$\bar{x} = \frac{\Sigma A x}{\Sigma A} = \frac{A_1 x_1 + A_2 x_2 + A_3 x_3 + \cdots}{A_1 + A_2 + A_3 + \cdots} \tag{7–7}$$

$$\bar{y} = \frac{\Sigma A y}{\Sigma A} = \frac{A_1 y_1 + A_2 y_2 + A_3 y_3 + \cdots}{A_1 + A_2 + A_3 + \cdots}$$

where A_1, A_2, A_3, etc., are the component areas, and $(x_1, y_1), (x_2, y_2), (x_3, y_3)$, etc., are the coordinates of the centroids of the component areas.

Distributed Line Load. In the general case of a distributed line load, the intensity q is a function of a distance x measured along the length of the beam. The loading diagram is a plot of q versus x. *A distributed line load can be replaced by its resultant having a magnitude equal to the area of the loading diagram and a line of action passing through the centroid of that area.*

Liquid Pressure. The gage pressure at a point having a depth h in a liquid with unit weight γ is

$$p = \gamma h \tag{7–8}$$

The loading diagram of a submerged rectangular area is a trapezoid. This loading diagram can be treated as the sum of a uniform load and a triangular load.

Flexible Cables. A cable that offers no resistance to bending is called a *flexible cable.* The cable is also assumed to be inextensible. A flexible cable subjected to a load uniformly distributed along its horizontal length is parabolic in shape. The tensile force in a cable is tangent to the cable. The magnitude and direction of the tensile force are

$$T = \sqrt{T_0^2 + q^2 x^2} \tag{7–10}$$

$$\tan \theta = \frac{qx}{T_0} \tag{7–11}$$

where q = the intensity of the uniform load per unit of horizontal length

T_0 = the minimum tension at the lowest point of the cable

The equation of the curve is

$$y = \frac{qx^2}{2T_0} \qquad (7\text{-}12)$$

From which we get

$$T_0 = \frac{qx_A^2}{2y_A} = \frac{qx_B^2}{2y_B} \qquad (7\text{-}13)$$

If the cable has equal sag h from the supports, the minimum tension T_0 is

$$T_0 = \frac{qL^2}{8h} \qquad (7\text{-}14)$$

If the cable has unequal sag, y_A and y_B, from the supports, the minimum tension is

$$T_0 = \frac{qL^2}{2(\sqrt{y_A} + \sqrt{y_B})^2} \qquad (7\text{-}16)$$

The maximum tension occurs at the support with greater sag, where the horizontal distance is a maximum from the lowest point.

The length s of the cable from the lowest point C to any point D (x, y) on the cable can be computed from:

$$s = x\left[1 + \frac{2}{3}\left(\frac{y}{x}\right)^2 - \frac{2}{5}\left(\frac{y}{x}\right)^4\right] \qquad (7\text{-}17)$$

PROBLEMS

Section 7–2   Center of Gravity and Centroid of a Body

7–1   Locate the centroid of a body consisting of the cylinder and circular disc shown in Fig. P7–1.

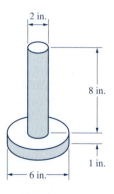

2 in.

8 in.

1 in.

6 in.

FIGURE P7–1

7-2 Locate the centroid of a body consisting of the cylinder and circular cone shown in Fig. P7–2.

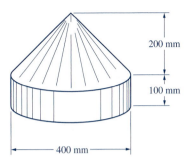

FIGURE P7-2

7-3 The machine member in Fig. P7–3 consists of a steel sphere and an aluminum rod. Locate the center of gravity of the member given $\gamma_{st} = 77$ kN/m³, and $\gamma_{al} = 27$ kN/m³.

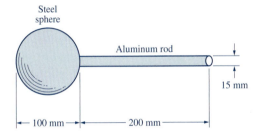

FIGURE P7-3

7-4 The mallet in Fig. P7–4 has a head made of steel and a cylindrical handle made of wood. The handle fits tightly into a hole over the entire width of the head. Locate the center of gravity of the mallet given $\gamma_{st} = 490$ lb/ft³, and $\gamma_{wd} = 40$ lb/ft³.

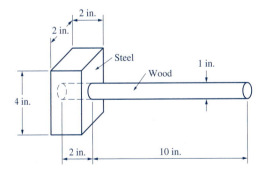

FIGURE P7-4

7-5 Refer to Fig. P7–5. Determine the centroid of the cylinder, which has a hole bored into one end as shown.

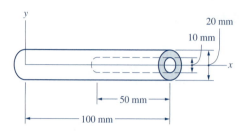

FIGURE P7-5

7-6   The homogeneous block in Fig. P7–6 has a uniform thickness of 1 in. The block is made of a plastic that has a specific weight of 55 lb/ft³. Determine the weight of the block and the location of its center of gravity.

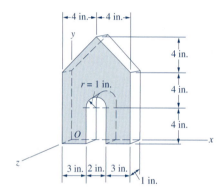

FIGURE P7-6

7-7   Determine the location of the center of gravity of the cast-aluminum block shown in Fig. P7–7.

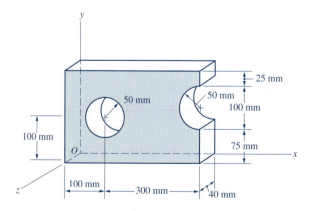

FIGURE P7-7

Section 7–3   Experimental Determination of the Center of Gravity

7–8   When a truck is placed in the horizontal position, its front and rear axle weights are measured to be 2040 lb and 1980 lb, respectively. When the front end is raised so that the truck is inclined at a 30° angle from the horizontal, its front axle weight is 1560 lb. Locate the center of gravity of the truck with respect to the rear axle if its wheel base is 12 ft.

7–9   Assume the car in Fig. 7–3 weighs 13.3 kN and has a wheel base of 3 m. When the car is placed in the horizontal position, the front axle weight is 6.67 N. When the front end is raised so that the car is inclined at a 30° angle from the horizontal, its front axle weighs 5.34 kN. Locate the center of gravity of the car with respect to the rear axle.

7–10   When the connecting rod in the reciprocal engine of Fig. P7–10 is weighed, the spring scales read 3 lb and 7 lb, respectively. Determine the location of its center of gravity along *AB*.

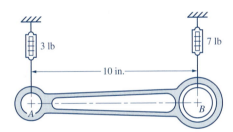

FIGURE P7–10

7–11   The machine bed in Fig. P7–11 weighs 25 kN. When its base *AB* is in the horizontal position, the scale at *B* reads 17.5 kN. When the right side is raised so that $\theta = 20°$, the scale reads 15 kN. Determine the location of its center of gravity.

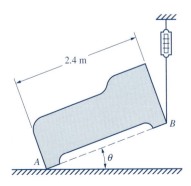

FIGURE P7–11

Section 7-4   Centroid of an Area

7-12 to **7-21** Locate the centroid of each plane area shown in Figs. P7-12 to P7-21.

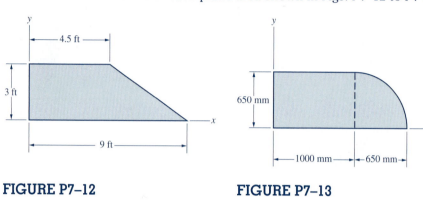

FIGURE P7-12   **FIGURE P7-13**

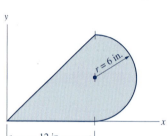

FIGURE P7-14   **FIGURE P7-15**

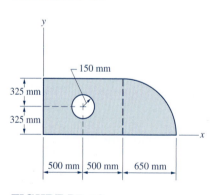
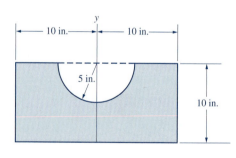

FIGURE P7-16   **FIGURE P7-17**

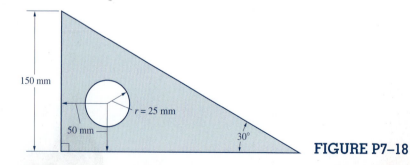

FIGURE P7-18

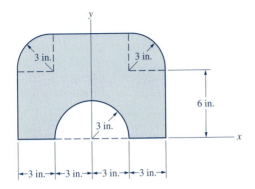

FIGURE P7–19

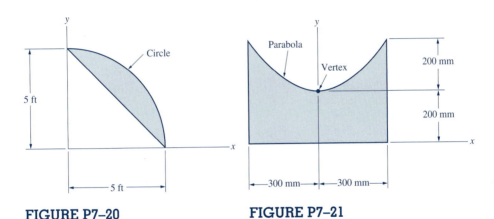

FIGURE P7–20   **FIGURE P7–21**

7–22 The homogeneous disk in Fig. P7–22 has a uniform thickness and weighs 30 lb. Determine the force P required to maintain the equilibrium of the disk in the position shown.

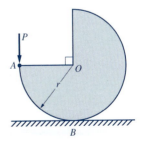

FIGURE P7–22

7–23 The semicircular disk in Fig. P7–23 is homogeneous and of uniform thickness. The mass of the disk is 10 kg. Determine (a) the force P required to maintain the equilibrium of the disk in the position shown, and (b) the minimum coefficient of static friction required to prevent the disk from slipping.

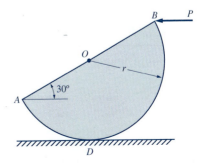

FIGURE P7–23

7–24   The triangular plate in Fig. P7–24 measures 3 ft, 4 ft, and 5 ft on the sides, and is attached to a support at corner *A* by a frictionless pin, as shown. Determine the angle θ that defines the equilibrium position of the plate.

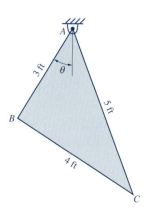

FIGURE P7–24

Section 7–5   General Types of Distributed Line Loads

7–25 to 7–28   Refer to Figs. P7–25 to P7–28. Determine the reactions at the supports of the beams for the loading shown in each figure.

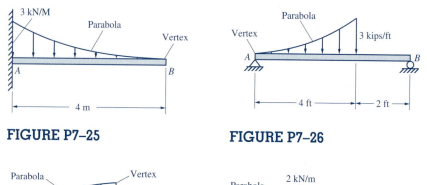

FIGURE P7–25   **FIGURE P7–26**

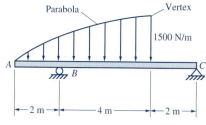

FIGURE P7–27   **FIGURE P7–28**

Section 7–6   Liquid Pressure

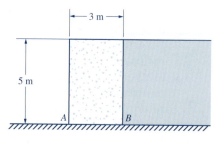

7–29   Consider a 1-m section of the gravity dam shown in Fig. P7–29. Determine the resultant of the water pressure and of the weight of the concrete dam, and the point to the right of *A* where the line of action of the resultant passes given $\gamma_w = 9.80$ kN/m³ and $\gamma_c = 23.6$ kN/m³. Is the dam safe?

FIGURE P7–29

7-30 Consider a 1-ft section of the gravity dam shown in Fig. P7–30. Determine the single resultant force of the water pressure and the weight of the dam given $\gamma_w = 62.4$ lb/ft^3 and $\gamma_c = 150$ lb/ft^3. Is the dam safe?

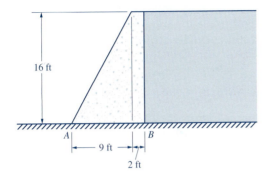

FIGURE P7–30

7-31 A gate, 2 m × 3 m, is placed in a wall below the water level, as shown in Fig. P7–31. The gate is hinged at A and supported by a roller at B. Determine the reaction at B given $\gamma_w = 9.80$ kN/m^3.

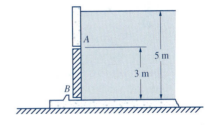

FIGURE P7–31

7–32 The automatic valve in Fig. P7–32 consists of a rectangular plate 3 ft × 4 ft, which is pivoted about a horizontal axis through C located at 2.5 ft below A. Determine the maximum depth h of water in the reservoir for which the valve will not open.

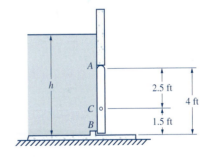

FIGURE P7–32

7–33 A freshwater container consists of an L-shaped gate ABC, as shown in Fig. P7–33. The gate is hinged at B. Determine the maximum depth of water h in the tank for which the gate will not open.

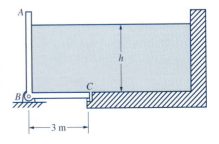

FIGURE P7–33

Section 7–7 Flexible Cables

7–34 A light cable supports a weight of 1 kN/m of horizontal length and is suspended between two supports, A and B, on the same level 200 m apart, as shown in Fig. P7–34. If the sag is 40 m, determine (*a*) the minimum tension at the midspan, (*b*) the magnitude and direction of the maximum tension at the supports, and (*c*) the total curve length of the cable.

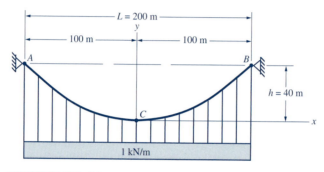

FIGURE P7–34

7–35 Refer to Fig. P7–35. The center span of a suspension bridge consists of a uniform roadway weighing 8 kips/ft suspended from two cables. The span between the two towers of equal height is 950 ft, and the sag of the cables at the midspan is 90 ft. Determine the minimum and maximum tensions in the cables and the length of each cable between the towers.

FIGURE P7–35

7–36 The cable shown in Fig. P7–36 supports a weight of 3 kN/m of horizontal length. The tangent to the cable at A is horizontal. Determine the maximum tension in the cable.

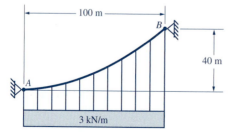

FIGURE P7–36

7–37 Cables AB and CD in Fig. P7–37 are both connected to the tower shown. The load is 2 kips/ft of horizontal length. The tangents to the cables at A and D are horizontal. Determine the resultant force exerted by the cables on the tower.

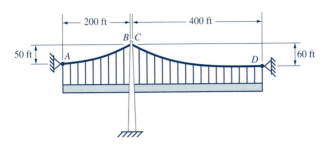

FIGURE P7–37

7–38 The ship in Fig. P7–38 is anchored in position with a cable attached to the stern at A, 10 ft above the water surface. The tangent to the cable at A is horizontal. The other end of the cable is tied to a fixed point B, which is 30 ft above the water surface. A tension of 100 kips is required in the cable at A. The weight of the cable is 40 lb/ft of cable. Determine (a) the horizontal distance X between A and B, and (b) the length of the cable. Assume that the weight of the cable is uniformly distributed along the horizontal length.

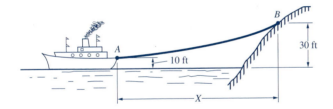

FIGURE P7–38

7–39 The portion of the pipeline above a river is suspended from a cable, as shown in Fig. P7–39. If the weight of the pipeline is 4 kN/m of pipe, determine (a) the maximum tension in the cable and (b) the length of the cable.

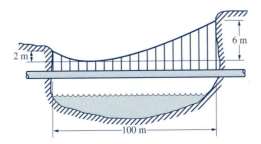

FIGURE P7–39

7-40   A load 500 N/m uniformly distributed along the horizontal length is supported by cable AB, as shown in Fig. P7-40. Determine the maximum and
minimum tensions in the cable.

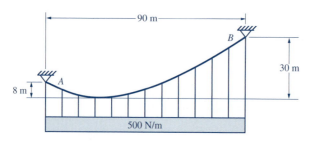

FIGURE P7-40

7-41   Refer to Fig. P7-41. A 20-lb homogeneous rod BC of uniform cross-section is
pinned to the wall at C and supported at B by a flexible cable AB, which
weighs 4 lb. The two ends of the cable are on the same level. Determine the
sag h of the cable at the midspan. Assume that the weight of the cable is
uniformly distributed along its horizontal length. (*Hint:* The tension in the
cable at B has a horizontal component equal to T_0 and a vertical component
equal to one-half of the cable weight.)

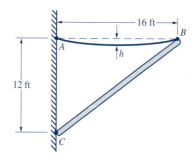

FIGURE P7-41

Computer Program Assignments

For each of the following problems, write a computer program using an appropriate programming language with which you are most familiar. Make the program user friendly by incorporating plenty of comments and input prompts so that the user will understand the input data to be entered and the limitations of their values. The output should include the data entered and the computed results, and they must be well labeled to identify each quantity. If a tabulated format is used, a proper heading must be included at the top of the table. Do not limit the program to any specific unit system. Indicate the consistent U.S. customary or SI units that can be used.

C7–1 Write a computer program that can be used to locate the centroid of a composite area. A proper rectangular coordinate system must be chosen first. The user input should include (1) the total number of component areas n, and (2) the area A_i and the coordinates x_i, y_i of the centroid of each area. Treat the area of a notch, hole, or cutoff as negative. Use this program to solve (*a*) Example 7–4, (*b*) Problem 7–16, and (*c*) Problem 7–19.

C7–2 Write a computer program that can be used to solve problems involving a light cable subjected to a load uniformly distributed along the horizontal length. The user input should include (1) the horizontal span L, (2) the intensity of the uniform load q, and (3) the vertical sags y_A and y_B of supports A and B from the lowest point of the cable. The output results must include (1) the minimum tension T_0, (2) the magnitude and direction of the maximum tension T_{max}, and (3) the total length of the cable. Use this program to solve (*a*) Example 7–8, (*b*) Problem 7–34, and (*c*) Problem 7–39.

CHAPTER 8

AREA MOMENTS OF INERTIA

8-1
INTRODUCTION

In addition to the centroid, the *area moment of inertia* is another property of an area used frequently in mechanics. For example, the moments of inertia of areas are used in calculating the stresses and deflections of beams, the torsion of shafts, and the buckling of columns. These topics will be discussed in later chapters in the second part of the book, the *strength of materials*.

Recall that the location of the centroid of an area involves the quantity $\Sigma x \Delta A$, which represents the *first moment* of the area. As you will see in the definition of area moment of inertia in the next section, it involves the quantity $\Sigma x^2 \Delta A$, which represents the *second moment* of an area (because x is squared). The moment of inertia is always computed with respect to an axis; its value is greatly affected by the distribution of the area relative to the axis.

The moment of inertia of an area arises in the mathematical derivation of many strengths of material formulas. Because of the frequent occurrence of area moments of inertia in later chapters, it is very important for us to learn the mathematical definition and the computations necessary for the determination of moments of inertia of a given area, although their physical meaning and the applications will take time for one to grasp fully.

8-2
MOMENTS OF INERTIA AND RADII OF GYRATION

Moments of Inertia. The moment of inertia of an area about an axis is the *second moment* of the area computed with respect to the axis. For the area in the x–y plane shown in Fig. 8–1, the moments of inertia of an area element ΔA about the x and y axes are $\Delta I_x = y^2 \Delta A$ and $\Delta I_y = x^2 \Delta A$, respectively. The *moments of inertia* for the entire area A with respect to the x and y axes are, respectively,

$$I_x = \Sigma y^2 \Delta A \qquad I_y = \Sigma x^2 \Delta A \qquad (8\text{–}1)$$

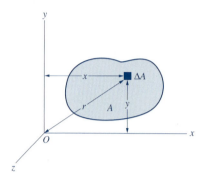

FIGURE 8–1

Where the summations are performed over the entire area, the value of the moment of inertia is always positive, regardless of the location of the axis. The units of moment of inertia are length raised to the fourth power, such as in.^{4} or m^4.

Polar Moment of Inertia. In a similar way, we can define the second moment of the area A in Fig. 8–1 with respect to the *pole O* or the *z* axis. This is referred to as the *polar moment of inertia*. For the area element ΔA, we write $\Delta J_O = r^2 \Delta A$, where r is the distance from the pole (or the z axis) to the element ΔA. The *polar moment of inertia* for the entire area with respect to the pole O or the z axis is

$$J_O = \Sigma r^2 \Delta A \qquad (8\text{–}2)$$

Since $r^2 = x^2 + y^2$, we have

$$J_O = \Sigma r^2 \Delta A = \Sigma (x^2 + y^2) \Delta A = \Sigma x^2 \Delta A + \Sigma y^2 \Delta A$$

Therefore, we conclude that, in general,

$$J_O = I_x + I_y \qquad (8\text{–}3)$$

Note that the moments of inertia are evaluated with respect to the axes on the plane of the area. The polar moments of inertia are computed with respect to the pole or the axes perpendicular to the plane of the area.

Radius of Gyration. The *radius of gyration* of an area with respect to the x axis is the distance at which the entire area could be located from the x axis to produce the same moment of inertia of the area with respect to the x axis. Let the radius of gyration with respect to the x axis be denoted by r_x. Then, by definition,

$$I_x = A r_x^2 \qquad (8\text{–}4)$$

or

$$r_x = \sqrt{\frac{I_x}{A}} \qquad (8\text{–}5)$$

Similarly, we have

$$I_y = Ar_y^{2} \qquad r_y = \sqrt{\frac{I_y}{A}} \qquad (8\text{–}6)$$

Centroidal Axes. The coordinate axes with the origin located at the centroid of an area are called the *centroidal axes* of the area.

Properties of Areas of Common Shapes. Formulas for moments of inertia for areas of simple geometric shapes can be derived by using integration. The properties of areas of some common shapes are given in Table 8–1, p. 300. The centroidal axes passing through the centroid C of the area are labeled with $\bar{x}$ and $\bar{y}$, and the properties with respect to the centroidal axes are denoted by a bar above the symbol of the quantity, such as $\bar{I}_x, \bar{I}_y, \bar{J}, \bar{r}_x$, etc. The noncentroidal axes are labeled by x and y, and properties with respect to the noncentroidal axes are denoted by a symbol without a bar, such as I_x, I_y, r_x, etc.

8–3
PARALLEL-AXIS THEOREM

The moment of inertia of an area with respect to a noncentroidal axis may be expressed in terms of the moment of inertia with respect to the parallel centroidal axis. Consider the area shown in Fig. 8–2, where the $\bar{x}$ axis is a centroidal axis through the centroid C of the area, and the x axis is a parallel noncentroidal axis located at a perpendicular distance d from the centroidal $\bar{x}$ axis. The coordinate of an area element ΔA from the centroidal $\bar{x}$ axis is denoted by y. By definition, the moment of inertia of the area about the noncentroidal x axis is

$$I_x = \Sigma(y + d)^2 \Delta A = \Sigma(y^2 + 2dy + d^2)\Delta A$$

or

$$I_x = \Sigma y^2 \Delta A + 2d\Sigma y\Delta A + d^2\Sigma\Delta A$$

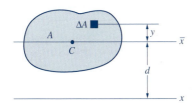

FIGURE 8–2

TABLE 8–1   Properties of Areas of Common Shapes

Rectangle	Triangle

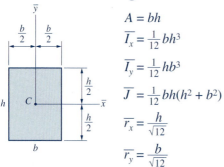

Rectangle

$$A = bh$$

$$\bar{I}_x = \tfrac{1}{12}bh^3$$

$$\bar{I}_y = \tfrac{1}{12}hb^3$$

$$\bar{J} = \tfrac{1}{12}bh(h^2 + b^2)$$

$$\bar{r}_x = \frac{h}{\sqrt{12}}$$

$$\bar{r}_y = \frac{b}{\sqrt{12}}$$

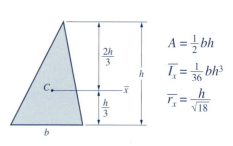

Triangle

$$A = \tfrac{1}{2}bh$$

$$\bar{I}_x = \tfrac{1}{36}bh^3$$

$$\bar{r}_x = \frac{h}{\sqrt{18}}$$

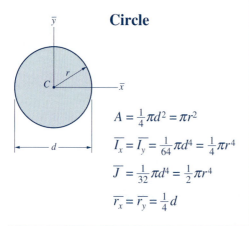

Circle

$$A = \tfrac{1}{4}\pi d^2 = \pi r^2$$

$$\bar{I}_x = \bar{I}_y = \tfrac{1}{64}\pi d^4 = \tfrac{1}{4}\pi r^4$$

$$\bar{J} = \tfrac{1}{32}\pi d^4 = \tfrac{1}{2}\pi r^4$$

$$\bar{r}_x = \bar{r}_y = \tfrac{1}{4}d$$

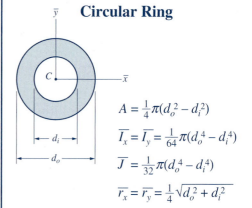

Circular Ring

$$A = \tfrac{1}{4}\pi(d_o^2 - d_i^2)$$

$$\bar{I}_x = \bar{I}_y = \tfrac{1}{64}\pi(d_o^4 - d_i^4)$$

$$\bar{J} = \tfrac{1}{32}\pi(d_o^4 - d_i^4)$$

$$\bar{r}_x = \bar{r}_y = \tfrac{1}{4}\sqrt{d_o^2 + d_i^2}$$

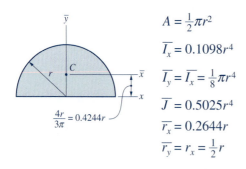

Semicircle

$$A = \tfrac{1}{2}\pi r^2$$

$$\bar{I}_x = 0.1098r^4$$

$$\bar{I}_y = \bar{I}_x = \tfrac{1}{8}\pi r^4$$

$$\bar{J} = 0.5025r^4$$

$$\bar{r}_x = 0.2644r$$

$$\bar{r}_y = r_x = \tfrac{1}{2}r$$

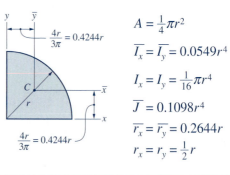

Quarter-Circle

$$A = \tfrac{1}{4}\pi r^2$$

$$\bar{I}_x = \bar{I}_y = 0.0549r^4$$

$$I_x = I_y = \tfrac{1}{16}\pi r^4$$

$$\bar{J} = 0.1098r^4$$

$$\bar{r}_x = \bar{r}_y = 0.2644r$$

$$r_x = r_y = \tfrac{1}{2}r$$

In the above equation, the first term represents moment of inertia $\bar{I}_x$ of the area about the centroidal $\bar{x}$ axis. The second term is zero since $\Sigma y \Delta A = A\bar{y}$, and $\bar{y}$ is zero with respect to the centroidal $\bar{x}$ axis. The third term is simply Ad^2. Thus, the expression for I_x becomes

$$I_x = \bar{I}_x + Ad^2 \tag{8-7}$$

This equation is the mathematical expression of the *parallel-axis theorem*. It states that the *moment of inertia of an area with respect to a noncentroidal axis is equal to the moment of inertia of the area with respect to the parallel centroidal axis plus the product of the area and the square of the perpendicular distance between the two axes.* Note that Equation 8–7 can be applied only to two *parallel* axes, and *one of the two axes must be a centroidal axis.*

─────── **EXAMPLE 8–1** ───

For the rectangular area shown in Fig. E8–1, calculate (*a*) the moments of inertia about the centroidal axes, (*b*) the centroidal polar moment of inertia, and (*c*) the moment of inertia of the area about the *x* axis.

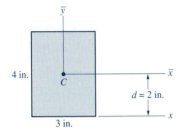

FIGURE E8–1

Solution.

(*a*) ***Centroidal Moments of Inertia.*** Using the formulas listed in Table 8–1, we get

$$\bar{I}_x = \frac{1}{12}bh^3 = \frac{1}{12}(3 \text{ in.})(4 \text{ in.})^3 = 16 \text{ in.}^4 \qquad \Leftarrow \textbf{Ans.}$$

$$\bar{I}_y = \frac{1}{12}hb^3 = \frac{1}{12}(4 \text{ in.})(3 \text{ in.})^3 = 9 \text{ in.}^4 \qquad \Leftarrow \textbf{Ans.}$$

(*b*) ***Centroidal Polar Moment of Inertia.*** From Equation 8–3, we get

$$\bar{J} = \bar{I}_x + \bar{I}_y = 16 \text{ in.}^4 + 9 \text{ in.}^4 = 25 \text{ in.}^4 \qquad \Leftarrow \textbf{Ans.}$$

(*c*) ***Moments of Inertia About the x Axis.*** Since the *x* axis is parallel to the centroidal $\bar{x}$ axis, we can use Equation 8–7 to transfer from the $\bar{x}$ axis to the *x* axis:

$$I_x = \bar{I}_x + Ad^2 = 16 \text{ in.}^4 + (12 \text{ in.}^4)(2 \text{ in.})^2$$

$$= 64 \text{ in.}^4 \qquad \Leftarrow \textbf{Ans.}$$

─────── **EXAMPLE 8–2** ───

For the semicircular area shown in Fig. E8–2, calculate its radius of gyration with respect to the *x* axis.

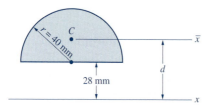

FIGURE E8–2

Solution. We must first find $\bar{I}_x$ with respect to the centroidal $\bar{x}$ axis, then calculate I_x with respect to the x axis by using the parallel-axis theorem. Finally the radius of gyration r_x can be determined by definition.

Centroidal Moment of Inertia. Using the formulas listed in Table 8–1, we get

$$\bar{I}_x = 0.1098r^4 = 0.1098(0.040 \text{ m})^4 = 2.81 \times 10^{-7} \text{ m}^4$$

Moment of Inertia About the x Axis. The distance d between the $\bar{x}$ axis and the x axis is

$$d = 0.028 \text{ m} + 0.4244(0.040 \text{ m}) = 0.0450 \text{ m}$$

The area of the semicircle is

$$A = \frac{1}{2}\pi(0.040 \text{ m})^2 = 0.00251 \text{ m}^2$$

Using the parallel-axis theorem, we get the moment of inertia of the area about the x axis:

$$I_x = \bar{I}_x + Ad^2$$
$$= 2.81 \times 10^{-7} \text{ m}^4 + (0.00251 \text{ m}^2)(0.0450 \text{ m})^2$$
$$= 5.37 \times 10^{-6} \text{ m}^4$$

Radius of Gyration About the x Axis. By definition, the radius of gyration about the x axis is

$$r_x = \sqrt{\frac{I_x}{A}} = \sqrt{\frac{5.37 \times 10^{-6} \text{ m}^4}{0.00251 \text{ m}^2}} = 0.0462 \text{ m}$$
$$= 46.2 \text{ mm} \qquad \qquad \Leftarrow \textbf{Ans.}$$

8-4
MOMENTS OF INERTIA OF COMPOSITE AREAS

It is frequently necessary to determine the moments of inertia of an area composed of several common geometric shapes, such as those listed in Table 8–1. Since a moment of inertia of an area with respect to an axis is the sum of the moments of inertia of the elements that comprise the area, it follows that *the moment of inertia of a composite area about an axis is the sum of the moments of inertia of the component parts about the same axis.* Before adding the moments of inertia, however, the parallel-axis theorem must be used to transfer the moment of inertia of each area to the desired axis. The moment of inertia of a composite area about the x axis may be computed from

$$I_x = \Sigma [I + Ay^2] \qquad (8–8)$$

where y is the distance from the centroid of a component area to the x axis.

Subsequent applications to the strength and deflection of beams require the determination of the moments of inertia of a composite beam section about its centroidal axes. Thus, the centroid of the section must be located first. The moment of inertia of the section about the centroidal $\bar{x}$ axis may be computed from

$$I_x = \Sigma [I + A(\bar{y} - y)^2] \qquad (8–9)$$

where y is the distance from the centroid of a component area to the reference x axis and $\bar{y}$ is the distance from the centroid of the entire section to the reference x axis.

When calculating the moment of inertia of an area that contains holes or notches, it is convenient to treat the values of their areas and their moments of inertia as negative, as illustrated in Example 8–4.

──────── **EXAMPLE 8–3** ────────

The area shown in Fig. E8–3(1) is a commonly used beam section. Determine the moment of inertia of the section with respect to the horizontal centroidal axis.

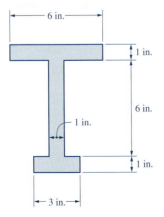

FIGURE E8–3(1)

Solution. The section is subdivided into three rectangular areas, designated by A_1, A_2, and A_3, as shown in Fig. E8–3(2). To locate the centroid of the section, a reference axis is chosen at the bottom of the section. We find

$$\bar{y} = \frac{A_1 y_1 + A_2 y_2 + A_3 y_3}{A_1 + A_2 + A_3}$$

$$= \frac{3(0.5) + 6(4) + 6(7.5)}{3 + 6 + 6}$$

$$= 4.70 \text{ in.}$$

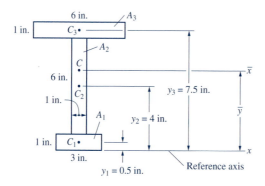

FIGURE E8–3(2)

The moment of inertia of each component area about its own horizontal centroidal axis is

$$I_1 = \frac{3(1)^3}{12} = 0.25$$

$$I_2 = \frac{1(6)^3}{12} = 18.0$$

$$I_3 = \frac{6(1)^3}{12} = 0.5$$

Now the moment of inertia of the section with respect to the centroidal $\bar{x}$ axis may be computed by using Equation 8–9:

$$\bar{I}_x = \Sigma[I + A(\bar{y} - y)^2]$$

$$= [0.25 + 3(4.70 - 0.5)^2] + [18.0 + 6(4.70 - 4.0)^2] + [0.5 + 6(4.70 - 7.5)^2]$$

From which we get

$$\bar{I}_x = 121.6 \text{ in.}^4 \qquad\qquad \Leftarrow \textbf{Ans.}$$

The above computations can be tabulated as follows.

(1) Part	(2) A (in.^{2})	(3) y (in.)	(4) Ay (in.^{3})	(5) $\bar{y} - y$ (in.)	(6) $A(\bar{y} - y)^2$ (in.^{4})	(7) I (in.^{4})
A_1	3.0	0.5	1.5	4.2	52.9	0.25
A_2	6.0	4.0	24.0	0.7	2.9	18.0
A_3	6.0	7.5	45.0	−2.8	47.0	0.5
Σ	15.0		70.5		102.8	18.8

Note that column (1) through column (4) of the table must be set up first. From the sums of column (2) and column (4), we find

$$\bar{y} = \frac{\Sigma Ay}{\Sigma A} = \frac{70.5}{15.0} = 4.70 \text{ in.}$$

Now column (6) and column (7) can be set up. From the sums of column (6) and column (7), we find

$$\bar{I}_x = \Sigma[I + A(\bar{y} - y)^2] = \Sigma I + \Sigma A(\bar{y} - y)^2$$

$$= 18.8 + 102.8 = 121.6 \text{ in.}^4$$

— **EXAMPLE 8–4** —

Determine the moment of inertia and the radius of gyration of the shaded area shown in Fig. E8–4(1) with respect to the x axis.

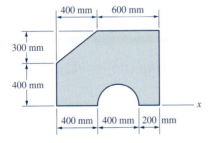

FIGURE E8–4(1)

Solution. The composite area can be divided into three components: the rectangular area A_1, the triangular area A_2, and the semicircular area A_3, as shown in Fig. E8–4(2). For each area, the distance from its own centroid to the x axis is indicated in the figure.

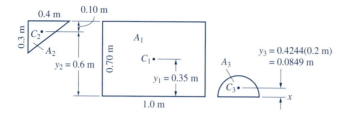

FIGURE E8–4(2)

Since the triangle and the semicircular notch are cutouts, both their areas and their moments of inertia are considered negative. The component areas are

$$A_1 = (1.00)(0.70) \quad = \quad 0.70 \ \text{m}^2$$

$$A_2 = -\frac{1}{2}(0.40)(0.30) = -0.60 \ \text{m}^2$$

$$A_3 = -\frac{1}{2}\pi(0.20)^2 \quad = -0.063 \ \text{m}^2$$

$$A = A_1 + A_2 + A_3 \quad = \quad 0.577 \ \text{m}^2$$

Using the formulas from Table 8–1, we can compute the moments of inertia of the component areas with respect to their own centroidal axes:

$$I_1 = \frac{bh^3}{12} = \frac{(1.0)(0.70)^3}{12} = 0.0286 \ \text{m}^4$$

$$I_2 = -\frac{bh^3}{36} = -\frac{(0.40)(0.30)^3}{36} = -0.0003 \ \text{m}^4$$

$$I_3 = -0.1098r^4 = -0.1098(0.20)^4 = -0.0002 \ \text{m}^4$$

Now the moment of inertia of the composite area with respect to the *x* axis may be computed by using Equation 8–8:

$$I_x = \Sigma[I + Ay^2]$$

$$= [0.0286 + (0.70)(0.35)^2] + [-0.0003 + (-0.06)(0.60)^2] + [-0.0002 + (-0.063)(0.0849)^2]$$

From which we get

$$I_x = 0.0918 \ \text{m}^4 \qquad\qquad \Leftarrow \textbf{Ans.}$$

By definition, the radius of gyration with respect to the *x* axis is

$$r_x = \sqrt{\frac{I_x}{A}} = \sqrt{\frac{0.0918}{0.577}} = 0.399 \ \text{m}$$

$$= 399 \ \text{mm} \qquad\qquad \Leftarrow \textbf{Ans.}$$

The computation may be tabulated as follows:

Part	A (m²)	y (m)	I (m⁴)	Ay^2 (m⁴)
A_1	0.70	0.35	0.0286	0.0858
A_2	−0.06	0.60	−0.0003	−0.0216
A_3	−0.063	0.0849	−0.0002	−0.0005
Σ	0.577		0.0281	0.0637

The moment of inertia of the composite area with respect to the reference x axis is

$$I_x = \Sigma\,[I + Ay^2] = \Sigma I + Ay^2$$

$$= 0.0281 + 0.0637 = 0.0918 \text{ m}^4$$

EXAMPLE 8–5

Refer to Fig. E8–5(1). For the cross-section of the timber beam made up of 1-in. thick planks, determine I_x with respect to the centroidal $\bar{x}$ axis.

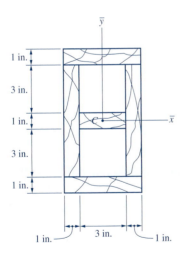

FIGURE E8–5(1)

Solution. Due to symmetry, the centroidal $\bar{x}$ and $\bar{y}$ axes are located at the middle of the section. The area is divided into three types designated by A_1, A_2, and A_3, as shown in Fig. E8–5(2).

The moments of inertia of the composite area with respect to the centroidal $\bar{x}$ axis may be computed by using Equation 8–8:

$$\overline{I}_x = 2[I_1 + A_1 y_1^{\,2}] + 2I_2 + I_3$$

$$= 2\left[\frac{5(1)^3}{12} + 5(4)^2\right] + 2\left[\frac{1(7)^3}{12}\right] + \frac{3(1)^3}{12}$$

$$= 218 \text{ in.}^4 \qquad\qquad \Leftarrow \textbf{Ans.}$$

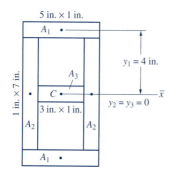

FIGURE E8–5(2)

Students are encouraged to find $\overline{I}_y$ for practice and to verify that $\overline{I}_y = 80.3$ in.^{4}

8–5
MOMENTS OF INERTIA OF BUILT-UP SECTIONS

Structural Steel Shapes. Structural steel is rolled into a wide variety of shapes and sizes. Structural steel sections are designated by letters that specify their shapes, followed by numbers that specify their sizes. Examples of this system follow:

1. A W21 × 83 is a wide-flange steel beam having a nominal depth of 21 in. and weighing 83 lb per linear foot.
2. An S20 × 75 is an American standard steel I-beam 20 in. deep and weighing 75 lb per linear foot.
3. A C12 × 30 is an American standard steel channel section 12 in. deep and weighing 30 lb per linear foot.
4. An L6 × 6 × $\frac{1}{2}$ is an equal-leg steel angle of legs 6 in. long and $\frac{1}{2}$ in. thick.

The American Institute of Steel Construction (AISC) manual provides detailed information for structural steel shapes. Selected structural steel shapes are listed in the tables in the appendix in both U.S. customary and SI units. These tables list dimensions, areas, locations of centroidal axes, centroidal moments of inertia, centroidal radii of gyration, etc., for selected structural steel shapes.

   Built-Up Structural Steel Sections.   A steel section may be composed of several steel shapes that are welded, riveted, or bolted together to form a single member. The centroidal moments of inertia are required in the analysis for strengths and deflections of beams. The radii of gyration are required for steel columns. Using the properties listed in the appendix tables, we can compute the moments of inertia of a built-up section in the same manner as that used for a composite area.

––––––––– **EXAMPLE 8–6** –––

Refer to Fig. E8–6(1). Determine the moment of inertia $\bar{I}_x$ with respect to the centroidal $\bar{x}$ axis for the built-up structural steel section. The W-shape is an SI designation.

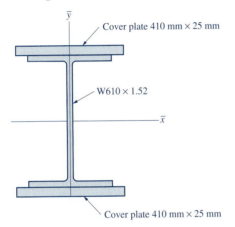

Cover plate 410 mm × 25 mm

W610 × 1.52

$\bar{x}$

Cover plate 410 mm × 25 mm

FIGURE E8–6(1)

   Solution. This section is symmetrical with respect to both the centroidal $\bar{x}$ and $\bar{y}$ axes. Therefore, there is no need for the computations required of centroidal axes. In Fig. E8–6(2), the wide-flange section is designated by A_1, and each cover plate is designated by A_2. From the appendix, Table A–1(b), the properties of a W610 × 1.52 section are:

$$d = 0.611 \text{ m} \qquad I_1 = 1.29 \times 10^{-3} \text{ m}^4$$

A_2 C_1

y_2

C

$d = 0.611$ m $\bar{x}$

A_1 y_2

A_2

C_2

FIGURE E8–6(2)

The moment of inertia of the built-up section with respect to the horizontal centroidal axis may be computed by using Equation 8–8:

$$\overline{I}_x = I_1 + 2[I_2 + A_2 y_2^2]$$

$$= 1.29 \times 10^{-3} + 2\left[\frac{0.410(0.025)^3}{12} + (0.410 \times 0.025)\left(\frac{0.611}{2} + \frac{0.025}{2}\right)^2\right]$$

$$= 3.36 \times 10^{-3} \ \text{m}^4 \qquad\qquad\qquad\qquad \Leftarrow \textbf{Ans.}$$

———— EXAMPLE 8–7 ————

Refer to Fig. E8–7(1). Determine the moment of inertia I_x of the built-up section shown with respect to the centroidal $\overline{x}$ axis.

C15 × 33.9

W18 × 50

FIGURE E8–7(1)

Solution. The C15 × 33.9 channel section is designated by A_1 in Fig. E8–7(2). The wide-flange section W18 × 50 is designated by A_2. For convenience, the reference x axis is taken at the top of the section. The distances below the x axis are treated as positive. From the appendix, Table A–3(a), the properties of a C15 × 33.9 channel section are:

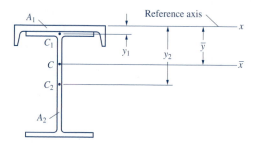

FIGURE E8–7(2)

$$A_1 = 9.96 \ \text{in.}^2$$

$$t_w = 0.400 \ \text{in.}$$

$$I_1 \ (I_{YY} \ \text{from the table}) = 8.13 \ \text{in.}^4$$

$$y_1 \ (\overline{x} \ \text{from the table}) = 0.787 \ \text{in.}$$

From the appendix, Table A–1(a), the properties of a W18 × 50 section are:

$$A_2 = 14.7 \text{ in.}^2$$

$$d = 17.99 \text{ in.}$$

$$I_2 = 800 \text{ in.}^4$$

$$y_2 = t_w \text{ (web thickness of the channel)} + \frac{d}{2}$$

$$= 0.400 \text{ in.} + \frac{(17.99 \text{ in.})}{2}$$

$$= 9.395 \text{ in.}$$

The centroid from the top of the section is located by

$$\bar{y} = \frac{A_1 y_1 + A_2 y_2}{A_1 + A_2}$$

$$= \frac{(9.96)(0.787) + (14.7)(9.395)}{9.96 + 14.7}$$

$$= 5.92 \text{ in.}$$

The moment of inertia of the built-up section about the horizontal centroidal axis can be computed by using Equation 8–9:

$$\bar{I}_x = [I_1 + A_1(\bar{y} - y_1)^2] + [I_2 + A_2(\bar{y} - y_1)^2]$$

$$= [8.13 + 9.96(5.92 - 0.787)^2] + [800 + 14.7(5.92 - 9.395)^2]$$

$$= 1248 \text{ in.}^4 \qquad\qquad \Leftarrow \textbf{Ans.}$$

The above computations may be tabulated as follows:

(1) Part	(2) A (in.²)	(3) y (in.)	(4) Ay (in.³)	(5) $\bar{y} - y$ (in.)	(6) $A(\bar{y} - y)^2$ (in.⁴)	(7) I (in.⁴)
A_1	9.96	0.787	7.84	5.13	262.4	8.13
A_2	14.7	9.395	138.1	−3.48	177.5	800
Σ	24.66		145.9		439.9	808.1

Note that column (1) through column (4) must be set up first. From the sums of column (2) and column (4), we find

$$\bar{y} = \frac{\Sigma Ay}{\Sigma A} = \frac{145.9}{24.66} = 5.92 \text{ in.}$$

Now columns (5), (6), and (7) can be set up. From the sums of columns (6) and (7), we find

$$\bar{I}_x = \Sigma[I + A(\bar{y} - y)^2] = \Sigma I + \Sigma A(\bar{y} - y)^2$$

$$= 808.1 + 439.9 = 1248 \text{ in.}^4$$

8–6
SUMMARY

Moment of Inertia. The area moment of inertia is a property of an area that represents the second moment of the area with respect to an axis. The moments of inertia for an area with respect to the x and y axes are, respectively,

$$I_x = \Sigma y^2 \Delta A \qquad I_y = \Sigma x^2 \Delta A \qquad\qquad (8\text{–}1)$$

Polar Moment of Inertia. The *polar moment of inertia* of an area with respect to pole O or the z axis is

$$J_O = \Sigma r^2 \Delta A \qquad\qquad (8\text{–}2)$$

Since $r^2 = x^2 + y^2$, we have

$$J_O = I_x + I_y \qquad\qquad (8\text{–}3)$$

Radius of Gyration. The *radius of gyration* of an area with respect to an axis is defined as

$$r_x = \sqrt{\frac{I_x}{A}} \qquad\qquad (8\text{–}5)$$

$$r_y = \sqrt{\frac{I_y}{A}} \qquad\qquad (8\text{–}6)$$

Formulas for the properties of areas of common geometric shapes are given in Table 8–1.

Parallel-Axis Theorem. The moment of inertia of an area with respect to a noncentroidal axis is equal to the moment of inertia of the area with respect to the parallel centroidal axis plus the product of the area and the square of the perpendicular distance between the two axes; that is

$$I_x = \overline{I}_x + Ad^2 \qquad\qquad (8\text{–}7)$$

Composite Area. The moment of inertia of a composite area composed of several common geometric shapes about an axis is the sum of the moments of inertia of the component parts about the same axis. The moment of inertia of a composite area about the x axis may be computed from

$$I_x = \Sigma[I + Ay^2] \qquad\qquad (8\text{–}8)$$

where y is the distance from the centroid of a component area to the x axis.

To compute the moments of inertia of a composite area about its cen-troidal axis, the centroid of the area must be located first. The moment of inertia of the section about the centroidal $\bar{x}$ axis may be computed from

$$\bar{I}_x = \Sigma[I + A(\bar{y} - y)^2] \qquad (8\text{--}9)$$

where y is the distance from the centroid of a component area to the refer-ence x axis and $\bar{y}$ is the distance from the centroid of the entire section to the reference x axis.

Built-Up Structural Steel Section. The centroidal moments of inertia of a structural steel section composed of several steel shapes can be com-puted in the same manner as a composite area.

PROBLEMS

Section 8–2　Moments of Inertia and Radii of Gyration

Section 8–3　Parallel-Axis Theorem

8–1　Refer to Fig. P8–1. Verify that the radii of gyration $\bar{r}_x$ and $\bar{r}_y$ of the rectangle shown with respect to its centroidal axes are $\bar{r}_x = h / \sqrt{12}$ and $\bar{r}_y = b / \sqrt{12}$.

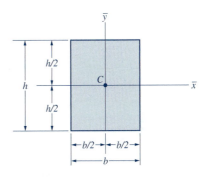

FIGURE P8–1

8–2　Verify that the radius of gyration for a circle of diameter d with respect to a centroidal axis is $\bar{r} = d / 4$.

8–3　Refer to Fig. P8–3. Determine the moment of inertia I_x and the radius of gyration r_x of the circular area about the x axis.

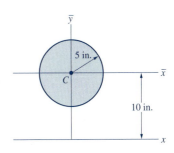

FIGURE P8–3

8–4 See Fig. P8–4. If the radius of gyration of the shaded area with respect to the y axis is $r_y = 12.4$ in., determine its centroidal moment of inertia $\bar{I}_y$ and the radius of gyration $\bar{r}_y$.

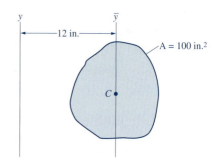

FIGURE P8–4

8–5 Determine the centroidal polar moment of inertia of a rectangle 100 mm wide by 200 mm high.

8–6 Refer to Fig. P8–6. If the moment of inertia I_x of the rectangular area about the x axis is 7320 in.^{4}, determine $I_{x'}$ of the area about the x' axis.

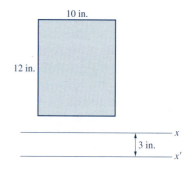

FIGURE P8–6

8–7 See Fig. P8–7. Determine the polar moment of inertia of the 18-in.-diameter circular area shown with respect to pole A on its circumference.

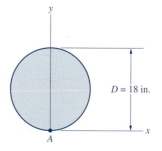

FIGURE P8–7

8–8 Refer to Fig. P8–8. Determine the radii of gyration r_x and r_y of the semicircular area shown with respect to the x and y axes, respectively.

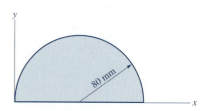

FIGURE P8–8

8–9 See Fig. P8–9. Determine the moments of inertia I_x and I_y and the radii of gyration r_x and r_y of the quarter-circle shown.

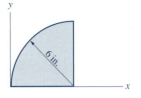

FIGURE P8–9

Section 8–4 Moments of Inertia of Composite Areas

8–10 to **8–17** For each composite area shown in Figs. P8–10 to P8–17, determine the moment of inertia of the area with respect to the horizontal centroidal axis.

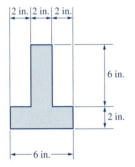

FIGURE P8–10

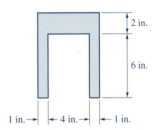

FIGURE P8–11

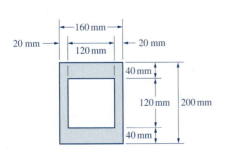

FIGURE P8–12

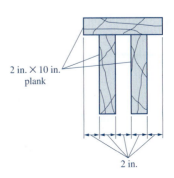

FIGURE P8–13

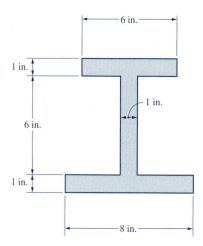

FIGURE P8–14

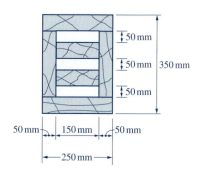

FIGURE P8–15

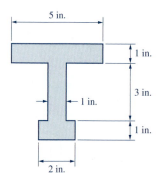

FIGURE P8–16

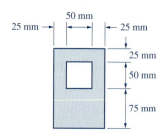

FIGURE P8–17

8–18   See Fig. P8–18. Determine the moment of inertia I_x and the radius of gyration r_x of the shaded area shown about the x axis.

8–19   Refer to Fig. P8–18. Determine the moment of inertia I_y and the radius of gyration r_y of the shaded area shown about the y axis.

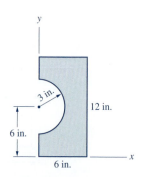

FIGURE P8–18

8–20 See Fig. P8–20. Determine the radius of gyration of the shaded area with respect to the x axis.

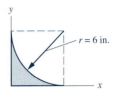

FIGURE P8–20

8–21 Refer to Fig. P8–21. Determine the polar moment of inertia of the shaded area with respect to the pole O.

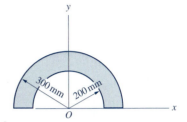

FIGURE P8–21

8–22 See Fig. P8–22. Determine the moment of inertia I_x of the shaded area about the x axis.

8–23 Refer to Fig. P8–22. Determine the moment of inertia I_y of the shaded area about the y axis.

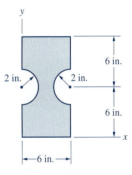

FIGURE P8–22

8–24 See Fig. P8–24. Determine the radius of gyration r_x of the shaded area about the x axis.

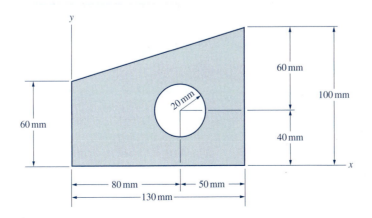

FIGURE P8–24

8–25   Refer to Fig. P8–24. Determine the moment of inertia I_y of the shaded area about the y axis.

Section 8–5   Moments of Inertia of Built-Up Sections

8–26 to **8–31** For each built-up section shown in Figs. P8–26 to P8–31, determine the moment of inertia and the radius of gyration of the section with respect to the horizontal centroidal axis.

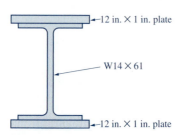

12 in. × 1 in. plate

W14 × 61

12 in. × 1 in. plate

FIGURE P8–26

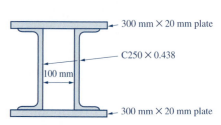

300 mm × 20 mm plate

C250 × 0.438

100 mm

300 mm × 20 mm plate

FIGURE P8–27

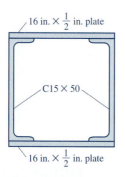

16 in. × $\frac{1}{2}$ in. plate

C15 × 50

16 in. × $\frac{1}{2}$ in. plate

FIGURE P8–28

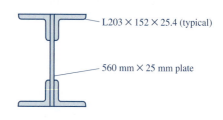

L203 × 152 × 25.4 (typical)

560 mm × 25 mm plate

FIGURE P8–29

C300 × 0.302

W410 × 0.53

FIGURE P8–30

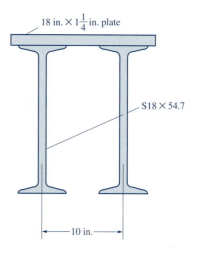

18 in. × $1\frac{1}{4}$ in. plate

S18 × 54.7

10 in.

FIGURE P8–31

8–32 See Fig. P8–32. Determine the radius of gyration of the built-up section shown with respect to the horizontal centroidal axis. The dashed line at the bottom of the section represents lacings for connecting the component parts at the open side. The areas of the lacings are not considered as effective areas when calculating the moment of inertia.

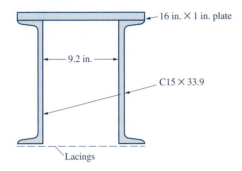

16 in. × 1 in. plate

9.2 in.

C15 × 33.9

Lacings

FIGURE P8–32

8–33 In Problem 8–32, determine the radius of gyration of the built-up section shown with respect to the vertical centroidal axis.

Computer Program Assignments

For each of the following problems, write a computer program using an appropriate programming language with which you are most familiar. Make the program user friendly by incorporating plenty of comments and input prompts so that the user will understand the input data to be entered and the limitations of their values. The output should include the data entered and the computed results, and they must be well labeled to identify each quantity. If a tabulated format is used, a proper heading must be included at the top of the table. Do not limit the program to any specific unit system. Indicate the consistent U.S. customary or SI units that can be used.

C8–1 Write a computer program that can be used to determine the moment of inertia I_x and the radius of gyration r_x of a composite area with respect to a given x axis (which may or may not be the centroidal axis). The user input should include (1) the total number of component areas n; and (2) the properties A_i, I_i, and y_i (the distance from the centroid of the component area to the x axis) of each component area. Treat the area and the moment of inertia of a notch, hole, or cutoff as negative. Use this program to solve (a) Example 8–4, (b) Problem 8–18, and (c) Problem 8–22.

C8–2 Modify the program in Problem C8–1 so that it can be used to determine the centroidal moment of inertia $\bar{I}_x$ and the centroidal radius of gyration $\bar{r}_x$ of a composite area with respect to its centroidal $\bar{x}$ axis. The user input should include (1) the total number of component areas n, and (2) the properties A_i, I_i, and y_i (the distance from the centroid of the component area to the reference axis) of each component area. Treat the area and the moment of inertia of a notch, hole, or cutoff as negative. Use this program to solve (a) Example 8–3, (b) Example 8–7, and (c) Problem 8–31.

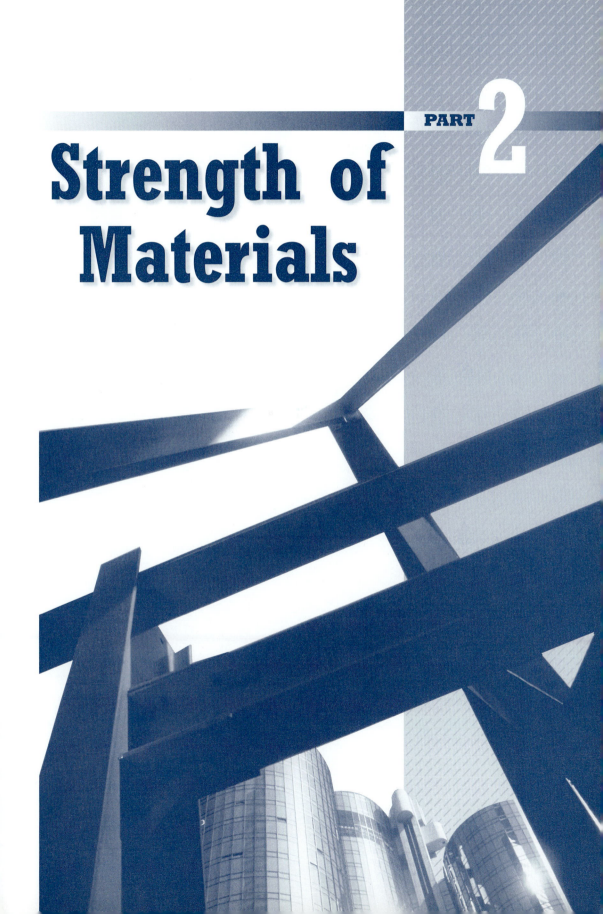

PART 2

Strength of
Materials

CHAPTER 9

SIMPLE STRESSES

9–1
INTRODUCTION TO STRENGTH OF MATERIALS

Statics was covered in the first eight chapters of this book. The remaining chapters are devoted to *strength of materials,* a subject dealing with the relationships between the external loads applied to an elastic body and the intensity of the internal forces acting within the body. The intensities of the internal resisting forces are called *stresses.* In the study of statics, all bodies are assumed to be rigid. In the study of strength of materials, however, bodies are considered deformable. Hence, the subject is also concerned with computing the *deformations* of the body. Deformation per unit length is called the *strain.*

The subject of strength of materials involves analytical methods for determining the *strength* (load-carrying capacity based on stresses inside a member), *stiffness* (deformation characteristics), and *stability* (the ability of a slender member to maintain its initial configuration without buckling while being subjected to compressive loading). The sizes of all structural or machine members must be properly designed according to the requirements for strength, stiffness, and/or stability. For example, the floor of a building must be strong enough to carry the design load; at the same time, it must be rigid enough not to deflect excessively under the load.

Strength of materials is one of the most fundamental subjects in the engineering and architectural curriculums. Comprehension of the topic is needed by structural engineers, mechanical engineers, and architects.

Keep in mind that, when relating internal resisting forces to external forces, the methods developed in statics still apply because the body or part of the body under consideration is only slightly deformed, and the small deformations have a negligible effect on equilibrium conditions. Therefore, free-body diagrams and application of the static equilibrium equations are essential to the determination of both the external reactions and the internal resisting forces in a body.

9–2
NORMAL AND SHEAR STRESSES

When a structural member is subjected to loads, internal resisting forces are generated within the member so that the external forces can be balanced and the body can hold itself together. The intensities of internal forces per unit area are called *stresses.* There are two types of stresses: normal stresses and shear stresses. *Normal stresses* are caused by internal forces normal (perpendicular) to the area in question, and *shear stresses* are caused by internal forces tangential (parallel) to the area under consideration.

Stresses are the most important concepts in the study of strength of materials. Whenever a body is subjected to external loads, stresses are induced within the body. Whether the material will fail and to what extent it will deform depend on the amount of stresses induced within the body.

In the U.S. customary system, the commonly used units of stresses are pounds per square inch (psi), pounds per square foot (psf), or kips (kilo-pound or 1000 lb) per square inch (ksi). The SI units of stresses are newtons per square meter (N/m^2), also designated pascal (Pa). When prefixes are used, the following SI units are frequently encountered:

$$1 \text{ kPa} = 10^3 \text{ Pa}$$

$$1 \text{ MPa} = 10^6 \text{ Pa}$$

$$1 \text{ GPa} = 10^9 \text{ Pa}$$

The conversion factors between stresses in U.S. customary units and in SI units are:

$$1 \text{ psi} = 6.895 \text{ kPa}$$

$$1 \text{ ksi} = 6.895 \text{ MPa}$$

$$1 \text{ psf} = 47.88 \text{ Pa}$$

9–3
DIRECT NORMAL STRESSES

A bar of uniform cross-section is called a *prismatic bar.* Figure 9–1a represents a prismatic bar subjected to a pair of equal and opposite pulling forces *P* acting along the axis of the rod. The forces applied along the axial direction of the member are called *axial loads.* A member subjected to axial loads is called an *axially loaded member.*

The pulling forces in Fig. 9–1a tend to elongate (stretch) the bar. The bar is said to be *in tension.* If the directions of the forces are reversed so that the bar is pushed, it tends to contract (shorten) and is said to be *in compression.* Any portion of the rod in equilibrium separated by imaginary transverse cutting planes must also be in equilibrium. Figure 9–1b shows the free-body diagrams of the two parts of the rod in Fig. 9–1a separated by the plane *m–m* perpendicular to the axis of the rod. In either one of the two free-body diagrams, the equilibrium condition requires that the internal

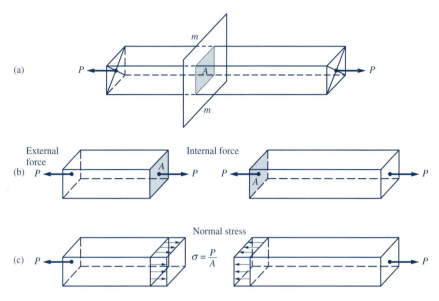

(a)

(b) External force

Internal force

(c) Normal stress

$$\sigma = \frac{P}{A}$$

FIGURE 9–1

force in the section be equal to the external force P. Since the internal force is normal to the section, the stress induced is the normal stress. The normal stresses due to axial loads through the centroids of the cross-sections are usually distributed uniformly over a cross-section. By definition, the uniform normal stress σ (the Greek lowercase letter sigma) in the section is

$$\sigma = \frac{P}{A} \tag{9–1}$$

where σ = the normal stress in the cross-section

P = the *internal* axial force at the section

A = the cross-sectional area of the rod

The normal stress caused by axial loads through the centroid of the section is called the *direct normal stress*. Equation 9–1 is referred to as the *direct normal stress formula*. There are two types of normal stress, namely, tensile stress and compressive stress. *Tensile stresses* are induced by tensile forces, while *compressive stresses* are induced by compressive forces.

Allowable Axial Load. Members are usually designed for a limited stress level called the *allowable stress* σ_{allow}, which is an upper limit of stress that must not be exceeded. A more comprehensive discussion of allowable stresses will be presented in Section 11–6. Using the allowable stress σ_{allow}, Equation 9–1 may be written in the following form for computing the allowable axial load P_{allow} that a member of cross-sectional area A can carry without being overstressed:

$$P_{allow} = \sigma_{allow} A \tag{9–2}$$

Required Area. The required minimum cross-sectional area A of a member designed to carry a maximum axial load P without exceeding the allowable stress σ_{allow} may be computed from

$$A = \frac{P}{\sigma_{\text{allow}}} \tag{9–3}$$

Equations 9–2 and 9–3 may be applied to a compression member only if the length of the member is relatively short compared to the lateral dimensions of the member. When a slender member is subjected to compressive axial loads, it tends to buckle before it fractures. Buckling of compression members will be presented in Chapter 19. For now, we will limit our discussion to short compression members that do not buckle.

Internal Axial Force Diagram. Variation of internal axial force along the length of a member can be depicted by an *internal axial force diagram* whose ordinate at any section of a member is equal to the value of the internal axial force at that section. When plotting an internal axial force diagram, we usually treat tensile force as positive and compressive force as negative. Example 9–2 illustrates the construction of an internal axial force diagram.

EXAMPLE 9–1

A compact car weighing 2500 lb is lifted by a crane, as shown in Fig. E9–1. The car is fastened to a steel cable of cross-sectional area $A = 0.25$ in.², which is attached to a steel hook. The upper part of the hook is a prismatic rod of uniform cross-sectional area $A' = 0.40$ in.². Neglecting the weights of the hook and the cable, calculate the normal stresses in the cable and the rod.

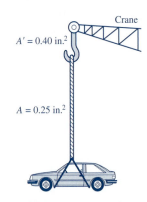

Crane
$A' = 0.40$ in.²
$A = 0.25$ in.²

FIGURE E9–1

Solution. The weight of the car exerts a 2500-lb tensile force on both the cable and the rod. Hence, the tensile stresses in the cable and rod are

$$\sigma_{\text{cable}} = \frac{W}{A} = \frac{2500 \text{ lb}}{0.25 \text{ in.}^2} = 10\ 000 \text{ psi (T)} \qquad \Leftarrow \textbf{Ans.}$$

$$\sigma_{\text{rod}} = \frac{W}{A'} = \frac{2500 \text{ lb}}{0.40 \text{ in.}^2} = 6250 \text{ psi (T)} \qquad \Leftarrow \textbf{Ans.}$$

EXAMPLE 9–2

A steel bar with 10-mm × 20 mm rectangular sections is subjected to four axial loads, as shown in Fig. E9–2(1). (a) Plot the internal axial force diagram of the rod, and (b) determine the normal stresses in segments *AB, BC,* and *CD.*

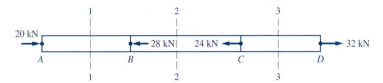

FIGURE E9–2(1)

Solution.

(a) The Internal Axial Force Diagram. Note that the algebraic sum of the four given axial forces is equal to zero; thus, the bar is in equilibrium. When the entire bar is in equilibrium, any segment of the bar must be in equilibrium. To determine the internal axial force in segment *AB*, pass section 1–1 through the segment. The free-body diagram of the part from *A* to the section is sketched in Fig. E9–2(2). The internal resisting force must be equal and opposite to the external force. Thus, the internal axial force in segment *AB* is

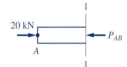

FIGURE E9–2(2)

$$P_{AB} = 20 \text{ kN (C)}$$

To determine the internal axial force in segment *BC,* pass section 2–2 through the segment. The free-body diagram of the part from A to the section is sketched in Fig. E9–2(3). The internal resisting force must be equal and opposite to the resultant external force. Thus, the internal axial force in segment *BC* is

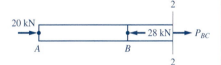

FIGURE E9–2(3)

$$P_{BC} = 8 \text{ kN (T)}$$

To determine the internal axial force in segment *CD,* pass section 3–3 through the segment. The free-body diagram of the part from the section to end *D* is sketched in Fig. E9–2(4). The internal resisting force must be equal and opposite to the external force. Thus, the internal axial force in segment *CD* is

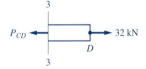

FIGURE E9–2(4)

$$P_{CD} = 32 \text{ kN (T)}$$

The internal axial force between section *A* and section *B* is constant; thus, the internal axial force diagram between the two sections is horizontal. Similarly, the internal axial force diagram between sections *B* and *C* and between sections *C* and *D* are also horizontal. The axial force diagram can be sketched as shown in Fig. E9–2(5). The axial forces that each segment is subjected to are also shown.

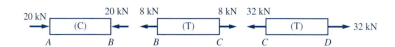

FIGURE E9–2(5)

(b) Normal Stresses in the Rod. The cross-sectional area of the bar is

$$A = (0.010 \text{ m})(0.020 \text{ m}) = 0.0002 \text{ m}^2$$

The normal stresses in the three segments are

$$\sigma_{AB} = \frac{P_{AB}}{A} = \frac{-20 \text{ kN}}{0.0002 \text{ m}^2} = -100\ 000 \text{ kN/m}^2$$

$$= 100 \text{ MPa (C)} \qquad \qquad \Leftarrow \textbf{Ans.}$$

$$\sigma_{BC} = \frac{P_{BC}}{A} = \frac{+8 \text{ kN}}{0.0002 \text{ m}^2} = +40\ 000 \text{ kN/m}^2$$

$$= 40 \text{ MPa (T)} \qquad \qquad \Leftarrow \textbf{Ans.}$$

$$\sigma_{CD} = \frac{P_{CD}}{A} = \frac{+32 \text{ kN}}{0.0002 \text{ m}^2} = +160\ 000 \text{ kN/m}^2$$

$$= 160 \text{ MPa (T)} \qquad \qquad \Leftarrow \textbf{Ans.}$$

———— EXAMPLE 9–3 ————

A 10-kip weight is supported by a rod and cables, as shown in Fig. E9–3(1). Neglecting the weights of the rod and the cables, determine the normal stresses in cable *AB* and in rod *AC.*

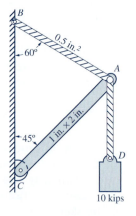

FIGURE E9–3(1)

Solution. Rod AC is a two-force member. Hence, the forces exerted on the rod must be along the axial direction of the member. To determine the axial forces in cable AB and rod AC, consider the equilibrium of joint A. The free-body diagram of the joint and the corresponding force triangle are sketched in Fig. E9–3(2). Note that the rod is assumed to be in compression, so it is shown to push joint A. Applying the law of sines to the force triangle, we write

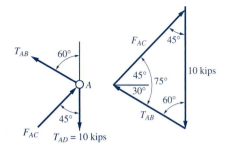

FIGURE E9–3(2)

$$\frac{T_{AB}}{\sin 45°} = \frac{F_{AC}}{\sin 60°} = \frac{10 \text{ kips}}{\sin 75°}$$

From which we get

$$T_{AB} = 7.32 \text{ kips (T)}$$
$$F_{AC} = 8.97 \text{ kips (C)}$$

The normal stresses in the cable and the rod are

$$\sigma_{AB} = \frac{T_{AB}}{A_{AB}} = \frac{7.32 \text{ kips}}{0.5 \text{ in.}^2} = 14.6 \text{ ksi (T)} \qquad \Leftarrow \textbf{Ans.}$$

$$\sigma_{AC} = \frac{F_{AC}}{A_{AC}} = \frac{8.97 \text{ kips}}{2 \text{ in.}^2} = 4.49 \text{ ksi (C)} \qquad \Leftarrow \textbf{Ans.}$$

EXAMPLE 9–4

Refer to Fig. E9–4(1). A load W is supported by a rigid beam AC and a tie rod BD. Using an allowable stress of $\sigma_{\text{allow}} = 140$ MPa and neglecting the weights of all members, determine (*a*) the maximum allowable load W that the tie rod can support if its cross-section is 20 mm $\times$ 30 mm, and (*b*) the required diameter of a circular tie rod to support a load of $W = 80$ kN.

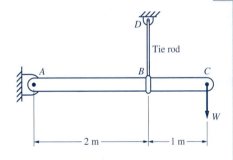

FIGURE E9–4 (1)

Solution. In both cases, we need to relate the tensile force T in the tie rod to the load W. To obtain this relationship, we will consider the equilibrium of the rigid beam AC. From the free-body diagram of the beam sketched in Fig. E9–4(2), we write the moment equation about A to get

$$\Sigma M_A = T(2 \text{ m}) - W(3 \text{ m}) = 0$$
$$2T = 3W$$

The allowable stress is

$$\sigma_{allow} = 140 \text{ MPa} = 140\,000 \text{ kN/m}^2$$

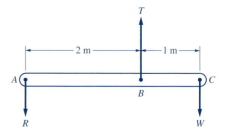

FIGURE E9–4(2)

(a) The Maximum Allowable Load W. For a tie rod of cross-sectional area $A = (0.020 \text{ m})(0.030 \text{ m}) = 0.0006 \text{ m}^2$, the allowable tensile force of the tie rod can be computed from Equation 9–2:

$$T_{allow} = \sigma_{allow} A$$
$$= (140\,000 \text{ kN/m}^2)(0.0006 \text{ m}^2)$$
$$= 84 \text{ kN}$$

$$W_{allow} = \frac{2}{3} T_{allow} = \frac{2}{3}(84 \text{ kN})$$
$$= 56 \text{ kN} \qquad \Leftarrow \textbf{Ans.}$$

(b) The Required Diameter of the Tie Rod. For a load $W = 80$ kN, the tensile force in the tie rod is

$$T = \frac{3}{2}W = \frac{3}{2}(80 \text{ kN}) = 120 \text{ kN}$$

The required cross-sectional area of the tie rod capable of carrying this force can be computed from Equation 9–3:

$$A = \frac{T}{\sigma_{allow}} = \frac{120 \text{ kN}}{140\,000 \text{ kN/m}^2} = 8.57 \times 10^{-4} \text{ m}^2$$

Since $A = \pi d^2/4 = 0.7854d^2$, the required diameter is

$$d = \sqrt{\frac{A}{0.7854}} = \sqrt{\frac{8.57 \times 10^{-4} \text{ m}^2}{0.7854}} = 0.0330 \text{ m}$$

$$= 33.0 \text{ mm} \qquad \Leftarrow \textbf{Ans.}$$

9–4
DIRECT SHEAR STRESSES

Shear stress has been defined in Section 9–2 as the intensity of internal force tangential (parallel) to the area in question. Shear stress differs from normal stress because shear stress is parallel to the area on which it acts, while normal stress is normal (perpendicular) to the area. Shear stress is similar to the stress on a paper exerted by a pair of scissors during cutting.

Consider a block with a protruded part, as shown in Fig. 9–2a. A horizontal force P applied to the protruded part tends to shear the part off the block along the shear plane *abcd*. The body resists the force P by developing resisting shear stresses in the shear plane. The resultant of the shear stresses must be equal to the applied force P, as shown in Fig. 9–2b. The shear stress may not be uniformly distributed over the shear area A_s. The average shear stress, however, can be calculated from

$$\tau_{avg} = \frac{P}{A_s} \tag{9-4}$$

where τ_{avg} = the average shear stress

P = the internal resisting shear force tangent to the shear plane

A_s = the area of the shear plane

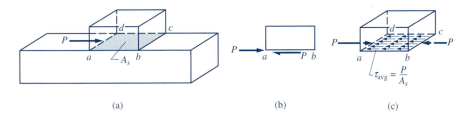

(a)   (b)   (c)

FIGURE 9–2

The shear stress caused directly by a shear force is called a *direct shear stress.* Indirect shear stresses, such as shear stresses on shafts and beams, will be discussed in later chapters. Direct shear stresses are found in bolts, rivets, pins, keys, etc., used to connect structural or machine members. A few examples are discussed in the following paragraphs.

The Lap Joint.   Figure 9–3a shows a lap joint connecting two overlapping tension plates with a rivet (or a bolt). The force on the rivet is shown in Fig. 9–3b. The rivet is subjected to shear stress through section *m–m,* as shown in Fig. 9–3c. The shear stress, in general, is not uniformly distributed in the section. The average shear stress is

$$\tau_{avg} = \frac{P}{A_s} \tag{9-5}$$

where A_s is the cross-sectional area of the rivet. Since the shear stress occurs only in one section of the rivet, it is said to be in *single shear.* If there are several rivets in the joint, the load is assumed to be shared equally by the rivets.

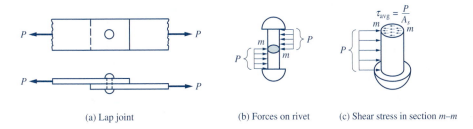

(a) Lap joint   (b) Forces on rivet   (c) Shear stress in section *m–m*

FIGURE 9–3

The Butt Joint. The butt joint connects nonoverlapping tension plates using connecting plates, as shown in Fig. 9–4a. Rivet A in the joint is subjected to shear stresses at sections m–m and n–n, as shown in Fig. 9–4b and c. Assuming that the shear force is shared equally by the two sections, then the average shear stress is

$$\tau_{avg} = \frac{P}{2A_s} \tag{9–6}$$

where A_s is the cross-sectional area of the rivet. Since the shear stresses occur in two sections of the rivet, the rivet is said to be in *double shear*. Rivet B is similarly loaded and is subjected to the same shear stress. If there are several rivets on each side of the joint, the load is assumed to be shared equally by the rivets.

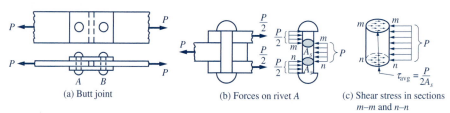

(a) Butt joint   (b) Forces on rivet A (c) Shear stress in sections m–m and n–n

FIGURE 9–4

The Shaft Key. The shaft key shown in Fig. 9–5a connects a gear to a shaft. The moment M on the gear is transmitted to the shaft through the key. The key is subjected to forces (labeled with P), shown in Fig. 9–5b. These forces are assumed to be concentrated on the rim of the shaft. The moment Pr of P about the center of the shaft must be equal to the transmitted moment M. Thus,

$$P = \frac{M}{r} \tag{9–7}$$

The average shear stress at section m–m of the key is

$$\tau_{avg} = \frac{P}{A_s} = \frac{M/r}{bL} = \frac{M}{rbL} \tag{9–8}$$

where b is the width of the key, L is the length of the key, and r is the radius of the shaft. For a square key, the width b is approximately equal to one-quarter of the shaft diameter.

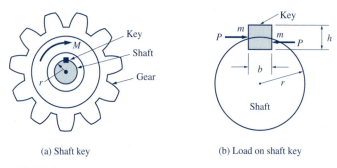

(a) Shaft key   (b) Load on shaft key

FIGURE 9–5

Required Shear Area. The *allowable shear stress* τ_{allow} is the upper limit of shear stress that must not be exceeded. From Equation 9–4, the minimum shear area A_s required to carry a design shear load P without exceeding the allowable shear stress τ_{allow} is

$$A_s = \frac{P}{\tau_{\text{allow}}} \tag{9–9}$$

9–5
BEARING STRESSES

When one body presses against another, *bearing stress* occurs between the two bodies. For example, Fig. 9–6a shows that the bottom of the block is pressed against the top of the pier by a compressive force P. Assuming that the bearing stress is uniformly distributed in the shaded contact area A_b shown in Fig. 9–6a, the bearing stress is

$$\sigma_b = \frac{P}{A_b} \tag{9–10}$$

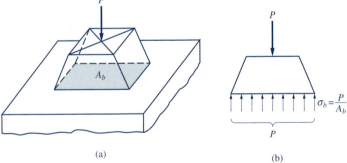

(a)   (b)

FIGURE 9–6

Bearing Stress in Shaft Key. Bearing stress occurs between the key and the gear and between the key and the shaft shown in Fig. 9–5. The compressive force P is assumed to be uniformly distributed over an area $(h/2)L$. The bearing stress is, therefore,

$$\sigma_b = \frac{P}{A_b} = \frac{M/r}{(h/2)L} = \frac{2M}{rhL} \tag{9–11}$$

where M = the moment transmitted by the key

        h = the height of the key

        L = the length of the key

        r = the radius of the shaft

Bearing Stress Between Rivet and Plate. In the lap joint and butt joint shown in Figs. 9–3 and 9–4, bearing stresses occur between rivets or bolts and the plates. The stress is distributed over a cylindrical surface, as shown in Fig. 9–7a and b. The maximum bearing stress is found to be approximately equal to the value obtained by dividing the compressive force by the projected area of the rivet onto the plate (the rectangular area with thickness t of the plate and diameter d of the rivet as its two sides, shown shaded in Fig. 9–7c). In engineering practice, the bearing stress between the rivet or the bolt and the plate is computed by

$$\sigma_b = \frac{P}{\text{projected area}} = \frac{P}{td} \qquad (9\text{–}12)$$

where P = force transmitted

t = thickness of the plate

d = diameter of the pin

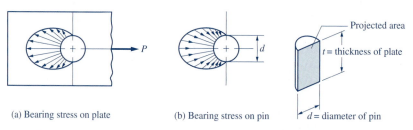

(a) Bearing stress on plate   (b) Bearing stress on pin d = diameter of pin

FIGURE 9–7

Required Bearing Area. The *allowable bearing stress* $(\sigma_b)_{\text{allow}}$ is the upper limit of compressive stress that must not be exceeded. From Equation 9–10, the minimum bearing area A_b required to carry a design bearing load P without exceeding the allowable bearing stress $(\sigma_b)_{\text{allow}}$ is

$$A_b = \frac{P}{(\sigma_b)_{\text{allow}}} \qquad (9\text{–}13)$$

―――― **EXAMPLE 9–5** ――――――――――――――――――――――――――――――

A circular blanking punch is shown in Fig. E9–5. It is operated by causing shear failure in the plate. Knowing that the thickness t of the steel plate is 10 mm and that the ultimate shear strength of the steel (the greatest shear stress a material can withstand before failure) is $\tau_u = 300$ MPa, determine the minimum force P required to punch a hole 50 mm in diameter.

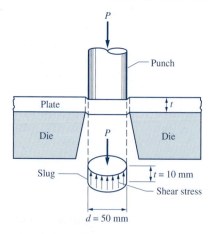

FIGURE E9–5

Solution. The shear area that resists the punch is the area on the side of the slug that is punched out. The shear area is

$$A_s = \pi d t = \pi(0.05 \text{ m})(0.01 \text{ m}) = 0.001\ 57 \text{ m}^2$$

The minimum force P is the force needed to cause shear failure over the shear area. Thus,

$$P_{min} = A_s \tau_u = (0.001\ 57 \text{ m}^2)(300\ 000 \text{ kN/m}^2)$$
$$= 471 \text{ kN} \qquad\qquad \Leftarrow \textbf{Ans.}$$

EXAMPLE 9–6

A rectangular key $b \times h \times L = \frac{3}{4}$ in. $\times \frac{1}{2}$ in. $\times$ 3 in. is used to connect a gear and a shaft of diameter $d = 3$ in., as shown in Fig. E9–6. The couple transmitted by the key is 15 kip · in. Determine (a) the shear stress in the key and (b) the bearing stress between the key and the shaft.

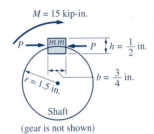

$M = 15$ kip-in.

$h = \frac{1}{2}$ in.

$b = \frac{3}{4}$ in.

$r = 1.5$ in.

Shaft

(gear is not shown)

FIGURE E9–6

Solution.
(a) *The Shear Stress.*   The shear force P on the key is

$$P = \frac{M}{r} = \frac{15 \text{ kip} \cdot \text{in.}}{1.5 \text{ in.}} = 10 \text{ kips}$$

The shear area in the key is

$$A_s = bL = \left(\frac{3}{4} \text{ in.}\right)(3 \text{ in.}) = 2.25 \text{ in.}^2$$

The shear stress in the key is

$$\tau = \frac{P}{A_s} = \frac{10 \text{ kips}}{2.25 \text{ in.}^2} = 4.44 \text{ ksi} \qquad\qquad \Leftarrow \textbf{Ans.}$$

(b) *The Bearing Stress.*   The bearing area is

$$A_b = \left(\frac{h}{2}\right)L = \left(\frac{1}{4} \text{ in.}\right)(3 \text{ in.}) = 0.75 \text{ in.}^2$$

The bearing stress is

$$\sigma_b = \frac{P}{A_b} = \frac{10 \text{ kips}}{0.75 \text{ in.}^2} = 13.3 \text{ ksi} \qquad\qquad \Leftarrow \textbf{Ans.}$$

EXAMPLE 9–7

The butt joint shown in Fig. E9–7(1) is fastened by four $\frac{3}{4}$-in.-diameter bolts. If the joint transmits a tensile force P of 24 kips, determine (a) the average shear stress in the bolts and (b) the bearing stress between the bolts and the plates.

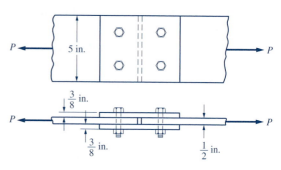

FIGURE E9–7(1)

Solution. The load P is transmitted by two bolts, as shown in the free-body diagram of the main plate and the two cover plates in Fig. E9–7(2).

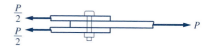

FIGURE E9–7(2)

(a) The Shear Stress. Assume that the force P is shared equally by the two bolts. Then the shear force on each bolt is $P/2$. The cross-sectional area of the bolt is

$$A_s = \frac{1}{4}\pi d^2 = \frac{1}{4}\pi\left(\frac{3}{4}\right)^2 = 0.442 \text{ in.}^2$$

For the bolt in double shear, the average shear stress is

$$\tau_{\text{avg}} = \frac{P/2}{2A_s} = \frac{(24 \text{ kips})/2}{2(0.442 \text{ in.}^2)} = 13.6 \text{ ksi} \qquad \Leftarrow \textbf{Ans.}$$

(b) The Bearing Stress. The compressive force transmitted by each bolt is $P/2$. From Equation 9–9, the bearing stress between the main plate and the bolt is

$$\sigma_b = \frac{P/2}{td} = \frac{(24 \text{ kips})/2}{\left(\frac{1}{2} \text{ in.}\right)\left(\frac{3}{4} \text{ in.}\right)} = 32 \text{ ksi} \qquad \Leftarrow \textbf{Ans.}$$

Note that the bearing stress between the cover plates and the bolt is smaller than the value calculated above because the total thickness of the cover plates is greater than the thickness of the main plate.

EXAMPLE 9–8

A clevis subjected to an axial load $P = 18$ kips is connected by a pin, as shown in Fig. E9–8. Determine (a) the required diameter d of the pin and (b) the required thickness t if the allowable stresses are $\tau_{\text{allow}} = 15$ ksi and $(\sigma_b)_{\text{allow}} = 48$ ksi.

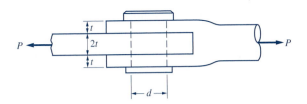

FIGURE E9–8

Solution.

(a) Diameter of the Pin. The pin is in double shear; thus, each area carries one-half the shear load. From Equation 9–8, the required cross-sectional area of the pin is

$$A_s = \frac{P/2}{\tau_{allow}} = \frac{9 \text{ kips}}{15 \text{ kips/in.}^2} = 0.600 \text{ in.}^2$$

Since $A_s = \pi d^2/4 = 0.7854d^2$, the required diameter of the pin is

$$d = \sqrt{\frac{A_s}{0.7854}} = \sqrt{\frac{0.600 \text{ in.}^2}{0.7854}} = 0.874 \text{ in.}$$

Use $d = \frac{7}{8}$ in. pin   ⇐ **Ans.**

(b) The Thickness t. From Equation 9–12, the required bearing area is

$$A_b = \frac{P}{(\sigma_b)_{allow}} = \frac{18 \text{ kips}}{48 \text{ kips/in.}^2} = 0.375 \text{ in.}^2$$

The projected bearing area is $A_b = d(2t)$; thus, the required thickness is

$$t = \frac{A_b}{2d} = \frac{0.375 \text{ in.}^2}{2(0.875 \text{ in.})} = 0.214 \text{ in.}$$

Use thickness $t = \frac{1}{4}$ in.   ⇐ **Ans.**

9–6
STRESSES ON INCLINED PLANES

For an axially loaded member, normal stress occurs on a plane perpendicular to the axis of the member. However, on an inclined plane, such as plane m–m shown in Fig. 9–8a, both normal and shear stresses exist. The equilibrium condition of the free-body diagram in Fig. 9–8b requires that the internal resisting force R in section m–m be equal to the applied force P. The force can be resolved into two components: the normal component R_n perpendicular to the inclined plane, and the tangential component R_s parallel to the inclined plane. The normal component R_n produces normal stress, and the tangential component R_s produces shear stress. These components are:

$$R_n = P \cos \theta$$
$$R_s = P \sin \theta$$

where θ is the angle between the inclined plane and the cross-section.

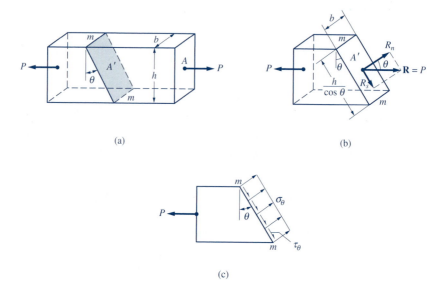

(a)   (b)

(c)

FIGURE 9–8

If the cross-sectional area A is a rectangle $b \times h$, then the area A' of the inclined plane is a rectangle $b \times (h\,/\cos\theta)$. By definition, the normal stress σ_θ, assumed to be distributed uniformly over the inclined plane, is

$$\sigma_\theta = \frac{R_n}{A'} = \frac{P\cos\theta}{b(h/\cos\theta)} = \frac{P}{A}\cos^2\theta$$

Since P/A is the normal stress σ over the cross-section, we write

$$\sigma_\theta = \sigma\cos^2\theta \tag{9–14}$$

The average shear stress τ_θ over the inclined plane is

$$\tau_\theta = \frac{R_s}{A'} = \frac{P\sin\theta}{b(h/\cos\theta)} = \frac{P}{A}\sin\theta\cos\theta$$

Using the trigonometric identity, $\sin 2\theta = 2\sin\theta\cos\theta$, we write

$$\tau_\theta = \frac{1}{2}\sigma\sin 2\theta \tag{9–15}$$

From Equation 9–14, we see that the maximum shear stress is

$$\tau_{max} = \frac{1}{2}\sigma \tag{9–16}$$

which occurs when $\sin 2\theta = 1$, or $\theta = 45°$, that is, on the 45° inclined plane.

The normal and shear stresses on any inclined plane in an axially loaded member can be computed by using Equations 9–13 and 9–14. These stresses can also be determined simply by definition, as demonstrated in the following two examples.

──────── **EXAMPLE 9–9** ────────

Refer to Fig. E9–9(1). Determine the normal and shear stresses on the inclined plane m–m of the axially loaded member shown.

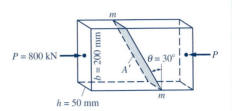

$P = 800$ kN $\theta = 30°$ P

$b = 200$ mm

A'

$h = 50$ mm

FIGURE E9–9(1)

Solution. The free-body diagram of the part of the member to the left of the section is sketched in Fig. E9–9(2). Equilibrium conditions along the axial direction require that

$$R = P = 800 \text{ kN}$$

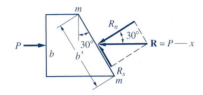

P $30°$ R_n $30°$ $R = P$ —x

b b' R_s

FIGURE E9–9(2)

The internal force **R** is resolved into two components, R_n and R_s:

$$R_n = R \cos 30° = (800 \text{ kN}) \cos 30° = 693 \text{ kN}$$

$$R_s = R \sin 30° = (800 \text{ kN}) \sin 30° = 400 \text{ kN}$$

The dimension b' along the incline is

$$b' = \frac{b}{\cos 30°} = \frac{200 \text{ mm}}{\cos 30°} = 231 \text{ mm} = 0.231 \text{ m}$$

Thus, the area A' of the inclined plane is

$$A' = b'h = (0.231 \text{ m})(0.050 \text{ m}) = 0.011\ 55 \text{ m}^2$$

By definition the normal and shear stresses on the inclined plane are

$$\sigma_\theta = \frac{R_n}{A'} = \frac{693 \text{ kN}}{0.011\ 55 \text{ m}^2} = 60\ 000 \text{ kN/m}^2$$

$$= 60 \text{ MPa (C)} \qquad \Leftarrow \textbf{Ans.}$$

$$\tau_\theta = \frac{R_s}{A'} = \frac{400 \text{ kN}}{0.011\ 55 \text{ m}^2} = 34\ 600 \text{ kN/m}^2$$

$$= 34.6 \text{ MPa} \qquad \Leftarrow \textbf{Ans.}$$

These stresses can also be determined from Equations 9–14 and 9–15. For $\theta = 30°$, we have

$$\sigma = \frac{P}{A} = \frac{-800 \text{ kN}}{0.200 \times 0.050 \text{ m}^2} = -80\ 000 \text{ kN/m}^2$$

$$\sigma_\theta = \sigma \cos^2\theta = (-80 \text{ MPa}) \cos^2 30°$$

$$= -60 \text{ MPa (C)} \qquad \Leftarrow \textbf{Ans.}$$

$$\tau_\theta = \frac{1}{2}\sigma \sin 2\theta = \frac{1}{2}(80 \text{ MPa}) \sin (2 \times 30°)$$

$$= 34.6 \text{ MPa}$$ ⇐ **Ans.**

EXAMPLE 9–10

The flanges of the member shown in Fig. E9–10(1) are connected by two bolts, one on each side. The diameter of each bolt is $\frac{1}{2}$ in. and the load P is 6 kips. Determine the normal and shear stresses in each bolt at the connecting plane.

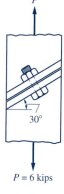

FIGURE E9–10(1)

Solution. A free-body diagram of the upper part of the member, with the bolts cut through the plane of the connection, is sketched in Fig. E9–10(2). The two bolts are subjected to normal and shear forces. Assume that the load is shared equally by the two bolts, and each bolt is thus subjected to one-half of the applied force. The normal and shear components of the load on each bolt are

$$R_n = \frac{1}{2}P\cos\theta = \frac{1}{2}(6 \text{ kips}) \cos 30° = 2.60 \text{ kips}$$

$$R_s = \frac{1}{2}P\sin\theta = \frac{1}{2}(6 \text{ kips}) \sin 30° = 1.50 \text{ kips}$$

The cross-sectional area of the bolt is

$$A = \frac{1}{4}\pi d^2 = \frac{1}{4}\pi \left(\frac{1}{2} \text{ in.}\right)^2 = 0.1963 \text{ in.}^2$$

By definition, the normal and shear stresses in the bolt are

FIGURE E9–10(2)

$$\sigma = \frac{R_n}{A} = \frac{2.60 \text{ kips}}{0.1963 \text{ in.}^2} = 13.2 \text{ ksi (T)}$$ ⇐ **Ans.**

$$\tau = \frac{R_s}{A} = \frac{1.50 \text{ kips}}{0.1963 \text{ in.}^2} = 7.64 \text{ ksi}$$ ⇐ **Ans.**

9–7
STRESSES IN THIN-WALLED PRESSURE VESSELS

Pressure vessels are leak-proof containers subjected to internal pressure. Boilers, fire extinguishers, and compressed air tanks are common examples. The primary purpose of pressure vessels is to contain liquids and/or gases under pressure. These containers are commonly shaped as cylinders or spheres. The discussion in this chapter is limited only to thin-walled pressure vessels, which have a wall thickness no greater than one-tenth of the internal radius of the vessel ($t \leq 0.1\, r_i$). For a thin-walled vessel, the tensile stresses caused by the internal pressure are assumed to be distributed uniformly throughout the thickness of the wall.

Cylindrical Pressure Vessel. When a thin-walled cylindrical pressure vessel, such as the one shown in Fig. 9–9a, is subjected to an internal pressure p, tensile stresses develop on the wall of the vessel, as shown on an element in Fig. 9–9b. The tensile stress σ_c acting along the circumferential direction is called the *circumferential stress* (also called the *hoop stress*). The normal stress σ_l acting along the longitudinal direction is called the *longitudinal stress*. Since the internal pressure p acting inside the vessel is much smaller than σ_c and σ_l, the pressure p acting on the element along the z direction can be neglected, and the element is subjected to normal stresses in only two directions. The element is said to be in a state of *biaxial stresses*.

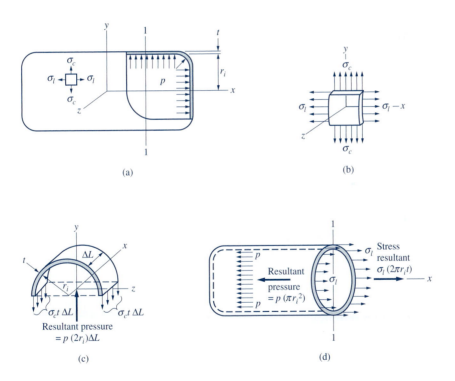

FIGURE 9–9

The formulas for σ_c can be derived by considering the equilibrium of one-half of the cylindrical wall segment with the enclosed fluid under a uniform internal gage pressure p (pressure above the atmospheric pressure). The free-body diagram is shown in Fig. 9–9c. The equilibrium condition along the y direction requires

$$\Sigma F_y = p(2r_i)\Delta L - 2\sigma_c t\, \Delta L = 0$$

which gives

$$\sigma_c = \frac{pr_i}{t} \qquad (9\text{--}17)$$

where σ_c = the circumferential tensile stress in the wall of the cylindrical vessel

p = the internal pressure

r_i = the inside radius of the cylinder

t = the wall thickness of the pressure vessel

The formula for σ_l may be derived by considering the equilibrium of the cylindrical vessel to the left of cross-section 1–1. The free-body diagram is shown in Fig. 9–9d. The equilibrium condition along the x direction requires

$$\Sigma F_x = \sigma_l(2\pi r_i t) - p(\pi r_i^2) = 0$$

From which the longitudinal stress σ_l is

$$\sigma_l = \frac{pr_i}{2t} \qquad (9\text{--}18)$$

From Equations 9–17 and 9–18, we see that the circumferential stress σ_c is twice the longitudinal stress σ_l. For design purposes, Equation 9–17 can be used to calculate a required wall thickness to resist a given internal pressure so that the tensile stress in the wall will not exceed an allowable tensile stress σ_{allow}. The equation can be written in the following form:

$$t_{req} = \frac{pr_i}{\sigma_{allow}} \qquad (9\text{--}19)$$

Spherical Pressure Vessel. A similar method can be used to derive an expression for the tensile stress in thin-walled spherical pressure vessels, as shown in Fig. 9–10a. By passing a section through the center of the sphere, a hemisphere with enclosed fluid under a uniform internal pressure p is isolated, as shown in Fig. 9–10b. The equilibrium condition along the x direction requires

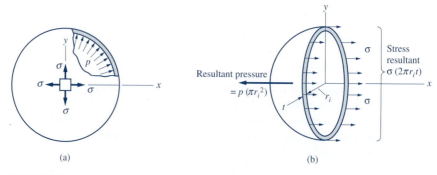

(a)   (b)

FIGURE 9–10

$$\Sigma F_x = \sigma(2\pi r_i t) - p(\pi r_i^2) = 0$$

From which the normal stress in the wall of the spherical pressure vessel is

$$\sigma = \frac{pr_i}{2t} \qquad (9\text{–}20)$$

where r_i is the inside radius of the spherical vessel. Since any section that passes through the center of the sphere yields the same result, the normal stress in the wall of the spherical vessel is the same value along *any* direction. This stress condition is called *all-around tension*. For design purposes, Equation 9–20 can be used to calculate a required wall thickness to resist a given internal pressure so that the tensile stress in the wall will not exceed an allowable tensile stress σ_{allow}. The equation can be written in the following form:

$$t_{req} = \frac{pr_i}{2\sigma_{allow}} \qquad (9\text{–}21)$$

──────── **EXAMPLE 9–11** ────────

A cylindrical, compressed air storage tank has a 10-in. inside radius and a wall thickness of $\frac{1}{2}$ in. The vessel is subjected to an internal gage pressure of 300 psi. Determine the circumferential and longitudinal stresses on the wall of the tank.

Solution. The tank can be regarded as a thin-walled cylindrical vessel since $t = 0.5$ in. $< 0.1 r_i = 1$ in. The circumferential stress σ_c in the wall of the tank can be computed from Equation 9–17:

$$\sigma_c = \frac{pr_i}{t} = \frac{(300 \text{ lb/in.}^2)(10 \text{ in.})}{\frac{1}{2} \text{ in.}} = 6000 \text{ psi}$$

$$= 6000 \text{ psi} \qquad \qquad \Leftarrow \textbf{Ans.}$$

The longitudinal stress σ_l is one-half the circumferential stress. Thus,

$$\sigma_l = 3000 \text{ psi} \qquad \qquad \Leftarrow \textbf{Ans.}$$

──────── **EXAMPLE 9–12** ────────

Find the maximum allowable internal gage pressure that can be sustained by a spherical steel vessel with a 500-mm inside radius and a wall thickness of 40 mm. Assume that the allowable tensile stress for steel is 60 MPa. (Note that the allowable stress is set at a lower value to account for corrosion effects.)

Solution. Since the thickness $t = 40$ mm is less than one-tenth of the inside radius of 500 mm, the vessel is a thin-walled spherical vessel. The allowable internal gage pressure may be obtained by solving p from Equation 9–20:

$$p = \frac{2t\sigma_c}{r_i} = \frac{2(0.040 \text{ m})(60 \text{ MPa})}{0.5 \text{ m}} = 9.6 \text{ MPa}$$

Thus, the allowable internal gage pressure (the pressure above atmosphere) is

$$p = 9.6 \text{ MPa} \qquad\qquad \Leftarrow \textbf{Ans.}$$

EXAMPLE 9–13

Refer to Fig. E9–13. The cylindrical storage tank with an open top has an inside diameter of 12 ft and a height of 40 ft. Assume that the allowable stress for the steel wall of the tank is 10 ksi, which provides an allowance for the corrosion effect. Determine the required wall thickness if the tank is filled with water to capacity. The specific weight of water is 62.4 lb/ft³.

Solution. From Section 7–6, we see that the liquid pressure varies linearly with respect to the depth; that is

$$p = \gamma h$$

FIGURE E9–13

where p is the gage pressure of a point at depth h below the free surface, and γ is the specific weight of the liquid. The maximum pressure occurs in the bottom of the tank, where the pressure is

$$p = (62.4 \text{ lb/ft}^3)(40 \text{ ft}) = 2496 \text{ psf} = 17.3 \text{ psi}$$

The required wall thickness can be computed from Equation 9–19:

$$t_{req} = \frac{pr_i}{\sigma_{allow}} = \frac{(17.3 \text{ psi})(72 \text{ in.})}{10\,000 \text{ psi}} = 0.125 \text{ in.}$$

$$= 0.125 \text{ in.}$$

Use a wall thickness of $t = \frac{1}{8}$ in. $\qquad \Leftarrow \textbf{Ans.}$

9–8
SUMMARY

Stresses. *Normal stresses* are caused by internal forces normal (perpendicular) to the area in question, and *shear stresses* are caused by internal forces tangential (parallel) to the area under consideration. The failure and deformation of structural or machine members depend on the amount of stresses induced within the member.

Normal Stresses. When a prismatic *axially loaded member* is subjected to pulling forces, the member is in *tension*; when it is subjected to pushing forces, the member is in *compression*. Normal stress due to axial loads is usually distributed uniformly over the cross-section. Normal stress can be computed by

$$\sigma = \frac{P}{A} \qquad\qquad (9\text{–}1)$$

There are two types of normal stress, namely, tensile stress and compressive stress. *Tensile stresses* are induced by tensile forces; *compressive stresses* are induced by compressive forces. The allowable axial load that a member of cross-sectional area A can carry without exceeding an allowable stress σ_{allow} may be computed from:

$$P_{allow} = \sigma_{allow} A \qquad\qquad (9\text{–}2)$$

The required minimum cross-sectional area A of a member designed to carry a maximum axial load P without exceeding the allowable stress σ_{allow} may be computed from

$$A = \frac{P}{\sigma_{allow}} \qquad\qquad (9\text{–}3)$$

Shear Stress. Shear stress is induced by a shear force over an area in the direction parallel to the area. An average shear stress can be calculated from

$$\tau_{avg} = \frac{P}{A_s} \qquad\qquad (9\text{–}4)$$

The shear stress caused directly by a shear force is called a *direct shear stress*. Direct shear stresses are found in bolts, rivets, pins, keys, etc., used to connect structural or machine members. The required minimum cross-sectional area A_s of a member to carry a maximum shear force P without exceeding the allowable shear stress τ_{allow} may be computed from

$$A_s = \frac{P}{\tau_{allow}} \qquad\qquad (9\text{–}9)$$

Bearing Stress. Bearing stress occurs between two bodies in contact pressing against each other. The bearing stress is assumed to be distributed uniformly in the contact area and it can be computed from

$$\sigma_b = \frac{P}{A_b} \qquad\qquad (9\text{–}10)$$

Stresses on an Inclined Plane. Both normal and shear stresses exist on an inclined plane in an axially loaded member. The internal force on the inclined plane can be resolved into two components: the normal component R_n perpendicular to the inclined plane and the tangential component R_s parallel to the inclined plane. If the area of the inclined plane is A', by definition, the normal stress σ_θ and the average shear stress τ_θ over the inclined plane are

$$\sigma_\theta = \frac{R_n}{A'}$$

$$\tau_\theta = \frac{R_s}{A'}$$

Thin-Walled Pressure Vessels. Pressure vessels are leak-proof containers subjected to internal pressure. The primary purpose of pressure vessels is to contain liquids and/or gases under pressure. Thin-walled pressure vessels have a wall thickness no greater than one-tenth of the internal radius of the vessel ($t \le 0.1\, r_i$). For a thin-walled vessel, the tensile stresses caused by internal pressure are assumed to be distributed uniformly throughout the thickness of the wall.

Cylindrical Pressure Vessel. When a thin-walled cylindrical pressure vessel of internal radius r_i is subjected to internal pressure p, tensile stresses develop in the wall of the vessel. The *circumferential stress* σ_c is

$$\sigma_c = \frac{pr_i}{t} \tag{9-17}$$

The longitudinal stress σ_l is

$$\sigma_l = \frac{pr_i}{2t} \tag{9-18}$$

Spherical Pressure Vessel. The tensile stress induced by the internal pressure in a thin-walled spherical pressure vessel is the same in all directions. It can be computed from

$$\sigma = \frac{pr_i}{2t} \tag{9-20}$$

PROBLEMS

Section 9–3   Direct Normal Stresses

9–1   A 1.5-ton crate is hoisted by three steel wires. Each wire is $\frac{1}{4}$ in. in diameter and each carries one-third of the load. Determine the stress in the wires.

9–2   An 80-kN hopper is supported by three steel wires. Each wire is 15 mm in diameter and each carries one-third of the load. Determine the stress in the wires.

9–3 to 9–5   Refer to Figs. P9–3 to P9–5. Plot the internal axial force diagram and determine the normal stresses in segments *AB, BC,* and *CD* of each member due to the axial loads shown.

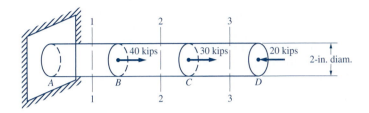

FIGURE P9–3

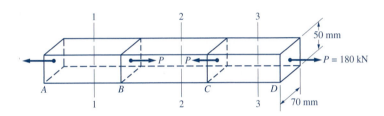

FIGURE P9–4

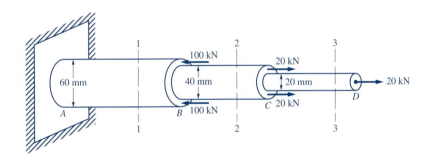

FIGURE P9–5

9–6   A short column composed of two standard steel pipes is subjected to a load *P* = 20 kips, as shown in Fig. P9–6. Determine the compressive stress in each pipe. Neglect the weight of the pipes.

9–7   Determine the size of steel rod, to the nearest sixteenth of an inch, needed to support a tensile load of 40 kips if the allowable tensile stress of steel is 22 ksi.

9–8   Determine the size of steel rod, to the nearest mm, needed to support a tensile load of 200 kN if the allowable tensile stress of steel is 150 MPa.

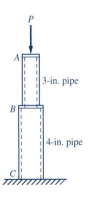

FIGURE P9–6

9–9 If rod *AB* in Fig. P9–9 has a diameter of 10 mm, determine the normal stress in the rod due to a weight $W = 15$ kN.

9–10 Refer to Fig. P9–9. Determine the diameter of the rod *AB,* to the nearest mm, needed to support a weight $W = 30$ kN if the allowable tensile stress is 150 MPa.

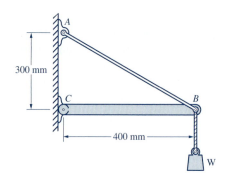

FIGURE P9–9

9–11 A 600-lb weight *W* is supported as shown in Fig. P9–11. Determine the normal stresses in cable *BC* and rod *AB* if their cross-sectional areas are 0.025 in.² and 0.5 in.², respectively.

9–12 Refer to Fig. P9–11. Determine the cross-sectional area required for rod *AB* to support a weight of $W = 1000$ lb if the allowable compressive stress for the member is 1200 psi.

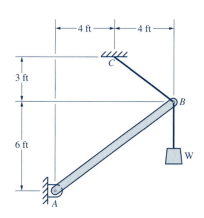

FIGURE P9–11

9–13 Determine the maximum hydraulic compression, in kN, that can be applied to the testing machine in Fig. P9–13. Each of the two posts, *A* and *B,* has a diameter $d = 80$ mm and an allowable tensile stress of 200 MPa.

9–14 The force applied to the brake pedal of a car is transmitted by lever *AD* and connecting rod *BC,* as shown in Fig. P9–14. If $P = 20$ lb, $a = 10$ in., $b = 2$ in., and $d = \frac{1}{4}$ in., determine the normal stress in rod *BC*.

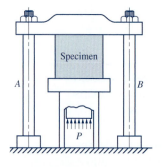

FIGURE P9–13

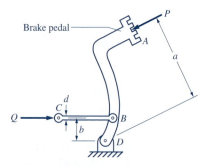

FIGURE P9–14

9–15 Refer to Fig. P9–15. Determine the required cross-sectional area in mm² of members *BD*, *BE*, and *CE* of the truss subjected to the forces shown. The allowable stresses are 140 MPa in tension and 70 MPa in compression.

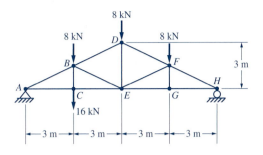

FIGURE P9–15

Section 9–4   Direct Shear Stresses
Section 9–5   Bearing Stresses

9–16 A schematic diagram of the apparatus for determining the ultimate shear strength (failure shear stress) of wood is sketched in Fig. P9–16. The test specimen is 4 in. high, 2 in. wide, and 2 in. deep. If the load required to shear the specimen into two pieces is 8000 lb, determine the ultimate shear strength of the specimen.

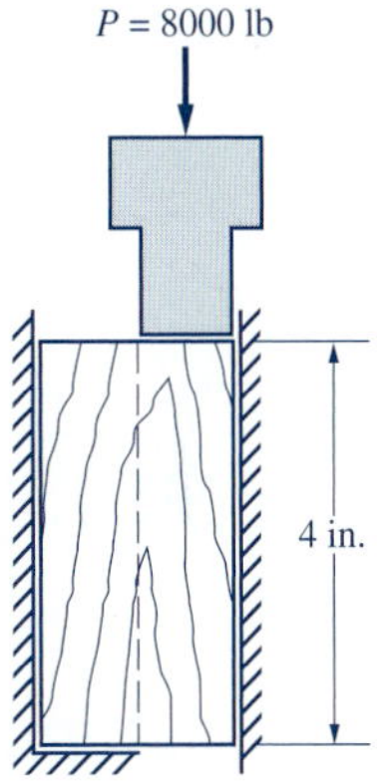

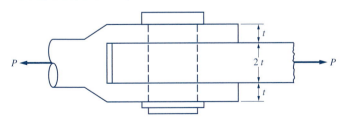

FIGURE P9–16

9–17 The lap joint shown in Fig. P9–17 is connected by four 20-mm-diameter rivets. Determine (*a*) the shear stress in the rivets and (*b*) the bearing stress between the rivets and the plates. Assume that the load $P =$ 120 kN is carried equally by the four rivets.

9–18 The clevis shown in Fig. P9–18 is connected by a pin of $\frac{3}{4}$-in. diameter. Determine the shear stress in the pin and the bearing stress between the pin and the plates if $P = 10$ kips and $t = \frac{1}{4}$ in.

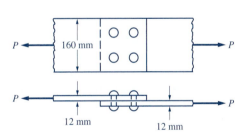

FIGURE P9–17

FIGURE P9–18

9–19 A 35-mm-diameter shaft transmits a torque of 950 N · m by means of a chain drive. The chain sprocket is fastened to the shaft by means of an 8-mm × 8-mm square key 50 mm long. Determine (*a*) the shear stress in the key and (*b*) the bearing stress between the key and the shaft.

9–20   The pulley shown in Fig. P9–20 is connected to an 80-mm-diameter shaft by a 20-mm square key that is 100 mm long. If the belt tensions are $T_1 = 40$ kN and $T_2 = 120$ kN, determine (a) the shear stress in the key and (b) the bearing stress between the key and the shaft.

9–21   A force $\mathbf{F} = 600$ lb is applied to a crank and is transmitted to a shaft through a steel key, as shown in Fig. P9–21. The key is $\frac{1}{2}$ in. square and $2\frac{1}{2}$ in. long. Determine (a) the shear stress in the key and (b) the bearing stress between the key and the shaft.

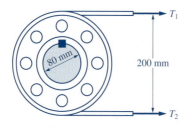

FIGURE P9–20

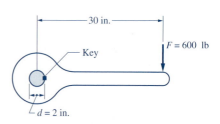

FIGURE P9–21

9–22   Determine the minimum force that must be exerted on a punch to shear a hole, having the shape shown in Fig. P9–22, through a steel plate 4 mm thick. The plate has an ultimate shear strength (failure shear stress) of 300 MPa.

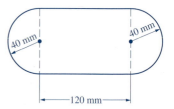

FIGURE P9–22

9–23   In the collar bearing shown in Fig. P9–23, the average bearing stress between the collar and the support is known to be 4000 psi. If $d = 2$ in., $D = 4$ in., and $t = \frac{1}{2}$ in., determine (a) the load P applied to the column and (b) the average shear stress on the area between the collar and the column.

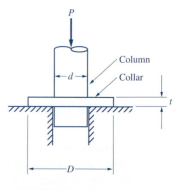

FIGURE P9–23

9–24   The dimensions in the wood joint shown in Fig. P9–24 are $a = 100$ mm, $b = 150$ mm, $c = 40$ mm, and $d = 90$ mm. Determine the shear stress and the bearing stress in the joint if $P = 50$ kN.

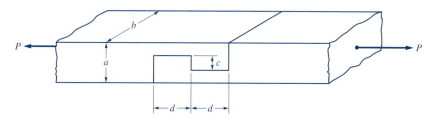

FIGURE P9–24

9–25   The bell crank mechanism in Fig. P9–25 is subjected to a vertical force of 10 kips applied at C. The force is resisted by a horizontal force P at A and a reaction at B. If the mechanism is in equilibrium and the allowable shear stress of the pin is 15 ksi, select the size of the pin at B.

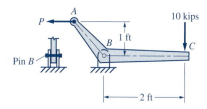

FIGURE P9–25

9–26   The structure shown in Fig. P9–26 is fastened to the support by bolts at A and B. The bolts are in double shear. If the allowable shear stress in the bolts is 120 MPa, select the sizes of the bolts at A and B.

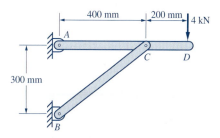

FIGURE P9–26

9–27   See Fig. P9–27. A tie rod of $\frac{1}{4}$-in. diameter is used to hold a plaster wall in place. The tensile stress in the rod caused by P is 20 ksi. Find the diameter d of the washer that keeps the bearing stress between the plaster and the washer from exceeding 300 psi.

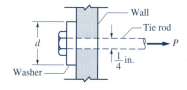

FIGURE P9–27

9–28   The control gate in Fig. P9–28 is operated by a wheel and shaft connected by a flat key, as shown. The allowable stresses in the key are 8000 psi in shear and 20 000 psi in bearing. If $d = 2\frac{1}{4}$ in., $D = 30$ in., $b = \frac{1}{2}$ in., $h = \frac{3}{8}$ in., and $F = 450$ lb, determine the length of the key.

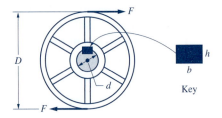

FIGURE P9–28

9–29   The wall bracket shown in Fig. P9–29 carries a load of $P = 12$ kips. The allowable tensile stress in the eye bar is 20 ksi, and the allowable shear stress in the pins is 12 ksi. Select (*a*) the diameter of the eye bar and (*b*) the diameter of the pin at A, which is in double shear.

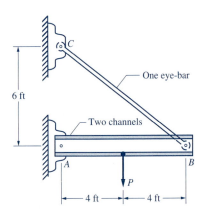

FIGURE P9–29

9–30   In the collar bearing shown in Fig. P9–30, the thickness of the collar is $\frac{1}{2}$ in. The load P is 50 kips. The allowable compressive stress in the column is 20 ksi, the allowable shear stress in the collar is 15 ksi, and the allowable bearing stress between the collar and the support is 5 ksi. Select the proper sizes for d and D.

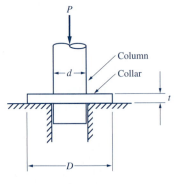

FIGURE P9–30

Section 9–6   Stresses on Inclined Planes

9–31   See Fig. P9–31. Determine the normal and shear stresses on the inclined plane *m–m* of the steel plate subjected to the axial load *P* shown.

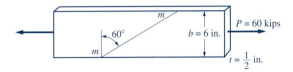

FIGURE P9–31

9–32   When subjected to an axial compressive load, bricks fail in shear on an approximately 45° inclined plane. Hence, the shear strength of bricks is less than one-half their compressive strength (refer to Equation 9–15). A brick of the dimensions shown in Fig. P9–32 is tested in compression. If its shear strength is 800 psi, determine the load *P* that will cause the brick to break.

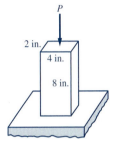

FIGURE P9–32

9–33   A short concrete post having a square section 100 mm × 100 mm is subjected to an axial load *P*. If the shear stress on an inclined plane at 30° from the cross-section is 2500 kPa, determine the value of the load *P*.

9–34   A flat plate $\frac{1}{2}$ in. thick is subjected to an axial force *P* of 40 kips, as shown in Fig. P9–34. Determine the normal and shear stresses on section *m–m*.

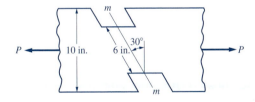

FIGURE P9–34

9–35   Dowels made of hard wood are used to connect the frame shown in Fig. P9–35. If the diameter of the dowel is 10 mm and the allowable shear stress of the dowels is 8 MPa, determine the maximum load *P* that can be applied. Neglect frictional effect.

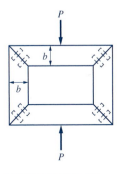

FIGURE P9–35

Section 9–7   Stresses in Thin-Walled Pressure Vessels

9–36   A cylindrical pressure vessel of 8-in. inside radius is made of $\frac{1}{8}$-in. steel plate. Determine the maximum permissible pressure within the vessel if the tensile stress must not exceed 8000 psi. (The allowable stress is set at a lower value to account for corrosion effects.)

9–37   A stainless steel cylindrical pressure vessel of 300-mm inside radius is subjected to an internal pressure of 3.5 MPa. If the allowable tensile stress is 140 MPa, determine the thickness of the wall.

9–38   A spherical pressure vessel with an inside diameter of 8 in. and a wall thickness of $\frac{1}{4}$ in. is made of a material having an allowable tensile stress of 6000 psi. Determine the maximum allowable gage pressure that the vessel can withstand.

9–39   Find the proper wall thickness t for the 30-kN hydraulic jack shown in Fig. P9–39 if the cylinder is made of steel having an allowable tensile stress of 140 MPa.

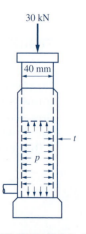

9–40   A steel pipe has an inside diameter of 15.0 in. and a wall thickness of 0.500 in. If the ultimate strength of steel is 65 ksi, determine the bursting pressure for the pipe.

FIGURE P9–39

Computer Program Assignments

For each of the following problems, write a computer program using an appropriate programming language with which you are most familiar. Make the program user friendly by incorporating plenty of comments and input prompts so that the user will understand the input data to be entered and the limitations of their values. The output should include the data entered and the computed results, and they must be well labeled to identify each quantity. If a tabulated format is used, a proper heading must be included at the top of the table. Do not limit the program to any specific unit system. Indicate the consistent U.S. customary or SI units that can be used.

C9–1 Refer to Fig. C9–1. Write a computer program that can be used to determine the average normal stress in each segment of the stepped, solid circular rod subjected to the axial forces shown. The user

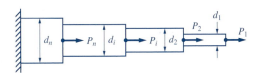

FIGURE C9–1

input should include (1) the total number of segments n, (2) the diameter d_i, and (3) the load P_i applied to the right end of each segment. The load P_i is treated as positive if its direction is the same as indicated in the figure. The output should include (1) the internal axial force in each segment, (2) the cross-sectional area of each segment, and (3) the average normal stress in each segment. Use this program to solve (*a*) Problem 9–3 and (*b*) Problem 9–5.

C9–2 Write a computer program that can be used to determine either the tensile stresses or the allowable internal pressure of a thin-walled pressure vessel. The user input should include (1) whether the vessel is cylindrical or spherical, (2) the internal diameter and the wall thickness of the vessel, and (3) the internal pressure if tensile stresses are required, or the allowable tensile stress if the allowable internal pressure is required. Before any computation is made, the program should verify whether the given vessel is indeed thin-walled. If not, a warning message should be printed to indicate the case. Use this program to solve (*a*) Example 9–11, (*b*) Problem 9–36, and (*c*) Problem 9–38.

CHAPTER 10

STRAINS

10-1
INTRODUCTION

Structural materials deform under the action of forces. There are three kinds of deformation: an increase in length is called an *elongation,* a decrease in length is called a *contraction,* and a change in shape is called an *angular distortion.* Deformation (either elongation or contraction) per unit length is called *linear strain.*

This chapter is devoted to the study of strain and the linear relationship between stress and strain. A formula for computing the deformations of axially loaded members will be established. This formula is not only useful for computing the amount of deformation, but it is also helpful in solving *statically indeterminate* problems for which the static equilibrium equation is not enough to solve for the unknown forces. Stresses caused by temperature changes are also discussed. Finally, shear strains are defined and related to shear stresses.

10-2
LINEAR STRAIN

Axial forces applied to a member tend to elongate or compress the member. Figure 10-1 shows a prismatic bar elongated by a tensile force. The original dimensions of the undeformed member are shown by dashed lines. The original length L of the member is elongated to a length $L + \delta$ after the tensile load P is applied. The total deformation is δ (the Greek lowercase letter delta).

Linear strain in a stressed member is defined as the deformation per unit of original length of the unstressed member. From this definition the linear strain is

$$\epsilon = \frac{\delta}{L} \qquad (10\text{-}1)$$

where ϵ = the linear strain (ϵ is the Greek lowercase letter epsilon)

δ = total axial deformation (elongation or contraction)

L = the original length of the member

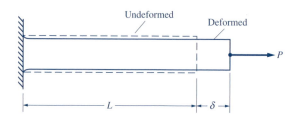

FIGURE 10–1

Since the linear strain is the ratio of two lengths, it is dimensionless and thus independent of unit systems. However, it is common practice to express strain in units such as in./in., ft/ft, or m/m. For a prismatic member of homogeneous material subjected to a constant load, the linear strain ϵ is a constant throughout the length of the member. If the section, material, or load varies, then the quantity δ/L represents the average linear strain along the length L.

10–3
HOOKE'S LAW

For most engineering materials, a linear relationship exists between stress and strain; that is, up to a certain limiting value of stress, the stress is proportional to the strain. Beyond this limit, stress will no longer be proportional to strain. This limiting value is called the *proportional limit* of the material. The linear relationship between stress and strain is known as *Hooke's law,* in honor of the English scientist Robert Hooke who first announced this property in 1676.

Hooke's law may be expressed by the equation

$$\frac{\sigma}{\epsilon} = E \tag{10–2a}$$

or

$$\sigma = E\epsilon \tag{10–2b}$$

where E is the constant of proportionality between stress and strain and is called the *modulus of elasticity.* Since ϵ is a dimensionless number, E must have the same units as those of stress. In U.S. customary units, E is usually expressed in psi or ksi. In SI units, E is expressed in GPa or MPa.

The modulus of elasticity is a definite property of a given material. For most of the common engineering materials, the modulus of elasticity of a material in compression is the same as that in tension. Physically, the modulus of elasticity of a material indicates its stiffness, which is the ability of a material to resist deformation. Typical moduli of elasticity are 30×10^3 ksi (or 210 GPa) for steel and 10×10^3 ksi (or 70 GPa) for aluminum. These values indicate that the stiffness of steel is much greater than that of aluminum. An aluminum bar would stretch three times more than a steel bar of the same length when subjected to the same stress. Average values of moduli of elasticity of some common engineering materials are given in the appendix, Table A–7.

10–4
AXIAL DEFORMATION

An axially loaded member elongates under a tensile load and contracts under a compressive load. If the normal stress in an axially loaded member is within the proportional limit of the material, Hooke's law applies and the axial deformation of the member can be computed.

Consider a homogeneous prismatic member of constant cross-sectional area A subjected to an axial tensile force P, as shown in Fig. 10–2. The tensile stress in the member is $\sigma = P/A$. By definition, the linear strain is the deformation per unit length, or $\epsilon = \delta/L$. Now if the tensile stress in the member is within the proportional limit, Hooke's law applies; that is

$$\sigma = E\epsilon \tag{10–3}$$

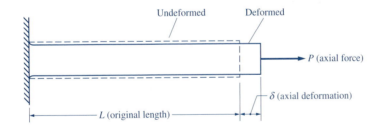

Undeformed   Deformed

P (axial force)

δ (axial deformation)

L (original length)

FIGURE 10–2

Substituting the expression for σ and ϵ into Equation 10–3, we get

$$\frac{P}{A} = E\frac{\delta}{L} \tag{10–4}$$

Solving for δ, we get

$$\delta = \frac{PL}{AE} \tag{10–5}$$

where δ = the total axial deformation

$\quad\quad P$ = the applied axial load

$\quad\quad L$ = the original length of the member

$\quad\quad A$ = the cross-sectional area of the member

$\quad\quad E$ = the modulus of elasticity of the material of the member

Equation 10–5 is valid only when the normal stress in the member does not exceed the proportional limit. For most structural materials, the moduli of elasticity for tension and for compression are the same. Thus, the formula is applicable for both tension and compression members. The stress, strain, and deformation caused by tensile forces are usually considered positive; those caused by compressive force are considered negative. For a member subjected to several axial forces, the axial deformation of each segment must be calculated. The total axial deformation of the member is the algebraic sum of the axial deformation of all the segments.

EXAMPLE 10–1

A 2-m long steel tie rod in a structure is subjected to a tensile axial load of 12 kN. The diameter of the rod is 10 mm. Determine the elongation of the rod if the proportional limit is 200 MPa and the modulus of elasticity is $E = 210$ GPa.

Solution. The cross-sectional area of the rod is

$$A = \frac{1}{4}\pi d^2 = \frac{1}{4}\pi(0.01 \text{ m})^2 = 7.85 \times 10^{-5}\,\text{m}^2$$

The tensile stress in the rod is

$$\sigma = \frac{P}{A} = \frac{12 \text{ kN}}{7.85 \times 10^{-5}\,\text{m}^2} = 153 \times 10^3 \text{ kPa}$$

$$= 153 \text{ MPa} < 200 \text{ MPa}$$

Thus, Hooke's law applies and Equation 10–5 can be used. We obtain

$$\delta = \frac{PL}{AE} = \frac{(12 \text{ kN})(2 \text{ m})}{(7.85 \times 10^{-5}\,\text{m}^2)(210 \times 10^6 \text{ kN/m}^2)}$$

$$= 0.00146 \text{ m} = 1.46 \text{ mm} \qquad\qquad \Leftarrow \textbf{Ans.}$$

EXAMPLE 10–2

See Fig. E10–2(1). Determine the total axial deformation of the steel bar between sections A and D. The bar has a cross-section $\frac{1}{2}$ in. $\times$ 1 in. and is subjected to the axial loads shown. The modulus of elasticity of steel is $E = 30 \times 10^3$ ksi and the proportional limit is 34 ksi.

FIGURE E10–2(1)

Solution. To find the internal axial force in each segment of the bar, the method of sections (as illustrated in Example 9–2) must be used. Passing a section through the bar between A and B, and using the part to the left of the section as a free-body diagram [see Fig. E10–2(2)],

FIGURE E10–2(2)

we find that the 5-kip load produces a compressive force of 5 kips for segment AB. Passing a section through the bar between B and C, and using the part to the left of the section as a free-body diagram [see Fig. E10–2(3)], we find that the 5-kip load produces a compressive force, but the 7-kip load produces a tensile force. The net effect of the two loads is a tensile force of 2 kips on segment BC. Passing a section through the bar between C and D, and using the part to the right of the section as a free-body diagram [see Fig. E10–2(4)], we find that the 8-kip load produces a tensile force of 8 kips for segment CD. The axial force diagram and the axial forces on each segment are shown in Figs. E10–2(5) and E10–2(6), respectively.

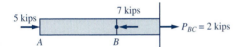

FIGURE E10–2(3)

FIGURE E10–2(4)

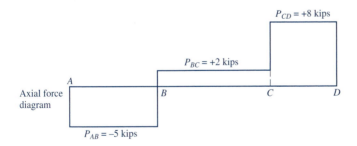

FIGURE E10–2(5)

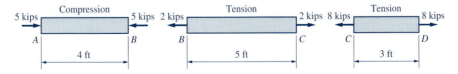

FIGURE E10–2(6)

The maximum load occurs in segment CD. The tensile stress in this segment is

$$\sigma = \frac{P}{A} = \frac{8 \text{ kips}}{0.5 \text{ in.}^2} = 16 \text{ ksi}$$

which is less than the proportional limit of 34 ksi. Thus, Hooke's law applies. The value of AE is

$$AE = \left(\frac{1}{2} \times 1 \text{ in.}^2\right)(30\ 000 \text{ kips/in.}^2) = 15\ 000 \text{ kips}$$

Thus, the axial deformation in each segment is

$$\delta_{AB} = \frac{P_{AB}L_{AB}}{AE} = \frac{(-5 \text{ kips})(4 \text{ ft})}{15\ 000 \text{ kips}} = -0.001\ 33 \text{ ft}$$

$$\delta_{BC} = \frac{P_{BC}L_{BC}}{AE} = \frac{(+2 \text{ kips})(5 \text{ ft})}{15\ 000 \text{ kips}} = +0.000\ 66 \text{ ft}$$

$$\delta_{CD} = \frac{P_{CD}L_{CD}}{AE} = \frac{(+8 \text{ kips})(3 \text{ ft})}{15\ 000 \text{ kips}} = +0.001\ 60 \text{ ft}$$

The total deformation of the bar is the algebraic sum of the axial deformations of the three segments. Thus,

$$\delta_{AD} = \delta_{AB} + \delta_{BC} + \delta_{CD}$$

$$= -0.001\ 33 \text{ ft} + 0.000\ 66 \text{ ft} + 0.001\ 60 \text{ ft}$$

$$= +0.000\ 93 \text{ ft}$$

$$= 0.0112 \text{ in. (elongation)} \qquad \Leftarrow \textbf{Ans.}$$

EXAMPLE 10–3

A circular steel bar 20 in. long is subjected to an axial tensile load of 4 kips. Determine the required diameter of the bar to the nearest sixteenth of an inch if the allowable tensile stress is 20 ksi and the total elongation is limited to 0.0085 in. The modulus of elasticity of steel is $E = 30 \times 10^3$ ksi.

Solution. From Equation 9–3, the required area A of the bar to keep the tensile stress to the allowable value of 20 ksi is

$$A = \frac{P}{\sigma_{\text{allow}}} = \frac{4 \text{ kips}}{20 \text{ kips/in.}^2} = 0.2 \text{ in.}^2$$

Solving A from Equation 10–5, the minimum area required to keep the elongation to a limiting value of 0.0085 in. is

$$A' = \frac{PL}{\delta E} = \frac{(4 \text{ kips})(20 \text{ in.})}{(0.0085 \text{ in.})(30 \times 10^3 \text{ kip/in.}^2)} = 0.314 \text{ in.}^2$$

Since A' is the larger one, the area A' must be provided and the requirement for stiffness controls. Equating A' with $\pi d^2/4 = 0.7854d^2$, we get

$$d = \sqrt{\frac{0.314 \text{ in.}^2}{0.7854}} = 0.632 \text{ in.}$$

$$\text{Use } d = \tfrac{11}{16} \text{ in. } (0.6875 \text{ in.}) \qquad \Leftarrow \textbf{Ans.}$$

10-5
STATICALLY INDETERMINATE PROBLEMS

When the unknown forces in structural members cannot be determined by the equilibrium equations alone, the structure is said to be *statically indeterminate*. Statically indeterminate problems involving axially loaded members can be analyzed by introducing the conditions of axial deformations. The following examples illustrate the use of the conditions of axial deformations, in addition to the equilibrium equations, when solving statically indeterminate problems.

──────── **EXAMPLE 10-4** ────────

Refer to Fig. E10-4(1). A bar is supported at both ends by fixed supports. Determine the reactions at the supports A and B caused by the axial force P, which acts at an intermediate point C.

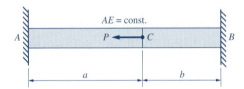

FIGURE E10-4(1)

Solution. The equilibrium condition of the free-body diagram of the bar shown in Fig. E10-4(2) requires that

$$R_A \xrightarrow{\quad} \boxed{A \qquad\qquad P \xleftarrow{\quad} C \qquad\qquad B} \xrightarrow{\quad} R_B$$

FIGURE E10-4(2)

$$\Sigma F_x = R_A + R_B - P = 0$$
$$R_A + R_B = P \tag{a}$$

Equation (a) alone is not sufficient to determine the two unknowns R_A and R_B; thus, the problem is statically indeterminate.

However, since the supports at A and B are fixed, the total deformation between A and B must be zero. This condition gives rise to another equation. From the free-body diagram of the bar in Fig. E10-4(2), we see that the AC part is subjected to a compressive load R_A, and the CB part is subjected to a tensile load R_B. Thus, we write

$$\delta_{AB} = \delta_{AC} + \delta_{CB} = \frac{(-R_A)(a)}{AE} + \frac{R_B(b)}{AE} = 0$$

Multiplying each term in the equation by AE, the equation is reduced to

$$-R_A a + R_B b = 0 \tag{b}$$

Solving Equation (a) for R_B and substituting into Equation (b) gives

$$-R_A a + (P - R_A)b = 0$$

From which we get

$$R_A = \frac{b}{a + b} P \qquad\qquad \Leftarrow \textbf{Ans.}$$

Substituting into Equation (a) gives

$$R_B = \frac{a}{a + b} P \qquad\qquad \Leftarrow \textbf{Ans.}$$

———— **EXAMPLE 10–5** ————————————————————

A steel cylinder fits loosely in a copper tube, as shown in Fig. E10–5. The length of the steel cylinder is 0.001 in. longer than the copper tube. Determine the stresses in the solid steel cylinder and in the copper tube caused by an axial force P of 55 kips applied on the rigid cap. The moduli of elasticity are: for steel, $E_{st} = 30 \times 10^3$ ksi; for copper, $E_{cu} = 17 \times 10^3$ ksi.

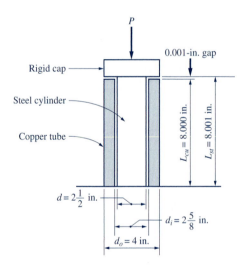

FIGURE E10–5

Solution. Assume that the given force P is large enough to close the gap. After the gap is closed, the copper tube will also be compressed and carry part of the load. Denote the forces carried by the steel cylinder and the copper tube by P_{st} and P_{cu}, respectively. The static equilibrium condition requires

$$\Sigma F_y = P_{st} + P_{cu} - 55 = 0$$

$$P_{cu} = 55 - P_{st} \qquad\qquad (a)$$

The amount of axial deformation in the steel cylinder is 0.001 in. more than that of the copper tube; that is

$$\delta_{st} = \delta_{cu} + 0.001$$

or

$$\frac{P_{st}L_{st}}{A_{st}E_{st}} = \frac{P_{cu}L_{cu}}{A_{cu}E_{cu}} + 0.001$$

The cross-sectional areas of the solid steel cylinder and the copper tube are

$$A_{st} = \frac{1}{4}\pi d^2 = \frac{1}{4}\pi(2.5)^2 = 4.91 \text{ in.}^2$$

$$A_{cu} = \frac{1}{4}\pi(d_o^2 - d_i^2) = \frac{1}{4}\pi(4^2 - 2.625^2) = 7.15 \text{ in.}^2$$

$$\frac{P_{st}(8.001)}{(4.91)(30\ 000)} = \frac{P_{cu}(8.000)}{(7.15)(17\ 000)} + 0.001$$

Multiplying each term by 10 000, we obtain

$$0.543P_{st} - 0.658P_{cu} = 10 \qquad\qquad \text{(b)}$$

Substituting Equation (a) into Equation (b) gives

$$0.543P_{st} - 0.658(55 - P_{st}) = 10$$

From which we get

$$P_{st} = 38.5 \text{ kips}$$

Substituting into Equation (a) gives

$$P_{cu} = 16.5 \text{ kips}$$

The stresses in the steel cylinder and in the copper tube are

$$\sigma_{st} = \frac{P_{st}}{A_{st}} = \frac{38.5 \text{ kips}}{4.91 \text{ in.}^2} = 7.84 \text{ ksi (C)} \qquad \Leftarrow \textbf{Ans.}$$

$$\sigma_{cu} = \frac{P_{cu}}{A_{cu}} = \frac{16.5 \text{ kips}}{7.15 \text{ in.}^2} = 2.31 \text{ ksi (C)} \qquad \Leftarrow \textbf{Ans.}$$

These stresses are well below the proportional limits of both materials. Hence the solution based on Hooke's law is justified.

——— **EXAMPLE 10–6** ———

The rigid beam AD is supported by a steel wire CF, a brass link BE, and a hinge at A, as shown in Fig. E10–6(1). The beam is in the horizontal position before the load P is applied. The moduli of elasticity are: for steel, $E_{st} = 210$

GPa; for brass, E_{br} = 105 GPa. The cross-sectional areas are: for the steel wire, A_{st} = 0.00015 m²; for the brass link, A_{br} = 0.0018 m². The dimensions indicated are the undeformed lengths before the load is applied. Find the stresses in the wire and the link.

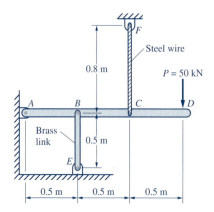

FIGURE E10–6(1)

Solution. Denote the tension in the wire by T and the compression in the link by F, as shown in the free-body diagram in Fig. E10–6(2). The moment equation about A is:

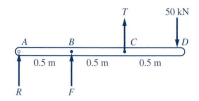

FIGURE E10–6(2)

$$\Sigma M_A = F(0.5) + T(1.0) - (50)(1.5) = 0$$

$$0.5F + T = 75 \tag{a}$$

Another equation can be established by the deformation condition. Since the beam is rigid, the deflected position of the beam is sketched, as shown in Fig. E10–6(3). We see that

$$\delta_{CF} = 2\,|\,\delta_{BE}\,|$$

or

$$\frac{TL_{CF}}{A_{st}E_{st}} = 2\,\frac{FL_{BE}}{A_{br}E_{br}}$$

$$\frac{T\,(0.8)}{(0.00015)(210 \times 10^6)} = \frac{2F\,(0.5)}{(0.0018)(105 \times 10^6)}$$

Multiplying both sides by 10^6, we obtain

$$25.4T = 5.29F$$

$$F = 4.80T \tag{b}$$

Substituting Equation (b) into Equation (a) gives

$$0.5(4.80T) + T = 75$$

$$T = 22.1 \text{ kN (T)}$$

Substituting into Equation (b) gives

$$F = 105.9 \text{ kN (C)}$$

The stresses in the steel wire and in the brass link are, respectively,

$$\sigma_{st} = \frac{T}{A_{st}} = \frac{22.1 \text{ kN}}{0.00015 \text{ m}^2} = 147\ 300 \text{ kPa}$$

$$= 147.3 \text{ MPa (T)} \qquad \Leftarrow \textbf{Ans.}$$

$$\sigma_{br} = \frac{F}{A_{br}} = \frac{105.9 \text{ kN}}{0.0018 \text{ m}^2} = 58\ 800 \text{ kPa}$$

$$= 58.8 \text{ MPa (C)} \qquad \Leftarrow \textbf{Ans.}$$

10–6
THERMAL STRESSES

Most materials expand when the temperature increases and contract when the temperature decreases. Ordinarily the expansion or contraction due to a temperature change is directly proportional to the amount of change of temperature and the length of the member. For homogeneous material, deformation due to temperature change can be calculated using the formula

$$\delta_T = \alpha L \Delta T \tag{10–6}$$

where δ_T = the change in length of the member due to temperature change

α = the *coefficient of thermal expansion per degree Fahrenheit* (/°F) or *per degree Celsius* (/°C)

ΔT = the change in temperature

L = the original length of the member

Average values of α for some common materials are given in the appendix, Table A–7. For a *statically determinate* member, the change in length caused by a temperature change can be computed readily from Equation 10–6. However, for a *statically indeterminate* member, deformations can be partially or fully constrained by supports, and the member cannot deform freely. Thus, significant stresses may be caused by a temperature change. Stresses produced by a temperature rise or drop are called *thermal stresses*.

Solutions of statically indeterminate problems involving temperature changes follow the general procedures discussed in the preceding section, as illustrated by the following three examples.

———— **EXAMPLE 10–7** ————————————————————————————

A steel rod having a uniform cross-sectional area of 3 in.^{2} is secured between rigid supports. The rod fits snugly between the supports at the temperature of 100°F. Determine the stress in the rod when the temperature drops to 32°F. For steel, $E = 30 \times 10^3$ ksi, $\alpha = 6.5 \times 10^{-6}/°F$.

Solution. The free contraction due to a temperature drop is

$$\delta_T = \alpha L \Delta T$$

The rigid supports resist the contraction by developing a reaction force P to pull the rod to its original length. The deformation caused by the pulling force P is

$$\delta_P = \frac{PL}{AE} = \frac{\sigma L}{E}$$

Equating the two values of δ, we write

$$|\delta_T| = \delta_P$$

$$\alpha L \Delta T = \frac{\sigma L}{E}$$

From which we get

$$\sigma = \alpha E \Delta T$$
$$= (6.5 \times 10^{-6}/°F)(30 \times 10^3 \text{ ksi})(100° F - 32° F)$$
$$= 13.26 \text{ ksi (T)} \qquad\qquad\qquad\qquad\qquad\qquad \Leftarrow \textbf{Ans.}$$

———— **EXAMPLE 10–8** ————————————————————————————

A steel rod 4.000 m long is fastened to a fixed support A, as shown in Fig. E10–8. There is a gap of 0.4 mm between the free end of the rod and the fixed wall B when the temperature is 10°C. Determine the thermal stress induced in the rod when the temperature rises to 38°C. For steel, $E = 210$ GPa, $\alpha = 12 \times 10^{-6}/°C$.

Solution. For a temperature rise of 38°C − 10°C = 28°C, the free expansion of the rod would be

$\delta_T = \alpha L \Delta T$

$\quad = (12 \times 10^{-6}/°C)(4 \text{ m})(28°C)$

$\quad = 0.001\ 34 \text{ m}$

$\quad = 1.34 \text{ mm}$

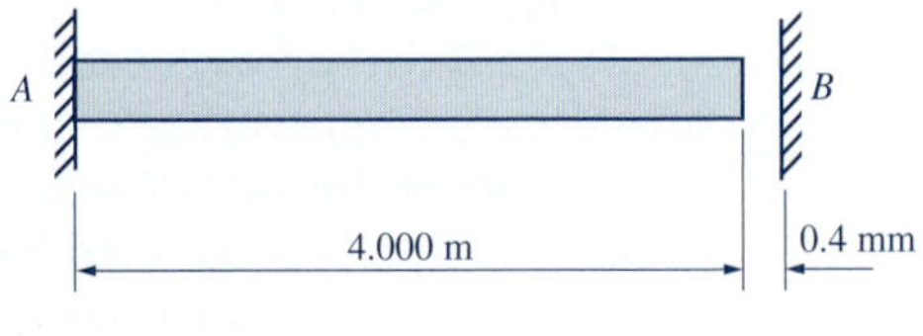

FIGURE E10–8

This is more than the dimension of the gap. Since the support and the wall are both fixed, the bar can expand only 0.4 mm to close the gap. The supports must exert a compressive force P large enough to deform the bar an amount equal to

$$\delta_p = 1.34 - 0.4 = 0.94 \text{ mm} = 0.000\ 94 \text{ m}$$

We write

$$\delta_p = \frac{PL}{AE} = \frac{\sigma L}{E}$$

From which we get

$$\sigma = \frac{\delta_p E}{L} = \frac{(0.00094 \text{ m})(210 \times 10^3 \text{ MN/m}^2)}{4 \text{ m}}$$

$$= 49.4 \text{ MPa (C)} \qquad \qquad \Leftarrow \textbf{Ans.}$$

EXAMPLE 10–9

Refer to Fig. E10–9(1). A steel bolt having a cross-sectional area of 0.442 in.^{2} is closely fitted in an aluminum sleeve having a cross-sectional area of 0.785 in.^{2}. At 60°F, the nut is hand-tightened until it fits snugly against the end of the sleeve. Determine the stresses in the bolt and the sleeve if the temperature rises to 170°F. For steel, $E_{st} = 30 \times 10^6$ psi and $\alpha_{st} = 6.5 \times 10^{-6}/°$F. For aluminum, $E_{al} = 10 \times 10^6$ psi and $\alpha_{al} = 13.0 \times 10^{-6}/°$F.

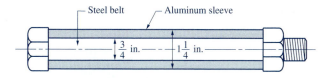

FIGURE E10–9(1)

Solution. Since α_{al} is greater than α_{st}, the thermal expansion of the aluminum sleeve is greater than that of the steel bolt. The final length of the bolt and the sleeve must be the same; therefore, the bolt is subjected to tension and the sleeve is subjected to compression.

From the static equilibrium condition, the tensile force in the bolt must be equal to the compressive force in the sleeve; that is,

$$P_{st} = |P_{al}| = P$$

From the deformation geometry shown in Fig. E10–9(2), we see that the thermal expansion $(\delta_{st})_T$ of the steel bolt plus the stretch $(\delta_{st})_P$ of the bolt is equal to the thermal expansion $(\delta_{al})_T$ of the aluminum sleeve minus the absolute value of the contraction $(\delta_{al})_P$ of the sleeve; that is,

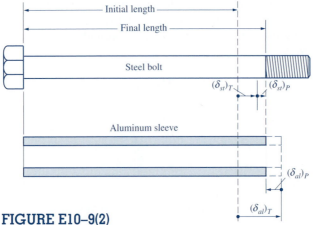

FIGURE E10–9(2)

$$(\delta_{st})_T + (\delta_{st})_P = (\delta_{al})_T - |(\delta_{al})_P|$$

or

$$\alpha_{st} L\Delta T + \frac{PL}{A_{st} E_{st}} = \alpha_{al} L\Delta T - \frac{PL}{A_{al} E_{al}}$$

After the length L is canceled from each term, the equation becomes

$$\alpha_{st}\Delta T + \frac{P}{A_{st} E_{st}} = \alpha_{al}\Delta T - \frac{P}{A_{al} E_{al}} \qquad\text{(a)}$$

in which the temperature change ΔT is

$$\Delta T = 170°\text{F} - 60°\text{F} = 110°\text{F}$$

When we substitute the numerical values, Equation (a) becomes

$$(6.5 \times 10^{-6})(110°) + \frac{P}{(0.442)(30 \times 10^6)}$$

$$= (13.0 \times 10^{-6})(110°) - \frac{P}{(0.785)(10 \times 10^6)}$$

Multiplying each term of the above equation by 10^6 and simplifying, we get

$$715 + 0.0754P = 1430 - 0.1274P$$

From which we get

$$P = 3526 \text{ lb}$$

The stresses for the steel bolt and the aluminum sleeve are, respectively,

$$\sigma_{st} = \frac{P}{A_{st}} = \frac{3526 \text{ lb}}{0.442 \text{ in.}^2} = 7980 \text{ psi (T)} \qquad \Leftarrow \textbf{Ans.}$$

$$\sigma_{al} = \frac{P}{A_{al}} = \frac{3526 \text{ lb}}{0.785 \text{ in.}^2} = 4490 \text{ psi (C)} \qquad \Leftarrow \textbf{Ans.}$$

10–7
POISSON'S RATIO

When a bar is subjected to an axial tensile load, it is elongated in the direction of the applied load; at the same time, its transverse dimension decreases, as shown in Fig. 10–3a. Similarly, if an axial compressive load is applied to the bar, the bar contracts along the axial direction while its transverse dimension increases, as shown in Fig. 10–3b.

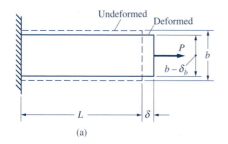
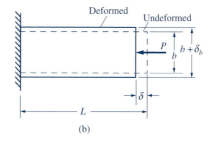

(a)   (b)

FIGURE 10–3

Experimental results show that the absolute value of the ratio of the transverse strain ϵ_t to the axial strain ϵ_a is a constant for a given material subjected to axial stresses within the proportional limit of the material. Thus,

$$\mu = \left| \frac{\text{transverse strain}}{\text{axial strain}} \right| = \left| \frac{\epsilon_t}{\epsilon_a} \right| \qquad (10\text{–}7)$$

where

$$\epsilon_a = \frac{\delta}{L} \qquad \epsilon_t = \frac{\delta_b}{b}$$

This relationship was established in the early nineteenth century by the French mathematician Poisson. The constant μ (the Greek lowercase letter mu) is called *Poisson's ratio*. Poisson's ratio is a distinct material constant. For most structural materials, Poisson's ratio ranges from 0.25 to 0.35.

Because ϵ_a and ϵ_t are always of opposite sign, Equation 10–7 can also be written as

$$\epsilon_t = -\mu\epsilon_a \qquad (10\text{–}8)$$

EXAMPLE 10–10

A steel rod 4 in. in diameter is subjected to an axial tensile force of 200 kips. Given $E = 30 \times 10^3$ ksi and $\mu = 0.29$, determine the change in diameter of the rod after the load is applied.

Solution. The tensile stress in the rod is

$$\sigma = \frac{P}{A} = \frac{200 \text{ kips}}{\frac{1}{4}\pi(4 \text{ in.})^2} = +15.9 \text{ ksi (T)}$$

This stress is within the proportional limit of steel, which is approximately 30 ksi. Hence, Hooke's law applies, and the axial strain is

$$\epsilon_a = \frac{\sigma}{E} = \frac{15.9 \text{ ksi}}{30 \times 10^3 \text{ ksi}} = +0.000\ 53 \text{ (elongation)}$$

From Equation 10–8, the transverse strain is

$$\epsilon_t = -\mu\epsilon_a = -0.29(+0.000\ 53)$$
$$= -0.000\ 154 \text{ (contraction)}$$

By definition,

$$\epsilon_t = \frac{\delta_D}{D}$$

Thus,

$$\delta_D = D\epsilon_t = (4 \text{ in.})(-0.000\ 154)$$
$$= -0.000\ 62 \text{ in. (contraction)} \qquad \Leftarrow \textbf{Ans.}$$

10–8
SHEAR STRAIN

A shear force causes shape distortion of a body. Figure 10–4 shows that a square element (shown by dashed lines) is distorted into a rhombus after the shear force F_s is applied. Total deformation δ_s occurs over a length a. The deformation per unit length δ_s/a is equivalent to tan γ. Since the angle γ (the Greek lowercase letter gamma) is very small, tan γ is equal to γ in radians; that is,

$$\gamma = \frac{\delta_s}{a} \qquad (10\text{–}9)$$

The shear strain γ is thus the change in radians in a right angle between two perpendicular lines.

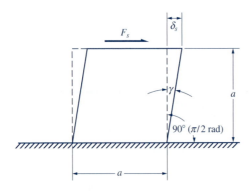

FIGURE 10–4

Hooke's Law.   When the shear stress is within the proportional limit of the material, the shear stress is proportional to shear strain for most materials. This is known as Hooke's law for shear stress and shear strain. Mathematical expression of this law is

$$\tau = G\gamma \tag{10–10}$$

where G is a constant of proportionality called the *shear modulus of elasticity* or the *modulus of rigidity.*

Like E and μ, G is a constant for a given material. Since γ is measured in radians (which are dimensionless), G is measured in the same units as those for stresses. It can be proved that the three elastic constants E, μ, and G are related by the equation

$$G = \frac{E}{2(1 + \mu)} \tag{10–11}$$

For example, if the modulus of elasticity and the modulus of rigidity of steel have been determined experimentally to be

$$E = 30 \times 10^6 \text{ psi} \quad \text{and} \quad G = 11.6 \times 10^6 \text{ psi}$$

Poisson's ratio for steel can be determined by solving μ from Equation 10–10:

$$\mu = \frac{E}{2G} - 1 = \frac{30 \times 10^6 \text{ psi}}{2(11.6 \times 10^6 \text{ psi})} - 1 = 0.29$$

EXAMPLE 10–11

An aluminum alloy rod 10 mm in diameter is subjected to an axial pull of 6 kN. Given $E = 70$ GPa and $G = 26.3$ GPa, determine the axial and transverse strain in the rod.

Solution. The axial tensile stress is

$$\sigma = \frac{P}{A} = \frac{6}{\frac{1}{4}\pi(0.01)^2} = 76\ 400 \text{ kN/m}^2 = 76.4 \text{ MPa}$$

This stress is within the proportional limit of aluminum, which is approximately 200 MPa. Thus, Hooke's law applies, and the axial strain is

$$\epsilon_a = \frac{\sigma}{E} = \frac{+76.4}{70 \times 10^3} = 0.001\ 09 \text{ m/m (elongation)}$$

Poisson's ratio of aluminum is

$$\mu = \frac{E}{2G} - 1 = \frac{70 \text{ GPa}}{2(26.3 \text{ GPa})} - 1 = 0.33$$

From Equation 10–8, the transverse strain is

$$\epsilon_t = -\mu\epsilon_a = -(0.33)(+0.001\ 09)$$

$$= -0.000\ 36 \text{ m/m (contraction)} \qquad \Leftarrow \textbf{Ans.}$$

10-9
SUMMARY

Linear Strain.   *Deformation* (δ) represents a change of dimension of a stressed member. *Linear strain* (ϵ) is defined as the deformation in a stressed member per unit of original length of the unstressed member, and it may be computed from

$$e = \frac{\delta}{L} \tag{10-1}$$

Hooke's Law.   When the stress in a member is within the *proportional limit* of the material, the stress is proportional to the strain. This linear relationship between stress and strain is called *Hooke's law* and it may be expressed as

$$\frac{\sigma}{\epsilon} = E \tag{10-2a}$$

where E is called the *modulus of elasticity*. It is a measure of resistance to deformation of a stressed member.

Axial Deformation.   If the normal stress in an axially loaded member is within the proportional limit, then Hooke's law applies and the total axial deformation of the member can be computed from:

$$\delta = \frac{PL}{AE} \tag{10-5}$$

Statically Indeterminate Problems.   When 'he unknown forces in structural members cannot be determined by the equilibrium equations alone, the structure is said to be *statically indeterminate*. Statically indeterminate problems involving axially loaded members can be analyzed by introducing the conditions of axial deformations.

Thermal Stresses.   The deformation due to a temperature change ΔT can be calculated using the formula

$$\delta_T = \alpha L \Delta T \tag{10-6}$$

where α is a property of the material referred to as the *coefficient of thermal expansion*. In a *statically indeterminate* structure, significant stresses may be caused by temperature change. Stresses produced by a temperature rise or drop are called *thermal stresses*. Solutions of statically indeterminate problems involving temperature changes follow the same general procedures as those of the statically indeterminate problems.

Poisson's Ratio.   Experimental results show that the ratio of the transverse strain ϵ_t to the axial strain ϵ_a is a constant for a given material subjected to axial stresses within the proportional limit of the material. This constant ratio (μ) is called *Poisson's ratio*. Because ϵ_a and ϵ_t are always of opposite sign, the transverse strain can be computed from

$$\epsilon_t = -\mu\epsilon_a \qquad (10\text{--}8)$$

Shear Strain. Shear stresses cause shear distortion in structural members. *Shear strain* (γ) is a measure of the amount of shear distortion and is defined as the change in angle that occurs between two lines that were originally perpendicular to each other. The shear strain is always measured in radians.

When the shear stress is within the proportional limit of the material in shear, the shear stress is proportional to shear strain. This is known as *Hooke's law* for shear stress and shear strain, and it can be expressed as

$$\tau = G\gamma \qquad (10\text{--}10)$$

where G is a constant of proportionality called the *shear modulus of elasticity* or the *modulus of rigidity*. The three elastic constants, E, μ, and G, are related by the equation

$$G = \frac{E}{2(1 + \mu)} \qquad (10\text{--}11)$$

PROBLEMS

Section 10–3   Hooke's Law

Section 10–4   Axial Deformation

10–1   A 10-ft steel bar is subjected to a tensile stress of 20 ksi. Determine (*a*) the linear strain and (*b*) the total deformation of the bar. The modulus of elasticity of steel is 30×10^3 ksi.

10–2   An aluminum rod of 20-mm diameter is elongated 3.5 mm along its longitudinal direction by a load of 25 kN. If the modulus of elasticity of aluminum is $E = 70$ GPa, determine the original length of the bar.

10–3   A 20-ft wrought-iron bar $\frac{1}{2}$ in. in diameter is subjected to a tensile force of 3 kips. Determine the stress, strain, and elongated length of the bar. The modulus of elasticity of wrought iron is $E = 29 \times 10^3$ ksi.

10–4   A metal wire is 10 m long and 2 mm in diameter. It is elongated 6.06 mm by a tensile force of 400 N. Determine the modulus of elasticity of the material and indicate a possible material for the wire.

10–5   A steel tape used in surveying is designed to be exactly 100 ft long when fully supported on a horizontal, frictionless plane and subjected to a tensile force of 10 lb. Determine the stretched length of the tape if it is subjected to an axial tensile force of 20 lb when supported in the same way. The tape is $\frac{1}{32}$ in. thick and $\frac{3}{8}$ in. wide.

10–6 Refer to Fig. P10–6. Determine the total elongation of strut AB due to a weight $W = 15$ kN if the strut has a 10-mm diameter and is made of steel with $E = 210$ GPa.

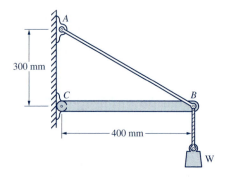

FIGURE P10–6

10–7 See Fig. P10–7. Determine the total elongation of cable BC due to a weight $W = 600$ lb if the cable has a cross-sectional area 0.025 in.^{2} and is made of steel with $E = 30 \times 10^6$ psi.

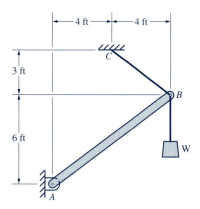

FIGURE P10–7

10–8 An aluminum bar 30 mm in diameter is suspended as shown in Fig. P10–8. Determine the total displacement of the lower end C after the loads are applied. The modulus of elasticity of aluminum is $E = 70$ GPa.

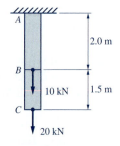

FIGURE P10–8

10–9 A brass bar having a uniform cross-sectional area of 2 in.^{2} is subjected to the forces shown in Fig. P10–9. Determine the total deformation of the bar. The modulus of elasticity of brass is $E = 17 \times 10^3$ ksi.

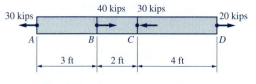

FIGURE P10–9

10–10 See Fig. P10–10. Determine the total elongation of the steel eye bar BC of 10-mm diameter due to the load $P = 8$ kN. The modulus of elasticity of steel is $E = 210$ GPa.

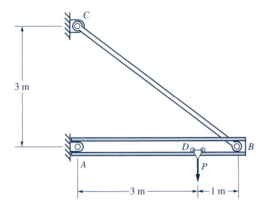

FIGURE P10–10

10–11 Determine the total deformation between points A and D of a stepped steel bar subjected to the axial forces shown in Fig. P10–11. The modulus of elasticity of steel is $E = 210$ GPa.

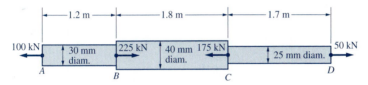

FIGURE P10–11

10–12 The two wires shown in Fig. P10–12 support a heavy bar weighing 900 lb. The wires AC and BD are identical, having the same $\frac{3}{8}$-in. diameter, the same 5-ft original length, and the same modulus of elasticity $E = 30 \times 10^6$ psi. Determine the deformation of each wire.

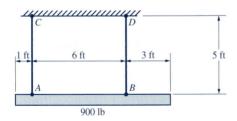

FIGURE P10–12

10–13 Determine the diameter of wire BD in Problem 10–12 so that the deformations of the two wires are equal. Other data remain unchanged.

10–14 A steel rod used in a control mechanism must transmit a tensile force of 10 kN without exceeding an allowable stress of 150 MPa or stretching more than 1 mm per 1 meter of length. The modulus of elasticity is $E = 210$ GPa. Find the proper diameter of the bar.

Section 10–5 Statically Indeterminate Problems

10–15 A stepped bar, with each portion made of the same material, is supported between two fixed supports as shown in Fig. P10–15. Determine the stress in each segment of the bar caused by the applied load of 100 kips.

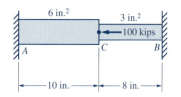

FIGURE P10–15

10–16 For the member in Fig. P10–16 supported between two fixed supports, determine the reactions at the supports due to the axial loads shown.

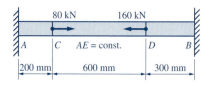

FIGURE P10–16

10–17 See Fig. P10–17. A bronze cylinder of 0.0065 m² cross-sectional area fits loosely inside an aluminum tube having 0.0045 m² cross-sectional area. The aluminum tube is 0.3 mm longer than the bronze cylinder before the load is applied. If the modulus of elasticity of bronze is 83 GPa and that of aluminum is 70 GPa, determine the stresses in the bronze cylinder and in the aluminum tube caused by an axial load P of 1000 kN.

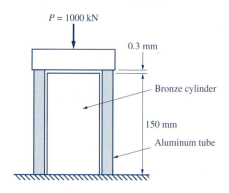

FIGURE P10–17

10–18 A timber post 10 in. × 10 in. is strengthened by four steel angles L2 × 2 × $\frac{3}{16}$, as shown in Fig. P10–18. If the moduli of elasticity are $E_{st} = 30 \times 10^3$ ksi and $E_{wd} = 1.5 \times 10^3$ ksi, and the allowable stresses are $(\sigma_{st})_{\text{allow}} = 23$ ksi and $(\sigma_{wd})_{\text{allow}} = 1.7$ ksi, determine the allowable load P.

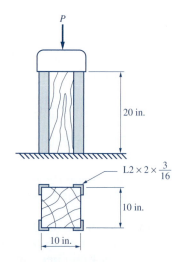

FIGURE P10–18

10–19 See Fig. P10–19. A 5-kip weight is lifted by two steel wires; one wire is initially 10.000 in. long and the other wire is initially 10.002 in. long. The cross-sectional area and modulus of elasticity of each wire are, respectively, $A = 0.25$ in.^{2} and $E = 30 \times 10^3$ ksi. Determine the stress in each wire.

5 kips

FIGURE P10–19

10–20 Refer to Fig. P10–20. A load $P = 50$ kN is applied to a rigid bar suspended by three wires of the same cross-sectional area. The outside wires are copper ($E_{cu} = 117$ GPa) and the inside wire is steel ($E_{st} = 210$ GPa). There is no slack or tension in the wires before the load is applied. Determine the load carried by each wire. (*Hint*: Due to symmetry, the axial deformation of each wire is identical.)

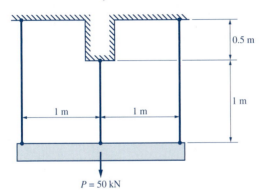

0.5 m

1 m

1 m

1 m

$P = 50$ kN

FIGURE P10–20

10–21 See Fig. P10–21. A rigid beam is supported by a hinge at A and two identical steel rods at C and E. Determine the forces in the rods due to the load $W = 40$ kN. Before the load is applied, the beam hangs in the horizontal position.

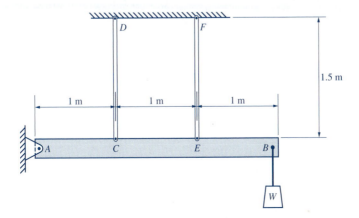

D

F

1.5 m

1 m

1 m

1 m

A

C

E

B

W

FIGURE P10–21

Section 10–6   Thermal Stresses

10–22 A steel structural member is supported between two fixed supports so that it cannot expand. At 60°F, there is no stress in the member. Determine the stresses in the member at 100°F. For steel, $E = 30 \times 10^6$ psi and $\alpha = 6.5 \times 10^{-6}/°F$.

10–23 If the member in Problem 10–22 is 40 ft long and is supported between two fixed supports, with a clearance of $\frac{1}{16}$ in. at 60°F, determine the stress in the member at 100°F.

10–24 A steel wire is held taut between two unyielding supports. At 10°C, the wire is tightened so that it has a tensile stress of 100 MPa. Determine the temperature at which the wire would become slack. For steel, $E = 210$ GPa and $\alpha = 12 \times 10^{-6}/°C$.

10–25 The composite steel and brass rod shown in Fig. P10–25 is attached to unyielding supports with no initial stress at 25°C. Determine the stress in each segment when the temperature is reduced to −5°C. For steel, $E_{st} = 210$ GPa and $\alpha_{st} = 12 \times 10^{-6}/°C$, and for brass, $E_{br} = 100$ GPa and $\alpha_{br} = 19 \times 10^{-6}/°C$.

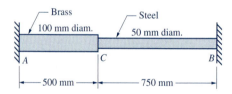

FIGURE P10–25

10–26 Three short posts of equal length support the 100-kN rigid concrete block shown in Fig. P10–26. Determine (a) the stress in each post and (b) the temperature decrease that will relieve the aluminum post of any stress. For steel, $E_{st} = 210$ GPa and $\alpha_{st} = 12 \times 10^{-6}/°C$, and for aluminum, $E_{al} = 70$ GPa and $\alpha_{al} = 23.4 \times 10^{-6}/°C$.

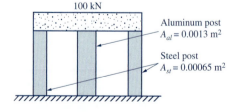

FIGURE P10–26

10–27 Solve Example 10–9 if the steel bolt is 1 in. in diameter, the aluminum sleeve has a $1\frac{1}{2}$-in. outside diameter, and the temperature rises to 150°F.

Section 10–7   Poisson's Ratio

Section 10–8   Shear Strain

10–28   Consider a carefully conducted tensile test of a copper specimen 10 mm in diameter and 50 mm in gage length. When an axial load of 10 kN is applied, the elastic deformation in the gage length is 0.0544 mm and the diameter is decreased by 0.0039 mm. Calculate the three elastic constants, E, μ, and G.

10–29   A steel tensile specimen 0.505 in. in diameter and 2 in. in gage length has stress and strain at the proportional limit equal to 42.0 ksi and 0.0014 in./in., respectively. The shear modulus is $G = 11.6 \times 10^3$ ksi. Determine the change in diameter of the specimen at the proportional limit.

10–30   An aluminum plate is subjected to an axial tensile force $P = 10$ kN. The plate has the following dimensions and characteristics: length $L = 100$ mm, width $b = 20$ mm, thickness $t = 5$ mm, $E = 70$ GPa, and $G = 26.3$ GPa. Determine the deformations in the length L, in the width b, and in the thickness t.

Computer Program Assignments

For each of the following problems, write a computer program using an appropriate programming language with which you are most familiar. Make the program user friendly by incorporating plenty of comments and input prompts so that the user will understand the input data to be entered and the limitations of their values. The output should include the data entered and the computed results, and they must be well labeled to identify each quantity. If a tabulated format is used, a proper heading must be included at the top of the table. Do not limit the program to any specific unit system. Indicate the consistent U.S. customary or SI units that can be used.

C10–1 Refer to Fig. C10–1. Write a computer program that can be used to determine the stress and strain in the stepped bar subjected to the axial loads shown. The user input should include (1) the total number of elements n; (2) the modulus of elasticity E of the material; and (3) the cross-sectional area A_i, the length L_i, and the load P_i applied to the right end of each segment. The load P_i is treated as positive if its direction is the same as shown in the figure. The output should include (1) the internal axial force and the average normal stress in each segment, indicating whether the stress is tension or compression, and (2) the axial deformation in each segment, as well as the total deformation over the entire length. Use this program to solve (a) Example 10–2 (b) Problem 10–8, (c) Problem 10–9, and (d) Problem 10–11.

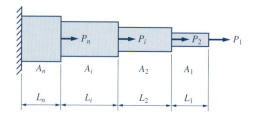

FIGURE C10–1

C10–2 Write a computer program that can be used to determine the reactions at A and B of the statically indeterminate, axially loaded rod AB shown in Fig. C10–2. The user input should include (1) the total number of elements n and (2) the cross-sectional area A_i, the length L_i, and the load P_i applied to the right end of each segment. The load P_i is treated as positive if its direction is the same as indicated in the figure. Note that, for the first segment, P_i is R_B, which is unknown, and there is no need to enter its value. The output should include (1) the reaction at A and B and (2) the internal force and the normal stress in each element, indicating whether the stress is in tension or compression. Use this program to solve (a) Problem 10–15; and (b) Problem 10–16, assuming that $A = 0.001$ m^2.

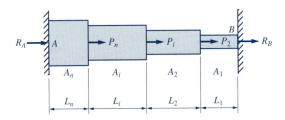

FIGURE C10–2

CHAPTER 11

MECHANICAL PROPERTIES OF MATERIALS

11–1
INTRODUCTION

Having discussed the basic concepts of stress and strain in the previous two chapters, we shall investigate in this chapter the relationship between stress and strain by conducting tension and/or compression tests of materials. These tests produce the stress–strain diagrams that reveal important properties of materials, including modulus of elasticity, yield and ultimate strengths, stiffness, elastic and plastic behavior, ductility, etc. Other related topics such as stress concentrations and ultimate strength design approach will also be introduced.

11–2
THE TENSION TEST

Information about the mechanical properties of materials is usually obtained through laboratory testing. *Tension tests* are the most commonly used tests for metals or other ductile materials. Figure 11–1 shows a universal testing machine used for this purpose. The universal testing machine is also capable of performing compression, shear, and bending tests.

In a *tension test*, a round test specimen, made to American Standard of Testing and Materials (ASTM) specifications, is clamped to the machine between the upper head and the lower head, as shown in Fig. 11–2. The lower head is stationary during the test, while the upper head is pushed upward by the hydraulic pressure. When the upper head moves upward, the specimen is stretched at a slow rate controlled by the load valve. The tensile force P acting on the specimen at any time during the test is indicated by the digital load indicator. While the tensile force P is recorded, the corresponding change in length between two punched marks A and B (Fig. 11–2) on the specimen is measured. The original distance between the marks is called the *gage length*. Commonly used gage lengths are 2 in. and 8 in. To measure elongation, a mechanical or electronic extensometer, capable of measuring deformation as small as 0.0001 in., is mounted on the specimen at marks A and B. The values of load P versus elongation δ at proper intervals are recorded. The corresponding values of stress versus strain can be calculated by

$$\sigma = \frac{P}{A} \qquad \epsilon = \frac{\delta}{L}$$

where A is the original cross-sectional area and L is the gage length.

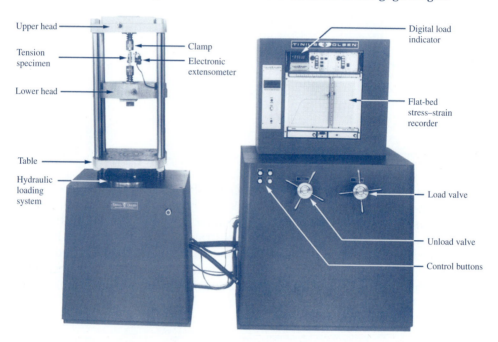

FIGURE 11–1   Universal Testing Machine (Courtesy of Tinius Olsen Testing Machine Co., Inc.)

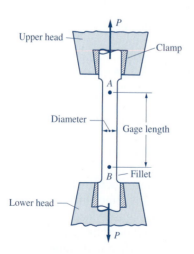

FIGURE 11–2   Tension Test

11–3
THE STRESS–STRAIN DIAGRAM

A *stress–strain diagram* is a graphical representation of the results of the tension test. From the data of the tension test, the corresponding values of stress versus strain can be plotted to produce a stress–strain diagram. A modern universal testing machine, such as the one shown in Fig. 11–1, can be set up to plot a stress–strain diagram automatically. The diagram establishes a relationship between stress and strain, and for most practical purposes, the relationship is independent of the cross-sectional area of the specimen and the gage length used.

It is customary to plot the stress–strain diagram by plotting stress as the ordinate and strain as the abscissa. A typical stress–strain diagram, obtained by testing a mild steel specimen, is shown in Fig. 11–3. (Mild steel has low carbon content and is widely used in construction.) The stress–strain diagram in Fig. 11–3 consists of four stages: elastic, yield, strain-hardening, and localized deformation.

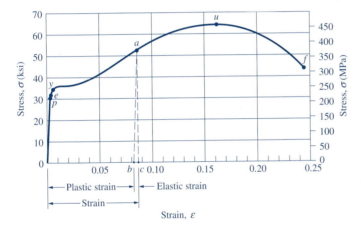

FIGURE 11–3

Elastic Stage. The diagram is a straight line up to point p, the *proportional limit*, because up to this point the stress is proportional to the strain. The stress at the proportional limit is denoted by σ_p. Beyond the proportional limit, the stress is no longer proportional to the strain.

Up to point e, the deformation is *elastic*, which means that the specimen will return to its initial size and shape on removal of the load. Point e is called the *elastic limit*, and the corresponding stress is denoted by σ_e. To determine the elastic limit, the load must be applied to the specimen and released; and a little greater load is applied again and released. The elastic limit is reached when a permanent deformation is detected in the specimen. For most materials, points p and e are very close, so for all practical purposes, the elastic limit and the proportional limit could be considered as the same point.

Beyond the elastic limit, only part of the deformation can be recovered after the load is removed; the remaining part of the deformation becomes *permanent set*. The deformation that can be recovered is called *elastic deformation*, and the permanent set is called *plastic deformation*.

Yield Stage.   At point y, the curve becomes horizontal or has a very mild slope. This means that the specimen continues to elongate without any significant increase in load. The material is said to have yielded, and the point is called the *yield point*. The corresponding stress is called the *yield strength* (*stress*) of the material and is denoted by σ_y.

Beyond the yield point, there will be an appreciable amount of plastic deformation. In machine parts, plastic deformation will affect their useful function. Therefore, the yield stress σ_y is an important index of the strength of the material.

Strain-Hardening Stage.   The ability of the material to resist deformation is regained after the yield stage has passed. Because of plastic deformation, the material *strain hardens*; thus, the stresses required to deform the specimen even more become greater. The stress–strain diagram reaches the highest point, u; the corresponding stress at this point is denoted by σ_u and is called the *ultimate strength*, that is, the maximum stress that a material can resist.

Stage of Localized Deformation.   While the specimen is being elongated, its lateral dimension contracts. The lateral contraction is so small for low stress levels that it can hardly be detected. Beyond point u, the lateral contraction becomes more pronounced. A drastic decrease in diameter occurs in a localized area, as shown in Fig. 11–4. This phenomenon is called *necking*. Only ductile metals exhibit this characteristic. After initial necking occurs, the cross-section at the necked-down section quickly decreases, and the tensile force required to produce additional stretch of the specimen also decreases. The tensile stress, which is computed based on the original cross-sectional area, decreases accordingly. When the stress–strain curve reaches point f, the specimen suddenly breaks into two parts. This point is called the *point of rupture*.

(a)   (b)

FIGURE 11–4

Because the ultimate strength σ_u is the maximum stress reached before failure, the ultimate strength is an important index of the strength of the material. In addition to the proportional limit, the yield strength, and the ultimate strength, the following indices can also be determined.

Modulus of Elasticity.   In the stress–strain diagram shown in Fig. 11–3, the part Op from the origin to the proportional limit is essentially a straight line, indicating that *Hooke's law* applies. Stress and strain are thus related by

$$\sigma = E\epsilon \qquad\qquad (11\text{--}1)$$

where E is the *modulus of elasticity*.
From Fig. 11–5, we see that

$$E = \frac{\sigma}{\epsilon} = \frac{\Delta\sigma}{\Delta\epsilon} = \tan\alpha \qquad\qquad (11\text{--}2)$$

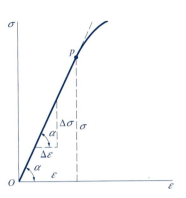

FIGURE 11–5

Geometrically, E can be interpreted as the slope of the straight line Op in the stress–strain diagram. The values of E for some common engineering materials are listed in the appendix, Table A–7.

Elastic and Plastic Behavior. Below the *elastic limit*, the material behaves *elastically*; that is, the strain induced by the load can be recovered totally if the load is removed. If the material is ductile and has a well-defined yield point, as in Fig. 11–3, the elastic limit, the proportional limit, and the yield point are essentially the same point. This means that the material behaves elastically and linearly up to the yield point. Beyond the yield point, the material behaves *plastically*. When a mild steel specimen is elongated to the plastic range at point a (Fig. 11–3) and then the load is gradually released, the unloading stress–strain curve ab has a slope that is essentially the same as that of line Op. When the load is removed completely, permanent deformation in the specimen remains. Thus, the strain at point a consists of two parts: part bc is the strain that is recovered on unloading; thus, it is the *elastic strain*. Part Ob is the permanent set in the material that cannot be recovered; thus, it is the *plastic strain*.

Yield Point by the Offset Method. When a material does not possess a well-defined yield point, the point can be determined by the *offset method*. A line parallel to the straight-line portion of the stress–strain diagram is drawn starting from a point on the abscissa with a given offset from the origin. The intersection of the line and the stress–strain diagram is the yield point. The most commonly used offset is 0.2 percent (i.e., $\epsilon = 0.002$). The stress at point y in Fig. 11–6 obtained by this method is called the yield strength of the material at 0.2 percent offset.

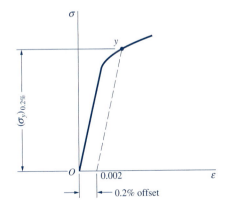

FIGURE 11–6

Percent Elongation.   When the specimen is removed from the testing machine and the two broken pieces are put together, the distance between the gage length marks becomes L'. The *percent elongation* is defined as the ratio of change of gage length to the original gage length L, expressed as a percentage. Thus,

$$\text{Percent elongation} = \frac{L' - L}{L} \times 100\% \qquad (11\text{--}3)$$

Percent Reduction in Area.   After the specimen breaks, the ratio of the reduction in cross-sectional area at the fractured section to the original cross-sectional area, expressed as a percentage, is called the percent reduction in area. Thus,

$$\text{Percent reduction in area} = \frac{A - A'}{A} \times 100\% \qquad (11\text{--}4)$$

where A is the original cross-sectional area and A' is the area at the fractured section.

─── **EXAMPLE 11–1** ───

The stress–strain diagram in Fig. E11–1 is the result of a tension test on a steel specimen. The specimen has an original diameter of 0.502 in. and a gage length of 2 in. between two punch marks. After rupture, the diameter at the rupture reduces to 0.405 in., and the length between the two punch marks stretches to 2.55 in. Determine (*a*) the stress at the proportional limit, (*b*) the modulus of elasticity, (*c*) the ultimate strength, (*d*) the yield stress at 0.2 percent offset, (*e*) the percent elongation, and (*f*) the percent reduction in area.

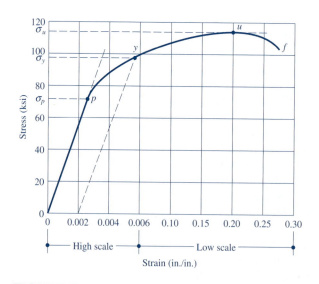

FIGURE E11–1

Solution.

(*a*) *The Stress at the Proportional Limit.* From the diagram, the stress at point *p* is

$$\sigma_p = 71 \text{ ksi} \qquad \Leftarrow \textbf{Ans.}$$

The corresponding strain at *p* is

$$\sigma_p = 0.0024 \text{ in./in.}$$

(*b*) *The Modulus of Elasticity.* The slope of the straight line, from the origin to the proportional limit, represents the value of *E*. Thus,

$$E = \frac{\delta_p}{\epsilon_p} = \frac{71 \text{ ksi}}{0.0024} = 30 \times 10^3 \text{ ksi} \qquad \Leftarrow \textbf{Ans.}$$

(*c*) *The Ultimate Strength.* The value of σ_u is the stress at the highest point *u* of the stress–strain diagram:

$$\sigma_u = 114 \text{ ksi} \qquad \Leftarrow \textbf{Ans.}$$

(*d*) *Yield Stress.* Since there is no well-defined yield point in the stress–strain diagram, the yield point is determined by the 0.2 percent offset method. From a point on the abscissa where the strain is equal to 0.002, draw a line parallel to the straight line. The point of intersection *y* of this line with the stress–strain diagram is the yield point corresponding to 0.2 percent offset. The yield stress at this point is

$$\sigma_y = 98 \text{ ksi} \qquad \Leftarrow \textbf{Ans.}$$

(*e*) *The Percent Elongation.* By definition, we find

$$\text{Percent elongation} = \frac{L' - L}{L} \times 100\%$$

$$= \frac{2.55 - 2.00}{2.00} \times 100\%$$

$$= 27.5\% \qquad \Leftarrow \textbf{Ans.}$$

Thus, the material is very ductile.

(*f*) *The Percent Reduction in Area.* By definition, we find

$$\text{Percent reduction in area} = \frac{A - A'}{A} \times 100\%$$

$$= \frac{\frac{1}{4}\pi(0.502)^2 - \frac{1}{4}\pi(0.405)^2}{\frac{1}{4}\pi(0.502)^2} \times 100\%$$

$$= 34.9\% \qquad \Leftarrow \textbf{Ans.}$$

11–4

MECHANICAL PROPERTIES OF MATERIALS

The stress–strain diagram of a material resulting from a tension test characterizes the mechanical behavior of the material. Different materials possess different characteristic curves. From a given stress–strain diagram for a material, the values of proportional limit, yield strength, and ultimate strength can be identified. In addition, the modulus of elasticity, percent elongation, and percent reduction in cross-sectional area are also obtained. These values are indices to the following mechanical properties of materials, which are significant in the selection of those materials.

1. **Strength.** *Strength* of a material is the greatest stress that it can withstand without failure. Depending on the type of material and nature of loading, it can be indicated by the proportional limit, yield strength, or ultimate strength of the material.

2. **Stiffness.** *Stiffness* is the ability of a material to resist deformation. A material with a high value of modulus of elasticity E is stiffer than materials with a lower value of E. From the appendix, Table A–7, we find that the moduli of elasticity for steel and aluminum are 30×10^3 ksi (210 GPa) and 10×10^3 ksi (70 GPa), respectively. This means that steel is three times stiffer than aluminum. For an aluminum rod of the same length and subjected to the same stress, the deformation is three times that of a steel rod.

3. **Elasticity.** *Elasticity* is the property of a material that enables it to regain its original, undeformed length once the load is removed. Most structural materials behave elastically up to the elastic limit of the material. If the material is ductile and has a well-defined yield point, as shown in Fig. 11–3, the elastic limit, proportional limit, and yield point are essentially the same. This means that the material behaves elastically and linearly up to the yield point.

4. **Ductility.** *Ductility* is the ability of a material to undergo a lot of plastic deformation before rupture. The stress–strain diagram shown in Fig. 11–3 is typical of ductile materials. Less ductile materials exhibit different stress–strain characteristics. Figure 11–7 shows the stress–strain diagrams of several common materials. Ductility is indicated by the *percent elongation* of the gage length as well as by the percent reduction in cross-sectional area at the fractured section. A material with a high percent of elongation indicates that the material has a high degree of plastic deformation and thus is more ductile than a material with a low percent of elongation. Mild steel, being very ductile, has a 20 to 30 percent elongation. Materials with a percent of elongation greater than 5 percent, such as steel, copper, and aluminum, are called *ductile* materials.

5. **Brittleness.** A material that undergoes very little plastic deformation before rupture is said to be *brittle*. A brittle material exhibits no yielding and does not exhibit the necking-down phenomenon. It ruptures suddenly and without warning at the ultimate strength. Brittle materials, such as cast iron, concrete, sandstone, glass, and ceramics, are generally weak and unreliable in tension.

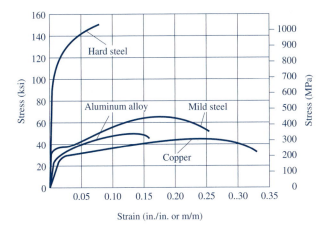

Strain (in./in. or m/m)

FIGURE 11–7

6. **Hardness.** *Hardness* is a measure of the resistance of a material to penetration. Hardness correlates well with strength and wearing of the material. Because tests are easy and inexpensive to perform and do not destroy the specimen, they are very useful for quality-control purposes.

7. **Machinability.** *Machinability* is the ease with which a material can be machined. The machinability of metal is essentially a function of hardness and ductility. It is frequently a critical factor for the selection of metals for parts made by automatic machine tools. Often an expensive metal that is readily machinable is a more economical selection than a lower-priced metal that is difficult to machine.

8. **Resilience.** The capacity of a material to absorb energy within the elastic range is called *resilience*. Resilience is measured by the triangular area under the elastic portion of the stress–strain curve. A material with greater resilience is capable of absorbing greater impact energy without any plastic deformation. Resilience should be considered when the material is subjected to shock loading.

9. **Toughness.** The capacity of a material to absorb energy without fracture is called *toughness*. Toughness is measured by the total area under the entire stress–strain curve up to the point of fracture. A material is said to be tough if it can absorb a lot of impact energy without fracture.

11–5
THE COMPRESSION TEST

The simplest and most informative test for establishing the mechanical properties of metals and other ductile materials is the tension test discussed in the previous sections. Brittle materials, such as concrete, are tested in compression due to low tensile strength and brittleness. In a *compression test*, the specimen used is generally a cylinder with a height $1\frac{1}{2}$ to 3 times the diameter, so that the specimen will not buckle under the compressive load.

The stress–strain diagram resulting from a compression test of mild steel is shown in Fig. 11–8. The modulus of elasticity E and yield stress σ_y in compression are the same as those determined from the tension test. After the specimen has yielded, it becomes flatter; its diameter increases continuously. Thus, the compressive load also increases continuously and the ultimate strength in compression cannot be obtained. Compression tests are rarely performed for ductile materials.

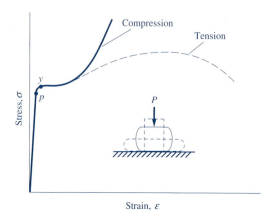

FIGURE 11–8

The compression characteristics of brittle materials differ greatly from the test results for ductile materials. Brittle materials rupture along a *cup–cone* section, at an angle of 45° from the axis of the specimen. This is evidence that the failure is caused by excessive shear stress (see Section 9–6). Figure 11–9 shows the stress–strain diagram of cast iron in a compression test. The ultimate strength of the cast iron in compression is four to five times higher than that in tension. Other brittle materials, such as concrete, brick, and ceramics, also display much higher resistance to compression than to tension. For these materials, a compression test is more important than a tension test.

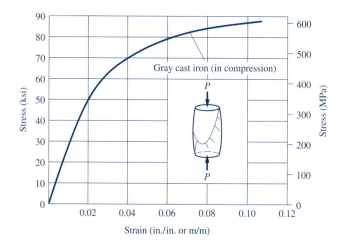

FIGURE 11–9

11–6
ALLOWABLE STRESSES AND FACTOR OF SAFETY

To provide a margin of safety in design, machine and structural members are designed for a limited stress level, called the *allowable stress*, which is the maximum stress considered to be safe when a member made of a given material is subjected to a known loading condition. Allowable stress is usually a fraction of the ultimate strength or the yield stress of the material. This ensures reserved strength in the load-carrying capacity of the member. The ratio of the failure stress to the allowable stress is defined as the *factor of safety* (F.S.). We write

$$F.S. = \frac{\text{failure stress}}{\text{allowable stress}} \tag{11–5}$$

In an elastic design with ductile materials, such as structural steel or aluminum, significant deformation may occur and render the structural element unusable when the yield strength is reached. Thus, the yield strength is taken as a failure stress, and the factor of safety is used to guard against yielding. On the other hand, a brittle material will fail suddenly at the ultimate strength without excessive deformation. Thus, for a brittle material, the ultimate strength is taken as the failure stress. Therefore, we write

$$F.S. = \frac{\text{yield strength}}{\text{allowable stress}} \quad \text{(for ductile materials)} \tag{11–6a}$$

$$F.S. = \frac{\text{ultimate strength}}{\text{allowable stress}} \quad \text{(for brittle materials)} \tag{11–6b}$$

A factor of safety of 2 means that the member can withstand a maximum load equal to twice the load for which the member is designed before failure or yield would occur.

Choosing an appropriate value for the factor of safety requires engineering judgment based on many considerations, including the following:

1. *Variation in material properties.* Materials of structural members are not entirely homogeneous or of uniform quality. For example, impurities such as carbon inclusions may cause nonuniformity in material properties.
2. *Uncertainty in the method of analysis.* The assumptions used in the analysis and design are often subject to appreciable error.
3. *Problems in manufacturing.* During manufacturing, uneven cooling in different portions of the metal often leaves some *residual stress* within a structural member.
4. *Environmental conditions.* A material may deteriorate due to rust, corrosion, or chemical attack.
5. *Uncertainty in loading conditions.* It is difficult to estimate the exact load to which a structure is subjected, and unexpectedly large loads might occur. Dynamic load may produce impact; a long-term sustained load may produce *creep*, which is the continuous deformation caused by a sustained load. Cyclic loading, applied repeatedly, will weaken the material and produce *fatigue* failure. In such cases, rupture may occur at a stress much lower than the static breaking strength.

6. *Risk and liability.* The failure of a member may produce loss of life or property damage.

Factors of safety for most structural components are specified by design specifications or building codes written by committees of experienced engineers working with professional societies, or with federal, state, and/or city agencies.

─────── **EXAMPLE 11–2** ───

According to the American Institute of Steel Construction (AISC) manual, the allowable stress for a steel tensile member is $(\sigma_t)_{allow} = 0.60\sigma_y$. What is the factor of safety used?

Solution. Steel tensile members are usually ductile. Using Equation 11–6a, we find

$$F.S. = \frac{\text{yield strength}}{\text{allowable stress}} = \frac{\sigma_y}{0.6\sigma_y} = 1.67 \qquad \Leftarrow \textbf{Ans.}$$

which means that there is a 67% margin of safety guarding against yielding of the tensile member.

─────── **EXAMPLE 11–3** ───

An A36 steel tie rod with $\sigma_y = 36$ ksi is designed to support a tensile load of 8 kips. Determine the required diameter of the rod to the nearest $\frac{1}{16}$ in. if the allowable tensile stress is $(\sigma_t)_{allow} = 0.60\sigma_y$ and the allowable elongation is 0.15 in. over a length of 10 ft.

Solution. The required area based on the allowable stress is

$$\text{Required } A = \frac{P}{(\sigma_t)_{allow}}$$

$$= \frac{8 \text{ kips}}{0.6(36 \text{ kips/in.}^2)}$$

$$= 0.370 \text{ in.}^2$$

Since the allowable tensile stress is well within the proportional limit, the axial deformation is elastic and the following formula for elastic deformation applies:

$$\delta = \frac{PL}{AE}$$

From which the required area based on the axial deformation is

$$\text{Required } A = \frac{PL}{\delta E}$$

$$= \frac{(8 \text{ kips})(120 \text{ in.})}{(0.15 \text{ in.})(30\ 000 \text{ kips/in.}^2)}$$

$$= 0.213 \text{ in.}^2$$

Comparing the two required areas, we see that the required area is controlled by the allowable tensile stress. Equating that area to $\frac{1}{4}\pi d^2 = 0.7854d^2$, we find

$$\text{Required diameter } d = \sqrt{\frac{0.370 \text{ in.}^2}{0.7854}} = 0.686 \text{ in.}$$

$$\text{Use an } \frac{11}{16} \text{ in. (0.6875 in.) rod} \qquad \Leftarrow \textbf{Ans.}$$

11–7
STRESS CONCENTRATIONS

As mentioned previously, when an axially loaded, prismatic member is subjected to central axial tension or compression, the normal stress is uniformly distributed in the cross-section. This uniform stress distribution will occur on all cross-sections except those in the vicinity of the points of application of the loads. Should abrupt changes in geometry (sometimes called *stress-raisers*) exist in a member, however, large irregularities in the stress distribution will develop. Examples of stress-raisers in flat, axially loaded members and the resulting stress distributions are shown in Fig. 11–10.

Figure 11–10a shows the presence of a circular hole centrally located in the member. Figure 11–10b demonstrates the presence of symmetrically placed, semicircular notches in the member. Figure 11–10c shows a member consisting of two segments of different widths connected by two quarter-circle fillets. In each case, a localized stress concentration develops immediately adjacent to the stress-raiser. In all these cases, the maximum stress may be several times greater than the average stress over the net cross-sectional area at section *m–m*. The stress distribution is uniform over sections at a considerable distance away from the region of abrupt geometric change, such as sections *n–n* and *q–q*. The abrupt increase in stress at localized regions is called *stress concentration*. The ratio of the maximum stress to the average stress over the *net* cross-sectional area at section *m–m* is called the *stress concentration factor*. Thus,

$$K = \frac{\sigma_{\text{max}}}{\sigma_{\text{avg}}} \qquad (11\text{–}7)$$

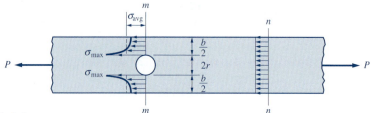

(a) Circular hole

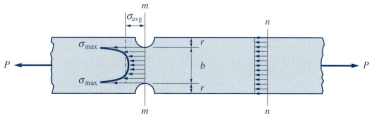

(b) Semicircular notches

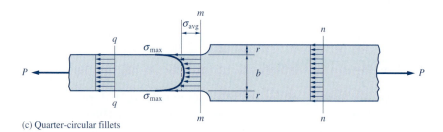

(c) Quarter-circular fillets

FIGURE 11–10

where K is the stress concentration factor and σ_{avg} is the average stress over the net cross-sectional area. If K is known, the maximum stress is

$$\sigma_{max} = K\sigma_{avg} = K\frac{P}{A_{net}} = K\frac{P}{bt} \tag{11–8}$$

where b is the net width at section m–m and t is the thickness of the plate.

Theoretical analysis as well as experimental results show that the values of K are a function of the ratio r/b in each of the three cases shown in Fig. 11–10. The variation of K with respect to the ratio r/b is plotted in Fig. 11–11 for the three cases. We see that a smaller ratio of r/b gives a larger value of K and, accordingly, a higher stress concentration. Therefore, the radii of holes, notches, or fillets should be reasonably large to avoid high stress concentrations but not so large that they leave a small net cross-sectional area.

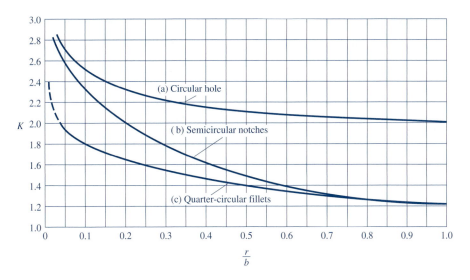

FIGURE 11–11

In some ductile metals, such as mild structural steel, the material at the point of maximum stress will yield when the yield point is reached. Additional load causes more points to yield while the maximum stress remains at σ_y, thereby distributing the load more evenly over the net cross-sectional area, as shown in Fig. 11–12. When every point in the critical section reaches the yield point, the stress distribution in the section is essentially uniform and equal to σ_y at every point. Therefore, the effect of stress concentrations due to static load is not an important design factor for ductile materials.

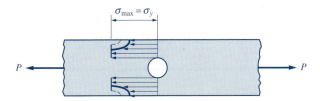

FIGURE 11–12

The lack of a yield point in brittle materials, however, causes a continuous increase in the maximum stress until σ_u is reached. The material will then begin to crack at the point of high localized stress due to its inability to deform plastically. For brittle materials, therefore, stress concentration is an important factor that must not be overlooked, even for static loading.

Stress concentrations are of particular importance in the design of machine parts subjected to cyclic stress variations or repetitive reversals of stress. Under these conditions, progressive cracks are likely to start gradually from the points of stress concentration for both ductile and brittle materials. These cracks may eventually lead to fatigue failure. Should stress concentrations be unavoidable in a member, a reduced allowable stress must be used.

━━━━ **EXAMPLE 11–4** ━━━━

Find the maximum stress in the $\frac{1}{2}$-in.-thick plate with the two semicircular notches as shown in Fig. E11–4.

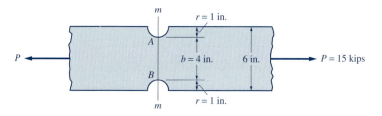

FIGURE E11–4

Solution. The ratio of the radius of the notch to the net width r/b is

$$\frac{r}{b} = \frac{1 \text{ in.}}{4 \text{ in.}} = 0.25$$

From Fig. 11–11, for semicircular notches and $r/b = 0.25$, we find

$$K = 1.9$$

The average stress in section m–m is

$$\sigma_{avg} = \frac{P}{bt} = \frac{15 \text{ kips}}{(4 \text{ in.})\left(\frac{1}{2}\right) \text{in.}} = 7.5 \text{ ksi}$$

The maximum stress due to stress concentration occurring at points A and B is

$$\sigma_{max} = K\sigma_{avg} = 1.9(7.5 \text{ ksi}) = 14.3 \text{ ksi} \qquad \Leftarrow \textbf{Ans.}$$

━━━━ **EXAMPLE 11–5** ━━━━

Determine the safe load P that can be applied to the 10-mm-thick plate in Fig. E11–5 without exceeding an allowable tensile stress of 100 MPa.

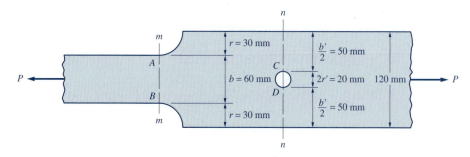

FIGURE E11–5

Solution. Consider the stress concentration at the fillets, section m–m:

$$\frac{r}{b} = \frac{30 \text{ mm}}{60 \text{ mm}} = 0.5$$

From Fig. 11–11, we find $K = 1.42$. The maximum stress at points A and B is

$$\sigma_{\max} = K\sigma_{\text{avg}} = K\frac{P}{bt} = \frac{1.42P}{(0.060)(0.010)} = 2370P$$

Consider the stress concentration at the circular hole, section n–n:

$$\frac{r'}{b'} = \frac{10 \text{ mm}}{100 \text{ mm}} = 0.10$$

From Fig. 11–11, we find $K = 2.5$. The maximum stress at points C and D is

$$\sigma_{\max} = K\sigma_{\text{avg}} = K\frac{P}{b't'} = \frac{2.5P}{(0.10)(0.010)} = 2500P$$

Equating the larger of the two maximum stresses to the allowable tensile stress, we find

$$2500P = 100\ 000$$

$$P = 40.0 \text{ kN} \qquad\qquad \Leftarrow \textbf{Ans.}$$

Thus, the load P must be no more than 40.0 kN.

11–8

ELASTIC DESIGN VERSUS PLASTIC DESIGN

Elastic Design. Up to this point, the analysis and design of axially loaded members are based on the allowable stress, which is set well within the elastic limit. The approach is called the *allowable stress design* or *elastic design*. Recall that for a ductile material, the allowable stress is computed by dividing the yield stress by the factor of safety. Therefore, the allowable load can be interpreted as the maximum elastic load P_y divided by the factor of safety, where the maximum elastic load P_y is the load that will cause the member to attain initial yielding without any plastic deformation. The *elastic design* approach assumes that, once a material starts to yield, it can no longer carry any additional load. The maximum elastic load P_y is reached when the maximum stress in the critical section reaches σ_y, as shown in Fig. 11–13. The allowable axial load is

$$[P_{\text{allow}}]_{\text{elastic}} = \frac{P_y}{F.S.} \qquad\qquad (11\text{–}9)$$

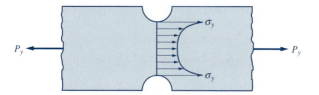

FIGURE 11–13   Maximum Elastic Load

Elastoplastic Materials. For some highly ductile metals, such as annealed low-carbon steel, Hooke's law is valid up to the yield point. Once the yield point is reached, the stress will remain constant at yield stress, and the stress–strain diagram can be modeled as shown in Fig. 11–14. A material that exhibits this behavior is referred to as *elastoplastic*. In reality, however, the material will, after some yielding, actually begin to strain-harden and stress will increase to a point above the yield stress. As a result, any design based on elastoplastic behavior will err on the safe side since strain-hardening provides the potential for the material to carry additional load.

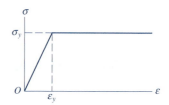

**FIGURE 11–14
Idealized Stress–Strain
Diagram**

Plastic Design. If the plate in Fig. 11–15 is elastoplastic, a load greater than P_y will cause more points to yield, while the maximum stress will remain at σ_y, as shown in Fig. 11–15a. The *maximum plastic load P_p* is reached when every point in section *m–m* has yielded, as shown in Fig. 11–15b. This is the ultimate load that the member can carry, and it can be calculated from

$$P_p = \sigma_y A \tag{11–10}$$

where A is the net cross-sectional area at section *m–m*. Thus, we see that members made of elastoplastic material can carry a load considerably higher than the maximum elastic load. An alternative design approach assumes that failure will not occur until every point in a critical section has yielded. When this occurs, the corresponding load is called the *ultimate load*. This approach is referred to as *ultimate strength design* or *limit design*. The allowable load in this approach is the ultimate load P_p divided by a specific load factor; that is,

$$[P_{\text{allow}}]_{\text{plastic}} = \frac{P_p}{\text{load factor}} \tag{11–11}$$

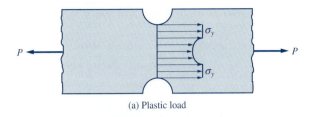

(a) Plastic load

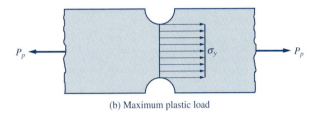

(b) Maximum plastic load

FIGURE 11–15

The ultimate strength design approach is supposedly more rational, resulting in greater economy and a more uniform factor of safety. This approach has been used in the field of reinforced concrete design for decades. In structural steel design, the approach was introduced in 1986 as *load and resistance factor design*. The following examples illustrate numerically elastic design versus ultimate strength design.

—————— **EXAMPLE 11–6** ——————————————————————

The bar shown in Fig. E11–6 is made of mild steel, which displays an elastoplastic behavior. The plate has a thickness of 3 mm and a yield strength of 250 MPa. Determine the maximum allowable load that the member can carry based on (*a*) the elastic design approach, using a factor of safety of 1.65, and (*b*) the ultimate strength design approach, with a load factor of 1.85.

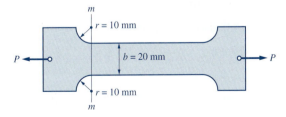

FIGURE E11–6

Solution. The ratio of the radius of the fillet to the net width r/b is

$$\frac{r}{b} = \frac{10 \text{ mm}}{20 \text{ mm}} = 0.5$$

From Fig. 11–11, for a quarter-circular fillet and $r/b = 0.5$, the stress concentration factor is

$$K = 1.4$$

(a) Allowable Load Based on Elastic Design. The maximum load, without causing any plastic deformation, occurs when $\sigma_{max} = \sigma_y$. We write

$$\sigma_{max} = \sigma_y = K\sigma_{avg} = K\frac{P_y}{bt}$$

From which the maximum elastic load is

$$P_y = \frac{bt\sigma_y}{K}$$

$$= \frac{(0.020)(0.003)(250\,000)}{1.4}$$

$$= 10.7 \text{ kN}$$

The maximum allowable load based on the elastic design approach is

$$[P_{allow}]_{elastic} = \frac{P_y}{F.S.} = \frac{10.7 \text{ kN}}{1.65}$$

$$= 6.49 \text{ kN} \qquad\qquad \Leftarrow \textbf{Ans.}$$

(b) Allowable Load Based on Ultimate Strength Design. The maximum plastic load that can be sustained by the plate requires all points at section *m–m* to reach the yield stress. Therefore, the maximum plastic load P_p is

$$P_p = A\sigma_y = bt\sigma_y$$

$$= (0.020)(0.003)(250\,000)$$

$$= 15.0 \text{ kN}$$

The maximum allowable load based on the ultimate strength design approach is

$$[P_{allow}]_{plastic} = \frac{P_p}{\text{load factor}} = \frac{15.0 \text{ kN}}{1.85}$$

$$= 8.11 \text{ kN} \qquad\qquad \Leftarrow \textbf{Ans.}$$

─────── **EXAMPLE 11–7** ───────

The steel wires in Fig. E11–7 are used to lift a load P. Wire AB has an unstretched length of 10.00 ft, and wire CD has an unstretched length of 10.01 ft. Each wire has a cross-sectional area of 0.04 in.^{2}, and a modulus of elasticity $E = 30\ 000$ ksi. If steel can be considered an elastoplastic material, determine the maximum allowable load that the member can carry based on (a) the elastic design approach, using a factor of safety of 1.65, and (b) the ultimate strength design approach, with a load factor of 1.85.

FIGURE E11–7

Solution.

(a) **Allowable Load Based on Elastic Design.** The maximum load, without causing wire AB to go into plastic deformation, occurs when $\sigma_{AB} = \sigma_y$. The strain in AB is

$$\epsilon_{AB} = \frac{\sigma_y}{E} = \frac{50 \text{ ksi}}{30\ 000 \text{ ksi}} = 0.00167$$

Elongation of wire AB is

$$\delta_{AB} = \epsilon_{AB} L = (0.00167)(10.00 \text{ ft}) = 0.0167 \text{ ft}$$

Elongation of wire CD is

$$\delta_{CD} = (0.0167 \text{ ft} - 0.01 \text{ ft}) = 0.0067 \text{ ft}$$

The strain in CD is

$$\epsilon_{CD} = \frac{\delta_{CD}}{L_{CD}} = \frac{0.0067 \text{ ft}}{10.01 \text{ ft}} = 0.00067$$

The stress in wire CD is

$$\sigma_{CD} = E\epsilon_{CD} = (30\ 000 \text{ ksi})(0.00067) = 20.1 \text{ ksi}$$

The maximum elastic load is

$$P_y = A\sigma_{AB} + A\sigma_{CD}$$
$$= (0.04 \text{ in.}^2)(50 \text{ ksi} + 20.1 \text{ ksi}) = 2.80 \text{ ksi}$$

The maximum allowable load based on the elastic design approach is

$$[P_{\text{allow}}]_{\text{elastic}} = \frac{P_y}{F.S.} = \frac{2.80 \text{ kips}}{1.65}$$

$$= 1.70 \text{ kips} \qquad \Leftarrow \text{Ans.}$$

(b) Allowable Load Based on Ultimate Strength Design. The maximum plastic load that can be sustained by the wires requires that both wires reach the yield stress. Therefore, the maximum plastic load P_p is

$$P_p = 2A\sigma_y = 2(0.04 \text{ in.}^2)(50 \text{ ksi}) = 4.00 \text{ kips}$$

The maximum allowable load based on the ultimate strength design approach is

$$[P_{\text{allow}}]_{\text{plastic}} = \frac{P_p}{\text{load factor}} = \frac{4.00 \text{ kips}}{1.85}$$

$$= 2.16 \text{ kips} \qquad\qquad \Leftarrow \textbf{Ans.}$$

11–9
SUMMARY

Tension Test.   *Tension tests* are the most commonly used tests for metals or other ductile materials. In a tension test, a standard test specimen is subjected to static tensile load until failure occurs. During the test, the applied loads and the corresponding elongations are measured and recorded. The corresponding values of stress versus strain can be calculated by

$$\sigma = \frac{P}{A} \qquad \epsilon = \frac{\delta}{L}$$

Stress–Strain Diagram.   The corresponding values of stress and strain obtained in the tension test can be plotted to produce a *stress–strain diagram*. The diagram establishes the relationship between stress and strain. From the stress–strain diagram, the following points can be identified:

Proportional Limit.   It is the upper limit where Hooke's law applies.

Elastic Limit.   It is the upper limit where the material behaves elastically. Beyond the elastic limit, the material deforms plastically. The *plastic deformation* cannot be recovered when the load is removed.

Yield Point.   At the *yield point*, the specimen continues to elongate without any significant increase in load.

Ultimate Strength.   The *ultimate strength* is the maximum stress that a material can resist before failure occurs.

Modulus of Elasticity.   In the stress–strain diagram, from the origin to the proportional limit, *Hooke's law* applies, and stress and strain are related by

$$\sigma = E\epsilon \qquad\qquad\qquad (11\text{–}1)$$

where E is the *modulus of elasticity*.

407

Plastic Behavior. Beyond the elastic limit, the material behaves *plastically*. At the plastic range, when the load is removed completely, permanent deformation in the specimen remains.

Percent Elongation. The *percent elongation* is defined as the ratio of change of gage length to the original gage length expressed as a percentage. Thus,

$$\text{Percent elongation} = \frac{L' - L}{L} \times 100\% \qquad (11\text{–}3)$$

Percent Reduction in Area. The ratio of the reduction in cross-sectional area at the fractured section to the original cross-sectional area, expressed as a percentage, is called the *percent reduction in area*. Thus,

$$\text{Percent reduction in area} = \frac{A - A'}{A} \times 100\% \qquad (11\text{–}4)$$

Yield Point by the Offset Method. When a material does not possess a well-defined yield point, it can be determined by the *offset method*. A line parallel to the straight-line portion of the stress–strain diagram is drawn, starting from a point on the abscissa with an offset of 0.2 percent (i.e., $\epsilon = 0.002$). The intersection of the line and the stress–strain curve is the yield point of the material at 0.2 percent offset.

Mechanical Properties of Materials. The stress–strain diagram obtained from tension characterizes many important properties of a material. These properties are:

1. ***Strength.*** The *strength* of a material is the greatest stress that the material can withstand without excessive deformation or failure.
2. ***Stiffness.*** The *stiffness* of a material is the ability of a material to resist deformation. A material with a high value for the modulus of elasticity E is stiffer than materials with a lower value of E.
3. ***Elasticity.*** *Elasticity* is the property of a material that enables it to regain its original undeformed length once the load is removed.
4. ***Ductility.*** *Ductility* is the ability of a material to undergo a lot of plastic deformation before rupture. Ductility is indicated by the *percent elongation* of the gage length as well as by the percent reduction in cross-sectional area at the fractured section. Materials with a percent of elongation greater than 5 percent are called *ductile* materials.
5. ***Brittleness.*** A material that undergoes very little plastic deformation before rupture is said to be *brittle*. A brittle material ruptures suddenly and without warning at the ultimate strength.
6. ***Hardness.*** *Hardness* is a measure of the resistance of a material to penetration. Hardness is very useful for quality control purposes.
7. ***Machinability.*** *Machinability* is the ease with which a material can be machined. It is frequently a critical factor for the selection of metals for parts made by automatic machine tools.
8. ***Resilience.*** The capacity of a material to absorb energy within the elastic range is called *resilience*. Resilience is measured by the triangular area under the elastic portion of the stress–strain curve.

9. ***Toughness.***   The capacity of a material to absorb energy without fracture is called *toughness*. Toughness is measured by the total area under the entire stress–strain curve up to the point of fracture.

Compression Test.   Brittle materials, such as concrete, are tested in compression due to low tensile strength and brittleness. The stress–strain diagram resulting from a compression test of mild steel shows that the modulus of elasticity E and yield stress σ_y in compression are the same as those determined in the tension test. For brittle materials such as cast iron, concrete, brick, ceramics, etc., a compression test is more important than a tension test.

Allowable Stress.   *Allowable stress* represents the maximum safe stress chosen as the limiting value that must not be exceeded by the load induced stresses. The ratio of failure stress to allowable stress is defined as the *factor of safety (F.S.)*. The failure stress can either be the yield strength or the ultimate strength, depending on whether the material is ductile or brittle. Thus,

$$F.S. = \frac{\text{yield strength}}{\text{allowable stress}} \quad \text{(for ductile materials)} \quad (11\text{--}6a)$$

$$F.S. = \frac{\text{ultimate strength}}{\text{allowable stress}} \quad \text{(for brittle materials)} \quad (11\text{--}6b)$$

Stress Concentration.   *Localized stress concentration* occurs at sections with abrupt changes in geometry (called *stress-raisers*). The *stress concentration factor* is the ratio of the maximum stress to the average stress over the *net* cross-sectional area of the section where there is a stress-raiser. Thus,

$$K = \frac{\sigma_{max}}{\sigma_{avg}} \quad (11\text{--}7)$$

where K is the stress concentration factor and σ_{avg} is the average stress over the net cross-sectional area.

   A high stress concentration may not be detrimental for ductile metals because of the occurrence of plastic yielding and subsequent stress redistribution. However, the inability to yield in brittle materials causes a continuous increase in the maximum stress, which will cause cracks to start at the point of high localized stress. Stress concentrations are of particular importance in the design of machine parts made of both ductile and brittle materials subjected to cyclic stress variations or repetitive reversals of stress.

Plastic Design.   A material is said to be *elastoplastic* if its stress–strain curve is such that Hooke's law is valid up to the yield point, and beyond the yield point the curve is a horizontal line. *Ultimate strength design* recognizes that a member made of elastoplastic material may be loaded beyond the initial yielding, and the ultimate strength of the member is reached when every point in a critical section has yielded. This load is called the *ultimate strength* or the *plastic strength* of the member. The allowable load that a member can carry is its ultimate strength divided by a certain specific *load factor*.

PROBLEMS

Section 11–2   The Tension Test

Section 11–3   The Stress–Strain Diagram

Section 11–4   Mechanical Properties of Materials

Section 11–5   The Compression Test

11–1   What is the difference between elastic deformation and plastic deformation?

11–2   What occurs when a mild steel bar is stretched to its yield point?

11–3   What is the necking phenomenon in a tension test?

11–4   What is the meaning and significance of the ultimate strength of a material?

11–5   Two bars, one made of aluminum and one made of copper, have the same length. The moduli of elasticity are 10×10^3 ksi for aluminum and 17×10^3 ksi for copper. Which bar stretches more if both bars are subjected to the same stress within their proportional limits?

11–6   How is the yield point at 0.2 percent offset determined?

11–7   Why is the tension test more important than the compression test for ductile materials?

11–8   Why is the compression test more important than the tension test for brittle materials?

11–9   In a tension test of a steel specimen with 0.50-in. diameter and 2.00-in. gage length, the maximum load is 15 200 lb, the final length is 2.59 in., and the final diameter at the necked-down section is 0.423 in. Calculate (*a*) the ultimate strength, (*b*) the percent of elongation, and (*c*) the percent reduction in area.

11–10   A metal bar with a 50-mm × 15-mm rectangular section is subjected to a tensile load of 80 kN. At this load, the bar stretches 0.305 mm over a 200-mm gage length. Assuming that the proportional limit of the material is 190 MPa, determine the modulus of elasticity of the material and indicate a possible type of metal comprising the bar.

11–11   The stress–strain diagram for a tension test of an alloy specimen is plotted as shown in Fig. P11–11. The following data are recorded:

Initial diameter   = 0.502 in.
Gage length   = 2.00 in.
Diameter at the fractured section = 0.412 in.
Final length after fracture   = 2.78 in.

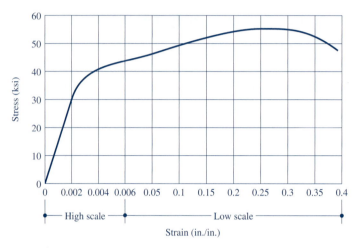

FIGURE P11–11

Determine (*a*) the stress at the proportional limit, (*b*) the modulus of elasticity, (*c*) the yield stress at 0.2 percent offset, (*d*) the ultimate strength, (*e*) the percent elongation, and (*f*) the percent reduction in area.

11–12 The data in Table P11–12 were obtained from a tension test of a steel specimen. The specimen had an original diameter of 0.505 in. and a gage length of 2.00 in. After the specimen ruptured, the gage length became 2.31 in. and the diameter at the section of rupture decreased to 0.450 in. Plot the stress–strain diagram. Determine (*a*) the stress at the proportional limit, (*b*) the modulus of elasticity, (*c*) the yield stress at 0.2 percent offset, (*d*) the ultimate strength, (*e*) the percent elongation, and (*f*) the percent reduction in area.

TABLE P11–12

Total Tensile Load (lb)	Total Elongation in 2-in. Gage Length (in.)	Total Tensile Load (lb)	Total Elongation in 2-in. Gage Length (in.)
200	0.0000	14 400	0.0118
1 000	0.0003	15 200	0.0167
2 000	0.0006	16 000	0.0212
4 000	0.0012	16 800	0.0263
6 000	0.0019	17 600	0.0327
8 000	0.0026	18 400	0.0380
10 000	0.0033	19 200	0.0440
12 000	0.0039	20 000	0.0507
13 400	0.0045	20 800	0.0580
13 600	0.0054	21 600	0.0660
13 800	0.0063	22 400	0.0780
14 000	0.0090	25 400	Specimen broke

Section 11–6 Allowable Stresses and Factor of Safety

11–13 A tie rod of A36 steel with $\sigma_y = 36$ ksi is used to support a tensile load of 2.50 kips. Select the diameter of the tie rod using a factor of safety of 2 to guard against yielding.

11–14 Two high-strength steel cables with $\sigma_y = 700$ MPa support a load of 800 kN. Assuming that the cables share the load equally and using a factor of safety of 2 to guard against yielding, select the diameter of the cable to the nearest mm.

11–15 A short, cast iron machine member of square section is subjected to a compressive load of 40 kips. If the compressive ultimate strength of cast iron is 90 ksi, select the cross-sectional dimensions of the member using a factor of safety of 4.

11–16 A wrought steel rod 500 mm long is subjected to a tensile load of 75 kN. The allowable tensile stress is 150 MPa, and the allowable elongation is not to exceed 0.25 mm. Select the diameter of the rod to the nearest mm. The modulus of elasticity is 210 GPa.

11–17 A 50-lb weight is lifted by a cable as shown in Fig. P11–17. If the rope has a tensile breaking strength of 180 lb, determine the maximum value of angle θ using a factor of safety of 3 to guard against breaking.

FIGURE P11–17

Section 11–7 Stress Concentrations

11–18 What is stress concentration? What is the stress concentration factor?

11–19 Why is the effect of stress concentration more important for brittle materials than for ductile materials when designing a member subjected to static load?

11–20 What is the significance of stress concentration when there are cyclic stress variations?

11–21 Determine the maximum stress in the $\frac{1}{2}$-in.-thick plate shown in Fig. P11–21 if (a) $r = \frac{1}{4}$ in., (b) $r = \frac{1}{2}$ in., and (c) $r = 1$ in.

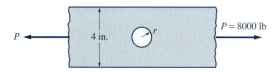

FIGURE P11–21

11–22 Refer to Fig. P11–22. Determine the maximum stress in the plate shown and indicate the points where the maximum stress occurs.

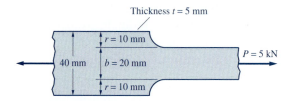

FIGURE P11–22

11–23 Determine the maximum permissible static load P that may be applied to the plate with semicircular notches shown in Fig. P11–23 if the tensile stress must not exceed 15 ksi.

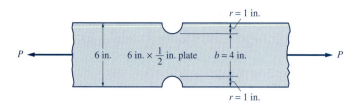

FIGURE P11–23

11–24 See Fig. P11–24. Determine the static axial load P that may be applied to the 10-mm-thick plate shown without causing the maximum stress in the plate to exceed 160 MPa.

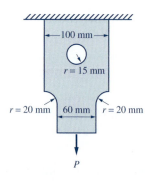

FIGURE P11–24

11–25 The two links shown in Fig. P11–25 have the same thickness $t = \frac{1}{4}$ in. and both are made of A36 steel with $\sigma_y = 36$ ksi. Determine the allowable load of each link using a factor of safety of 3 to guard against yielding. Assume that the links are subjected to cyclic stress variation, so the stress concentration is an important factor to consider.

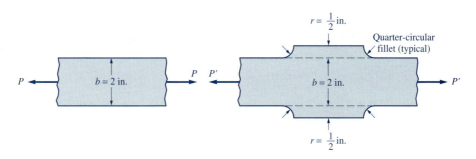

FIGURE P11–25

11–26 A plate 5 mm thick has two semicircular notches, as shown in Fig. P11–26. The material of the plate is mild steel with a yield stress $\sigma_y = 250$ MPa. Determine the maximum stress at section m–m if (a) $P = 40$ kN and (b) $P = 70$ kN. In each case, sketch the stress distribution in the section.

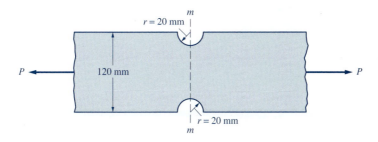

FIGURE P11–26

Section 11–8   Elastic Design Versus Plastic Design

11–27 A plate 5 mm thick has two semicircular notches, as shown in Fig. P11–27. The material of the plate is mild steel with a yield stress $\sigma_y = 250$ MPa. Determine the maximum allowable load that the member can carry based on (a) the elastic design approach, using a factor of safety of 1.65, and (b) the ultimate strength design approach, using a load factor of 1.85.

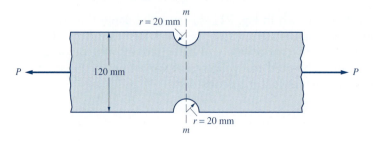

FIGURE P11–27

11–28 A rigid beam is symmetrically supported by three steel rods, as shown in Fig. P11–28. The steel rods have a yield stress $\sigma_y = 36$ ksi and a diameter of $\frac{1}{2}$ in. Determine the allowable distributed load w that can be applied to the beam based on (*a*) the elastic design approach, using a factor of safety of 1.65, and (*b*) the ultimate strength design approach, using a load factor of 1.85.

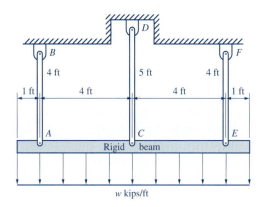

FIGURE P11–28

CHAPTER 12

TORSION OF CIRCULAR SHAFTS

12–1
INTRODUCTION

A member subjected to *twisting moments* (also referred to as *torques* or *torsional loads*) is called a *shaft*. Many technical applications involve the twisting of circular shafts. A torsion bar used in the front suspension of some cars is an example of a stationary shaft. The impact force on the front wheels caused by bumps on the road will make the torsion bar twist and untwist, thereby absorbing some of the impact and improving the riding comfort. Rotating shafts are frequently used to transmit power in machines. A power transmission shaft in an automobile with rear wheel drive transmits power from the engine to the driving wheels.

Only solid and hollow circular shafts will be considered in this chapter. We will develop basic equations that can be used to determine stress distribution and the angle of twist in circular shafts subjected to torsional loads. Most important applications in engineering involve shafts of solid or hollow circular sections. The formulas developed in this chapter will be useful in many engineering applications.

12–2
EXTERNAL AND INTERNAL TORQUES

External Torque.   To see the action of an external torque, consider the steel shaft *AB* and the handle *CD* shown in Fig. 12–1. Two equal and opposite forces of magnitude *F* are applied to the handle. The moment of the couple is $T = Fa$. This moment has a twisting effect to shaft *AB*. The applied torque at *A* and the torsional reaction at *B* are the *external torques*. The torques are vector quantities and may be represented either by curved arrows, as in Fig. 12–1b, or by vector representation, as in Fig. 12–1c. In Fig. 12–1c, the torque vector is represented by an arrow using the *right-hand rule*:

Right-Hand Rule. *Curl the fingers of your right hand in the direction of a torque; the vector representation of the torque is now indicated by the direction of the thumb.*

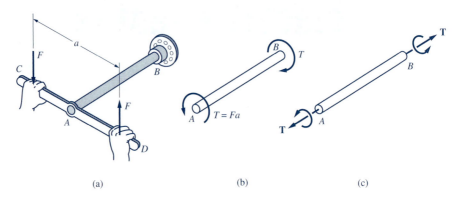

(a)   (b)   (c)

FIGURE 12–1

Internal Torque. The *internal torque* at a section of the shaft is the torque within the section required to resist the external torque. The method of sections is used in determining internal torque. First, the entire system is analyzed using equilibrium conditions. The value of an unknown reaction can be determined by equating the sum of the moments about the axis of the shaft to zero. Next, pass an imaginary cutting plane through the section where the internal torque is to be determined. The cutting plane separates the shaft into two parts. Either part can be used as a free body to determine the internal torque at the section.

For example, Fig. 12–2a shows a shaft subjected to three balanced external torques. To determine the internal torque in segment *AB,* a plane *m–m* is passed through the segment, separating the shaft into two parts. If we

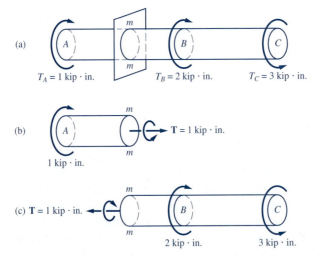

(a)

$T_A = 1$ kip · in. $T_B = 2$ kip · in. $T_C = 3$ kip · in.

(b)

1 kip · in. $T = 1$ kip · in.

(c) $T = 1$ kip · in.

2 kip · in.   3 kip · in.

FIGURE 12–2

consider the equilibrium of the shaft to the left of section *m–m* (Fig. 12–2b), the internal torque in section *m–m* is found to be 1 kip · in. If the equilibrium of the shaft to the right of the section is considered (Fig. 12–2c), the internal torque is also found to be 1 kip · in., but acting in the opposite direction. Since the internal torque at a section viewed from two different directions must have the same sign, a special sign convention must be used. We will define the positive internal torque in a section as follows:

Positive Internal Torque. The internal torque is considered positive if the vector representation of the internal torque in the section is directed outward from the section, as shown in Fig. 12–3.

Positive internal torque

FIGURE 12–3

According to this sign convention, the internal torque at section *m–m*, viewed from either direction as shown in Fig. 12–2b and c, is positive.

Internal Torque Diagram. An axial force diagram has been discussed in Chapter 9 to show the variation of internal axial force along an axially loaded member. Similarly, an internal torque diagram can be plotted to show the variation of internal torque along a shaft. Example 12–2 demonstrates the construction of an internal torque diagram.

EXAMPLE 12–1

Determine the internal resisting torque in segments *AB* and *BC* of the shaft shown in Fig. E12–1(1).

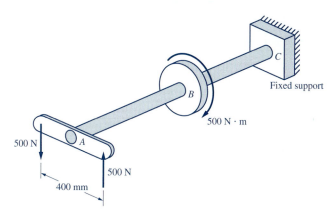

FIGURE E12–1(1)

Solution. The moment of the couple produced by the two 500-N forces is

$$(500 \text{ N})(0.4 \text{ m}) = 200 \text{ N} \cdot \text{m}$$

This moment is a torsion acting at A.

Internal Torque T_{AB}. To determine the internal resisting torque in segment AB, pass section 1–1 through any section within the segment. The free-body diagram of the part from A to the section is sketched in Fig. E12–1(2). The internal resisting torque must be equal and opposite to the external torque. The vector representation of T_{AB} is directed *outward* from the section and is thus considered *positive*. Hence,

$$T_{AB} = +200 \text{ N} \cdot \text{m} \qquad\qquad \Leftarrow \textbf{Ans.}$$

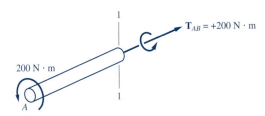

FIGURE E12–1(2)

Internal Torque T_{BC}. To determine the internal resisting torque in segment BC, pass section 2–2 through any section within the segment. The free-body diagram of the part from A to the section is sketched as in Fig. E12–1(3). The internal resisting torque must be equal and opposite to the resultant external torque. The vector representation of T_{BC} is directed *inward* toward the section and is thus considered *negative*. Hence,

$$T_{BC} = -300 \text{ N} \cdot \text{m} \qquad\qquad \Leftarrow \textbf{Ans.}$$

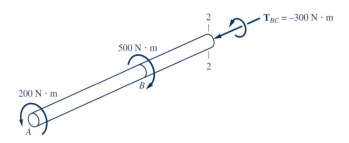

FIGURE E12–1(3)

EXAMPLE 12–2

Refer to Fig. E12–2(1). Determine the internal resisting torque in each segment and draw the internal torque diagram for the shaft subjected to the external torques shown.

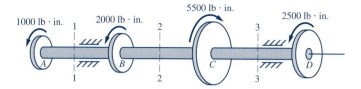

FIGURE E12–2(1)

Solution. Since the algebraic sum of the given external torques about the axis of the shaft is zero, the shaft is in equilibrium and every segment of the shaft must also be in equilibrium.

Internal Torques. To determine the internal resisting torque in segment *AB*, pass section 1–1 through the segment. The free-body diagram of the part from *A* to the section is sketched as in Fig. E12–2(2). The internal resisting torque must be equal and opposite to the external torque. The vector representation of T_{AB} is directed *inward* toward the section and is thus considered *negative*. Hence,

1000 lb · in.

$T_{AB} = -1000$ lb · in.

FIGURE E12–2(2)

$$T_{AB} = -1000 \text{ lb} \cdot \text{in.} \qquad \Leftarrow \textbf{Ans.}$$

To determine the internal resisting torque in segment *BC*, pass section 2–2 through the segment. The free-body diagram of the part from *A* to the section is sketched as in Fig. E12–2(3). The internal resisting torque must be equal and opposite to the resultant external torque. The vector representation of T_{BC} is directed *inward* toward the section and is thus considered *negative*. Hence,

$$T_{BC} = -3000 \text{ lb} \cdot \text{in.} \qquad \Leftarrow \textbf{Ans.}$$

1000 lb · in.   2000 lb · in.

2000 lb · in.

$T_{BC} = -3000$ lb · in.

FIGURE E12–2(3)

To determine the internal resisting torque in segment *CD*, pass section 3–3 through the segment. The free-body diagram of the part of the shaft to the right of the section is sketched in Fig. E12–2(4). The internal resisting torque must be equal and opposite to the external torque. The vector representation of T_{CD} is directed *outward* from the section and is thus considered *positive*. Hence,

$$T_{CD} = +2500 \text{ lb} \cdot \text{in.}$$ ⇐ **Ans.**

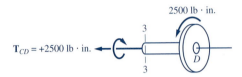

FIGURE E12–2(4)

Internal Torque Diagram. From the internal torques determined above, the internal torque diagram can be plotted as shown in Fig. E12–2(5). The torques acting on the three segments of the shaft can also be sketched as shown.

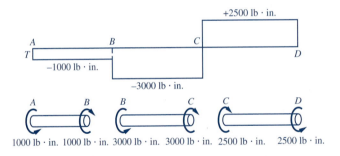

FIGURE E12–2(5)

12–3
THE TORSION FORMULA

Shear Distortion of Shafts. Figure 12–4a shows a circular member fixed to a support at the left end and free at the right end. Figure 12–4b shows the deformation of the member after a torque T is applied to the free end B. We see that the longitudinal line AB on the surface of the shaft is twisted into a helix AB'. The radius OB is rotated through an angle ϕ (the Greek lowercase letter phi) to a new position OB', but the radius remains a straight line.

The entire section at the free end rotates through the same angle ϕ, while the size and shape of the section and distance to the adjacent section are unchanged. A square element bounded by the adjacent longitudinal and circumferential lines on the surface of the shaft deforms into a rhombus. This deformation is evidence that the element is subjected to shear

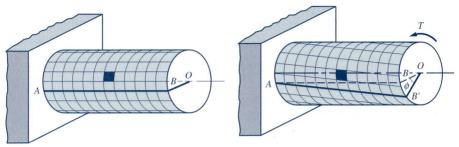

(a) Circular member before torque is applied   (b) Circular member after torque is applied

FIGURE 12–4

stresses. Since the dimensions of all sides of the element are unchanged, there are no normal stresses in the element along the longitudinal and transverse directions. Thus, the element is subjected only to shear stresses and is said to be in *pure shear.*

   Torsion Formula. If the maximum shear stress due to the torque in the circular shaft is within the elastic range of the shaft material, shear stresses vary linearly from the axis of the shaft to the outside surface. Figure 12–5 shows the variation of shear stresses at points along a radius. These shear stresses are perpendicular to the radial direction. The maximum shear stress occurs at points on the periphery of a section, such as point C. We denote the radius of the shaft by c and the maximum shear stress by τ_{max}. Then, by virtue of the linear stress variation, the shear stress τ on an element ΔA located at a distance ρ (the Greek lowercase letter rho) from the center is

$$\tau = \frac{\rho}{c}\,\tau_{max} \tag{a}$$

The shear force on the element ΔA is

$$\Delta F = \tau \Delta A = \frac{\rho}{c}\,\tau_{max}\Delta A$$

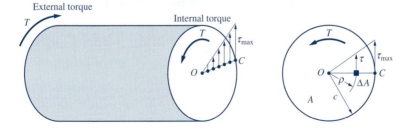

FIGURE 12–5

The resisting torque produced by the shear force ΔF about the axis of the shaft is

$$\Delta T = \rho\,\Delta F = \frac{\tau_{max}}{c}\,\rho^2\,\Delta A$$

The total resisting torque produced by the shear forces on all the elements over the entire section must be equal to the internal torque T in the section; thus,

$$T = \Sigma \, \Delta T = \Sigma \frac{\tau_{max}}{c} \rho^2 \, \Delta A$$

At any given section, c and τ_{max} are both constant. When the constant τ_{max}/c is factored from each term, the equation becomes

$$T = \frac{\tau_{max}}{c} \Sigma \, \rho^2 \, \Delta A$$

where $\Sigma \rho^2 \, \Delta A$ is, by the definition given in Chapter 8, the polar moment of inertia of the cross-sectional area, which is a constant for a given section. We denote the polar moment of inertia by J and solve for τ_{max} to get

$$\tau_{max} = \frac{Tc}{J} \tag{12–1}$$

where τ_{max} = the maximum shear stress on the outer surface of the shaft

T = the internal torque in the shaft

c = the distance from the centroidal axis to the outer surface, or simply the radius of the shaft

J = the polar moment of inertia of the cross-section of the shaft

Substituting Equation 12–1 into Equation (a) gives

$$\tau = \frac{T\rho}{J} \tag{12–2}$$

Equations 12–1 and 12–2 are two forms of the well-known *torsion formula*. These formulas may be used for both solid and hollow circular shafts. The formulas for computing the appropriate values of J for solid or hollow shafts, from Table 8–1, are

For solid circular shaft: $\quad J = \dfrac{\pi d^4}{32}$ $\qquad$ (12–3)

For hollow circular shaft: $\quad J = \dfrac{\pi}{32} (d_o^{\,4} - d_i^{\,4})$ $\qquad$ (12–4)

where d is the diameter of the solid circular shaft, and d_o and d_i are the outside and inside diameters of the hollow shaft, respectively.

Allowable Torque.   If the allowable shear stress τ_{allow} for a shaft is known, Equation 12–1 can be rewritten for computing the allowable torque that a shaft can resist:

$$T_{allow} = \frac{\tau_{allow} J}{c} \tag{12–5}$$

where the allowable shear stress τ_{allow} may be computed from

$$\tau_{allow} = \frac{\tau_y}{F.S.} \tag{12–6}$$

If τ_y is unknown, we will take its value as being equal to 57% of the yield stress in tension; that is

$$\tau_y = 0.57\sigma_y \tag{12–7}$$

Required Diameter for Solid Circular Shaft.   For purposes of the design of solid circular shafts, Equation 12–1 can be expressed in terms of the diameter. Substituting $c = d/2$ and J from Equation 12–3 into Equation 12–5, we get

$$T = \frac{\tau_{\text{allow}}}{d/2}\left(\frac{\pi d^4}{32}\right) = \frac{\pi\tau_{\text{allow}}d^3}{16}$$

Solving this equation for the diameter gives

$$d_{\text{req}} = \sqrt[3]{\frac{16T}{\pi\tau_{\text{allow}}}} \tag{12–8}$$

Required Diameter for Hollow Circular Shaft.   For hollow shafts, the polar moment of inertia J is given by Equation 12–4. In terms of the ratio of diameters $k = d_i/d_o$, the equation becomes

$$J = \frac{\pi}{32}\left[d_o^4 - \left(kd_o\right)^4\right] = \frac{\pi d_o^{\,4}}{32}\left(1-k^4\right)$$

Substituting this expression into Equation 12–5 and solving for d_o, we get

$$(d_o)_{\text{req}} = \sqrt[3]{\frac{16T}{\pi\tau_{\text{allow}}\,(1-k^4)}} \tag{12–9}$$

where k is the ratio of diameters $k = d_i\,/\,d_o$.

───── **EXAMPLE 12–3** ─────

A solid steel shaft 40 mm in diameter is subjected to the torsional loads shown in Fig. E12–3. Determine the maximum shear stress in the shaft.

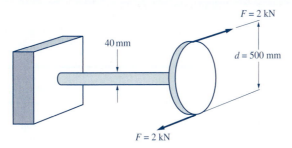

FIGURE E12–3

Solution. The torque applied to the shaft is

$$T = Fd = (2\text{ kN})(0.5\text{ m}) = 1\text{ kN}\cdot\text{m}$$

From Equation 12–3, the polar moment of inertia of the cross-sectional area is

$$J = \frac{\pi d^4}{32} = \frac{\pi (0.040 \text{ m})^4}{32} = 2.51 \times 10^{-7} \text{m}^4$$

From Equation 12–1, the maximum shear stress on the periphery of the shaft is

$$\tau_{max} = \frac{Tc}{J} = \frac{(1 \text{ kN} \cdot \text{m})(0.020 \text{ m})}{2.51 \times 10^{-7} \text{m}^4}$$

$$= 79\ 600 \text{ kPa}$$

$$= 79.6 \text{ MPa} \qquad \Leftarrow \textbf{Ans.}$$

EXAMPLE 12–4

A shaft of hollow circular section, with outside diameter $d_o = 3\frac{1}{2}$ in. and inside diameter $d_i = 3\frac{1}{4}$ in., is subjected to a torque of 30 kip $\cdot$ in. Determine the maximum and minimum shear stresses in the shaft.

Solution. From Equation 12–4, the polar moment of inertia of the hollow section is

$$J = \frac{\pi}{32}(d_o^4 - d_i^4) = \frac{\pi}{32}(3.5^4 - 3.25^4) = 3.78 \text{ in.}^4$$

From Equation 12–1, the maximum shear stress on the outside surface of the shaft is

$$\tau_{max} = \frac{Tc}{J} = \frac{(30 \text{ kip} \cdot \text{in.})\left[\frac{1}{2}(3.5 \text{ in.})\right]}{3.78 \text{ in.}^4}$$

$$= 13.9 \text{ ksi} \qquad \Leftarrow \textbf{Ans.}$$

The minimum shear stress on the inner surface can be calculated from Equation 12–2 with $\rho = d_i/2$.

$$\tau_{min} = \frac{T\left(\frac{1}{2}d_i\right)}{J} = \frac{(30 \text{ kip} \cdot \text{in.})\left[\frac{1}{2}(3.25 \text{ in.})\right]}{3.78 \text{ in.}^4} = 12.9 \text{ ksi} \qquad \Leftarrow \textbf{Ans.}$$

EXAMPLE 12–5

Determine the required diameter of a solid circular shaft subjected to a torque load of 10 kN $\cdot$ m. The yield stress in shear of the shaft material is 170 MPa. Use a factor of safety of 2.5.

Solution. From Equation 12–6, the allowable shear stress is

$$\tau_{allow} = \frac{\tau_y}{F.S.} = \frac{170 \text{ MPa}}{2.5} = 68 \text{ MPa} = 68\ 000 \text{ kN/m}^2$$

Using Equation 12–8, we get

$$\text{Required } d = \sqrt[3]{\frac{16\,T_{max}}{\pi\tau_{allow}}} = \sqrt[3]{\frac{16(10 \text{ kN} \cdot \text{m})}{\pi(68\,000 \text{ kN/m}^2)}}$$

From which we get

$$\text{Required } d = 0.0908 \text{ m} = 90.8 \text{ mm} \qquad \Leftarrow \textbf{Ans.}$$

──── **EXAMPLE 12–6** ────

A machine shaft made of plain carbon steel, having a tensile yield stress of 47.2 ksi, is to transmit a torque of 30 kip · in. Use a factor of safety of 2. (a) Determine the size of the solid circular shaft. (b) Determine the size of the hollow circular shaft if the inside diameter is to be three-quarters that of the outside diameter. (c) Compare the weights of the shafts selected in (a) and (b).

Solution. The yield stress in shear is not given, but its value may be determined approximately from Equation 12–7 as

$$\tau_y = 0.57\sigma_y = 0.57 \,(47.2 \text{ ksi}) = 26.9 \text{ ksi}$$

The allowable shear stress is

$$\tau_{allow} = \frac{\tau_y}{F.S.} = \frac{26.9 \text{ ksi}}{2} = 13.5 \text{ ksi}$$

(a) Solid Shaft. From Equation 12–8, we get

$$d_{req} = \sqrt[3]{\frac{16T}{\pi\tau_{allow}}} = \sqrt[3]{\frac{16(30 \text{ kip} \cdot \text{in.})}{\pi\,(13.5 \text{ kip/in.}^2)}}$$

From which we get

$$d = 2.25 \text{ in.}$$
$$\text{Use } d = 2\tfrac{1}{4} \text{ in.} \qquad \Leftarrow \textbf{Ans.}$$

(b) Hollow Shaft. The ratio of the inside and outside diameters is

$$k = d_i/d_o = 0.75$$

From Equation 12–9, we get

$$(d_o)_{req} = \sqrt[3]{\frac{16T}{\pi\tau_{allow}(1 - k^4)}}$$

$$= \sqrt[3]{\frac{16(30 \text{ kip} \cdot \text{in.})}{\pi\,(13.5 \text{ kip/in.}^2)(1 - 0.75^4)}}$$

From which we get

$$(d_o)_{req} = 2.55 \text{ in.}$$

The required inside diameter is

$$d_i = 0.75 \, d_o = 0.75(2.55 \text{ in.}) = 1.91 \text{ in.}$$

Use $d_o = 2\frac{9}{16}$ in. and $d_i = 1\frac{7}{8}$ in.   ⇐ **Ans.**

Note that the outside diameter must be a little larger than the required value, and the inside diameter must be a little smaller than the required value.

(c) *Comparison of Weights.* The cross-sectional areas of the solid and hollow shafts are, respectively,

$$A_{\text{solid}} = \frac{\pi(2.25 \text{ in.})^2}{4} = 3.98 \text{ in.}^2$$

$$A_{\text{hollow}} = \frac{\pi}{4}\left[\left(2\frac{9}{16}\right)^2 - \left(1\frac{7}{8}\right)^2\right] = 2.40 \text{ in.}^2$$

The weight of a shaft is equal to its cross-sectional area multiplied by its length and its specific weight. For the same length and the same material, the ratio of the weights of the shafts is equal to the ratio of the areas. Thus,

$$\frac{W_{\text{hollow}}}{W_{\text{solid}}} = \frac{A_{\text{hollow}}}{A_{\text{solid}}} = \frac{2.40}{3.98} = 0.60$$

Thus, we see that the weight of the hollow shaft is only 60% of the weight of the solid shaft of the same material. By using a hollow shaft for this problem, the weight will be cut down by 40% from the solid shaft. For structures such as aircraft, this weight reduction could be a crucial factor. The drive shafts in automobiles are hollow because of the desire for fuel economy.

EXAMPLE 12–7

Refer to Fig. E12–7(1). A machine shaft of uniform cross-section having a diameter of $d = 3$ in. is subjected to the torsional loads shown. Find the maximum shear stress in the shaft.

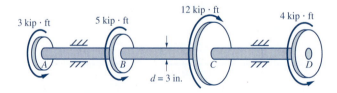

FIGURE E12–7(1)

Solution. From the free-body diagrams sketched in Fig. E12–7(2) and using the sign convention for internal torque presented in Section 12–2, we find the internal torque in each segment as

$$T_{AB} = -3 \text{ kip} \cdot \text{ft}$$
$$T_{BC} = -8 \text{ kip} \cdot \text{ft}$$
$$T_{CD} = +4 \text{ kip} \cdot \text{ft}$$

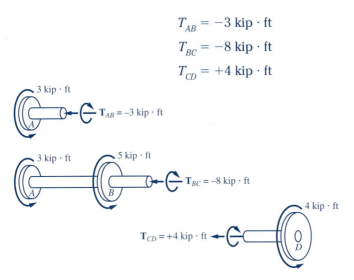

FIGURE E12–7(2)

Since the magnitude of the shear stress in a shaft is independent of the sign of the internal shear force, the maximum shear stress occurs at the outside surface of the segment with a maximum absolute value of internal torque. The internal torque at segment BC has the greatest absolute value of

$$\left| T_{BC} \right| = 8 \text{ kip} \cdot \text{ft} = 96 \text{ kip} \cdot \text{in.}$$

The polar moment of inertia of the cross-section of the shaft is

$$J = \frac{\pi d^4}{32} = \frac{\pi (3 \text{ in.})^4}{32} = 7.95 \text{ in.}^4$$

From Equation 12–1, the maximum shear stress on the outside surface of segment BC is

$$\tau_{max} = \frac{Tc}{J} = \frac{(96 \text{ kip} \cdot \text{in.})(1.5 \text{ in.})}{7.95 \text{ in.}^4}$$

$$= 18.1 \text{ ksi} \qquad \qquad \Leftarrow \textbf{Ans.}$$

12–4
SHEAR STRESSES ON MUTUALLY PERPENDICULAR PLANES

In the preceding section, we learned that shear stresses exist in the cross-section of a shaft. In this section, we shall show that shear stresses exist in the longitudinal planes of a shaft as well.

Consider a small rectangular element $abcd$ on the surface of the shaft shown in Fig. 12–6a. Assume that the dimensions of the element are $\Delta x \times \Delta y \times \Delta z$, as shown in the free-body diagram of the element in Fig. 12–6b. An element isolated from an equilibrium body must also be in equilibrium. The shear stress τ on side ab can be determined from the torsion formula. The equilibrium condition along the y direction requires that the shear stress on side cd be equal to the shear stress on side ab, but acting in the opposite direction. The equilibrium condition along the x direction requires that the shear stresses on the opposite sides ad and bc be equal and opposite; thus, the shear stresses on these two sides are denoted by τ'. The equilibrium equation $\Sigma M_z = 0$ of the element is

$$\Sigma M_z = (\tau)(\Delta y \Delta z)(\Delta x) - (\tau')(\Delta x \, \Delta z)(\Delta y) = 0$$

where the moment is obtained by multiplying the shear stress (inside the first set of parentheses) by the area (inside the second set of parentheses) and by the moment arm (inside the third set of parentheses). When we divide both terms by the common factor $\Delta x \, \Delta y \, \Delta z$, the equation above is simplified to

$$\tau' = \tau$$

Thus, we conclude that *the shear stresses on the adjacent faces of an element in a stressed body must have equal magnitude and must be directed either toward or away from a corner of the element,* as shown in Fig. 12–6c.

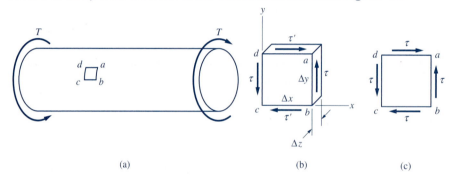

(a)   (b)   (c)

FIGURE 12–6

Consequently, a typical element on the surface of the shaft in Fig. 12–6a is subjected to the shear stress shown in Fig. 12–6c. Thus, shear stresses exist on the longitudinal plane of the shaft. The shear stress variation along a radius is shown in Fig. 12–7, both in the cross-section and in the longitudinal plane. Note that the direction of the shear stress in the cross-section must coincide with the direction of the internal torque.

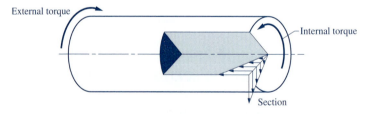

FIGURE 12–7

Materials whose properties are the same in all directions are said to be *isotropic* materials. Most structural materials are isotropic. Some materials such as wood, however, exhibit drastically different properties in different directions. The shear strength of wood on planes parallel to the grain is much smaller than that on planes perpendicular to the grain. Therefore, wooden shafts tend to split along the axial plane parallel to the grain when overloaded with torsion.

──────── **EXAMPLE 12–8** ────────────────────────────────────

A 4-in.-diameter Douglas fir shaft has an allowable shear stress parallel to the grain of 95 psi and an allowable shear stress perpendicular to the grain of 120 psi. Determine the allowable torque that the shaft can resist.

Solution. The direction of wood grain is usually parallel to the longitudinal axis of the shaft. The allowable torque that the shaft can resist is governed by the allowable shear stress parallel to the grain. For a 4-in.-diameter shaft, the polar moment of inertia of the cross-section of the shaft is

$$J = \frac{\pi d^4}{32} = \frac{\pi (4 \text{ in.})^4}{32} = 25.1 \text{ in.}^4$$

From Equation 12–5, we find

$$T_{\text{allow}} = \frac{\tau_{\text{allow}} J}{c} = \frac{(95 \text{ lb/in.}^2)(25.1 \text{ in.}^4)}{2 \text{ in.}}$$

$$= 1194 \text{ lb} \cdot \text{in.} \qquad \Leftarrow \textbf{Ans.}$$

12–5
POWER TRANSMISSION

The main purpose of a machine shaft is to transmit power. The shaft is usually rotating at a constant speed and the bearings are assumed to be frictionless. A shaft may have several gears (or pulleys) mounted on it. The power is delivered to (input) and transported out (output) through the gears (or pulleys). When transmitting power, a shaft is subjected to torques that depend on the power transmitted and the angular velocity of the shaft.

Power is defined as the work done per unit time. Work done by a torque acting on a rotating shaft is equal to the torque T multiplied by the angular displacement θ of the shaft; that is,

$$\text{Work done} = T\theta$$

Since power is defined as work done per unit time, we write the expression for power P as

$$P = \frac{\text{work done}}{\text{time}} = T\frac{\theta}{t} = T\omega$$

Solving for T, we get

$$T = \frac{P}{\omega} \qquad\qquad (12\text{--}10)$$

where T = the torque in the shaft, $\text{lb} \cdot \text{in.}$ or $\text{N} \cdot \text{m}$

P = the power transmitted by the shaft, $\text{lb} \cdot \text{in./s}$ or $\text{N} \cdot \text{m/s}$ (watt or W)

ω = the angular velocity of the shaft, rad/s

U.S. Customary System.   In U.S. customary units, the commonly used unit for power is horsepower (hp), and the angular velocity is commonly expressed in revolutions per minute (rpm), which is denoted by n. The conversion factors are

$$1 \text{ hp} = 6600 \text{ lb} \cdot \text{in./s}$$

$$1 \text{ rpm} = (2\pi/60) \text{ rad/s}$$

When horsepower is used for P and n (rpm) is used to express the angular velocity in Equation 12–10, the conversion factors listed above must be used. After simplification, the equation becomes:

$$T = \frac{63\ 000P}{n} \tag{12–11}$$

where T = the torque in the shaft, lb · in.

P = the power transmitted by the shaft, hp

n = the angular velocity of the shaft, rpm

SI Units.   In SI units, the commonly used unit for power is kilowatt (kW), and the angular velocity is commonly expressed in revolutions per minute (rpm), which is denoted by n. The conversion factors are

$$1 \text{ kW} = 1000 \text{ W}$$

$$1 \text{ rpm} = (2\pi/60) \text{ rad/s}$$

When kW is used for P and n (rpm) is used to express the angular velocity in Equation 12–10, the conversion factors listed above must used. After simplification, the equation becomes:

$$T = \frac{9550P}{n} \tag{12–12}$$

where T = the torque in the shaft, N · m

P = the power transmitted by the shaft, kW

n = the angular velocity of the shaft, rpm

The conversion factor between horsepower and kilowatt is

$$1 \text{ hp} = 0.7457 \text{ kW}$$

─── **EXAMPLE 12–9** ───────────────────────────────────────

Find the maximum stress in a 2-in.-diameter solid steel shaft that transmits 120 horsepower operating at 600 rpm.

Solution. The torque that the shaft is subjected to can be computed from Equation 12–11 as

$$T = \frac{63\ 000\ (120)}{600} = 12\ 600 \text{ lb} \cdot \text{in.}$$

The polar moment of inertia is

$$J = \frac{\pi d^4}{32} = \frac{\pi(2 \text{ in.})^4}{32} = 1.571 \text{ in.}^4$$

The maximum shear stress on the outside surface of the shaft is

$$\tau_{max} = \frac{Tc}{J} = \frac{(12\ 600 \text{ lb} \cdot \text{in.})(1 \text{ in.})}{1.571 \text{ in.}^4}$$

$$= 8020 \text{ psi} \qquad\qquad\qquad \Leftarrow \textbf{Ans.}$$

EXAMPLE 12–10

A hollow circular shaft is to transmit 100 kW rotating at a speed of 1800 rpm. If the allowable shear stress is 60 MPa, select the size of the hollow shaft. Assume a ratio of $k = d_i/d_o = 0.7$.

Solution. The torque that the shaft is subjected to can be computed from Equation 12–12 as

$$T = \frac{9550P}{n} = \frac{9550(100)}{1800} = 531 \text{ N} \cdot \text{m}$$

From Equation 12–9, we find

$$(d_o)_{req} = \sqrt[3]{\frac{16T}{\pi\tau_{allow}(1 - k^4)}}$$

$$= \sqrt[3]{\frac{16\ (531 \text{ N} \cdot \text{m})}{\pi\ (60 \times 10^6 \text{ N/m}^2)(1 - 0.7^4)}}$$

From which we get

$$d_o = 0.0390 \text{ m}$$

$$d_i = 0.7d_o = 0.7(0.0390 \text{ m}) = 0.0273 \text{ m}$$

$$\text{Use } d_o = 39 \text{ mm and } d_i = 27 \text{ mm} \qquad \Leftarrow \textbf{Ans.}$$

EXAMPLE 12–11

The 2-in.-diameter solid steel line shaft in Fig. E12–11(1) is used for power transmission in a manufacturing plant. A motor inputs 100 hp to a pulley at A, which is transmitted by the shaft to pulleys at B, C, and D. The output horsepower from pulleys located at B, C, and D are 45 hp, 25 hp, and 30 hp, respectively. Determine the maximum shear stress in the shaft.

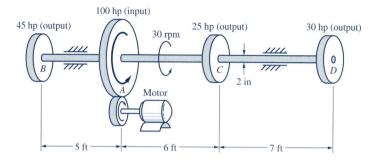

FIGURE E12–11(1)

Solution. The torque exerted on each gear can be computed from Equation 12–11 as

$$T_A = \frac{63\,000\,(100)}{300} = 21\,000\ \text{lb} \cdot \text{in.}$$

$$T_B = \frac{63\,000\,(45)}{300} = 9450\ \text{lb} \cdot \text{in.}$$

$$T_C = \frac{63\,000\,(25)}{300} = 5250\ \text{lb} \cdot \text{in.}$$

$$T_D = \frac{63\,000\,(30)}{300} = 6300\ \text{lb} \cdot \text{in.}$$

These torques act on the pulley as shown in Fig. E12–11(2). From the free-body diagrams shown in Fig. E12–11(3) and using the sign convention presented in Section 12–2, the internal torque at each segment is

$$T_{BA} = +9450\ \text{lb} \cdot \text{in.}$$

$$T_{AC} = -11\,550\ \text{lb} \cdot \text{in.}$$

$$T_{CD} = -6300\ \text{lb} \cdot \text{in.}$$

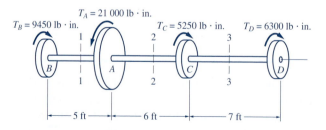

FIGURE E12–11(2)

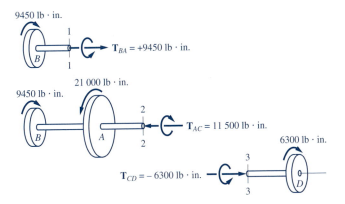

FIGURE E12–11(3)

The internal torque diagram is sketched in Fig. E12–11(4). Since the shaft is of uniform cross-section, the maximum shear stress occurs on the outside surface of segment AC, where the absolute value of the internal torque is a maximum. Thus,

$$\tau_{max} = \frac{Tc}{J} = \frac{(11\ 550\ \text{lb} \cdot \text{in.})(1\ \text{in.})}{\frac{\pi}{32}(2\ \text{in.})^4}$$

$$= 7350\ \text{psi} \qquad\qquad \Leftarrow \textbf{Ans.}$$

FIGURE E12–11(4)

12–6
ANGLE OF TWIST

When a shaft is subjected to torque, two end sections rotate through an angular displacement relative to each other. This relative angular displacement between two sections in the shaft is called the angle of twist. Figure 12–8 shows a shaft of length L that is fixed at the left end and subjected to a torque T at the free end. The longitudinal line AB on the surface of the shaft is twisted into a helix AB by the torque T. The radius OB at the free end rotates through an angle ϕ (the Greek letter phi) to OB'. The angle ϕ is called the angle of twist of the shaft over the length L.

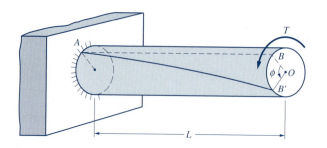

FIGURE 12–8

Attention is now directed to a short length ΔL isolated from the shaft, as shown in Fig. 12–9. The longitudinal line PQ assumes a new position PQ' after the torque is applied. At the same time, the radius OQ rotates through a small angle $\Delta\phi$ to a new position OQ'. The angle $\angle QPQ'$ in radians represents the angular distortion between two lines that were perpendicular before twisting. This angle is identified as the shear strain, according to the definition given in Section 10–8. This shear strain occurs on the periphery of the shaft, where the shear stress is a maximum; accordingly, it is a maximum shear strain and is thus denoted by γ_{max}. Both γ_{max} and $\Delta\phi$ are measured in radians. Since γ_{max} in the elastic range is very small, we have

$$\text{arc } QQ' = \gamma_{max}\,\Delta L = c\Delta\phi$$

From which we get

$$\Delta\phi = \frac{\gamma_{max}\,\Delta L}{c} \tag{a}$$

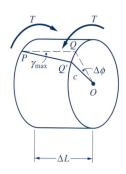

FIGURE 12–9

For elastic distortion of the shaft, the maximum shear stress is within the elastic range and Hooke's law applies. Then, according to Equation 10–9,

$$\gamma_{max} = \frac{\tau_{max}}{G} = \frac{Tc}{JG} \tag{b}$$

Substituting Equation (b) into Equation (a) gives

$$\Delta\phi = \frac{T}{JG}\,\Delta L$$

which gives the angle of twist of the shaft over a short length ΔL. The total angle of twist of the shaft of uniform cross-section, subjected to a constant torque T over a length L, is

$$\phi = \Sigma \Delta \phi = \Sigma \frac{T}{JG} \Delta L = \frac{T}{JG} \Sigma \Delta L$$

But $\Sigma \Delta L = L$; thus,

$$\phi = \frac{TL}{JG} \tag{12-13}$$

where ϕ = the angle of twist of the shaft in radians

T = the constant internal torque in the shaft over the length L

L = the length of the shaft

J = the polar moment of inertia of the cross-section of the shaft

G = the shear modulus of the shaft material

This equation is valid for both solid and hollow circular shafts. Consistent units must be used for quantities on the right-hand side of the equation so that the expression is dimensionless, since ϕ (in radians) must be a dimensionless quantity.

We will consider the sign of the angle of twist ϕ to be the same as the sign of the internal torque T, as defined in the sign convention presented in Section 12-2. Therefore, we state the sign convention for the angle of twist as: *A positive angle of twist is produced by a positive internal torque, and a negative angle of twist is produced by a negative internal torque.*

───── **EXAMPLE 12-12** ─────

A steel shaft 50 mm in diameter is subjected to a torque of 2000 N · m. Determine the maximum shear stress in the shaft and the angle of twist of the shaft over a 2-m length. The shear modulus of steel is $G = 84$ GPa.

Solution. The polar moment of inertia of the cross-sectional area is

$$J = \frac{\pi}{32} (0.050 \text{ m})^4 = 6.14 \times 10^{-7} \text{ m}^4$$

The maximum shear stress is

$$\tau_{max} = \frac{Tc}{J} = \frac{(2000 \text{ N} \cdot \text{m})(0.025 \text{ m})}{6.14 \times 10^{-7} \text{ m}^4}$$

$$= 81.5 \times 10^6 \text{ N/m}^2 = 81.5 \text{ MPa} \qquad \Leftarrow \textbf{Ans.}$$

which is within the elastic range of steel, so the twisting of the shaft is elastic. The given shear modulus is

$$G = 84 \text{ GPa} = 84 \times 10^9 \text{ N/m}^2$$

From Equation 12–13, the angle of twist is

$$\phi = \frac{TL}{JG} = \frac{(2000 \text{ N} \cdot \text{m})(2 \text{ m})}{(6.14 \times 10^{-7} \text{ m}^4)(84 \times 10^9 \text{ N/m}^2)}$$

$$= 0.0776 \text{ rad} = 4.45° \qquad \Leftarrow \textbf{Ans.}$$

EXAMPLE 12–13

Determine the relative angle of twist between D and B of the shaft in Example 12–11. The shear modulus of steel is $G = 12 \times 10^6$ psi.

Solution. From the solution to Example 12–11, the internal torques in the three segments are $T_{BA} = +9450$ lb · in., $T_{AC} = -11\,550$ lb · in., and $T_{CD} = -6300$ lb · in.

Since the shaft has a uniform cross-section and is made of the same material, the value of JG is a constant throughout the length of the shaft.

$$JG = \frac{\pi}{32} (2 \text{ in.})^4 (12 \times 10^6 \text{ lb/in.}^2) = 1.885 \times 10^7 \text{ lb} \cdot \text{in.}^2$$

From Equation 12–13, the angle of twist of each segment is

$$\phi_{BA} = \frac{T_{BA} L_{BA}}{JG} = \frac{(+9450)(5 \times 12)}{1.885 \times 10^7} = +0.0301 \text{ rad}$$

$$\phi_{AC} = \frac{T_{AC} L_{AC}}{JG} = \frac{(-11\,550)(6 \times 12)}{1.885 \times 10^7} = -0.0441 \text{ rad}$$

$$\phi_{CD} = \frac{T_{CD} L_{CD}}{JG} = \frac{(-6300)(7 \times 12)}{1.885 \times 10^7} = -0.0281 \text{ rad}$$

The relative angle of twist of the shaft between D and B is the algebraic sum of the angle of twist of each of the three segments between D and B. Thus,

$$\phi_{D/B} = \phi_{BA} + \phi_{AC} + \phi_{CD}$$

$$= +0.0301 - 0.0441 - 0.0281$$

$$= -0.0421 \text{ rad} = -2.41° \qquad \Leftarrow \textbf{Ans.}$$

EXAMPLE 12–14

Select the size of a solid steel shaft necessary to transmit 150 hp at 300 rpm without exceeding an allowable shear stress of 8000 psi, or without having a relative angle of twist beyond an allowable value of 0.3° per ft length of the shaft. The shear modulus of steel is $G = 12 \times 10^6$ psi.

Solution. The torque that the shaft is subjected to can be computed from Equation 12–11:

$$T = \frac{63\,000P}{n} = \frac{63\,000(150)}{300} = 31\,500 \text{ lb} \cdot \text{in.}$$

Required Diameter for Strength. From Equation 12–8, we get

$$d = \sqrt[3]{\frac{16T}{\pi \tau_{allow}}} = \sqrt[3]{\frac{16(31\ 500\ \text{lb} \cdot \text{in})}{\pi\ (8000\ \text{lb/in.}^2)}} = 2.72\ \text{in.}$$

Required Diameter for Stiffness. The angle twist in radians is

$$\phi = (0.3°)\left(\frac{\pi\ \text{rad}}{180°}\right) = 0.00524\ \text{rad}$$

Solving for J from Equation 12–13, we get

$$J = \frac{TL}{\phi G} = \frac{(31\ 500\ \text{lb} \cdot \text{in.})(12\ \text{in.})}{(0.00524\ \text{rad})(12 \times 10^6\ \text{lb/in.}^2)} = 6.011\ \text{in.}^4$$

Expressing J in terms of d and equating the expression to the required J computed above, we write

$$J = \frac{\pi d^4}{32} = 0.09817 d^4 = 6.011\ \text{in.}^4$$

From which we get

$$d = \sqrt[4]{\frac{6.011\ \text{in.}^4}{0.09817}} = 2.80\ \text{in.}$$

The larger value of d controls. Thus,

$$d_{req} = 2.80\ \text{in.} \qquad \Leftarrow \textbf{Ans.}$$

EXAMPLE 12–15

If the shaft in Example 12–14 is rotating 16 times faster, at 4800 rpm, while transmitting the same horsepower, and all the other conditions remain unchanged, select the size of the solid steel shaft.

Solution. The torque that the shaft is subjected to can be computed from Equation 12–11:

$$T = \frac{63\ 000P}{n} = \frac{63\ 000(150)}{4800} = 1969\ \text{lb} \cdot \text{in.}$$

Required Diameter for Strength. Using Equation 12–6, we get

$$d = \sqrt[3]{\frac{16T}{\pi \tau_{allow}}} = \sqrt[3]{\frac{16\ (1969\ \text{lb} \cdot \text{in.})}{\pi(8000\ \text{lb/in.}^2)}} = 1.08\ \text{in.}$$

Required Diameter for Stiffness. The angle of twist in radians is

$$\phi = (0.3°)\left(\frac{\pi\ \text{rad}}{180°}\right) = 0.00524\ \text{rad}$$

Solving for J from Equation 12–13, we get

$$J = \frac{TL}{\phi G} = \frac{(1969\ \text{lb} \cdot \text{in.})(12\ \text{in.})}{(0.00524\ \text{rad})(12 \times 10^6\ \text{lb/in.}^2)} = 0.376\ \text{in.}^4$$

Expressing J in terms of d and equating the expression to the required J computed above, we write

$$J = \frac{\pi d^4}{32} = 0.09817 d^4 = 0.376 \text{ in.}^4$$

From which we get

$$d = \sqrt[4]{\frac{0.376 \text{ in.}^4}{0.09817}} = 1.40 \text{ in.}$$

Thus,

$$d_{\text{req}} = 1.40 \text{ in.} \qquad\qquad \Leftarrow \textbf{Ans.}$$

Remark. The required diameter above is one-half of the required diameter in the preceding example. The weights of solid shafts are proportional to their diameters squared; hence, the shaft weight in this example is only one-quarter of the shaft weight in Example 12–14. There is a 75 percent saving in weight. This is the reason for the increased use of high-speed shafts in modern machinery.

12–7
SUMMARY

Shafts. Members subjected to twisting moments are referred to as *torsional members.* Power transmission shafts are examples of torsional members. Only solid and hollow circular shafts were discussed in this chapter.

Torsion Formula. Torsional shear stresses are developed in a shaft as a result of the applied torques. The shear stress in a circular shaft varies linearly from zero at the axis to a maximum at the outer surface of the shaft. The maximum shear stress can be computed from

$$\tau_{\text{max}} = \frac{Tc}{J} \qquad\qquad (12\text{--}1)$$

The shear stresses inside the shaft can be computed from

$$\tau = \frac{T\rho}{J} \qquad\qquad (12\text{--}2)$$

The polar moments of inertia are

For solid circular shaft: $= \dfrac{\pi d^4}{32}$ $(12\text{--}3)$

For hollow circular shaft: $J = \dfrac{\pi}{32}(d_o^4 - d_i^4)$ $(12\text{--}4)$

The allowable torque that a shaft can resist is

$$T_{\text{allow}} = \frac{\tau_{\text{allow}} J}{c} \qquad\qquad (12\text{--}5)$$

where the allowable shear stress is

$$\tau_{\text{allow}} = \frac{\tau_y}{F.S.}$$ (12–6)

If τ_y is unknown, its value can be taken as

$$\tau_y = 0.57 \, \sigma_y$$ (12–7)

The required diameter of a solid circular shaft can be computed from

$$d_{\text{req}} = \sqrt[3]{\frac{16T}{\pi\tau_{\text{allow}}}}$$ (12–8)

The required outside diameter of a hollow circular shaft with a given ratio of diameters $k = d_i/d_o$ can be computed from

$$(d_o)_{\text{req}} = \sqrt[3]{\frac{16T}{\pi\tau_{\text{allow}}(1 - k^4)}}$$ (12–9)

Power Transmission. Rotating shafts are often used to transmit power. In U.S. customary units, the torque in lb · in. developed in a shaft transmitting P hp and rotating at n rpm can be computed from

$$T = \frac{63\,000P}{n}$$ (12–11)

In SI units, the torque T in N · m developed in a shaft transmitting P kW and rotating at n rpm can be computed from

$$T = \frac{9550\,P}{n}$$ (12–12)

Angle of Twist. Within the elastic range, the angle of twist, in radians, can be computed from

$$\phi = \frac{TL}{JG}$$ (12–13)

PROBLEMS

Section 12–2   External and Internal Torques

12–1 to 12–6   For the shaft subjected to the external torque in each of Figs. P12–1 to P12–6, find the internal torque in each segment and plot the internal torque diagram of the shaft.

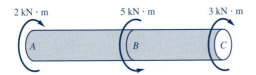

FIGURE P12–1

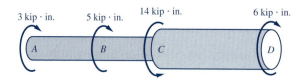

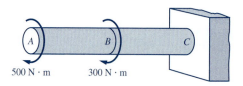

FIGURE P12–2

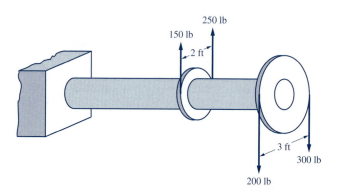

FIGURE P12–3

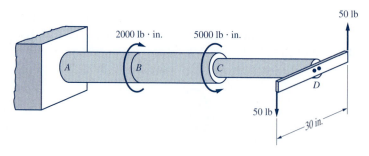

FIGURE P12–4

FIGURE P12–5

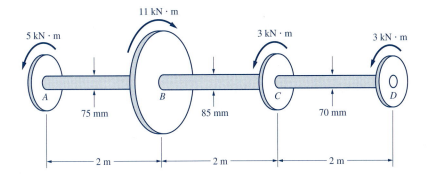

FIGURE P12–6

Section 12–3 The Torsion Formula

12–7 Calculate the maximum shear stress developed in a solid circular shaft with a $4\frac{1}{4}$-in. diameter subjected to a torque of 10 kip · ft.

12–8 Calculate the maximum shear stress developed in a solid circular shaft with a 90-mm diameter subjected to a torque of 8.5 kN · m.

12–9 Calculate the maximum shear stress developed in a hollow circular shaft with a 4-in. outside diameter and a 3-in. inside diameter subjected to a torque of 80 kip · in.

12–10 Calculate the maximum shear stress developed in a hollow circular shaft with a 50-mm outside diameter and a 40-mm inside diameter subjected to a torque of 890 N · m.

12–11 A hollow steel shaft has an outside radius of 6 in. and an inside radius of 4 in. If the maximum shear stress due to a torsional load is 9000 psi, find the minimum shear stress in the section.

12–12 A hollow steel shaft with a 4-in. outside diameter and a 3-in. inside diameter is subjected to the torsional loads shown in Fig. P12–12. Determine the maximum and minimum shear stresses in the shaft.

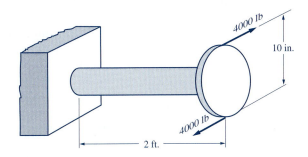

FIGURE P12–12

12–13 See Fig. P12–13. Determine the maximum shear stress in the steel shaft subjected to the external torques shown.

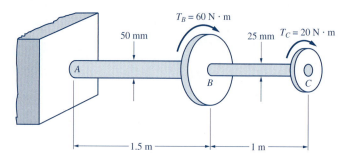

FIGURE P12–13

12–14 Refer to Fig. P12–14. Determine the maximum shear stress in the shaft subjected to the external torques shown.

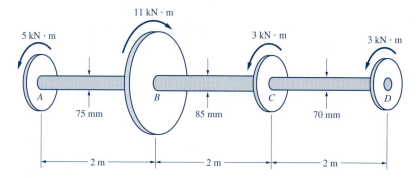

FIGURE P12–14

12–15 Find the allowable torque that may be applied to a solid circular shaft with a $2\frac{1}{2}$-in. diameter if the allowable shear stress is 8000 psi.

12–16 Find the allowable torque that may be applied to a solid circular shaft with a 75-mm diameter if the allowable shear stress is 60 MPa.

12–17 A hollow shaft is produced by boring a 200-mm diameter concentric core in a 300-mm diameter solid circular shaft. Compute the percentage of reduction of the torsional strength and the percentage reduction of weight.

12–18 Show that the torsional strength of a solid circular shaft is reduced by only about 25% if an axial hole is bored to remove half the shaft material.

12–19 A solid circular shaft is to transmit a torque of 1500 N · m. Select the size of the shaft to the nearest mm if the allowable shear stress is 50 MPa.

12–20 A solid circular steel shaft is subjected to a torque of 30 kip · ft. Select the size of the shaft to the nearest sixteenth of an inch if the allowable shear stress is 12 ksi.

12–21 A shaft of A36 steel having a tensile yield stress of 36 ksi is to transmit a torque of 20 kip · in. Using a factor of safety of 3, select the size of the hollow circular shaft to the nearest sixteenth of an inch if the inside diameter is to be three-quarters of the outside diameter.

12–22 A hollow shaft having the ratio of inside diameter to outside diameter of 0.75 is to transmit a torque of 12.5 kN · m. Select the size of the shaft to the nearest mm if the allowable shear stress is 70 MPa.

Section 12–4   Shear Stresses on Mutually Perpendicular Planes

12–23 In the segment of shaft shown in Fig. P12–23, sketch the shear stress distribution on the cross-section and the longitudinal axial planes $OACO'$ and $OBDO'$ along the radii OA and OB. The external torque T is applied as shown.

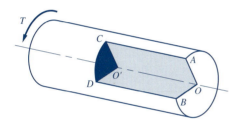

FIGURE P12–23

12–24 A 6-in.-diameter oak shaft, with the grain parallel to the longitudinal axis, is used in a water mill. If the allowable shear stress is 140 psi parallel to the grain and 300 psi perpendicular to the grain, determine the allowable torque of the shaft.

Section 12–5   Power Transmission

12–25 Calculate the torque developed by a motor that delivers 5 hp rotating at 1500 rpm.

12–26 Calculate the torque developed by a motor that delivers 3.75 kW rotating at 1200 rpm.

12–27 Find the maximum shear stress in a $1\frac{3}{4}$-in.-diameter steel shaft transmitting 100 hp at 1000 rpm.

12–28 Find the maximum shear stress in a 35-mm-diameter steel shaft transmitting 75 kW at 1260 rpm.

12–29 Find the horsepower that a solid steel shaft 1 in. in diameter rotating at 1000 rpm can transmit safely without exceeding a maximum allowable shear stress of 8000 psi.

12–30 A 4-in.-diameter solid steel shaft is transmitting 200 hp at 100 rpm. Determine the maximum shear stress in the shaft and the reduction in the maximum shear stress that would occur if the speed of the shaft were increased to 300 rpm.

12–31 Determine the allowable power in kW that a hollow shaft, with an outside diameter of 50 mm and an inside diameter of 35 mm, can transmit at 250 rpm without exceeding an allowable shear stress of 50 MPa.

12–32 A hydraulic turbine generates 30 000 kW of electric power when rotating at 250 rpm. Determine the maximum shear stress in the hollow generator shaft with an outside diameter of 550 mm and an inside diameter of 300 mm.

12–33 A motor inputs 112 kW to gear A and drives a line shaft as shown in Fig. P12–33. The solid steel shaft has a uniform cross-section of 50-mm diameter. The shaft rotates at 500 rpm, and delivers 60 kW to gear B and 52 kW to gear C. Determine the maximum shear stress in the shaft.

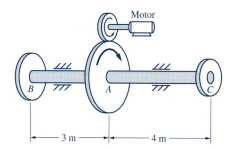

FIGURE P12–33

12–34 The solid steel shafts shown in Fig. P12–34 transmit an input power of 35 kW at pulley C to pulleys A and B. Pulley A outputs 15 kW and pulley B outputs 20 kW. Determine the maximum shear stress in the shafts.

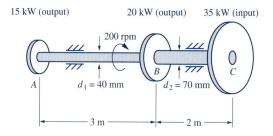

FIGURE P12–34

12–35 The solid steel shafts, with the diameters indicated as shown in Fig. P12–35, are driven by a 50-hp motor. The output horsepower at A, C, and D is 10 hp,

20 hp, and 20 hp, respectively. The shafts rotate at a constant speed of 200 rpm. Determine the maximum shear stress in the shafts.

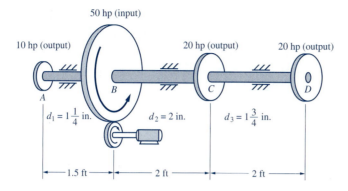

FIGURE P12–35

Section 12–6 Angle of Twist

12–36 Determine the angle of twist in degrees per foot of a 2-in.-diameter steel shaft subjected to a torque of 1200 lb · ft. The shear modulus of steel is $G = 12 \times 10^6$ psi.

12–37 Calculate the angle of twist over a 2-ft length of a tubular steel shaft subjected to a torque of 40 kip · in. The outside and inside diameters are 4 in. and 3 in., respectively. The shear modulus of steel is $G = 12 \times 10^6$ psi.

12–38 Find the angular displacement, in degrees, of pulley C due to the applied torques shown in Fig. P12–38. The shear modulus of steel is $G = 84$ GPa.

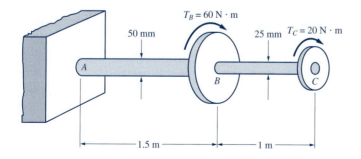

FIGURE P12–38

12–39 Find the total angle of twist, in degrees, between A and D of the shafts shown in Fig. P12–39. The shear modulus of steel is $G = 84$ GPa.

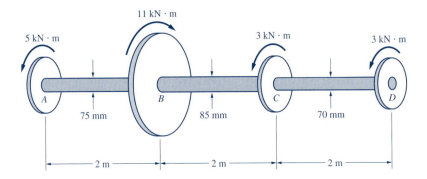

FIGURE P12–39

12–40 Find the angle of twist, in degrees, of section C relative to section B of the shaft in Problem 12–33. The shear modulus of steel is $G = 84$ GPa.

12–41 Find the angle of twist, in degrees, of section C relative to section A of the shaft in Problem 12–34. The shear modulus of steel is $G = 84$ GPa.

12–42 Determine the relative angle of twist of end D relative to end A of the shaft in Problem 12–35. The shear modulus of steel is $G = 12 \times 10^6$ psi.

12–43 Select the size of a solid steel shaft that will not twist more than $1°$ per meter of length when subjected to a torque of 4 kN · m. The shear modulus of steel is $G = 84$ GPa.

12–44 Select the size of a solid steel shaft necessary to transmit 100 hp at 250 rpm without exceeding an allowable shear stress of 8000 psi, or having a relative angle of twist beyond an allowable value of $0.24°$ per 1-ft length. The shear modulus of steel is $G = 12 \times 10^6$ psi.

12–45 Select the size of a solid steel shaft necessary to transmit 186 kW at 450 rpm without exceeding an allowable shear stress of 70 MPa, or having a relative angle of twist beyond an allowable value of $1°$ per 1-m length. The shear modulus of steel is $G = 84$ GPa.

12–46 Select the size of a hollow steel shaft necessary to transmit 100 hp at 250 rpm without exceeding an allowable shear stress of 8000 psi, or having a relative angle of twist beyond an allowable value of $0.24°$ per 1-ft length. The shear modulus of steel is $G = 12 \times 10^6$ psi. Let the ratio $k = d_i/d_o$ be 0.8.

Computer Program Assignments

For each of the following problems, write a computer program using an appropriate programming language with which you are most familiar. Make the program user friendly by incorporating plenty of comments and input prompts so that the user will understand the input data to be entered and the limitations of their values. The output should include the data entered and the computed results, and they must be well labeled to identify each quantity. If a tabulated format is used, a proper heading must be included at the top of the table. Do not limit the program to any specific unit system. Indicate the consistent U.S. customary or SI units that can be used.

C12–1 Refer to Fig. C12–1. Write a computer program that can be used to determine the shear stresses and the angles of twist in the shaft shown. The user input should include (1) the total number of segments n; (2) the value of the modulus of rigidity G; (3) the outside and inside diameters, DO_i and DI_i (note that $DI_i = 0$ for a solid segment); (4) the length L_i; and (5) the torque T_i applied to the right end of each segment. Each torque is treated as positive if it is in the same direction as that indicated in the figure. The output should include (1) the maximum shear stress in each segment, (2) the angle of twist of each segment, and (3) the angle of twist of the entire shaft. Use this program to solve (*a*) Example 12–13, (*b*) Problem 12–38, and (*c*) Problem 12–39.

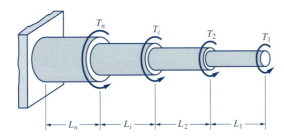

FIGURE C12–1

C12–2 Write a computer program that can be used to determine the required size of a solid or hollow shaft. The shaft transmits a given power and rotates at a given angular velocity. The shear stress in the shaft must be within a given allowable shear stress, and the angle of twist per unit length must be within a specified limit. The user input should include (1) the power P transmitted by the shaft, (2) the angular velocity N of the shaft, (3) the value of the shear modulus G, (4) the value of the allowable shear stress τ_{allow}, (5) the allowable angle of twist in degrees per unit length ϕ_{allow}, and (6) the ratio k $= d_i/d_o$ for a hollow shaft (the ratio $k = 0$ for a solid shaft). Use this program to solve (*a*) Example 12–14, (*b*) Problem 12–45, and (*c*) Problem 12–46.

CHAPTER 13

SHEAR FORCES AND BENDING MOMENTS IN BEAMS

13–1
INTRODUCTION

Beams are members that carry transverse loads and are subjected to bending. Beams are among the most important structural members. Any member subjected to bending may be referred to as a beam. A machine shaft designed to transmit power is also subjected to bending; hence, a shaft also acts as a beam. Depending on the functions and types of structures in which beams are used, they may be called by different names, such as girders, stringers, floor beams, joists, etc.

The main objective of this chapter is to determine the internal forces at various sections along a beam. First, the types of beam supports and beam loadings and the calculation of beam reactions are reviewed. Next, the internal forces in the beam, the shear forces, and the bending moments are determined. The graphical representations, called the shear force and bending moment diagrams, will be sketched.

The beams considered in this book are limited to those that are:

1. Straight and of uniform cross-section, and that possess a vertical plane of symmetry, as shown in Fig. 13–1.

2. Horizontal, although in actual situations beams may be inclined or in vertical positions.

3. Subjected to forces applied in the vertical plane of symmetry, as shown in Fig. 13–1.

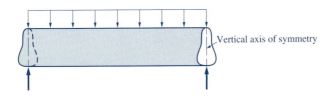
Vertical axis of symmetry

FIGURE 13–1

13–2
TYPES OF BEAMS

Types of Beam Support. Supports for structures have been discussed previously in Section 3–4. In this section, three types of beam supports will be reviewed.

Roller Supports. A roller (or link) support resists motion of the beam only along the direction perpendicular to the plane of the support (or along the axis of the link). Hence, the reaction at a roller support acts along the known direction, as shown in Fig. 13–2.

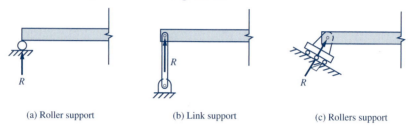

(a) Roller support   (b) Link support   (c) Rollers support

FIGURE 13–2

Hinge Supports. A hinge support resists motion of the beam at the support in any direction on the plane of loading. Hence, the reaction at a pin support consists of two components, usually represented by horizontal and vertical components, as shown in Fig. 13–3a. Figure 13–3b shows a "knife-edge" support, which is a schematic drawing of a hinge support.

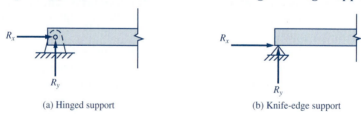

(a) Hinged support   (b) Knife-edge support

FIGURE 13–3

Fixed Supports. At a fixed support, a beam is either built in as an integral part of a concrete column or welded to a steel column. The end of the beam at the fixed support is prevented from displacement in any direction and also from rotation. In general, the reaction at a fixed support consists of three unknowns, that is, two unknown components of force and one unknown moment, as shown in Fig. 13–4.

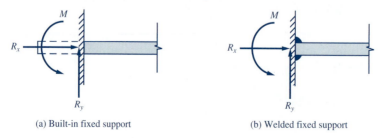

(a) Built-in fixed support   (b) Welded fixed support

FIGURE 13–4

At a roller or hinge support, a beam is free to rotate. Hence, roller and hinge supports are termed *simple supports* to differentiate them from fixed supports.

Types of Beams. Beams can be classified into the types shown in Fig. 13–5, according to the kind of support used.

Simple Beam. A beam supported at its ends with a hinge and a roller, as shown in Fig. 13–5a, is called a *simple beam.*

Overhanging Beam. A simply supported beam with an overhang from one or both ends, as shown in Fig. 13-5b, is called an *overhanging beam.*

Cantilever Beam. A beam that is fixed at one end and free at the other, as shown in Fig. 13–5c, is called a *cantilever beam.*

Propped Cantilever Beam. A beam that is fixed at one end and simply supported at the other, as shown in Fig. 13–5d, is called a *propped cantilever beam.*

Fixed Beam. When both ends of a beam are fixed to supports, as shown in Fig. 13–5e, the beam is called a *fixed beam.*

Continuous Beam. A *continuous beam* is supported on a hinge support and two or more roller supports, as shown in Fig. 13–5f.

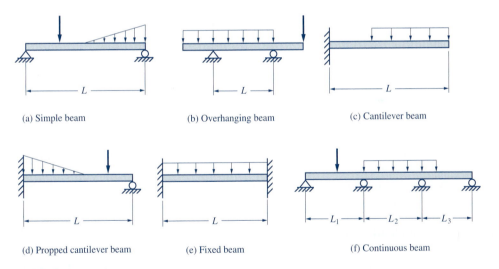

(a) Simple beam (b) Overhanging beam (c) Cantilever beam

(d) Propped cantilever beam (e) Fixed beam (f) Continuous beam

FIGURE 13–5

Statically Determinate Beams. In the first three types of beams, shown in Fig. 13–5a, b, and c, there are three unknown reaction components that may be determined from the static equilibrium equations. Such beams are said to be *statically determinate.*

Statically Indeterminate Beams. When the number of unknown reaction components exceeds three, as in the beams shown in Fig. 13–5d, e, and f, the three equilibrium equations are insufficient for determining the unknown reaction components. Such beams are said to be *statically indeterminate.* Statically indeterminate beams will be discussed in Chapter 17.

13–3
TYPES OF LOADING

Beams are subjected to various loads. Only the concentrated, uniform, and linearly varying loads will be discussed here.

Concentrated Loads. A *concentrated load* is applied at a specific point on the beam and is considered as a discrete force acting at the point, as shown in Fig. 13–6a. For example, a weight fastened to a beam by a cable applies a concentrated load to the beam.

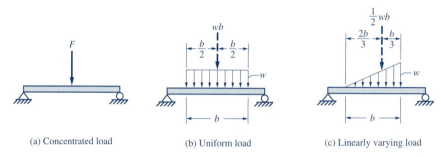

(a) Concentrated load   (b) Uniform load   (c) Linearly varying load

FIGURE 13–6

Uniform Loads. When a load is distributed over a part or the entire length of the beam, it is called a *distributed load*. If the intensity of a distributed load is a constant value, it is called a *uniform load*. The load intensity is expressed as force per unit length of the beam, such as lb/ft or N/m. For computing the reactions, the distributed load may be replaced by its equivalent force. The equivalent force of a uniform load is equal to the load intensity w multiplied by the length of distribution b, and the line of action of the equivalent force passes through the midpoint of the length b, as shown in Fig. 13–6b. The weight of a beam is an example of a uniformly distributed load.

Linearly Varying Loads. A *linearly varying load* is a distributed load with a uniform variation of intensity. Such a load condition occurs on a vertical or inclined wall due to liquid pressure. Figure 13–6c shows a linearly varying load, with intensity varying linearly from zero to a maximum value w.

From Section 7–5, we recall that a distributed force may be replaced by an equivalent concentrated force having a magnitude equal to the area of the load diagram and a line of action passing through the centroid of that area. For the linearly varying load in Fig. 13–6c, the equivalent concentrated force has a magnitude equal to the area of the load triangle, $\frac{1}{2}wb$, and a line of action passing through the centroid of the load triangle at a distance $\frac{1}{3}b$ from the point with maximum load intensity w.

13–4
BEAM REACTIONS

Since the subsequent computation of internal forces needs the beam reactions, it is important that they are determined correctly. The determination of reactions has been discussed in Chapter 3. In this section, we will review the procedure.

As stated in Chapter 3, for a structure in a plane, we can write three independent equilibrium equations to solve for three unknowns. The three equations could be any combination of force-component equations or moment equations, as long as the three equations are independent. In this chapter, we assume that there are no horizontal forces acting on the beam. The equilibrium equation $\Sigma F_x = 0$ requires that the horizontal component of the reaction be zero. Hence, the horizontal component of reaction will not be shown in the following examples.

———— EXAMPLE 13–1 ————

Refer to Fig. E13–1(1). Determine the external reactions in the overhanging beam due to the loading shown.

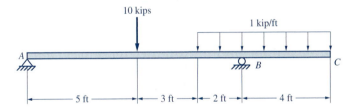

FIGURE E13–1(1)

Solution. The free-body diagram of the beam is sketched in Fig. E13–1(2), with all the externally applied forces and the unknown reaction components drawn as shown. Note that the uniform load is replaced by its resultant force in calculating the reactions. Without the horizontal component of reaction at A (there is no applied horizontal load), we have only two unknown reactions to solve. The two unknowns can be solved by the following static equilibrium equations. Note that counterclockwise (c.c.w.) moments are considered positive in the moment equations.

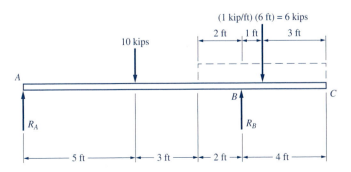

FIGURE E13–1(2)

$$\Sigma M_B = -R_A(10 \text{ ft}) + (10 \text{ kips})(5 \text{ ft}) - (6 \text{ kips})(1 \text{ ft}) = 0$$
$$R_A = +4.4 \text{ kips} \uparrow \qquad\qquad \Leftarrow \textbf{Ans.}$$

$$\Sigma M_A = +R_B(10 \text{ ft}) - (10 \text{ kips})(5 \text{ ft}) - (6 \text{ kips})(11 \text{ ft}) = 0$$
$$R_B = +11.6 \text{ kips} \uparrow \qquad\qquad \Leftarrow \textbf{Ans.}$$

Check. The equation $\Sigma F_y = 0$ has not been used in the solution; it can be used as a check. Thus,

$$\Sigma F_y = 4.4 - 10 + 11.6 - 6 = 0 \qquad\qquad \text{(Checks)}$$

Remark. An alternative solution would be to solve R_A from the moment equation $\Sigma M_B = 0$ and to compute R_B from the component equation $\Sigma F_y = 0$. However, if an error is made in computing R_A, the error would be carried over to the calculation of R_B, and that would lead to an incorrect result for R_B also.

EXAMPLE 13–2

Refer to Fig. E13–2(1). Determine the external reactions at the fixed support of the cantilever beam due to the loading shown.

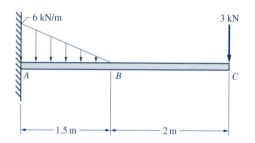

FIGURE E13–2(1)

Solution. The free-body diagram of the beam is sketched in Fig. E13–2(2). Note that the triangular load is replaced by its equivalent resultant force. The unknown reactions at the fixed support are the vertical component R_A and the moment M_A. Note that again the horizontal component of reaction is not shown because there is no horizontal force applied. The two unknown reactions can be determined from the following equilibrium equations.

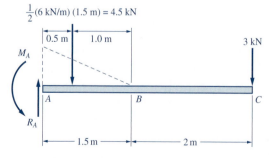

FIGURE E13–2(2)

$$\Sigma F_y = R_A - 4.5 \text{ kN} - 3 \text{ kN} = 0$$

$$R_A = 7.5 \text{ kN} \uparrow \qquad\qquad \Leftarrow \textbf{Ans.}$$

$$\Sigma M_A = +M_A - (4.5 \text{ kN})(0.5 \text{ m}) - (3 \text{ kN})(3.5 \text{ m}) = 0$$

$$M_A = 12.75 \text{ kN} \circlearrowright \qquad\qquad \Leftarrow \textbf{Ans.}$$

Check. The moment equation $\Sigma M_B = 0$ has not been used in the solution, so it can be used for a check. Thus,

$$\Sigma M_B = +12.75 - 7.5(3.5) + 4.5(3) = 0 \qquad\qquad \text{(Checks)}$$

Note that the moment reaction M_A is a couple. Recall that the moment of a couple is independent of the moment center, so the value of M_A about B or any other point remains the same.

13–5
SHEAR FORCE AND BENDING MOMENT IN BEAMS

Internal shear force and bending moment are developed in a beam to resist the external forces and to maintain equilibrium. Consider the beam of Fig. 13–7a, which is subjected to the two concentrated loads shown. The reactions R_A and R_B are determined by considering the equilibrium of the entire beam. The results are shown in Fig. 13–7a. To find the internal forces at section 1–1, pass a plane through the section so that the beam is separated into two parts. Since the entire beam is in equilibrium, each part of the beam separated by section 1–1 must also be in equilibrium.

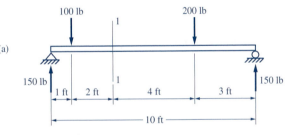

(a)

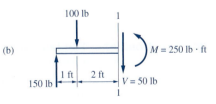

(b)

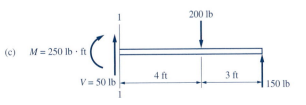

(c)

FIGURE 13–7

Figure 13–7b shows the free-body diagram of the beam to the left of section 1–1. The resultant external force on this segment of the beam is 50 lb, acting upward. The internal resisting force at section 1–1, denoted by V, must be equal and opposite to the resultant external force. Hence, $V = 50$ lb is shown acting downward. Since this force has a shearing effect on the section, it is called a *shear force.* The resultant moment of the external forces about the section is 250 lb · ft acting clockwise ($150 \times 3 - 100 \times 2 = 250$). The internal resisting moment at section 1–1, denoted by M, must be equal and opposite to the resultant moment of the external forces. Hence, $M = 250$ lb · ft is shown acting counterclockwise. Since this moment has a bending effect on the section, it is called a *bending moment.*

If the beam to the right of section 1–1 is isolated as a free body, as shown in Fig. 13–7c, the shear force and the bending moment in the section can also be determined from this segment. The resultant external force on this segment is 50 lb acting downward. The internal resisting shear force at the section is $V = 50$ lb acting upward. The resultant moment of the external forces about the section is 250 lb · ft acting counterclockwise ($150 \times 7 - 200 \times 4 = 250$). The internal resisting bending moment at the section is $M = 250$ lb · ft acting clockwise. We see that the shear force V and the bending moment M at the section shown in Fig. 13–7b and c are of equal magnitude but of opposite directions. Therefore, it is obvious that the algebraic sign conventions do not apply. We cannot say that the upward shear force is positive or the counterclockwise bending moment is positive. Consequently, the beam sign conventions presented below must be adopted.

Beam Sign Conventions. The signs for the internal shear forces and bending moments are based on the effects that they produce:

1. ***Positive Shear.*** The shear force at a section is positive if the external forces on the beam produce a shear effect that tends to cause the left side of the section to move up relative to the right side, as shown in Fig. 13–8a.

2. ***Positive Moment.*** The bending moment at a section is positive if the external forces on the beam produce a bending effect that causes the beam to bend concave upward (i.e., the center of curvature is above the curve) at the section, as shown in Fig. 13–8b.

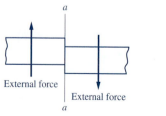

(a) Effect of positive shear

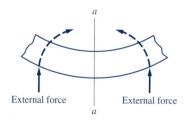

(b) Effect of positive moment

FIGURE 13–8 Beam Sign Convention

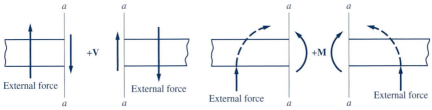

(c) Direction of positive internal shear force **V**   (d) Direction of positive internal bending moment **M**

FIGURE 13–8 (continued)

The positive internal shear force V at a given section of a beam viewed from both directions is shown in Fig. 13–8c, and the positive internal bending moment M at a given section of a beam viewed from both directions is shown in Fig. 13–8d. According to the sign conventions stated above, the shear force and bending moment shown in Fig. 13–7b and c are both positive.

From the beam sign conventions stated above, we derive the following two rules for finding the internal shear force and bending moment in a beam:

Rule 1 *(For Finding Shear Forces) The internal shear force at any section of a beam is equal to the algebraic sum of the external forces on either segment separated by the section. If the summation is from the left end of the beam to the section, treat the upward forces as positive. If the summation is from the right end of the beam to the section, treat the downward forces as positive.*

$$V = \Sigma \text{ Ext. Forces} \begin{cases} \text{From left:} \quad \text{Upward force as positive} \\ \text{From right: Downward force as positive} \end{cases} \quad (13\text{–}1)$$

Rule 2 *(For Finding Bending Moments) The internal bending moment at any section of a beam is equal to the algebraic sum of the moments about the section due to the external forces on either segment separated by the section. In either case, treat the moment produced by upward forces as positive.*

$$M = \Sigma \text{ Moments of Ext. Forces} \begin{cases} \text{From either side: Moment due to} \\ \text{upward force as positive} \end{cases} \quad (13\text{–}2)$$

— EXAMPLE 13–3 —

Calculate the shear forces and bending moments at sections C and D of the beam in Fig. E13–3.

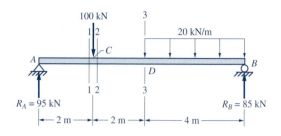

FIGURE E13–3

Solution. If we consider the equilibrium of the entire beam, the reactions are found to be $R_A = 95$ kN and $R_B = 85$ kN.

Shear Force. Because the concentrated load is applied at C, we must choose a section a little to the left and another section a little to the right of C. Let section 1–1 be just to the left and section 2–2 be just to the right of C. The difference between the two sections is that, between A and section 1–1, the 100-lb force is *not* included, while between A and section 2–2, the 100-lb force *is included*. Therefore, the shear force at C has two different values, depending on whether the section is a little to the left or a little to the right of C. We will denote the shear force at section 1–1 to the left of C by V_{C^-} and the shear force at section 2–2 to the right of C by V_{C^+}. Note that the shear force right at the section where a concentrated force is applied is undefined. Hence, V_C has no meaning. Using Rule 1, we find the algebraic sum of the external forces to the left of the appropriate section, treating upward forces as positive. We get

$$V_{C^-} = V_{1-1} = +95 \text{ kN} \qquad\qquad \Leftarrow \textbf{Ans.}$$

$$V_{C^+} = V_{2-2} = +95 - 100 = -5 \text{ kN} \qquad\qquad \Leftarrow \textbf{Ans.}$$

Note that these values indicate that the shear force decreases abruptly from +95 kN on the section just to the left of C to −5 kN on the section just to the right of C. It is generally true that the shear force changes abruptly in the section where a concentrated force is applied. *An upward force causes an abrupt increase in shear force, and a downward force causes an abrupt decrease in shear force.*

To determine the internal forces at D, pass section 3–3 through D. Although a uniform load is applied to the right of D, it does not make any difference whether section 3–3 is a little to the left or a little to the right of D because the amount of distributed load between the two sections at a small distance apart is too small to cause any change in the shear force. Using Rule 1, we can find the shear force at D by taking the algebraic sum of the external forces from A to D, considering upward forces as positive. We get

$$V_D = V_{3-3} = +95 - 100 = -5 \text{ kN} \qquad\qquad \Leftarrow \textbf{Ans.}$$

We could find the shear force at D by taking the algebraic sum of the external forces on segment BD. From Rule 1, we have to treat the downward forces as positive. We get

$$V_D = V_{3-3} = -85 + 20(4) = -5 \text{ kN}$$

Bending Moment.   To find the bending moment at a section, we will apply Rule 2. According to this rule, we need to compute the algebraic sum of the moments about the section produced by the external forces either to the left or to the right of the section. Either way, the moment produced by an upward force will be treated as positive. Starting from the left end of the beam, we write

$$M_C = +95(2) = +190 \text{ kN} \cdot \text{m} \qquad\qquad \Leftarrow \textbf{Ans.}$$

$$M_D = +95(4) - 100(2) = 180 \text{ kN} \cdot \text{m} \qquad\qquad \Leftarrow \textbf{Ans.}$$

We could start from the right end of the beam to get the same results:

$$M_C = +85(6) - (20 \times 4)(4) = +190 \text{ kN} \cdot \text{m}$$
$$M_D = +85(4) - (20 \times 4)(2) = +180 \text{ kN} \cdot \text{m}$$

Note that, although the shear force is not defined at a section where a concentrated force is applied (such as at C), the moment at the section is well defined. And it does not make any difference whether section 1–1 or section 2–2 is considered. The difference of the two sections is a very small difference in moment arm; its effect on the moment is negligible.

EXAMPLE 13–4

Calculate the internal shear force and bending moment at sections A, B, C, and D of the overhanging beam shown in Fig. E13–4(1).

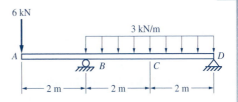

FIGURE E13–4(1)

Solution. The reactions can be determined by considering the equilibrium of the entire beam. The results are $R_B = 15$ kN and $R_D = 3$ kN, as shown in the loading diagram in Fig. E13–4(2).

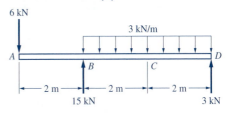

FIGURE E13–4(2)

Shear Force.   Using Rule 1 and starting from the left side of the beam, we find

$$V_{A^-} = 0 \qquad\qquad \Leftarrow \textbf{Ans.}$$

$$V_{A^+} = -6 \text{ kN} \qquad\qquad \Leftarrow \textbf{Ans.}$$

$$V_{B^-} = -6 \text{ kN} \qquad \qquad \Leftarrow \textbf{Ans.}$$

$$V_{B^+} = -6 + 15 = +9 \text{ kN} \qquad \qquad \Leftarrow \textbf{Ans.}$$

The following shear forces are computed from the right side of the beam.

$$V_C = -3 + 3 \times 2 = +3 \text{ kN} \qquad \qquad \Leftarrow \textbf{Ans.}$$

$$V_{D^-} = -3 \text{ kN} \qquad \qquad \Leftarrow \textbf{Ans.}$$

$$V_{D^+} = 0 \qquad \qquad \Leftarrow \textbf{Ans.}$$

Bending Moment. Using Rule 2 and starting from the left side of the beam, we find

$$M_A = 0 \qquad \qquad \Leftarrow \textbf{Ans.}$$

$$M_B = -6 \times 2 = -12 \text{ kN} \cdot \text{m} \qquad \qquad \Leftarrow \textbf{Ans.}$$

The following bending moments are computed from the right side of the beam.

$$M_C = 3 \times 2 - (3 \times 2)(1) = 0 \qquad \qquad \Leftarrow \textbf{Ans.}$$

$$M_D = 0 \qquad \qquad \Leftarrow \textbf{Ans.}$$

13–6
SHEAR FORCE AND BENDING MOMENT DIAGRAMS

Shear force and bending moment diagrams depict the variation of shear force and bending moment along a beam. To construct such diagrams, points with ordinates equal to the computed values of shear forces or bending moments are plotted from a baseline equal to the length of the beam. Beam sign conventions must be used for plotting the shear force and bending moment diagrams. A positive shear or moment is plotted above the baseline; a negative shear or moment is plotted below the baseline. When a series of points is plotted and interconnected, a shear force or bending moment diagram results. It is convenient to make the baselines of the diagrams directly below the loading diagram, which is the diagram of the beam with the applied loads and computed reactions shown.

Shear force and bending moment diagrams are important in beam design. With the aid of these diagrams, the magnitudes and locations of the maximum shear force and the maximum bending moment becomes immediately apparent.

──────── **EXAMPLE 13–5** ────────

Draw the shear force and bend-ing moment diagrams of the overhanging beam in Example 13–4. The beam is shown in Fig. E13–5(1).

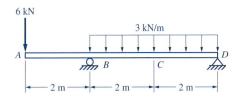

FIGURE E13–5(1)

Solution. The loading diagram, the shear forces, and the bending moments at sections A, B, C, and D have been computed in Example 13–4.

These values are used as ordinates for the shear force and bending moment diagrams at the corresponding sections. When points are plotted and connected by lines, the shear force and the bending moment dia-grams are obtained, as shown in Fig. E13–5(2). The two diagrams should be plotted on the same horizontal scale as the loading diagram of the beam. Usually the corre-sponding horizontal posi-tion lies on the same ver-tical line.

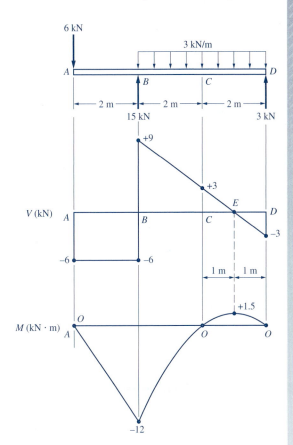

From the shear dia-gram, we see that the zero shear force occurs at E. By inspection, E is the midpoint between C and D. As will be shown in the next section, the moment is either a maxi-mum or a minimum at the section where the shear force is zero. Using Rule 2 and starting from the right, the moment at E is

FIGURE E13–5(2)

$$M_E = +3 \times 1 - (3 \times 1)\left(\frac{1}{2}\right) = +1.5 \text{ kN} \cdot \text{m}$$

From the moment diagram, we see that this moment is a maximum positive moment.

13–7
RELATIONSHIPS AMONG LOAD, SHEAR, AND MOMENT

Certain relationships exist among the loading diagram, the shear diagram, and the moment diagram. To establish the relationships, let us consider a beam element with incremental length Δx shown in Fig. 13–9a. The free-body diagram of the element isolated from the beam is shown in Fig. 13–9b. The shear force and bending moment on section 1–1 are denoted by V and M, respectively. Both are assumed to act in the positive direction. The shear and moment on section 2–2 is $V + \Delta V$ and $M + \Delta M$, respectively, where ΔV and ΔM represent the change in shear and moment between the two sections. The length of the element Δx can be as small as we can imagine, and the distributed load on the element can be considered as having a uniform intensity w. Upward load is considered positive.

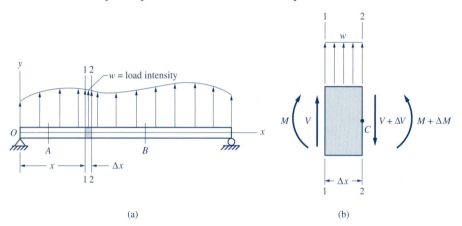

(a)   (b)

FIGURE 13–9

Relationship Between Load and Shear.   The equilibrium equation $\Sigma F_y = 0$ for the element in Fig. 13–9b gives

$$\Sigma F_y = V + w\Delta x - (V + \Delta V) = 0$$

From which we get

$$\Delta V = w\Delta x \qquad (13\text{–}3)$$

Dividing both sides by Δx, we obtain

$$\frac{\Delta V}{\Delta x} = w \qquad (13\text{–}4)$$

which means that *the slope of the shear diagram (the rate of change of the shear force per unit length of beam) at any section is equal to the load intensity at that section.* Note that an upward load is considered positive.

Equation 13–3, $\Delta V = w\Delta x$, means that the incremental change ΔV is equal to $w\Delta x$, which represents the load on the beam over the incremental length Δx. The difference of shear force between sections A and B (Fig. 13–9a) is equal to the sum of all the incremental change ΔV between the two sections:

$$V_B - V_A = \Sigma \Delta V = \Sigma w \Delta x = \text{Total loads between } A \text{ and } B$$

or

$$V_B = V_A + [\text{Load}]_A^B \qquad (13\text{–}5)$$

which means that *the shear force at a section is equal to the shear force at the previous section plus the total load between the two sections.* Note that this relationship is also valid when a concentrated load is applied. The shear force at the section immediately to the right of the concentrated load is equal to the shear force immediately to the left plus the load. Thus, *the shear diagram has an abrupt change at the concentrated load, an upward force will cause an abrupt increase in the shear force, and a downward force will cause an abrupt decrease in the shear force.*

Relationship Between Shear and Moment.   The equilibrium equation $\Sigma M_C = 0$ for the element in Fig. 13–9b gives

$$\Sigma M_C = -M - V \Delta x - (w \Delta x)(\Delta x / 2) + (M + \Delta M) = 0$$

From which we get

$$\Delta M = V \Delta x + w(\Delta x)^2 / 2$$

Dividing both sides by Δx and remembering that Δx could be as small as we can imagine, the term containing Δx vanishes. We obtain

$$\frac{\Delta M}{\Delta x} = V \qquad (13\text{–}6)$$

which means that *the slope of the moment diagram (the rate of change of moment per unit length of beam) at any section is equal to the value of the shear force at that section.*

    Equation 13–6 can be written as $\Delta M = V \Delta x$, which means that the incremental change ΔM is equal to the narrow strip of area in the shear diagram over the incremental length Δx. The difference in the bending moment between sections A and B is equal to the sum of all the incremental changes ΔM between the two sections:

$$M_B - M_A = \Sigma \Delta M = \sum_A^B V \Delta x = \text{Total area under the } V\text{-diagram between } A \text{ and } B$$

or

$$M_B = M_A + [\text{Area under the V-diagram}]_A^B \qquad (13\text{–}7)$$

which means that *the moment at a section is equal to the moment at the previous section plus the area under the shear diagram between the two sections.* Note that the area under the positive shear curve is considered positive, and the area above the negative shear curve is considered negative.

13–8
SKETCHING SHEAR AND MOMENT DIAGRAMS
USING THEIR RELATIONSHIPS

The relationships established in the previous section may be used to facilitate the sketching of shear force and bending moment diagrams.

Loading Diagram. Show all the applied forces and reactions on the beam, including all the relevant dimensions along the beam. *Never* replace a distributed load by its equivalent concentrated force.

Shear Diagram. The following procedure may be followed for sketching the shear diagram:

1. For convenience and clarity, the shear diagram should be drawn directly below the loading diagram. A horizontal baseline for the shear diagram is drawn at a proper location below the loading diagram. Draw lines vertically downward from controlling sections, including the sections at the supports, sections at the concentrated forces, and the beginning and end of a distributed load.
2. Starting at the left end, compute the shear at the controlling sections using Equation 13–5. Note that at the section where a concentrated force is applied, the shear force diagram has an abrupt change of values equal to the concentrated load. An upward concentrated load causes an abrupt increase; a downward load causes an abrupt decrease.
3. Plot points on the shear diagram using the shear force of each controlling section as the ordinate. A positive value is plotted above the baseline; a negative value is plotted below the baseline.
4. Connect the adjacent points plotted, and keep in mind that the slope of the shear diagram is equal to the load intensity. The shear diagram is horizontal for the segment of the beam that is not loaded. At the segment of the beam where a downward uniform load is applied, the shear diagram is an inclined line with a downward slope. If the inclined line intersects the baseline, the shear force at the point is zero. Find the location of this point.

Moment Diagram. The following procedure may be followed for sketching the moment diagram:

1. The moment diagram is usually drawn directly under the shear diagram using the same horizontal scale. A horizontal baseline for the moment diagram is drawn at a proper location below the shear diagram. The controlling sections for the moment diagram include those used in sketching the shear diagram plus the section where the shear is zero or where the shear changes sign.
2. Calculate all the areas under the shear diagram between the adjacent controlling sections.
3. Note that the moments at the free end or the ends of a simple beam are always equal to zero. Starting at the left end, compute the moment at the controlling sections using Equation 13–7.

4. Plot points on the moment diagram using the moment of each controlling section as the ordinate. A positive value is plotted above the baseline; a negative value is plotted below the baseline.
5. Connect the adjacent plotted points with proper straight lines or curves. Keep in mind that the slope of the moment diagram at a section is equal to the shear force at the section. For the part of the beam where the shear is a constant, the moment diagram is a straight line. For the part of the beam where the shear diagram is an inclined straight line, the moment diagram is a parabola.
6. The maximum or minimum moment occurs at the section where the shear is zero or where the shear changes sign.

––––––––––– **EXAMPLE 13–6** –––––––––––

Refer to Fig. E13–6(1). Draw the shear force and bending moment diagram for the simple beam due to the concentrated force shown.

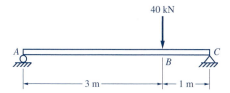

FIGURE E13–6(1)

Solution. Considering the entire beam as a free body, as shown in Fig. E13–6(2), we determine the following reactions:

$$\Sigma M_C = -R_A(4 \text{ m}) + (40 \text{ kN})(1 \text{ m}) = 0$$
$$R_A = +10 \text{ kN} \uparrow$$

$$\Sigma M_A = +R_C(4 \text{ m}) + (40 \text{ kN})(3 \text{ m}) = 0$$
$$R_C = +30 \text{ kN} \uparrow$$

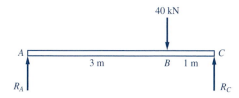

FIGURE E13–6(2)

Loading Diagram. The reactions are shown as concentrated forces in the loading diagram, as shown in Fig. E13–6(3).

Shear Diagram. See Fig. E13–6(3). At point A, the shear force increases abruptly from 0 to +10 kN due to the upward reaction of 10 kN. Between A and B, the shear force remains constant because this segment is not loaded. At point B, the 40 kN downward force makes the shear force decrease abruptly from 10 kN to -30 kN. Between B and C, the shear force remains constant at -30 kN since this segment is not loaded. At point C, the shear force increases abruptly from -30 kN to 0 because of the upward reaction of 30 kN. The fact that the shear force goes back to zero on the right end of the beam provides a useful check. Note that the shear force to the left of B is $+10$ kN; to the right of B, it is -30 kN. The shear force exactly at B is undefined.

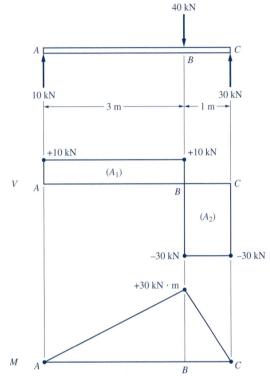

FIGURE E13–6(3)

Moment Diagram.   Refer to Fig. E13–6(3). The two rectangular areas in the shear diagram are

$$A_1 = (10 \text{ kN})(3 \text{ m}) = 30 \text{ kN} \cdot \text{m}$$

$$A_2 = (-30 \text{ kN})(1 \text{ m}) = -30 \text{ kN} \cdot \text{m}$$

The moments at sections A, B, and C are

$$M_A = 0 \text{ (end of a simple beam)}$$

$$M_B = M_A + A_1 = 0 + 30 = +30 \text{ kN} \cdot \text{m}$$

$$M_C = M_B + A_2 = 30 - 30 = 0$$

The result $M_C = 0$ provides a check to the computations because the moment at the end of a simple beam is supposed to be zero. Now three points on the moment diagram can be plotted using the values above as ordinates. Recall that the slope of the moment diagram at a section is equal to the shear at the section. Since the shear is constant between A and B, the moment diagram in this segment is a straight line. For the same reason, the moment diagram between B and C is also a straight line.

─────── **EXAMPLE 13–7** ───────

Refer to Fig. E13–7(1). Draw the shear force and bending moment diagram for the simple beam subjected to the uniform load shown.

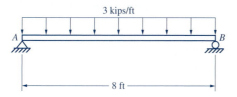

FIGURE E13–7(1)

Solution. The resultant of the uniform load is

$$(3 \text{ kips/ft})(8 \text{ ft}) = 24 \text{ kips}$$

Since the loading is symmetric, the reactions at A and B are each equal to one-half the load, or 12 kips.

Loading Diagram. Refer to the top part of Fig. E13–7(2). Note that the uniform load must be shown as a uniform load, not its equivalent concentrated force. In general, *a distributed load can never be replaced by a concentrated force in the loading diagram.*

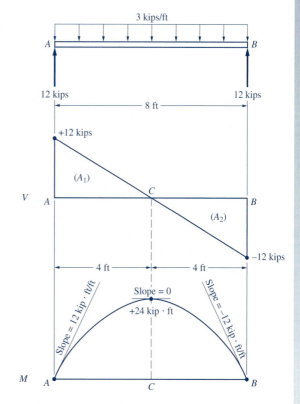

Shear Diagram. See Fig. E13–7(2). Just to the right of A, the shear force is +12 kips. The uniform load will make the shear force decrease at a uniform rate. The total decrease from A to B is equal to the total uniform load of 24 kips. Thus,

$$V_B = V_A - (3 \text{ kip/ft})(8 \text{ ft})$$

$$= +12 \text{ kips} - 24 \text{ kips}$$

$$= -12 \text{ kips}$$

FIGURE E13–7(2)

Between A and B, the slope of the shear diagram is a constant because the load intensity is constant. So the shear diagram is a straight inclined line. The two triangles in the shear diagram are congruent; hence, the point with zero shear occurs at the midspan C. This conclusion can also be obtained by dividing the change of shear force between A and C by the intensity of the uniform load:

$$AC = \frac{12 \text{ kips}}{3 \text{ kips/ft}} = 4 \text{ ft}$$

Moment Diagram. Refer to Fig. E13–7(2). The areas of the two triangles in the shear diagram are

$$A_1 = \frac{1}{2}(+12 \text{ kips})(4 \text{ ft}) = +24 \text{ kip} \cdot \text{ft}$$

$$A_2 = \frac{1}{2}(-12 \text{ kips})(4 \text{ ft}) = -24 \text{ kip} \cdot \text{ft}$$

The moments at sections A, C, and B are

$$M_A = 0 \text{ (end of a simple beam)}$$

$$M_C = M_A + A_1 = 0 + 24 = +24 \text{ kip} \cdot \text{ft}$$

$$M_B = M_C + A_2 = 24 - 24 = 0$$

The result $M_B = 0$ provides a check on the computations. Now three points on the moment diagram can be plotted using the above values as ordinates. Since the load is uniform, the three points are connected by a parabolic curve. Recall that the slope of the moment diagram at a point is equal to the shear at the point. Thus, the slope of the moment diagram at points A, C, and B are $+12 \text{ kip} \cdot \text{ft/ft}$, 0, and $-12 \text{ kip} \cdot \text{ft/ft}$, respectively. The zero slope at C means that the tangent to the moment diagram at C is horizontal. Thus, the moment diagram has a maximum value at C.

EXAMPLE 13–8

Refer to Fig. E13–8(1). Draw the shear force and bending moment diagrams for the overhanging beam due to the loading shown.

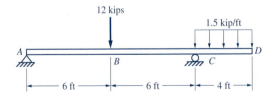

FIGURE E13–8(1)

Solution. Referring to the free-body diagram of the entire beam, as shown in Fig. E13–8(2), we determine the following reactions:

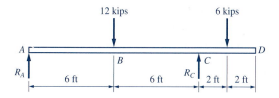

FIGURE E13–8(2)

$$\Sigma M_C = -R_A(12 \text{ ft}) + (12 \text{ kips})(6 \text{ ft}) - (6 \text{ kips})(2 \text{ ft}) = 0$$
$$R_A = +5 \text{ kips} \uparrow$$

$$\Sigma M_A = R_C(12) - (12 \text{ kips})(6 \text{ ft}) - (6 \text{ kips})(14 \text{ ft}) = 0$$
$$R_C = +13 \text{ kips} \uparrow$$

Loading Diagram.
See Fig. E13–8(3). The reactions are shown as concentrated forces, and the uniform load is shown as it is, not its equivalent concentrated force.

Shear   Diagram.
See Fig. E13–8(3). Just to the left of A, the shear is zero. The shear at other points is

$V_{A^+} = 0 + 5 = +5 \text{ kips}$

$V_{B^-} = +5 + 0 = +5 \text{ kips}$

$V_{B^+} = +5 - 12 = -7 \text{ kips}$

$V_{C^-} = -7 + 0 = -7 \text{ kips}$

$V_{C^+} = -7 + 13 = +6 \text{ kips}$

$V_D = +6 - (1.5)(4) = 0$

(Checks)

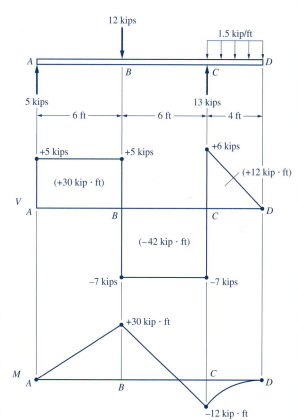

FIGURE E13–8(3)

Points on the shear diagram can now be plotted using the values above as ordinates at the appropriate sections. The shear is constant between concentrated forces. Thus, the shear diagram between A and B and between B and C is composed of horizontal lines. Under a uniform load, the shear diagram has a constant slope, so the shear diagram between C and D is an inclined straight line.

Moment Diagram. Refer to Fig. E13–8(3). The area of each portion of the shear diagram is computed and is indicated in parentheses on the shear diagram. The moments at the controlling points are

$$M_A = 0$$
$$M_B = M_A + 30 = 0 + 30 = +30 \text{ kip} \cdot \text{ft}$$
$$M_C = M_B - 42 = +30 - 42 = -12 \text{ kip} \cdot \text{ft}$$
$$M_D = M_C + 12 = -12 + 12 = 0 \qquad \text{(Checks)}$$

Points are plotted on the moment diagram using the values above as ordinates at the appropriate sections. Between the concentrated forces, the shear is constant, and the corresponding slope of the moment diagram is constant. Thus, between A and B and between B and C, the moment diagram is drawn by connecting the known points with straight lines. Between C and D, the moment diagram is a parabola tangent to the baseline at D, where the shear is zero.

EXAMPLE 13–9

Refer to Fig. E13–9(1). Draw the shear force and bending moment diagrams for the beam subjected to the loading shown. Find the maximum moments along the beam.

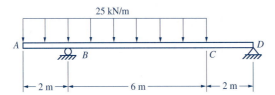

FIGURE E13–9(1)

Solution. Considering the entire beam as a free body, we obtain the following reactions:

$$R_B = 150 \text{ kN} \uparrow$$

$$R_D = 50 \text{ kN} \uparrow$$

These reactions are shown on the loading diagram in Fig. E13–9(2).

Shear Diagram. The shear force at A is zero. The shear forces at other sections are

$$V_{B^-} = 0 - (25)(2) = -50 \text{ kN}$$
$$V_{B^+} = -50 + 150 = +100 \text{ kN}$$
$$V_C = +100 - (25)(6) = -50 \text{ kN}$$
$$V_{D^-} = -50 + 0 = -50 \text{ kN}$$
$$V_{D^+} = -50 + 50 = 0$$

Points on the shear diagram can be plotted using the values above as ordinates at the appropriate sections, as shown in Fig. E13–9(2). The slope of the shear diagram is constant where the load is uniform. Hence, between A and B and between B and C, the shear diagram is drawn by connecting the known points with straight lines. Between C and D, the shear diagram is a horizontal line. The distance from B to point E, of zero shear, can be computed by dividing V_{B^+} by the intensity of the uniform load. We obtain

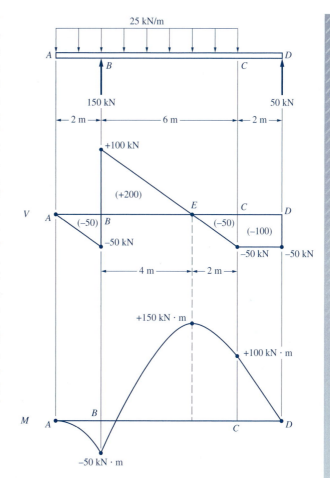

FIGURE E13–9(2)

$$BE = \frac{100 \text{ kN}}{25 \text{ kN/m}} = 4 \text{ m}$$

Moment Diagram. Refer to Fig. E13–9(2). The area of each portion of the shear diagram is computed and is indicated in parentheses inside the shear diagram. The moments at the controlling points are

$$M_A = 0$$
$$M_B = M_A - 50 = 0 - 50 = -50 \text{ kN} \cdot \text{m}$$
$$M_E = M_B + 200 = -50 + 200 = +150 \text{ kN} \cdot \text{m}$$
$$M_C = M_E - 50 = +150 - 50 = +100 \text{ kN} \cdot \text{m}$$
$$M_D = M_C - 100 = +100 - 100 = 0$$

Points are plotted on the moment diagram using the values above as ordinates at the appropriate sections. The moment diagram between A and B is a parabola tangent to the baseline at A, where the shear is zero. The moment diagram between B and C is also a parabola that has a horizontal tangent at E, where the shear is zero. The moment diagram between C and D is a straight line with a constant slope because the shear force is constant between C and D.

From the moment diagram, we find

$$\text{Maximum positive moment} = +150 \text{ kN} \cdot \text{m} \qquad \Leftarrow \textbf{Ans.}$$

$$\text{Maximum negative moment} = -50 \text{ kN} \cdot \text{m} \qquad \Leftarrow \textbf{Ans.}$$

13–9
SHEAR AND MOMENT FORMULAS

Using the method presented in the preceding section, the shear force and bending moment diagrams can be plotted for some simple but typical loading conditions. The results are shown in Table 13–1 for simple beams and cantilever beams subjected to either a concentrated or a uniform load. The expressions for maximum shear and maximum moment listed in the table can be used as formulas. For example, to find the maximum moment in a simple beam subjected to a uniform load of 4 kips/ft over the entire 10-ft span, we can use the expression in case 4 of Table 13–1 to find

$$M_{max} = \frac{wL^2}{8} = \frac{(4 \text{ kips/ft})(10 \text{ ft})^2}{8} = 50 \text{ kip} \cdot \text{ft}$$

If the maximum shear or the maximum moment is required for a beam subjected to a loading consisting of several forces, the *method of superposition* can be used. Using this method, the effect of each load is computed separately and the combined effect is added algebraically. For example, in addition to the uniform load on the simple beam of 10-ft span cited above, there is a concentrated load of 6 kips applied at the midspan. Using case 1 in Table 13–1, the maximum moment caused by the concentrated force is

$$M_{max} = \frac{PL}{4} = \frac{(6 \text{ kips})(10 \text{ ft})}{4} = 15 \text{ kip} \cdot \text{ft}$$

When the uniform load and the concentrated load are both applied, the maximum moment in the beam would be the algebraic sum of the above quantities. If both loads are downward, the maximum moments are both positive. We have

$$M_{max} = 50 + 15 = 65 \text{ kip} \cdot \text{ft}$$

TABLE 13–1   Shear and Moment Formulas for Some Simple Loadings

1. Simple beam with a concentrated load at the center

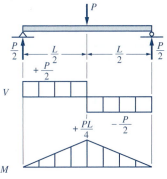

2. Simple beam with a concentrated load at any point

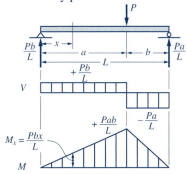

3. Simple beam with two equal concentrated loads symmetrically placed

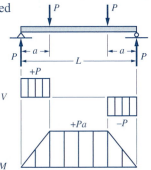

4. Simple beam with a uniform load

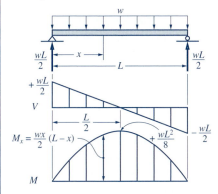

5. Cantilever beam with a concentrated load at any point

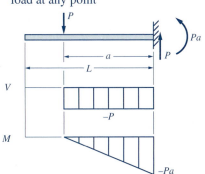

6. Cantilever beam with a uniform load

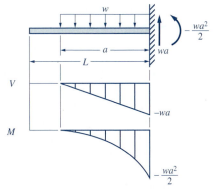

In the beam design problems to be discussed in Chapter 15, the maximum shear force and the maximum bending moment are required. These maximum values can be identified from the shear force and the bending moment diagrams. For some simple cases, it is easier to find these maximum values by using Table 13–1 and the method of superposition, as illustrated in the following examples.

──── **EXAMPLE 13–10** ────────────────────────────────────

Refer to Fig. E13–10. Find the maximum shear force and the maximum bending moment in the simple beam due to the loading shown.

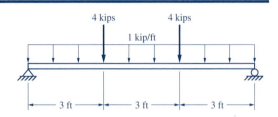

FIGURE E13–10

Solution. The loading represents a combination of cases 3 and 4 in Table 13–1. Using the method of superposition, we will consider the two cases separately and then take the algebraic sum.

Maximum Shear. From cases 3 and 4 in Table 13–1, we see that the maximum shear occurs at the end of the beam with the same sign. Using the expression for the maximum shear for the two cases, we get

$$V_{max} = P + \frac{wL}{2} = 4 \text{ kips} + \frac{(1 \text{ kip/ft})(9 \text{ ft})}{2}$$
$$= 8.5 \text{ kips} \qquad\qquad \Leftarrow \textbf{Ans.}$$

Maximum Moment. From cases 3 and 4 in Table 13–1, we see that the maximum moment occurs at the midspan of the beam and both are positive. Using the expression for the maximum moment for the two cases, we get

$$M_{max} = Pa + \frac{wL^2}{8}$$
$$= (4 \text{ kips})(3 \text{ ft}) + \frac{(1 \text{ kip/ft})(9 \text{ ft})^2}{8}$$
$$= 22.1 \text{ kip} \cdot \text{ft} \qquad\qquad \Leftarrow \textbf{Ans.}$$

──── **EXAMPLE 13–11** ────────────────────────────────────

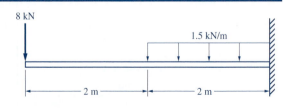

Refer to Fig. E13–11. Find the maximum shear force and the maximum bending moment in the cantilever beam due to the loading shown.

FIGURE E13–11

Solution. The loading represents a combination of cases 5 and 6 in Table 13–1. Using the method of superposition, we will consider the two cases separately and then take the algebraic sum.

Maximum Shear. From cases 5 and 6 in Table 13–1, we see that the maximum shear occurs at the fixed end of the beam and both cases are negative. Using the expression for the maximum shear for the two cases, we get

$$\left| V_{max} \right| = P + wa = 8 \text{ kN} + (1.5 \text{ kN/m})(2 \text{ m})$$

$$= 11 \text{ kN} \qquad \qquad \Leftarrow \textbf{Ans.}$$

Maximum Moment. From cases 5 and 6 in Table 13–1, we see that the maximum moment occurs at the fixed end of the beam and both cases are negative. Using the expression for the maximum moment from these cases, we get

$$\left| M_{max} \right| = PL + \frac{wa^2}{2}$$

$$= (8 \text{ kN})(4 \text{ m}) + \frac{(1.5 \text{ kN/m})(2 \text{ m})^2}{2}$$

$$= 35 \text{ kN} \cdot \text{m} \qquad \qquad \Leftarrow \textbf{Ans.}$$

13–10
SUMMARY

Beams. *Beams* are members that carry transverse loads and are subjected to bending.

Types of Beams. Beams can be classified according to the kind of support used. Types of beams commonly used are the *simple beam,* the *overhanging beam,* and the *cantilever beam.* These beams are all *statically determinate,* and the three static equilibrium equations are sufficient for determining the unknown reaction components.

Types of Loads. Loads on beams are either *concentrated* or *distributed.* Only two special types of distributed loads, the *uniform load* and the *linearly varying load,* were discussed.

Reactions. External beam reactions may be computed by writing the equilibrium equation for the entire beam.

Internal Shear Force and Bending Moment. The internal shear force and bending moment in a beam section are the internal reactions required to resist the external forces to maintain equilibrium. The *method of section* is used to find the internal forces at a section by passing a plane through the section so that the beam is separated into two parts. Since the entire beam is in equilibrium, each part of the beam separated by the section must also be in equilibrium.

Beam Sign Conventions. The following sign conventions were used for the internal shear force and bending moment:

1. *Positive Shear.* The shear force at a section is positive if the external forces on the beam produce a shear effect that tends to cause the left side of the section to move up relative to the right side.

2. *Positive Moment.* The bending moment at a section is positive if the external forces on the beam produce a bending effect that causes the beam to bend *concave upward.*

The following two rules may be used to find the internal shear force and bending moment in a beam:

Rule 1 *(For Finding the Shear Force)*   The internal shear force at any section of a beam is equal to the algebraic sum of the external forces on either segment separated by the section. If the summation is from the left end of the beam to the section, treat the upward forces as positive. If the summation is from the right end of the beam to the section, treat the downward forces as positive.

Rule 2 *(For Finding the Bending Moment)*   The internal bending moment at any section of a beam is equal to the algebraic sum of the moments about the section due to the external forces on either segment separated by the section. In either case, treat the moment produced by the upward forces as positive.

Shear Force and Bending Moment Diagrams.   The shear force diagram is a graphical representation showing the variation of the shear force along the beam. It is drawn directly below the loading diagram. The bending moment diagram is a graphical representation showing the variation of the bending moment along the beam. It is drawn directly below the shear diagram.

Relationships Between the Load and Shear Diagrams. (1) The slope of the shear diagram at a section along a beam is equal to the load intensity at that section.   (2) The change in shear force between two sections along a beam is equal to the total load between the two sections.

Relationships Between the Shear and Moment Diagrams.   (1) The slope of the moment diagram at a section along a beam is equal to the value of the shear force at that section.   (2) The change in bending moment between two sections along a beam is equal to the area under the shear diagram between the two sections.

The above relationships may be used to facilitate the sketching of shear force and bending moment diagrams.

Shear Diagram. The following procedure may be followed for sketching the shear diagram:

1. Draw the shear diagram directly below the loading diagram and use the same horizontal scale.
2. Starting at the left end, compute the shear at various controlling sections by adding the load on the next segment to the shear of the previous section.
3. Plot points on the shear diagram using the shear of each controlling section as the ordinate.
4. Connect the adjacent points plotted. Keep in mind that the slope of the shear diagram is equal to the load intensity. Find the location of the point of zero shear or the point where shear force changes sign.

Moment Diagram. The following procedure may be followed for sketching the moment diagram:

1. The moment diagram is usually drawn directly under the shear diagram using the same horizontal scale.
2. Calculate the areas under the shear diagram.
3. Starting from the left end, add the area of the shear diagram between the controlling sections to the moment of the previous section to get the moment of the next section.
4. Plot points on the moment diagram using the moment of each controlling section as the ordinate.
5. Connect the adjacent points plotted with proper straight lines or curves. Keep in mind that the slope of the moment diagram at a section is equal to the shear force at the section.
6. The maximum or minimum moment occurs at the section where the shear force is zero or where the shear force changes sign.

Shear and Moment Formulas. The expressions for maximum shear and maximum moment listed in Table 13–1 can be used to compute the maximum shear and maximum moment for some loading combinations.

PROBLEMS

Section 13–4   Beam Reactions

13–1 to **13–6**   Determine the external reactions on each beam in Figs. P13–1 to P13–6 due to the loading shown.

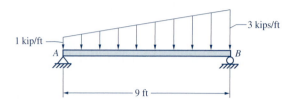

FIGURE P13–1

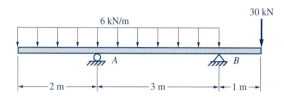

FIGURE P13–2

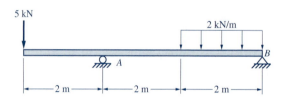

FIGURE P13–3

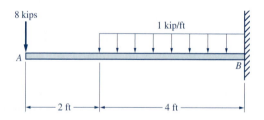

FIGURE P13–4

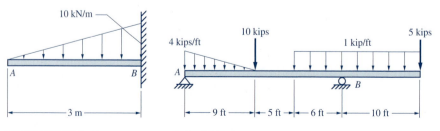

FIGURE P13–5   **FIGURE P13–6**

Section 13–5 Shear Force and Bending Moment in Beams

13–7 to 13–12 Refer to Figs. P13–7 to P13–12. Use the rules for finding shear forces and bending moments to determine the shear forces and bending moments in each figure at sections 1–1, 2–2, and 3–3.

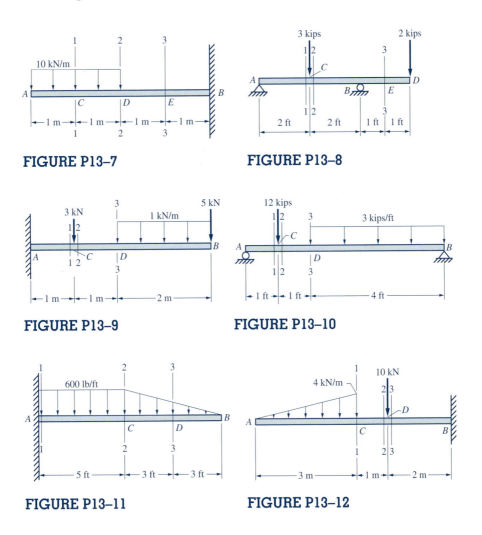

FIGURE P13–7

FIGURE P13–8

FIGURE P13–9

FIGURE P13–10

FIGURE P13–11

FIGURE P13–12

13–13 to 13–15 Determine the shear forces and bending moments at sections A, B, C, D, E, and F in Figs. P13–13 to P13–15.

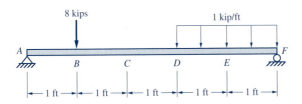

FIGURE P13–13

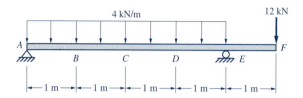

FIGURE P13–14

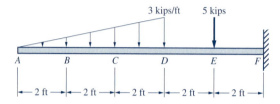

FIGURE P13–15

Section 13–6   Shear Force and Bending Moment Diagrams

13–16 to **13–21**   Refer to Figs. P13–16 to P13–21. Draw the shear force and bending moment diagrams for each beam. Locate the section with zero shear force (if any) and determine the moment at the section.

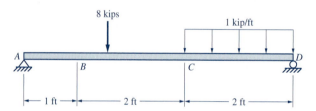

FIGURE P13–16

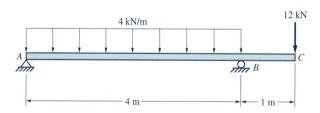

FIGURE P13–17

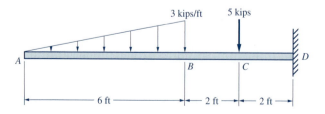

FIGURE P13–18

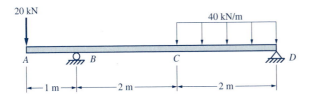

FIGURE P13–19

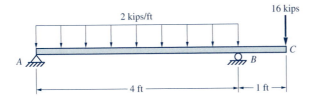

FIGURE P13–20

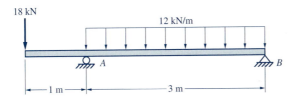

FIGURE P13–21

Section 13–7 Relationships Among Load, Shear, and Moment

Section 13–8 Sketching Shear and Moment Diagrams
Using Their Relationships

13–22 to 13–37 Refer to Figs. P13–22 to P13–37. Construct the shear force and the bending moment diagrams for the beam in each figure due to the loading shown by using the relationships among the load, shear, and moment diagrams.

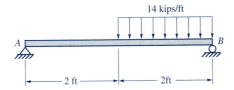

FIGURE P13–22

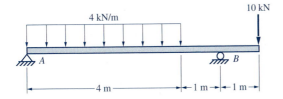

FIGURE P13–23

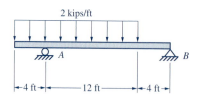

FIGURE P13–24

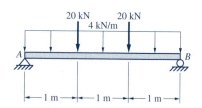

FIGURE P13–25

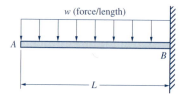

FIGURE P13–26

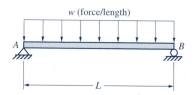

FIGURE P13–27

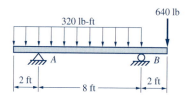

FIGURE P13–28

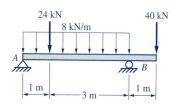

FIGURE P13–29

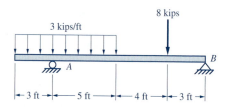

FIGURE P13–30

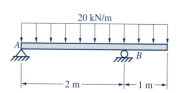

FIGURE P13–31

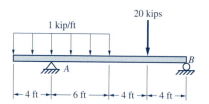

FIGURE P13–32

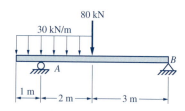

FIGURE P13–33

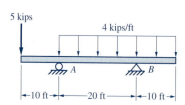

FIGURE P13–34

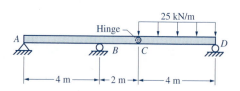

FIGURE P13–35

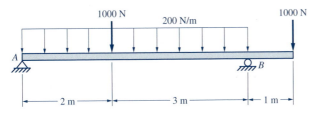

FIGURE P13–36

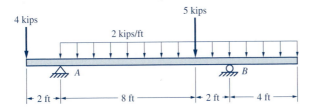

FIGURE P13–37

Section 13–9 Shear and Moment Formulas

13–38 to 13–45 Refer to Figs. P13–38 to P13–45. Find the maximum shear force and the maximum bending moment in the beam in each figure due to the loading shown by using the formulas in Table 13–1.

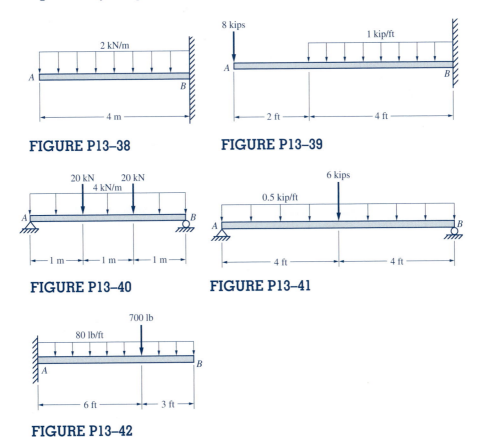

FIGURE P13–38

FIGURE P13–39

FIGURE P13–40

FIGURE P13–41

FIGURE P13–42

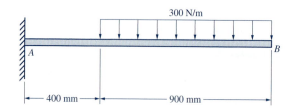

FIGURE P13–43

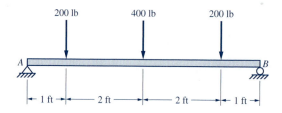

FIGURE P13–44

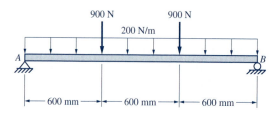

FIGURE P13–45

Computer Program Assignments

For each of the following problems, write a computer program using an appropriate programming language with which you are most familiar. Make the program user friendly by incorporating plenty of comments and input prompts so that the user will understand the input data to be entered and the limitations of their values. The output should include the data entered and the computed results, and they must be well labeled to identify each quantity. If a tabulated format is used, a proper heading must be included at the top of the table. Do not limit the program to any specific unit system. Indicate the consistent U.S. customary or SI units that can be used.

C13–1 Write a computer program that can be used to produce a table showing the shear force and the bending moment at every tenth point of a simply supported beam subjected to a uniform load over the entire span. The user input should be the span length L and the load intensity w. Apply this program to Example 13–7.

C13–2 Refer to Fig. C13–2. Write a computer program that can be used to calculate the shear force and the bending moment of the simple beam subjected to the concentrated loads shown. The user input should include (1) the span length L, (2) the total number of loads n, and (3) the magnitude and location of each load P_i and a_i. The load P_i is treated as positive if it acts in the same direction as that shown in the figure. The output should include (1) the value of the shear force between the adjacent loads and (2) the value of the bending moment at each load. Use this program to solve (a) Example 13–6 and (b) Problem 13–44.

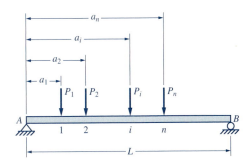

FIGURE C13–2

CHAPTER 14

STRESSES IN BEAMS

14–1
INTRODUCTION

In this chapter, the stresses caused by bending moments and shear forces are considered. Normal stresses along the longitudinal direction are caused by bending moments, and shear stresses are caused by shear forces. The purposes of this chapter are to study the distribution of normal and shear stresses in a beam, and to relate these stresses to the internal bending moment and shear force in the beam.

Normal stresses in beams due to bending (also called *flexural stresses*) are discussed first, followed by a study of shear stress variation in beams. Inelastic bending of beams will also be studied in this chapter so that ultimate strength design of beams can be presented in the next chapter.

14–2
NORMAL STRESSES IN BEAMS DUE TO BENDING

For a straight beam having a constant cross-sectional area with a vertical axis of symmetry, as shown in Fig. 14–1a, a line through the centroid of all cross-sections is referred to as the *axis of the beam*. Consider the two cross-sections, *ab* and *cd*, in the beam. Before the application of loads, the cross-sections are in the vertical direction and the beam axis is a straight horizontal line. Assume that the beam segment between the two sections is subjected to a positive bending moment $+M$. The beam bends and the cross-sections *ab* and *cd* tilt slightly. Note that lines *ab* and *cd* remain straight and perpendicular to the axis of the beam, as shown in Fig. 14–1b. Experimental results indicate that a plane section before bending remains a plane after bending.

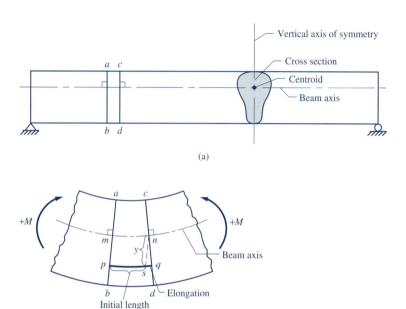

(a)

(b)

FIGURE 14–1

Imagine that the beam is composed of an infinite number of fibers along the longitudinal direction. The fibers along *bd* become longer and those along *ac* become shorter. Hence, the fibers along *bd* are subjected to tension and those along *ac* are subjected to compression.

Neutral Surface and Neutral Axis. The fibers *mn* along the beam axis do not undergo any change of length due to bending; hence, these fibers are not stressed. The surface *mn* is called the *neutral surface*. The intersection of the neutral surface with a cross-section is called a *neutral axis. For a beam subjected to pure bending (no axial force), the neutral axis is a horizontal line that passes through the centroid of the cross-sectional area.*

Variation of Linear Strains. In Fig. 14–1b, consider a typical fiber *pq* parallel to the neutral surface and located at a distance *y* from it. From point *n*, draw a line *ns* parallel to *mp*; then

$$ps = mn = \text{initial undeformed length of the fiber}$$

and the fiber *pq* elongates by an amount *sq*. From triangle *nsq*, we see that the elongation (or contraction) of a fiber varies linearly as the distance *y* to the neutral surface. Since the initial length of all fibers between the sections is the same, it follows that *the linear strains of the longitudinal fibers due to bending vary linearly from zero at the neutral surface to the maximum value at the outer fibers.*

Variation of Flexural Stresses. For elastic bending of the beam, Hooke's law applies; that is, stress is proportional to strain. For most materials, the moduli of elasticity in tension and in compression are equal. Under these conditions, we also conclude that *the flexural stresses in a beam section vary linearly from zero at the neutral axis to the maximum value at the outer fibers.*

Figure 14–2 shows the normal stress distribution in a beam section. The stresses vary from zero at the neutral axis, to a maximum tensile stress at the bottom outer fiber and a maximum compressive stress at the top outer fiber. Note that the resultant T of the tensile stresses below the neutral axis and the resultant C of the compressive stress above the neutral axis must be equal and opposite (in the absence of axial forces) and form a couple equal to the internal resisting moment M.

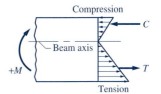

FIGURE 14–2

14–3
THE FLEXURE FORMULA

Derivation of the Flexure Formula. Consider a beam segment subjected to a positive bending moment $+M$, as shown in Fig. 14–3a. At section m–m, the applied moment is resisted by normal stresses that vary linearly from the neutral axis. The maximum normal stresses occur at points on the bottom of the section that are most remote from the neutral axis. Denote the maximum normal stress by σ_{max} and the distance from the neutral axis to the bottom of the section by c. Then the normal stress σ at the narrow strip of area ΔA located at distance y from the neutral axis is, by proportion,

$$\sigma = \frac{y}{c}\sigma_{max} \qquad (14\text{–}1)$$

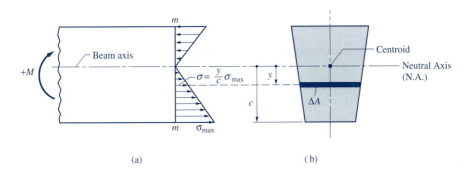

FIGURE 14–3

The force on the incremental area ΔA is $\sigma\Delta A$. The incremental moment of this force about the neutral axis is

$$\Delta M = (\sigma\Delta A)y$$

The total internal resisting moment developed by the normal stresses is the sum of the incremental moment over the entire section. This internal moment must be equal to the external moment M to satisfy the equilibrium condition. We write

$$M = \Sigma \; \Delta M = \Sigma \; (\sigma \; \Delta A)y$$

Substituting $\sigma = (y/c)\sigma_{max}$ from Equation 14–1, we have

$$M = \Sigma \left(\frac{y}{c} \; \sigma_{max} \; \Delta A \right) y$$

Since σ_{max}/c is a constant factor in each term, it can be factored out of the summation. Thus,

$$M = \frac{\sigma_{max}}{c} \; \Sigma y^2 \; \Delta A$$

According to the definition given in Chapter 8, the expression $\Sigma y^2 \Delta A$ is the moment of inertia of the cross-sectional area with respect to the neutral axis. Its value is a constant and depends on the size and shape of the cross-sectional area. Denote it by I, and the equation can be written as

$$M = \frac{\sigma_{max}}{c} I$$

From which we get

$$\sigma_{max} = \frac{Mc}{I} \tag{14–2}$$

where σ_{max} = the maximum normal stress due to bending (also called the flexural stress)

  M = the internal resisting moment at the section under consideration

  c = the distance from the neutral axis to the outermost fiber

  I = the moment of inertia of the cross-sectional area about the neutral axis

Substituting in Equation 14–1, we have

$$\sigma = \frac{My}{I} \tag{14–3}$$

where σ = the flexural stress at any point in the section

  y = the distance from the neutral axis to the point where the flexural stress is desired

Equations 14–2 and 14–3 are two forms of the *flexure formula.*

  The moment of inertia of simple geometric shapes can be computed by using the formulas listed in Table 8–1 in Chapter 8. (The table is also printed inside the front cover for easy reference.) The moments of inertia of composite areas or built-up steel sections can be determined by the method discussed in Sections 8–4 and 8–5.

The sketches in Fig. 14–4a and b are helpful for determining whether a fiber is in tension or in compression due to a given bending moment.

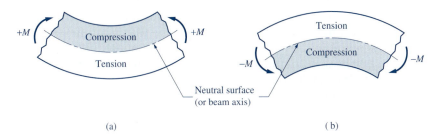

(a)   (b)

FIGURE 14–4

Section Modulus. Note that the moment of inertia *I* and the distance *c* are both geometric properties of the cross-section and their values are constant for a given section. Their ratio *I/c* is also a geometric property and is constant for a given section. The quantity *I/c* is called the *section modulus*. Denoting the section modulus by *S*, we write

$$S = \frac{I}{c} \tag{14–4}$$

Expressed in terms of the section modulus, Equation 14–2 becomes

$$\sigma_{max} = \frac{M}{S} \tag{14–5}$$

This equation is widely used in engineering practice because of its simplicity. The equation is especially useful in the design of beams, as we will see in the next chapter. The section modulus for a rectangular section of width *b* and height *h* is

$$S = \frac{I}{c} = \frac{bh^3/12}{h/2} = \frac{bh^2}{6} \tag{14–6}$$

The section modulus for a circular section of diameter *d* is

$$S = \frac{I}{c} = \frac{\pi d^4/64}{d/2} = \frac{\pi d^3}{32} \tag{14–7}$$

To facilitate computations, section moduli for manufactured sections are tabulated in handbooks. Values of section moduli for selected structural steel shapes are given in the appendix, Tables A–1 to A–5, and those for structural timber sections are given in Table A–6.

The moment of inertia has the units in.^{4} or m^4. The section modulus has the units in.^{3} or m^3. It should be emphasized that when numerical values are substituted into Equations 14–2, 14–3, or 14–5, consistent units must be used for each of the quantities. To avoid using incorrect units, units must be written for each quantity when substituting into the flexure formula.

Maximum Tensile and Compressive Stresses. For a beam with sections that are symmetrical with respect to the horizontal neutral axis, the maximum tensile and compressive stresses in the beam occur at the extreme fibers of the section where the absolute value of the bending moment is a maximum. For a beam with sections that are not symmetrical with respect to the neutral axis, such as the beam with an inverted T-section in Fig. 14–5, the maximum tensile and compressive stresses in a given section are not equal. Expressions for the maximum flexure stresses are shown in the figure for a positive bending moment. In this case, the maximum tensile and compressive stresses must be calculated at both the section with the maximum positive moment and the section with the maximum negative moment, as illustrated in Example 14–3.

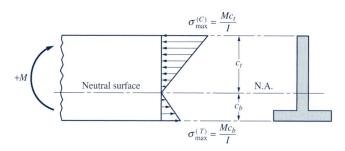

FIGURE 14–5

EXAMPLE 14–1

A timber section with a nominal 4 in. × 10 in. rectangular section is used on a simple span of 10 ft. The beam supports a uniformly distributed load of 450 lb/ft (which includes the weight of the beam). Determine the maximum flexural stress due to bending.

Solution. From Table 13–1 for a uniform load on a simple span (case 4), the maximum moment at the midspan is

$$M_{max} = \frac{wL^2}{8} = \frac{(450\ \text{lb/ft})(10\ \text{ft})^2}{8} = 5625\ \text{lb} \cdot \text{ft}$$

From the appendix, Table A–6(a), we find that the dressed size for a 4-in. × 10-in. timber section is $3\frac{1}{2}$ in. × $9\frac{1}{4}$ in. The table also lists the values of the moment of inertia of the section about the neutral axis and the section modulus. From the table, we find

$$I = 231\ \text{in.}^4 \qquad S = 49.9\ \text{in.}^3$$

These values can be verified as follows:

$$I = \frac{bh^3}{12} = \frac{(3.5\ \text{in.})(9.25\ \text{in.})^3}{12} = 231\ \text{in.}^4$$

$$S = \frac{bh^2}{6} = \frac{(3.5\ \text{in.})(9.25\ \text{in.})^2}{6} = 49.9\ \text{in.}^3$$

The maximum flexural stress in the beam is

$$\sigma_{max} = \frac{M_{max}c}{I} = \frac{(5625 \times 12 \text{ lb} \cdot \text{in.})(9.25 \text{ in.}/2)}{231 \text{ in.}^4}$$

$$= 1350 \text{ psi} \qquad\qquad \Leftarrow \textbf{Ans.}$$

or

$$\sigma_{max} = \frac{M_{max}}{S} = \frac{5625 \times 12 \text{ lb} \cdot \text{in.}}{49.9 \text{ in.}^3} = 1350 \text{ psi}$$

EXAMPLE 14–2

A cantilever beam with a 4-m span and a solid circular cross-section of 100-mm diameter is subjected to a concentrated load $P = 2$ kN applied at the free end, as shown in Fig. E14–2. Determine the maximum flexural stress in the beam caused by the load.

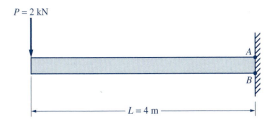

FIGURE E14–2

    Solution. From Table 13–1 for a concentrated load on a cantilever span (case 5), the maximum moment at the fixed end is

$$M_{max} = -PL = -(2 \text{ kN})(4 \text{ m}) = -8 \text{ kN} \cdot \text{m}$$

The section modulus of the circular section is

$$S = \frac{\pi d^3}{32} = \frac{\pi (0.1 \text{ m})^3}{32} = 9.82 \times 10^{-5} \text{ m}^3$$

The maximum flexural stress at the fixed end of the beam is

$$\sigma_{max} = \frac{M_{max}}{S} = \frac{8 \text{ kN} \cdot \text{m}}{9.82 \times 10^{-5} \text{ m}^3} = 81.5 \times 10^3 \text{ kN/m}^2$$

$$= 81.5 \text{ MPa} \qquad\qquad \Leftarrow \textbf{Ans.}$$

Since the maximum moment is negative, the bending at the fixed end is concave downward. The maximum tensile stress occurs at the top fiber at A, and the maximum compressive stress occurs at the bottom fiber at B. These stresses have the same magnitude because the section is symmetrical with respect to the neutral axis.

─── **EXAMPLE 14–3** ───

The overhanging beam in Fig. E14–3(1) is built up with two full-size timber planks, 2 in. × 6 in., glued together to form a T-section, as shown in Fig. E14–3(2). The beam is subjected to a uniform load of 400 lb/ft, which includes the weight of the beam. Determine the maximum tensile and compressive flexural stresses in the beam.

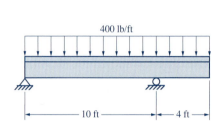

FIGURE E14–3(1)   **FIGURE E14–3(2)**

Solution. The reactions are determined first from the equilibrium conditions of the entire beam. The shear force and bending moment diagrams are then drawn as shown in Fig. E14–3(3). From the moment diagram, we find

$$M_{max}^{(+)} = +3530 \text{ lb} \cdot \text{ft} \qquad M_{max}^{(-)} = -3200 \text{ lb} \cdot \text{ft}$$

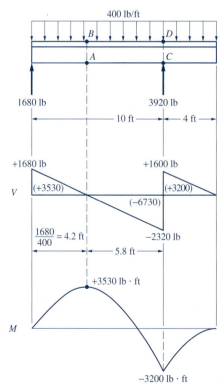

FIGURE E14–3(3)

Next, the properties of the cross-section must be determined. The centroid of the section is determined first. Refer to Fig. E14–3(4). The distance $\bar{y}$ of the centroid from the bottom of the section can be computed from

$$\bar{y} = \frac{\Sigma Ay}{\Sigma A} = \frac{A_1 y_1 + A_2 y_2}{A_1 + A_2}$$

$$= \frac{(6 \text{ in.}^2)(3 \text{ in.}) + (6 \text{ in.}^2)(7 \text{ in.})}{6 \text{ in.}^2 + 6 \text{ in.}^2} = 5 \text{ in.}$$

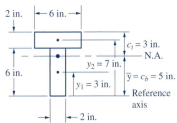

FIGURE E14–3(4)

Thus, the neutral axis is located 5 in. above the bottom of the section. The distance c for the bottom and top fibers are, respectively,

$$c_b = 5 \text{ in.} \qquad c_t = 3 \text{ in.}$$

The moment of inertia of the section about the neutral axis is computed from

$$I_{NA} = \Sigma [I + A (\bar{y} - y)^2]$$

$$= \left[\frac{(2)(6)^3}{12} + (12)(5 - 3)^2 \right] + \left[\frac{(6)(2)^3}{12} + (12)(5 - 7)^2 \right]$$

$$= 136 \text{ in.}^4$$

Since the cross-section is not symmetrical about the neutral axis, the flexural stresses must be computed at the section where the maximum positive moment occurs as well as the section where the maximum negative moment occurs. At the section where the maximum positive moment occurs, the maximum tensile stress occurs at the bottom fiber at A, and the maximum compressive stress occurs at the top fiber at B. The values are

$$\sigma_A = \frac{M_{max}^{(+)} c_b}{I} = \frac{(3530 \times 12)(5)}{136} = 1560 \text{ psi (T)}$$

$$\sigma_B = \frac{M_{max}^{(+)} c_t}{I} = \frac{(3530 \times 12)(3)}{136} = 934 \text{ psi (C)}$$

At the section where the maximum negative moment occurs, the maximum tensile stress occurs at the top fiber at D, and the maximum compressive stress occurs at the bottom fiber at C. The values are

$$\sigma_D = \frac{M_{max}^{(-)} c_t}{I} = \frac{(3200 \times 12)(3)}{136} = 953 \text{ psi (T)}$$

$$\sigma_C = \frac{M_{max}^{(-)} c_b}{I} = \frac{(3200 \times 12)(5)}{136 \text{ in.}^4} = 1410 \text{ psi (C)}$$

Thus, the maximum tensile stress in the beam is

$$\sigma_{max}^{(T)} = \sigma_A = 1560 \text{ psi} \qquad \Leftarrow \textbf{Ans.}$$

and the maximum compressive stress in the beam is

$$\sigma_{max}^{(C)} = \sigma_C = 1410 \text{ psi} \qquad \Leftarrow \textbf{Ans.}$$

14–4
ALLOWABLE MOMENT

Solving Equation 14–2 for the moment M and using the allowable flexural stress σ_{allow} for σ_{max}, we get the formula for computing the allowable moment of a beam:

$$M_{allow} = \frac{I\sigma_{allow}}{c} \qquad (14\text{–}8)$$

where M_{allow} = the allowable moment of a beam, which is the maximum moment that the beam can resist without causing the maximum flexure stress in the beam to exceed the allowable stress

I = the moment of inertia of the cross-sectional area about the neutral axis

σ_{allow} = the allowable flexural stress of the beam, which depends on the type of material and the specifications used (as will be discussed in more detail in the next chapter)

c = the distance from the neutral axis to the outermost fiber

Since the section modulus of the cross-section is $S = I/c$, Equation 14–8 may be written as

$$M_{allow} = S\,\sigma_{allow} \qquad (14\text{–}9)$$

Allowable moment is calculated mainly for the purpose of computing the allowable load that can be applied safely to the beam without causing overstress of the beam. Expressing the maximum moment along the beam as a function of the allowable load and equating the expression to the allowable moment, we can solve for the allowable load that the beam is capable of carrying. The following two examples illustrate this process.

──── **EXAMPLE 14–4** ────

Determine the allowable uniform load that a structural steel W14 × 38 beam can support over a simple span of 12 ft without exceeding an allowable flexural stress of 24 ksi.

Solution. From the appendix, Table A–1(a), the section modulus of W14 $\times$ 38 about the x axis is found to be $S_x = 54.6$ in.^{3}. Substituting this value and the given allowable flexural stress into Equation 14–9, we find

$$M_{allow} = S\sigma_{allow} = (54.6 \text{ in.}^3)(24 \text{ kip/in.}^2)$$
$$= 1310 \text{ kip} \cdot \text{in.} = 109.2 \text{ kip} \cdot \text{ft}$$

For a uniform load on a simple span (from Table 13–1, case 4), the maximum moment at the midspan is

$$M_{max} = \frac{wL^2}{8}$$

Solving for w and substituting M_{allow} for M_{max}, we get

$$w_{allow} = \frac{8M_{allow}}{L^2} = \frac{8(109.2 \text{ kip} \cdot \text{ft})}{(12 \text{ ft})^2}$$
$$= 6.067 \text{ kip/ft} = 6067 \text{ lb/ft}$$

The weight of the beam $w = 38$ lb/ft must be subtracted from the load. Thus, the allowable uniform load that the beam can carry is

$$w_{allow} = 6067 \text{ lb/ft} - 38 \text{ lb/ft} = 6030 \text{ lb/ft} \qquad \Leftarrow \textbf{Ans.}$$

EXAMPLE 14–5

Refer to Fig. E14–5. Determine the allowable concentrated load P in kN that can be applied to the structural steel W410 $\times$ 1.46 (SI designation) beam without exceeding an allowable flexural stress of 165 MPa.

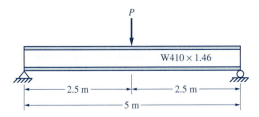

FIGURE E14–5

Solution. From the appendix, Table A–1(b), the section modulus of a W410 $\times$ 1.46 shape about the x axis is found to be $S_x = 2.87 \times 10^{-3}$ m^3. Substituting this value and the allowable flexural stress of

$$\sigma_{allow} = 165 \text{ MPa} = 165 \times 10^3 \text{ kN/m}^2$$

into Equation 14–9, we find

$$M_{allow} = S\sigma_{allow}$$
$$= (2.87 \times 10^{-3} \text{ m}^3)(165 \times 10^3 \text{ kN/m}^2)$$
$$= 474 \text{ kN} \cdot \text{m}$$

The maximum moment at the midspan due to the weight of the beam is

$$M_w = \frac{wL^2}{8} = \frac{(1.46 \text{ kN/m})(5 \text{ m})^2}{8} = 4.6 \text{ kN} \cdot \text{m}$$

The moment due to the concentrated force P must be the difference of the two moments above.

$$M_P = 474 \text{ kN} \cdot \text{m} - 4.6 \text{ kN} \cdot \text{m} = 469 \text{ kN} \cdot \text{m}$$

For a centrally placed, concentrated load on a simple span (from Table 13–1, case 1), the maximum moment at the midspan is

$$M_{max} = \frac{PL}{4}$$

Solving for P and substituting M_P for M_{max}, we get

$$P_{allow} = \frac{4M_P}{L} = \frac{4(469 \text{ kN} \cdot \text{m})}{5 \text{ m}} = 376 \text{ kN} \qquad \Leftarrow \textbf{Ans.}$$

14–5
SHEAR STRESS FORMULA FOR BEAMS

   Derivation of Shear Stress Formula.   Internal shear forces exist in beam sections. The shear forces cause shear stresses in the cross-sections. Since equal shear stresses exist on mutually perpendicular planes at a point (see Section 12–4), shear stresses also exist in the longitudinal sections of a beam.
   The existence of shear stresses in the longitudinal sections can be seen in Fig. 14–6. Figure 14–6a shows a simple beam subjected to a concentrated load. The shear force and bending moment diagrams of the beam are shown in Fig. 14–6b and c. Consider the forces acting on element *abcd*, which is between two adjacent cross-sections, 1–1 and 2–2, at a small distance Δx apart, and above the longitudinal section *bc*. Since M_2 (the moment at section 2–2) is greater than M_1 (the moment at section 1–1), the resultant of the normal stresses on side *cd* is greater than the resultant of the normal stresses on side *ab*. The difference of the resultant forces on the two sides is resisted by the shear force acting on the longitudinal section *bc*.

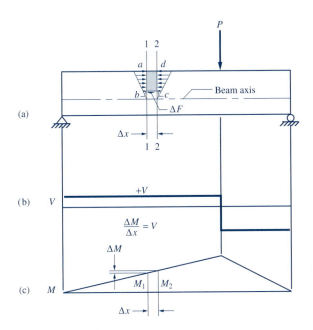

FIGURE 14–6

The free-body diagram of the element *abcd* is shown in Fig. 14–7a. The cross-section of the beam is shown in Fig. 14–7c.

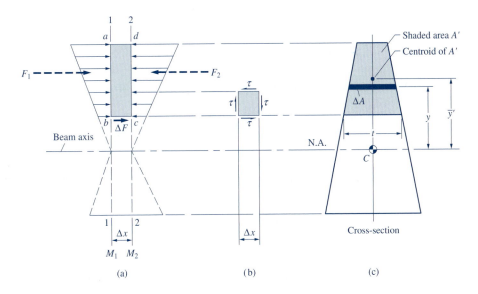

FIGURE 14–7

The resultant F_1 of the flexural stresses that act on the shaded area A' of section 1–1 due to the moment M_1 is

$$F_1 = \Sigma \, \sigma \Delta A = \Sigma \, \frac{M_1 y}{I} \, \Delta A = \frac{M_1}{I} \, \Sigma \, y \Delta A$$

where the summation is performed over the shaded area A' of the cross-section. Similarly, the resultant F_2 of the flexural stresses on the shaded area A' of section 2–2 due to the moment M_2 is

$$F_2 = \Sigma\, \sigma\Delta A = \Sigma\, \frac{M_2 y}{I}\, \Delta A = \frac{M_2}{I}\, \Sigma\, y\Delta A$$

Equilibrium of element $abcd$ along the horizontal direction requires that the shear force ΔF acting on side bc be equal to

$$\Delta F = F_2 - F_1 = \frac{M_2 - M_1}{I}\, \Sigma\, y\Delta A = \frac{\Delta M}{I}\, \Sigma\, y\Delta A$$

Assume that the shear stress τ is uniformly distributed across the section at bc, where the width of the section is t. Then the shear stress in the area may be obtained by dividing ΔF by the area $t\Delta x$. This gives the horizontal shear stress τ. It was shown in Section 12–4 that for a small element, numerically equal shear stresses act on the mutually perpendicular planes, as indicated in Fig. 14–7b. Hence, the shear stress τ in both the longitudinal plane and the vertical section is

$$\tau = \frac{\Delta F}{t\Delta x} = \frac{\Delta M}{\Delta x}\frac{\Sigma y\Delta A}{It}$$

From Section 13–7, $\Delta M/\Delta x = V$. Denoting $\Sigma y\Delta A$ by Q, the above equation becomes

$$\tau = \frac{VQ}{It} \qquad\qquad (14\text{--}10)$$

where τ = the shear stress at a point in a given section of the beam

V = the shear force at the given section

Q = the first moment of the area A' about the neutral axis, $Q = A'\,\overline{y}'$

   A' = the part of the area in the cross-section above (or below) the horizontal line where the shear stress is to be calculated

   $\overline{y}'$ = the distance from the neutral axis to the centroid of the area A'

I = the moment of inertia of the *entire* section with respect to the neutral axis (the same I as in the flexure formula)

t = the width of the cross-section at the horizontal line where the shear stress is being calculated

Equation 14–10 is the shear formula for beams. This formula can be used to calculate the shear stresses either on the vertical section or on the longitudinal planes.

   In Equation 14–10, the shear force V and the moment of inertia I are constant for a given section; it follows that the shear stresses in a section vary in accordance with the variation of Q/t.

Maximum Shear Stress in a Rectangular Section.   In a rectangular section (Fig. 14–8), the width is a constant ($t = b$). The maximum value of Q occurs at the neutral axis and is equal to

$$Q = A'\bar{y}' = \left(b \times \frac{h}{2}\right)\left(\frac{h}{4}\right) = \frac{bh^2}{8}$$

The maximum shear stress is

$$\tau_{max} = \frac{VQ}{It} = \frac{V(bh^2/8)}{(bh^3/12)(b)} = \frac{3V}{2bh}$$

where bh = the rectangular area A of the section. We write

$$\tau_{max} = 1.5\frac{V}{A} \tag{14–11}$$

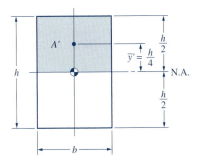

FIGURE 14–8

Equation 14–11 means that *the maximum shear stress in a rectangular section is 1.5 times the average shear stress in the section.*
   Since beams of rectangular cross-section, especially timber beams, are frequently used in engineering practice, Equation 14–11 is very useful. Timber beams have a tendency to split along the neutral surface because the maximum shear stress occurs at the neutral surface, and the shear strength of wood in longitudinal planes parallel to the grain is weaker than the shear strength perpendicular to the grain.

Maximum Shear Stress in a Circular Section.   For a member of circular cross-section, the maximum shear stress also occurs at the neutral axis, despite the fact that the width is the greatest at the neutral axis. To find the maximum stress at the neutral axis in a circular section, the area A' is a semicircle, as shown in Fig. 14–9. Referring to the properties of a semicircle in Table 7–2 and expressing all the quantities in terms of diameter d, we obtain

$$A' = \frac{\pi d^2}{8} \qquad \bar{y}' = \frac{2d}{3\pi}$$

$$Q = A'\bar{y}' = \left(\frac{\pi d^2}{8}\right)\left(\frac{2d}{3\pi}\right) = \frac{d^3}{12}$$

$$\tau_{max} = \frac{VQ}{It} = \frac{V(d^3/12)}{(\pi d^4/64)(d)} = \frac{16V}{12(\pi d^2/4)} = \frac{4V}{3A}$$

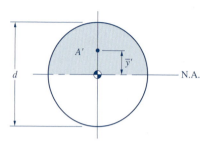

FIGURE 14–9

From which we find the maximum shear stress in a circular section as

$$\tau_{max} = \frac{4V}{3A} \tag{14–12}$$

which means that *the maximum shear stress in a beam of circular cross-section occurs at the neutral axis and is equal to $\frac{4}{3}$ times the average shear stress in the section.*

───── **EXAMPLE 14–6** ─────

The simple beam in Fig. E14–6(1) is subjected to a concentrated load $P = 20$ kN at the midspan. The beam has the rectangular section shown. Determine the shear stresses at points along line 1, line 2, and the neutral axis. Sketch the shear stresses distribution in the section.

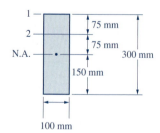

FIGURE E14–6(1)

Solution. For a simple beam subjected to a concentrated load at the midspan (from case 1 in Table 13–1), the maximum shear force is

$$V_{max} = \frac{P}{2} = +10 \text{ kN}$$

The moment of inertia of the rectangular section with respect to the neutral axis is

$$I = \frac{bh^3}{12} = \frac{(0.1 \text{ m})(0.3 \text{ m})^3}{12} = 2.25 \times 10^{-4} \text{ m}^4$$

Along Line 1.   The line is located at the top of the section. There is no area above the line. Thus, $A' = 0$ and

$$Q = A'\bar{y}' = 0$$

Therefore, the shear stress at points along line 1 is

$$\tau_1 = 0 \qquad\qquad \Leftarrow \textbf{Ans.}$$

If we take the area below line 1 as A', the centroid of the area is located at the neutral axis. Hence, $\bar{y}' = 0$ and we have $Q = A'\bar{y}' = 0$ also.

Along Line 2.   The area A' and $\bar{y}'$ are indicated in Fig. E14–6(2). We have

$$Q = A'\bar{y}' = (0.1 \text{ m} \times 0.075 \text{ m})(0.1125 \text{ m})$$
$$= 8.44 \times 10^{-4} \text{ m}^3$$

$$\tau_2 = \frac{VQ}{It} = \frac{(10 \text{ kN})(8.44 \times 10^{-4}\text{m}^3)}{(2.25 \times 10^{-4}\text{m}^4)(0.1 \text{ m})}$$

$$= 375 \text{ kPa} \qquad\qquad \Leftarrow \textbf{Ans.}$$

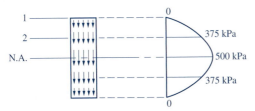

FIGURE E14–6(2)

Along the Neutral Axis.   For a rectangular section, the maximum stress at the neutral axis can be computed from Equation 14–11:

$$\tau_{max} = 1.5\frac{V}{A}$$

$$= 1.5 \frac{10 \text{ kN}}{(0.1 \text{ m})(0.3 \text{ m})}$$

$$= 500 \text{ kPa} \qquad\qquad \Leftarrow \textbf{Ans.}$$

The shear stresses at the lines below the neutral axis can be calculated in the same way, except that for convenience, the area A' is taken below the line where the shear stress is to be computed. The magnitudes of the shear stresses are symmetrical with respect to the neutral axis. The distribution of the shear stresses in the section is plotted as shown in Fig. E14–6(3). Additional investigation will show that the curve is parabolic. Note that the direction of the shear stresses coincides with the direction of the shear force on the section.

FIGURE E14–6(3)

━━ **EXAMPLE 14–7** ━━

The overhanging beam in Fig. E14–7(1) is subjected to a uniform load as shown. The beam is built of three steel plates welded together to form an integral section, as shown in Fig. E14–7(2). Such a built-up section is called a *plate girder*. (*a*) Calculate the maximum shear stress and indicate this stress on a rectangular element at the point where it occurs. (*b*) At the section where the maximum shear stress occurs, calculate the shear stress at the junction of the flange and the web. (*c*) Plot the distribution of the shear stress for the section where the maximum shear stress occurs.

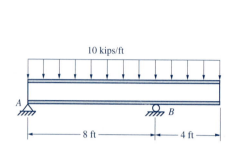

FIGURE E14–7(1)   **FIGURE E14–7(2)**

Solution. The shear force diagram is first plotted as shown in Fig. E14–7(3), from which the maximum absolute value of the shear force is 50 kips at section 1–1 just to the left of the right support *B*.

Treating the cross-sectional area as a rectangle 5 in. × 12 in. minus a rectangle 4.5 in. × 11 in., the moment of inertia of the section about the neutral axis is

$$I = \bar{I}_1 - \bar{I}_2$$

$$= \frac{(5)(12)^3}{12} - \frac{(4.5)(11)^3}{12}$$

$$= 221 \text{ in.}^4$$

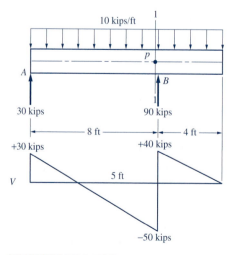

FIGURE E14–7(3)

(a) The Maximum Shear Stress. The maximum shear stress occurs at point p located at the neutral axis of section 1–1 just to the left of the support at B, where the absolute value of the shear force is a maximum. The shaded area A' above the neutral axis is divided into areas A_1' and A_2', as shown in Fig. E14–7(4). The first moments of the two areas about the neutral axis are

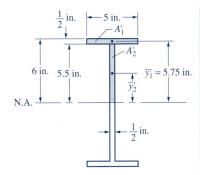

FIGURE E14–7(4)

$$Q_1 = A_1'\bar{y}_1' = \left(5 \times \frac{1}{2}\right)(5.75) = 14.38 \text{ in.}^3$$

$$Q_2 = A_2'\bar{y}_2' = \left(\frac{1}{2} \times 5.5\right)\left(\frac{5.5}{2}\right) = 7.56 \text{ in.}^3$$

$$Q = Q_1 + Q_2 = 21.94 \text{ in.}^3$$

Using the shear stress formula, the maximum shear stress at point p is

$$\tau_{max} = \frac{VQ}{It} = \frac{(50 \text{ kips})(21.91 \text{ in.}^3)}{(221 \text{ in.}^4)\left(\frac{1}{2} \text{ in.}\right)} = 9.91 \text{ ksi}$$

The maximum shear stress on the element at p is shown in Fig. E14–7(5). Pay special attention to the direction of the shear stresses. Note that the element is at the neutral surface, where the flexural stress is zero; the element is thus under pure shear.

(b) The Shear Stress at the Junction. Note that, at the junction of the flange and the web, the value of t may be either 5 in. or 0.5 in., depending on whether the stress to be computed is at the flange or the web. The magnitude of the shear stress at the junction, therefore, undergoes an abrupt change. The area A' is the area of the flange, A_1'; thus,

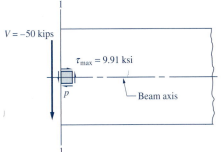

FIGURE E14–7(5)

$$Q = Q_1 = 14.38 \text{ in.}^3$$

The shear stress in the flange where $t = 5$ in. is

$$\sigma_{flange} = \frac{VQ}{It_f} = \frac{(50 \text{ kips})(14.38 \text{ in.}^3)}{(221 \text{ in.}^4)(5 \text{ in.})} = 0.651 \text{ ksi}$$

The stress in the web where $t = \frac{1}{2}$ in. is

$$\sigma_{web} = \frac{VQ}{It_w} = \frac{(50 \text{ kips})(14.38 \text{ in.}^3)}{(221 \text{ in.}^4)\left(\frac{1}{2} \text{ in.}\right)} = 6.51 \text{ ksi}$$

(c) Shear Stress Distribution.   The shear stress distribution in the section is plotted as shown in Fig. E14–7(6). Note that the shear stresses in the flange are very small. Therefore, the flanges resist only a small portion of the shear force. The web resists most of the shear force in a wide-flange section or an I-beam section. The flanges, on the other hand, resist most of the bending moment.

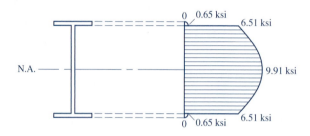

FIGURE E14–7(6)

Average Web Shear.   From the above example, we see that most of the shear force is resisted by the web in a wide-flange section. For this reason, design codes such as American Institute of Steel Construction (AISC) specifications allow the use of an *average web shear* to calculate the maximum shear stress in fabricated or hot-rolled wide-flange or I-beam sections. The average shear stress is approximately equal to

$$\tau_{avg} = \frac{V_{max}}{dt_w} \qquad\qquad (14\text{--}13)$$

where τ_{avg} = the average shear stress in the web

V_{max} = the maximum shear force along the beam

d = the *full* depth of the beam

t_w = the web thickness of the beam section

For the plate girder in Example 14–7, the average web stress is

$$\tau_{avg} = \frac{V_{max}}{dt_w} = \frac{50 \text{ kips}}{(12 \text{ in.})\left(\frac{1}{2} \text{ in.}\right)} = 8.33 \text{ ksi}$$

which is about 16% below the maximum shear stress $\tau_{\max} = 9.91$ ksi calculated by using the shear formula. The average shear stress seems to be too low. However, this approximation is widely used in engineering practice because, in most beams, the maximum shear stress is well within the allowable shear stress.

14–6
COMPOSITE BEAMS

Beams made of several materials are called *composite beams*. For example, timber beams are often strengthened by metal plates, and concrete beams are almost always reinforced with steel reinforcing bars. Consider the rectangular wooden beam reinforced with steel plate shown in Fig. 14–10a. Assume that the steel plate is fastened properly to the wooden section so that there is no sliding between steel and wood during bending. The basic deformation assumption used in deriving the flexure formula remains valid; that is, plane sections at right angles to the beam axis remain planes, and the strains of the longitudinal fibers vary linearly from the neutral axis, as shown in Fig. 14–10b. For elastic deformations, stress is proportional to strain. Thus, when the section is subjected to a positive moment, the flexural stress is distributed as shown in Fig. 14–10c. Note that since $E_{st} > E_{wd}$, the flexural stress in the steel plate is greater than the flexural stress in the wooden section. At the interface (level 3), the stress in steel is

$$(\sigma_{st})_3 = E_{st}\epsilon_3$$

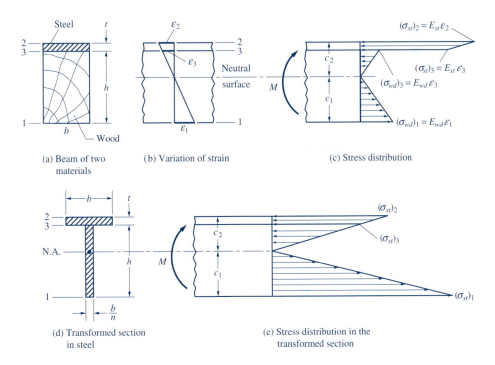

(a) Beam of two materials

(b) Variation of strain

(c) Stress distribution

(d) Transformed section in steel

(e) Stress distribution in the transformed section

FIGURE 14–10

and the stress in wood at level 3 is

$$(\sigma_{wd})_3 = E_{wd}\epsilon_3$$

The ratio of the above stresses is

$$\frac{(\sigma_{st})_3}{(\sigma_{wd})_3} = \frac{E_{st}}{E_{wd}}$$

Let the ratio of the moduli of elasticity of steel and wood, E_{st}/E_{wd}, be denoted by n. Then

$$(\sigma_{st})_3 = n\,(\sigma_{wd})_3$$

The flexural stresses in the section can be determined by using the transformed section technique, which consists of constructing a section of one material on which the resisting moment is the same as that on the composite section. The section can be transformed into either one of the component materials. To obtain a transformed section in steel, as shown in Fig. 14–10d, the dimensions of the steel section remain unchanged. The width b of the wooden section is reduced to b/n, while its depth remains unchanged. On the transformed section, the normal stresses vary linearly from the neutral axis, as shown in Fig. 14–10e, and thus the flexure formula applies. The normal stress at level 1 is

$$(\sigma_{st})_1 = E_{st}\epsilon_1 = nE_{wd}\epsilon_1 = n(\sigma_{wd})_1$$

Thus, we see that while the width of the wooden section is reduced to b/n in the transformed section, the stress is increased by n times. Therefore, the resisting force of the composite section is equivalent to that of the transformed section. Since there is no change in the vertical dimensions, moment arms of the forces are unchanged. Hence, the resisting moment of the composite section is equivalent to that of the transformed section.

Using the flexure formula, the maximum normal stresses in steel and in wood are, respectively,

$$(\sigma_{st})_{max} = \frac{Mc_2}{I_{st}} \tag{14–14}$$

$$(\sigma_{wd})_{max} = \frac{1}{n}\left[\frac{Mc_1}{I_{st}}\right] \tag{14–15}$$

where I_{st} is the moment of inertia of the transformed section in steel (Fig. 14–10d) with the respect to the neutral axis.

An alternative solution is to transform the section into wood, as shown in Fig. 14–11a, where the width of steel section is multiplied by n. The stress distribution in this transformed section is shown in Fig. 14–11b. In this case the maximum normal stresses in wood and in steel are, respectively,

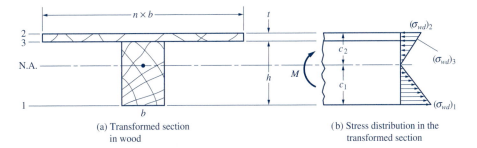

(a) Transformed section in wood

(b) Stress distribution in the transformed section

FIGURE 14–11

$$(\sigma_{wd})_{max} = \frac{Mc_1}{I_{wd}} \qquad (14\text{–}16)$$

$$(\sigma_{st})_{max} = n\left[\frac{Mc_2}{I_{wd}}\right] \qquad (14\text{–}17)$$

where I_{wd} is the moment of inertia of the transformed section in wood (Fig. 14–11a) with respect to the neutral axis.

──────── **EXAMPLE 14–8** ────────

The composite beam shown in Fig. E14–8(1) consists of a wooden section reinforced with a steel plate at the bottom. If the beam is subjected to a maximum positive moment of 50 kN · m, determine the maximum normal stresses in steel and in wood. The moduli of elasticity are: $E_{st} = 210$ GPa, and $E_{wd} = 10.5$ GPa.

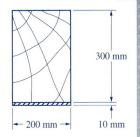

FIGURE E14–8(1)

Solution. The ratio of the moduli of elasticity is

$$n = \frac{E_{st}}{E_{wd}} = \frac{210 \text{ GPa}}{10.5 \text{ GPa}} = 20$$

For the transformed section in steel, the width of the wood section is divided by $n = 20$, as shown in Fig. E14–8(2).
  The neutral axis is located by

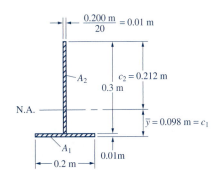

FIGURE E14–8(2)

$$\bar{y} = \frac{A_1 y_1 + A_2 y_2}{A_1 + A_2}$$

$$= \frac{(0.2 \times 0.01)(0.005) + (0.01 \times 0.3)(0.16)}{0.2 \times 0.01 + 0.01 \times 0.3}$$

$$= 0.098 \text{ m}$$

The moment of inertia of the transformed section about the neutral axis is

$$I_{st} = \Sigma[I + A(\bar{y} - y)^2]$$

$$= \left[\frac{0.2 \times 0.01^3}{12} + (0.2 \times 0.01)(0.098 - 0.005)^2\right]$$

$$+ \left[\frac{0.01 \times 0.3^3}{12} + (0.01 \times 0.3)(0.098 - 0.16)^2\right]$$

$$= 5.14 \times 10^{-5} \text{ m}^4$$

From Equations 14–15 and 14–16, the maximum normal stresses in steel and in wood are, respectively,

$$(\sigma_{st})_{max} = \frac{Mc_1}{I_{st}} = \frac{(50 \text{ kN} \cdot \text{m})(0.098 \text{ m})}{5.14 \times 10^{-5} \text{ m}^4}$$

$$= 95.3 \times 10^3 \text{ kN/m}^2 = 95.3 \text{ MPa (T)} \qquad \Leftarrow \textbf{Ans.}$$

$$(\sigma_{wd})_{max} = \frac{1}{n}\left[\frac{Mc_2}{I_{st}}\right] = \frac{1}{20}\left[\frac{(50 \text{ kN} \cdot \text{m})(0.212 \text{ m})}{5.14 \times 10^{-5} \text{ m}^4}\right]$$

$$= 10.3 \times 10^3 \text{ kN/m}^2 = 10.3 \text{ MPa (C)} \qquad \Leftarrow \textbf{Ans.}$$

14–7
REINFORCED CONCRETE BEAMS

The transformed section method used in Section 14–6 can be applied to the reinforced concrete beams shown in Fig. 14–12a. The reinforced concrete section is usually transformed into an equivalent concrete area. Concrete is very weak in resisting tension and it may crack in the tension zone. Therefore, concrete is not reliable in resisting tension. The transformed section for the reinforced beam is shown in Fig. 14–12b, where the area of concrete in the compression zone remains unchanged, but it does not include any concrete area in the tension zone. The tensile force is assumed to be carried by steel reinforcing bars alone. The transformed area of steel bars into concrete is nA_{st}, where

A_{st} = the total cross-sectional area of the bars in the section

$n = E_{st}/E_{cn}$

E_{st} = modulus of elasticity of steel

E_{cn} = modulus of elasticity of concrete

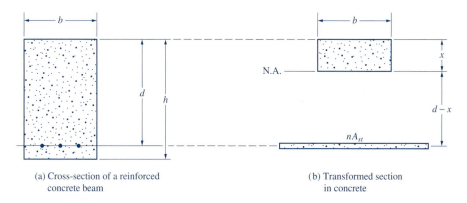

(a) Cross-section of a reinforced concrete beam

(b) Transformed section in concrete

FIGURE 14–12

For computation purposes, the area is assumed to be concentrated at the centerline of the steel bars.

Since the neutral axis must pass through the centroid of the transformed section, the first moment of the concrete area in the compression zone about the neutral axis must be equal to the first moment of the transformed steel area about the same axis. Thus,

$$(bx)\left(\frac{x}{2}\right) = nA_{st}(d - x) \tag{14–18}$$

from which x can be solved. The moment of inertia of the transformed section about the neutral axis is then

$$I = \frac{bx^3}{3} + nA_{st}(d - x)^2 \tag{14–19}$$

The maximum compressive stress in concrete and the tensile stress in the steel bars are, respectively,

$$(\sigma_{cn})_{max} = \frac{M_{max}x}{I} \tag{14–20}$$

$$(\sigma_{st})_{max} = n\left[\frac{M_{max}(d - x)}{I}\right] \tag{14–21}$$

––––––– EXAMPLE 14–9 –––

The cross-section of the simply supported reinforced concrete beam in Fig. E14–9(1) is shown in Fig. E14–9(2). Determine the maximum compressive stress in concrete and the maximum tensile stress in the steel bars. Assume that $n = E_{st}/E_{cn} = 10$ and that the concrete weighs 150 lb/ft³.

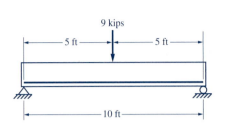

FIGURE E14–9(1)

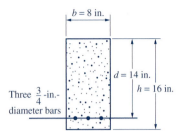

FIGURE E14–9(2)

Solution. The transformed section of the beam is shown in Fig. E14–9(3). The total cross-sectional area of the $\frac{3}{4}$-in. diameter bars is

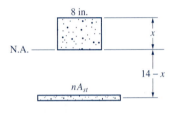

FIGURE E14–9(3)

$$A_{st} = 3\left[\frac{1}{4}\pi d^2\right] = 3\left[\frac{1}{4}\pi\left(\frac{3}{4}\text{ in.}\right)^2\right] = 1.32 \text{ in.}^2$$

Thus, the transformed area of steel (into concrete) is

$$nA_{st} = 10(1.32 \text{ in.}^2) = 13.2 \text{ in.}^2$$

The neutral axis of the transformed section can be located by solving Equation 14–18. We have

$$8x\left(\frac{x}{2}\right) = 13.2(14 - x)$$

or

$$4x^2 + 13.2x - 185 = 0$$

The coefficients of the quadratic equation (in standard form, $ax^2 + bx + c = 0$) are

$$a = 4 \qquad b = 13.2 \qquad c = -185$$

From the quadratic formula, the roots are

$$x = \frac{-b \pm \sqrt{b^2 - 4ac}}{2a}$$

$$= \frac{-13.2 \pm \sqrt{(13.2)^2 - 4(4)(-185)}}{2(4)}$$

$$= \frac{-13.2 \pm 56.0}{8}$$

$$= 5.35 \text{ in.} \qquad \text{(The negative root is dropped.)}$$

From Equation 14–19, the moment of inertia of the transformed section about the neutral axis is

$$I = \frac{8(5.35)^3}{3} + 13.2(14 - 5.35)^2 = 1400 \text{ in.}^4$$

The weight of the beam per foot of length is

$$w = \left[\frac{8 \times 16}{144} \text{ ft}^2\right](150 \text{ lb/ft}^3) = 133 \text{ lb/ft} = 0.133 \text{ kip/ft}$$

The maximum moment at the midspan of the beam is

$$M_{max} = \frac{PL}{4} + \frac{wL^2}{8} = \frac{(9 \text{ kips})(10 \text{ ft})}{4} + \frac{(0.133 \text{ kip/ft})(10 \text{ ft})^2}{8}$$

$$= 24.2 \text{ kip} \cdot \text{ft}$$

From Equation 14–20, the maximum compressive stress in concrete is

$$(\sigma_{cn})_{max} = \frac{M_{max}x}{I} = \frac{(24.2 \times 12 \text{ kip} \cdot \text{in.})(5.36 \text{ in.})}{1400 \text{ in.}^4}$$

$$= 1.11 \text{ ksi (C)} \qquad\qquad \Leftarrow \textbf{Ans.}$$

From Equation 14–21, the maximum tensile stress in the steel bar is

$$(\sigma_{st})_{max} = n\left[\frac{M_{max}(d - x)}{I}\right] = 10\left[\frac{(24.2 \times 12 \text{ kip} \cdot \text{in.})(14 \text{ in.} - 5.36 \text{ in.})}{1400 \text{ in.}^4}\right]$$

$$= 18.0 \text{ ksi (T)} \qquad\qquad \Leftarrow \textbf{Ans.}$$

14–8
ELASTIC ANALYSIS OF BEAMS

Beam problems can be divided into two major categories, namely, analysis problems and design problems. Design of beams will be discussed in the next chapter. In the beam analysis problems, the beam material and size are given. The usual types of problems are to find the maximum flexural and shearing stresses, the maximum load-carrying capacity, or the maximum possible span length. In these problems, the common goal is to keep the maximum stress computed from the flexure formula and shear stress formula from exceeding the allowable limits, which are usually well within the proportional limit of the material. Thus, this analysis is based on the elastic theory, which is different from plastic analysis (to be discussed in the next section). The following examples illustrate the elastic analysis of beams.

―――― **EXAMPLE 14–10** ――――

The simply supported beam in Fig. E14–10 is subjected to two symmetrically placed, concentrated loads as shown. The beam is made of Douglas fir having an allowable flexural stress of 1450 psi and an allowable shear stress parallel to the grain of 95 psi. Determine whether a 4 × 10 nominal section is satisfactory for bending and shearing strength.

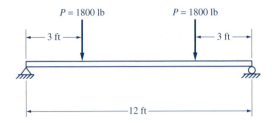

FIGURE E14–10

Solution. From the appendix, Table A–6(a), we find the following properties for a 4×10 section:

$$A = 32.4 \text{ in.}^2$$

$$S_x = 49.9 \text{ in.}^3$$

$$w = 8.93 \text{ lb/ft}$$

The beam is subjected to two symmetrically placed, concentrated loads and a uniform load w (the weight of the beam). This loading is the combination of cases 3 and 4 from Table 13–1. The maximum moment at the midspan is

$$M_{max} = Pa + \frac{wL^2}{8}$$

$$= (1800)(3) + \frac{(8.93)(12)^2}{8}$$

$$= 5561 \text{ lb} \cdot \text{ft}$$

From Equation 14–5, we find

$$\sigma_{max} = \frac{M_{max}}{S_x} = \frac{5561 \times 12 \text{ lb} \cdot \text{in.}}{49.9 \text{ in.}^3} = 1340 \text{ psi}$$

which is less than the allowable flexural stress of 1450 psi; hence, the bending strength of the beam is satisfactory.
    The maximum shear near the left support is

$$V_{max} = P + \frac{wL}{2} = 1800 + \frac{(8.93)(12)}{2} = 1854 \text{ lb}$$

From Equation 14–11, we find

$$\tau_{max} = 1.5 \frac{V_{max}}{A} = 1.5 \frac{1854 \text{ lb}}{32.4 \text{ in.}^2} = 85.8 \text{ psi}$$

which is less than the allowable shear stress of 95 psi; hence, the shear strength of the beam is satisfactory. We conclude that the beam is satisfactory for bending and shearing strength.

―――――― **EXAMPLE 14–11** ――――――

Refer to Fig. E14–11. Determine the allowable uniform load w that can be applied to the structural steel W460 × 0.73 (SI designation) beam that is also subjected to a pair of 50-kN loads as shown. The allowable flexural stress is 165 MPa, and the allowable shear stress is 100 MPa.

Solution. From the appendix, Table A–1(b), the properties of a W460 × 0.73 shape are found to be

$A = 9.48 \times 10^{-3} \text{ m}^2$

$S_x = 1.46 \times 10^{-3} \text{ m}^3$

$d = 457 \text{ mm}$

$t_w = 9.02 \text{ mm}$

wt. $= 0.73 \text{ kN/m}^3$
    (from the designation)

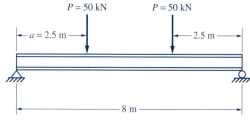

FIGURE E14–11

The allowable moment is

$$M_{allow} = S\sigma_{allow}$$

$$= (1.46 \times 10^{-3} \text{ m}^3)(165 \times 10^3 \text{ kN/m}^2)$$

$$= 241 \text{ kN} \cdot \text{m}$$

The moment at the midspan due to the concentrated forces is

$$M_p = Pa = (50 \text{ kN})(2.5 \text{ m}) = 125 \text{ kN} \cdot \text{m}$$

The moment at the midspan due to the uniform load w is

$$M_w = M_{allow} - M_p = 241 - 125 = 116 \text{ kN} \cdot \text{m}$$

Solving w from $M_w = wL^2/8$, we get

$$w = \frac{8M_w}{L^2} = \frac{8(116 \text{ kN} \cdot \text{m})}{(8 \text{ m})^2} = 14.5 \text{ kN/m}$$

Subtracting the weight of the beam from w, we get

$$w_{allow} = 14.5 - 0.73 = 13.8 \text{ kN/m}$$

This load is calculated based on the allowable flexural stress, which is usually the governing factor for steel beams. Now we need to check if the shear stress is satisfactory. The maximum shear force near the left support is

$$V_{max} = P + \frac{wL}{2}$$

$$= 50 \text{ kN} + \frac{(14.5 \text{ kN/m})(8 \text{ m})}{2}$$

$$= 108 \text{ kN}$$

From Equation 14–12, the average web stress is

$$\tau_{\text{avg}} = \frac{V_{\text{max}}}{dt_w} = \frac{108 \text{ kN}}{(0.457 \text{ m})(0.00902 \text{ m})}$$

$$= 26\ 200 \text{ kPa} = 26.2 \text{ MPa}$$

which is well within the allowable shear stress of 100 MPa. Therefore the allowable uniform load is

$$w_{\text{allow}} = 13.8 \text{ kN/m} \qquad \Leftarrow \textbf{Ans.}$$

EXAMPLE 14–12

Refer to Fig. E14–12(1). The floor system of a residential building is supported by 2 × 8 Southern pine joists spaced 2 ft on centers and having a simple span length of 12 ft. (*a*) Calculate the allowable uniform load *w* for each joist in lb/ft. (*b*) Calculate the allowable live load in lb/ft² (psf) on the floor. Assume that the floor system is 2 in. thick at a specific weight of 45 lb/ft³. The allowable flexural stress is 1600 psi and the allowable shear stress is 90 psi.

Solution. From the appendix, Table A–6(a), the properties of a 2 × 8 timber section are found to be

$A = 10.9 \text{ in.}^2$

$S = 13.1 \text{ in.}^3$

wt. = 3.02 lb/ft

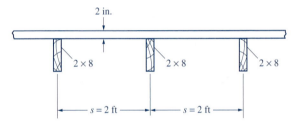

FIGURE E14–12(1)

(a) The Allowable Load on the Joist. The allowable moment is

$$M_{\text{allow}} = S\sigma_{\text{allow}}$$

$$= (13.1 \text{ in.}^3)(1600 \text{ lb/in.}^2)$$

$$= 20\ 960 \text{ lb} \cdot \text{in.} = 1747 \text{ lb} \cdot \text{ft}$$

Solving *w* from $M = wL^2/8$, we get

$$w = \frac{8M}{L^2} = \frac{8(1747 \text{ lb} \cdot \text{ft})}{(12 \text{ ft})^2} = 97.0 \text{ lb/ft}$$

The allowable shear force can be obtained by solving Equation 14–11 for *V*:

$$V_{\text{allow}} = \frac{A\tau_{\text{allow}}}{1.5} = \frac{(10.9 \text{ in.}^2)(90 \text{ lb/in.}^2)}{1.5} = 654 \text{ lb}$$

Equating the allowable shear force to $V = wL/2$ due to the uniform load and solving for *w*, we get

$$w = \frac{2V_{\text{allow}}}{L} = \frac{2(654 \text{ lb})}{12 \text{ ft}} = 109 \text{ lb/ft}$$

The smaller of the two w's computed above is $w = 97.0$ lb, which is controlled by bending. Subtracting the weight of the joist from this load, we obtain the allowable load on the joist:

$$w_{\text{allow}} = 97.0 - 3.02 = 94.0 \text{ lb/ft} \qquad \Leftarrow \textbf{Ans.}$$

(b) The Live Load on the Floor.   As shown in Fig. E14–12(2), each joist supports a width of floor equal to the spacing of the joist $s = 2$ ft. Note that the allowable load w_{allow} on the joist is the load per unit length of the joist. The allowable load per unit of floor area is equal to w_{allow} divided by the spacing s:

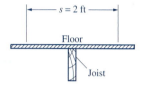

Allowable load per unit area $= \dfrac{w_{\text{allow}}}{s}$

$$= \frac{94.0 \text{ lb}}{2 \text{ ft}} = 47 \text{ lb/ft}^2 \qquad \textbf{FIGURE E14–12(2)}$$

The dead weight of the floor system per unit area can be obtained by multiplying the specific weight of the floor system by the thickness of the floor in ft:

$$(45 \text{ lb/ft}^3)\left(\frac{2}{12} \text{ ft}\right) = 7.5 \text{ lb/ft}^2$$

Subtracting the dead weight of the floor from the total allowable load, the superimposed live load on the floor is

$$\text{Allowable live load} = 47 - 7.5 = 39.5 \text{ lb/ft}^2 \qquad \Leftarrow \textbf{Ans.}$$

14–9
PLASTIC ANALYSIS OF BEAMS

The analysis of beams discussed in the preceding section is based on the elastic behavior of materials. Tests have shown that ductile materials can carry much higher loads than those allowed by the elastic theory. An inherent property of structural steel is its ability to resist large deformations without failure. These large deformations occur primarily in the plastic range, where the large deformations occur without significant increase in stress. Because of this behavior of ductile materials, the plastic design approach, called the *ultimate strength design,* has been developed.

Plastic Behavior. Figure 14–13 represents the stress–strain curve of a ductile steel. It shows that, up to the yield point y, the stress is proportional to strain. Beyond the yield point, the material deforms plastically without an increase in stress. This portion of the curve is called the *plastic range.* Beyond this range, *strain hardening* begins, where the material loses part of its ductility and additional deformation can occur only at an increased stress.

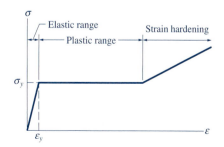

FIGURE 14–13   Ductile Material

Omitting the strain hardening effect, a ductile steel is assumed to be an idealized *elastoplastic* material that has the ideal stress–strain curve shown in Fig. 14–14. Steels with a yield stress no more than 65 psi (450 MPa) may be considered elastoplastic.

FIGURE 14–14   Elastoplastic Material

Maximum Elastic Moment. Beams made of ductile steel deform elastically when the maximum stress in the beam is not greater than the yield stress. Figure 14–15a shows the elastic bending where the maximum stress in the extreme fibers is less than the yield stress. When the maximum stress in the extreme fibers reaches the yield stress, as shown in Fig. 14–15b, the corresponding resisting moment M_y in the section is called the *maximum elastic moment*. At this point, the material will start to yield but there is no plastic deformation yet. The stress distribution in the beam section is still linear and the elastic theory applies. Hence, from Equation 14–5, we have

$$M_y = S\sigma_y \qquad\qquad (14\text{–}22)$$

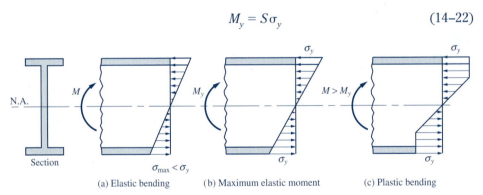

(a) Elastic bending   (b) Maximum elastic moment   (c) Plastic bending

FIGURE 14–15

Plastic Bending Moment. When the moment is greater than M_y, the stress on the extreme fiber remains at σ_y, and a greater area of the cross-section is also stressed to σ_y. The stress distribution is shown in Fig. 14–15c.

Now imagine that the load is increased further. More and more points in the cross-section will reach σ_y. This process is called *plastification* of the cross-section. If the material is elastoplastic, eventually all the fibers will be stressed to σ_y. This idealized plastic stress distribution is shown in Fig. 14-16. The section is said to be *fully plastified,* and the corresponding bending moment M_p is called the *plastic bending moment.* The beam acts as if it were hinged at the section where the plastic moment occurs. When a *plastic hinge* is formed, no additional moment can be resisted. Hence, the plastic moment represents the limiting moment that the beam can resist.

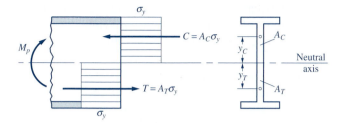

FIGURE 14–16   Ultimate Strength

To find M_p, consider the full plastification of the wide-flange section shown in Fig. 14–16. The area A_C above the *plastic neutral axis* is subjected to uniform compression σ_y, and the area A_T below the neutral axis is subjected to uniform tension σ_y. The resultant T of the tensile stresses and the resultant C of the compressive stresses are, respectively,

$$T = A_T\sigma_y \qquad C = A_C\sigma_y$$

Assuming that there is no axial force acting on the beam, then the equilibrium condition along the axial direction requires that

$$T = C$$

$$A_T\sigma_y = A_C\sigma_y$$

From which we get

$$A_T = A_C$$

The equation shows that *the plastic neutral axis divides the cross-section into equal areas.* This is apparent in symmetrical sections, but it applies to non-symmetrical sections as well.

The two stress resultants T and C, being equal and opposite, form a couple representing the internal resisting moment M_p. Thus, we obtain

$$M_p = (A_T\sigma_y)\,y_T + (A_C\sigma_y)\,y_C = (A_T\,y_T + A_C\,y_C)\sigma_y$$

or

$$Z = A_T y_T + A_C y_C \tag{14–23}$$

$$M_p = Z\sigma_y \tag{14–24}$$

where Z = the plastic section modulus of the cross-section, equal to the sum of the (absolute value of) the first moments with respect to the neutral axis of the cross-sectional areas above and below the neutral axis

A_T = the area of the part of the cross-section that is under tension

A_C = the area of the part of the cross-section that is under compression

y_T = the distance of the centroid of area A_T from the neutral axis

y_C = the distance of the centroid of area A_C from the neutral axis

M_p = the plastic moment

Note that for a symmetrical section, the elastic neutral axis and the plastic neutral axis are located at the horizontal line of symmetry. For a nonsymmetrical section, however, they are two distinct lines.

Computation of the Plastic Section Modulus.   Equation 14–23 may be used to determine the plastic section modulus and the plastic moment. For the rectangular section shown in Fig. 14–17a, we have

$$Z = A_T y_T + A_C y_C = \left(\frac{bh}{2}\right)\left(\frac{h}{4}\right) + \left(\frac{bh}{2}\right)\left(\frac{h}{4}\right)$$

$$Z = \frac{bh^2}{4} \tag{14–25}$$

For the circular section shown in Fig. 14–17b, we have

$$Z = A_T y_T + A_C y_C = \left(\frac{\pi d^2}{8}\right)\left(\frac{2d}{3\pi}\right) + \left(\frac{\pi d^2}{8}\right)\left(\frac{2d}{3\pi}\right)$$

$$Z = \frac{d^3}{6} \tag{14–26}$$

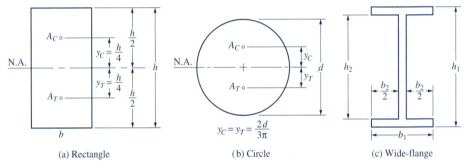

(a) Rectangle   (b) Circle   (c) Wide-flange

FIGURE 14–17

The wide-flange section shown in Fig. 14–17c can be treated as the difference of two rectangles; thus,

$$Z = Z_1 - Z_2 = \frac{b_1 h_1^2}{4} - \frac{b_2 h_2^2}{4} \tag{14–27}$$

The plastic section modulus for a wide-flange steel section or an I-beam steel section can be found in the appendix, Tables A–1 and A–2. The plastic section modulus for a nonsymmetrical section can be determined by the definition given in Equation 14–23.

Shape Factor.   The ratio of the plastic moment to the maximum elastic moment is called the *shape factor, k,* of a beam. Thus,

$$k = \frac{M_p}{M_y} \tag{14–28}$$

Since

$$M_p = Z\sigma_y \quad \text{and} \quad M_y = S\sigma_y$$

we see that the shape factor is also equal to the ratio of the plastic section modulus to the elastic section modulus. Thus,

$$k = \frac{Z}{S} \qquad (14\text{-}29)$$

The shape factor for a rectangular section is

$$\frac{Z}{S} = \frac{\left(\dfrac{bd^2}{4}\right)}{\left(\dfrac{bd^2}{6}\right)} = 1.5$$

This value means that the plastic moment for a rectangular section is 50% greater than its maximum elastic moment. Thus, for a beam of rectangular section, it takes 50% more bending load to produce a plastic failure from the point where yield begins.

For steel sections such as W and S shapes, the shape factor is approximately 1.13 for bending about the strong axis of the section. This means that it takes only about 13% more bending load to produce a plastic failure after the section started to yield. This value indicates that a wide-flange beam is very efficient in resisting an elastic moment.

EXAMPLE 14–13

The wide-flange steel beam in Fig. E14–13 has the dimensions shown. If it is made of an elastoplastic material having a tensile and compressive yield stress of $\sigma_y = 250$ MPa, determine (*a*) the maximum elastic moment M_y, (*b*) the plastic moment M_p, (*c*) the shape factor of the beam, and (*d*) the uniformly distributed load that the beam can carry at yield and at full plastification for a simple span of 10 m.

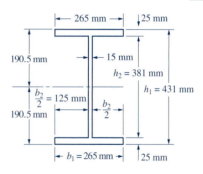

FIGURE E14–13

Solution. The section is symmetrical with respect to the horizontal centroidal axis. Therefore, the elastic and plastic neutral axes are located at the horizontal axis of symmetry.

(*a*) **The Maximum Elastic Moment.** The cross-section may be considered as a rectangle $b_1 \times h_1$ less the rectangle $b_2 \times h_2$. Therefore, the moment of inertia of the section is

$$I = I_1 - I_2 = \frac{b_1 h_1^3}{12} - \frac{b_2 h_2^3}{12}$$

$$= \frac{(0.265)(0.431)^3}{12} - \frac{(0.250)(0.381)^3}{12}$$

$$= 6.16 \times 10^{-4} \text{ m}^4$$

The distance from the extreme fibers to the neutral axis is

$$c = \frac{h_1}{2} = 0.2155 \text{ m}$$

The elastic section modulus is

$$S = \frac{I}{c} = \frac{6.16 \times 10^{-4}}{0.2155} = 2.86 \times 10^{-3} \text{ m}^3$$

The maximum elastic moment is

$$M_y = S\sigma_y = (2.86 \times 10^{-3} \text{ m}^3)(250 \times 10^3 \text{ kN/m}^2)$$

$$= 715 \text{ kN} \cdot \text{m} \qquad\qquad \Leftarrow \textbf{Ans.}$$

(b) The Plastic Moment. Again we consider the cross-section to be a rectangle $b_1 \times h_1$ less the rectangle $b_2 \times h_2$. Thus, we compute the plastic section modulus as

$$Z = Z_1 - Z_2 = \frac{b_1 h_1^2}{4} - \frac{b_2 h_2^2}{4}$$

$$= \frac{(0.265)(0.431)^2}{4} - \frac{(0.250)(0.381)^2}{4}$$

$$= 3.23 \times 10^{-3} \text{ m}^3$$

The plastic moment is

$$M_p = Z\sigma_y = (3.23 \times 10^{-3} \text{ m}^3)(250 \times 10^3 \text{ kN/m}^2)$$

$$= 807 \text{ kN} \cdot \text{m} \qquad\qquad \Leftarrow \textbf{Ans.}$$

(c) The Shape Factor. By definition, the shape factor is either the ratio of M_p/M_y or Z/S. Thus,

$$k = \frac{M_p}{M_y} = \frac{807}{715} = 1.13 \qquad\qquad \Leftarrow \textbf{Ans.}$$

or

$$k = \frac{Z}{S} = \frac{3.23 \times 10^{-3}}{2.86 \times 10^{-3}} = 1.13$$

(d) Load at Yield. Solving w_y from $M_y = w_y L^2/8$, we get

$$w_y = \frac{8\,M_y}{L^2} = \frac{8(715 \text{ kN} \cdot \text{m})}{(10 \text{ m})^2} = 57.2 \text{ kN/m} \qquad\qquad \Leftarrow \textbf{Ans.}$$

Load at Full Plastification. Solving w_p from $M_p = w_p L^2/8$, we get

$$w_p = \frac{8M_p}{L^2} = \frac{8(807 \text{ kN} \cdot \text{m})}{(10 \text{ m})^2} = 64.6 \text{ kN/m} \qquad\qquad \Leftarrow \textbf{Ans.}$$

────── **EXAMPLE 14–14** ──────

The cross-section of the T-beam in Fig. E14–14(1) has the dimensions shown. If it is made of an elastoplastic material having a tensile and compressive yield stress of $\sigma_y = 36$ ksi, determine (*a*) the maximum elastic moment M_y, (*b*) the plastic moment M_p, and (*c*) the shape factor of the beam.

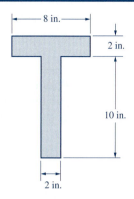

Solution. The section is nonsymmetrical with respect to the horizontal centroidal axis. The elastic and plastic neutral axes are at different locations.

FIGURE E14–14(1)

(*a*) *The Maximum Elastic Moment.* Refer to Fig. E14–14(2). The neutral axis for elastic bending passes through the centroid of the section. Its location from the bottom of the section is

$$\bar{y} = \frac{A_1 y_1 + A_2 y_2}{A_1 + A_2} = \frac{(2 \times 10)(5) + (8 \times 2)(11)}{2 \times 10 + 8 \times 2} = 7.67 \text{ in.}$$

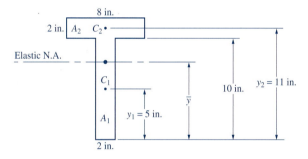

FIGURE E14–14(2)

The moment of inertia of the section is

$$I = \Sigma \, [I + A(\bar{y} - y)^2]$$

$$= \left[\frac{2(10)^3}{12} + 20(7.67 - 5)^2\right] + \left[\frac{8(2)^3}{12} + 16(7.67 - 11)^2\right]$$

$$= 492 \text{ in.}^4$$

The distance from the extreme fibers at the bottom of the section to the neutral axis is

$$c = \bar{y} = 7.67 \text{ in.}$$

The elastic section modulus is

$$S = \frac{I}{c} = \frac{492}{7.67} = 64.1 \text{ in.}^3$$

The maximum elastic moment is

$$M_y = S\sigma_y \ (64.1 \text{ in.}^3)(36 \text{ kip/in.}^2)$$

$$= 2310 \text{ kip} \cdot \text{in.} \qquad\qquad \Leftarrow \textbf{Ans.}$$

(b) The Plastic Moment.   Refer to Fig. E14–14(3). The neutral axis for plastic bending divides the section into two equal areas. Assume the plastic neutral axis is at a distance x above the bottom of the section. We require that

$$A_1 = A_2 + A_3$$

$$2x = 2(10 - x) + 2(8)$$

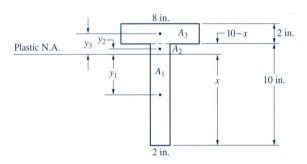

Plastic N.A.

FIGURE E14–14(3)

From which we get

$$x = 9 \text{ in.}$$

The plastic section modulus may be computed by taking the sum of the absolute values of the first moment of the areas about the plastic neutral axis:

$$Z = A_1 y_1 + A_2 y_2 + A_3 y_3$$

$$= (2 \times 9)(4.5) + (2 \times 1)(0.5) + (8 \times 2)(2)$$

$$= 114 \text{ in.}^3$$

The plastic moment is

$$M_p = Z\sigma_y = (114 \text{ in.}^3)(36 \text{ kip/in.}^2)$$

$$= 4104 \text{ kip} \cdot \text{in.} \qquad\qquad \Leftarrow \textbf{Ans.}$$

(c) The Shape Factor.   By definition, the shape factor is either the ratio of M_p/M_y or Z/S. Thus,

$$k = \frac{M_p}{M_y} = \frac{4014}{2310} = 1.78 \qquad\qquad \Leftarrow \textbf{Ans.}$$

or

$$k = \frac{Z}{S} = \frac{114}{64.1} = 1.78$$

14–10
SUMMARY

Flexural Stresses.   The bending moment in a beam causes the beam to bend and produces normal stresses called *flexural stresses.* The surface on which the flexural stresses are zero is called the *neutral surface.* The intersection of the neutral surface with a cross-section is called the *neutral axis.* The flexural stresses vary linearly from zero at the neutral axis to the maximum value at the outer fibers.

Flexure Formula.   The maximum flexural stress occurs at the outer fibers, those most remote from the neutral axis. Its value can be computed from the flexure formula:

$$\sigma_{max} = \frac{Mc}{I} \qquad (14\text{–}2)$$

Using the *section modulus* $S = I/c$, the flexure formula can be written as

$$\sigma_{max} = \frac{M}{S} \qquad (14\text{–}5)$$

Allowable Moment.   The allowable moment that a beam can resist without causing the maximum stress in the beam to exceed the allowable flexural stress σ_{allow} of the material can be computed from:

$$M_{allow} = \frac{I\sigma_{allow}}{c} = S\sigma_{allow} \qquad (14\text{–}8, 14\text{–}9)$$

Shear Stresses in Beams.   Horizontal and vertical shear stresses developed in beams subjected to transverse loads may be computed from the shear stress formula:

$$\tau = \frac{VQ}{It} \qquad (14\text{–}10)$$

The shear stress is zero at the outer fibers and is almost always a maximum at the neutral axis.

Maximum Shear Stress in a Rectangular Section.   The maximum shear stress at the neutral axis of a rectangular cross-section is

$$\tau_{max} = 1.5\frac{V}{A} \qquad (14\text{–}11)$$

Maximum Shear Stress in a Circular Section.   The maximum shear stress at the neutral axis of a circular cross-section is

$$\tau_{max} = \frac{4V}{3A} \qquad (14\text{–}12)$$

Average Web Shear.   The AISC specifications allow the use of an *average web shear* to calculate the maximum shear stress in fabricated or hot-rolled wide-flange or I-beam sections by using:

$$\tau_{avg} = \frac{V_{max}}{dt_w} \qquad (14\text{–}13)$$

Composite Beams.   Beams made of several materials are called *composite beams.* The normal stresses in a composite beam can be determined by using the *transformed section,* which is an equivalent section transformed into one of the component materials. On the transformed section, the normal stresses vary linearly from the neutral axis, and the flexure formula applies.

Reinforced Concrete Beams.   The transformed section method can be applied to reinforced concrete beams. The reinforced concrete section is usually transformed into an equivalent concrete area. Concrete is very weak in resisting tension, so the transformed section for the reinforced concrete beam contains only the area of concrete in the compression zone. The tensile force is assumed to be carried by steel reinforcing bars alone.

Elastic Beam Analysis.   In elastic beam analysis, the main goal is to keep the maximum stresses computed from the flexure formula and shear stress formula within the allowable limits of the material.

The Maximum Elastic Moment.   The resisting moment in a beam when only the extreme fibers of a beam are stressed to the yield point, called the *maximum elastic moment,* may be computed from

$$M_y = S\sigma_y \tag{14–22}$$

Plastic Moment.   When all the fibers are stressed to the yield point, the section is said to be *fully plastified.* The resisting moment, called the *plastic moment,* represents the limiting moment that the beam can resist and its value may be computed from

$$M_p = Z\sigma_y \tag{14–24}$$

where Z is the *plastic section modulus* equal to the sum of the first moments with respect to the neutral axis of the cross-sectional areas above and below the neutral axis. The neutral axis for plastic bending divides the cross-section into two equal areas. The plastic section modulus for a rectangular shape is

$$Z = \frac{bh^2}{4} \tag{14–25}$$

The plastic section modulus for a circular section is

$$Z = \frac{d^3}{6} \tag{14–26}$$

Shape Factor.   The ratio of the plastic moment to the maximum elastic moment, or the ratio of the plastic section modulus to the elastic section modulus, is called the *shape factor, k,* of a beam; that is,

$$k = \frac{M_p}{M_y} = \frac{Z}{S} \tag{14–28, 14–29}$$

PROBLEMS

Section 14–2   Normal Stresses in Beams Due to Bending

Section 14–3   The Flexure Formula

14–1   A timber beam has a 10-ft simple span and a full rectangular section 4 in. wide and 6 in. deep. Determine the maximum flexural stress in the beam due to a concentrated load of 800 lb applied at the midspan.

14–2   Rework Problem 14–1 using the properties of the dressed size listed in the appendix, Table A–6(a), for a nominal 4 × 6 timber section.

14–3   A cantilever beam has a 3-m span and a circular section of 100-mm diameter. Determine the maximum flexural stress in the beam due to a 5-kN concentrated load applied at the free end.

14–4   Verify the section moduli tabulated in the appendix, Tables A–1 and A–2, for the following sections:

(a) S_x for W18 × 35

(b) S_y for W250 × 0.71 (SI designation)

(c) S_x and S_y for S12 × 31.8

14–5   A simple beam has a 24-ft span and a W16 × 50 section. Determine the maximum flexural stress due to two concentrated loads of 20 kips each applied at the third points along the beam.

14–6   A simple beam has an 8-m span and a W410 × 0.73 section. Determine the maximum flexural stress due to two concentrated loads of 90 kN each applied at the third points along the beam.

14–7   A standard-weight steel pipe of 50-mm nominal diameter is used as a post for a clothesline. The pipe is firmly embedded in a concrete base. Determine the maximum normal stress in the pipe caused by a horizontal force of 400 N applied at the section 2 m above the base.

14–8   A log of 10-in. average diameter is used in a simple span of 15 ft. Determine the maximum normal stress in the log caused by a uniform load of 200 lb/ft.

14–9   A simple beam has an 18-ft span and a W18 × 50 section. Determine the maximum flexural stress due to two concentrated loads of 10 kips each applied at the third points of the span and a uniform load of 3 kips/ft (including the weight of the beam) over the entire length of the beam.

14–10   A timber beam has a 5-m simple span and a rectangular section of nominal size 150 × 410. Determine the maximum flexural stress due to a concentrated load of 16 kN applied at the midspan and a uniform load of 4.5 kN/m (including the weight of the beam) over the entire length of the beam.

14–11 The overhanging beam shown in Fig. P14–11 has a timber section of nominal size 100×300. Determine the maximum flexural stress due to a uniform load of 20 kN/m over the entire length of the beam.

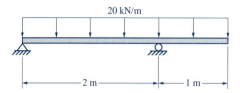

FIGURE P14–11

14–12 A beam with an inverted T-section is subjected to the two concentrated loads shown in Fig. P14–12. Determine the maximum tensile and compressive stresses in the beam. Neglect the weight of the beam.

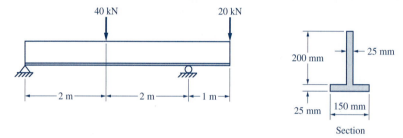

FIGURE P14–12

14–13 See Fig. P14–13. Determine the maximum tensile and compressive stresses in the beam due to the loading shown.

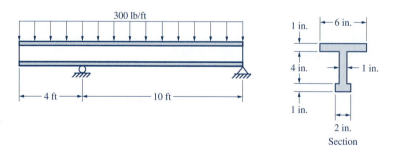

FIGURE P14–13

Section 14–4   Allowable Moment

14–14 Determine the allowable moment about the horizontal neutral axis that a timber beam with a nominal size 50×100 can resist without exceeding the allowable stress of 10 MPa.

14–15 Determine the allowable moment about the horizontal neutral axis that a $W16 \times 50$ section can resist without exceeding the allowable stress of 24 ksi.

14–16 Determine the allowable moment about the horizontal neutral axis that the beam with the built-up section shown in Fig. P14–16 can resist without exceeding the allowable stress of 24 ksi.

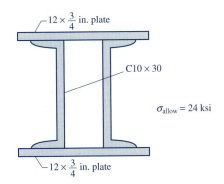

12 × $\frac{3}{4}$ in. plate

C10 × 30

σ_{allow} = 24 ksi

12 × $\frac{3}{4}$ in. plate

FIGURE P14–16

14–17 A cast-iron machine part has a channel section, as shown in Fig. P14–17. Determine the allowable positive moment about the horizontal neutral axis that the section can resist without exceeding the allowable stress of 21 MPa in tension and 84 MPa in compression.

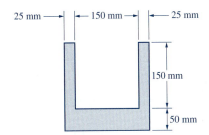

25 mm — |← 150 mm →| ← 25 mm

150 mm

50 mm

FIGURE P14–17

14–18 Refer again to Fig. P14–17. Determine the maximum negative moment that the section can resist without exceeding the allowable stresses in tension and compression given in Problem 14–17.

14–19 Determine the allowable load P that can be applied to the midspan of the simply supported beam shown in Fig. P14–19. The beam has a structural steel W14 × 82 section and an allowable flexural stress of 33 ksi. Neglect the weight of the beam.

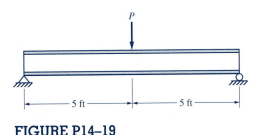

P

5 ft —|— 5 ft

FIGURE P14–19

14–20 See Fig. P14–20. Determine the allowable uniform load w in lb/ft that the structural steel S15 × 50 cantilever beam can carry without exceeding an allowable flexural stress of 24 ksi.

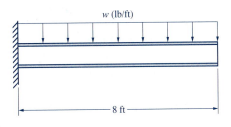

w (lb/ft)

8 ft

FIGURE P14–20

Section 14–5   Shear Stress Formula for Beams

14–21   The beam of rectangular section in Fig. P14–21 is subjected to a maximum shear force of 1900 lb. Determine the shear stresses at points A, B, and C.

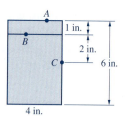

FIGURE P14–21

14–22   The beam of circular section in Fig. P14–22 is subjected to a maximum shear force of 15 kN. Determine the shear stresses at points A, B, and C.

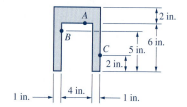

FIGURE P14–22

14–23   A beam having the channel section shown in Fig. P14–23 is subjected to a maximum shear force of 10 kips. Determine the shear stresses at points A, B, and C.

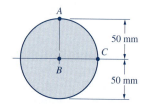

FIGURE P14–23

14–24   A cantilever timber beam having the full-size rectangular section shown in Fig. P14–24 is subjected to the concentrated load P at its free end. Determine the maximum allowable load P if the allowable flexural stress is 10 MPa and the allowable shear stress in the beam is 800 kPa.

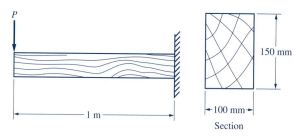

FIGURE P14–24

14–25   A simple beam having the full-size rectangular section shown in Fig. P14–25 carries a uniform load. The beam is made of oak with an allowable flexural stress of 1900 psi and an allowable longitudinal shear stress (parallel to the grain) of 145 psi. Determine the maximum superimposed uniform load w in lb/ft that can be applied to the beam.

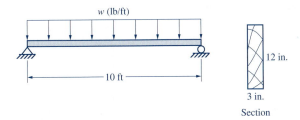

FIGURE P14–25

14–26 Determine the maximum load P in kN that can be applied to the circular log shown in Fig. P14–26. The beam has an allowable flexural stress of 9 MPa and an allowable shear stress parallel to the grain of 850 kPa.

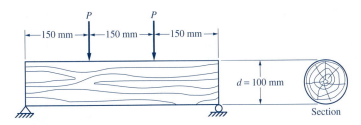

FIGURE P14–26

14–27 Determine the maximum shear stress in the beam shown in Fig. P14–12.

14–28 Determine the maximum shear stress in the beam shown in Fig. P14–13.

14–29 An overhanging beam having a T-section ($I = 136$ in.^{4}) is subjected to the concentrated load shown in Fig. P14–29. Determine the shear stresses in section A–A at the levels indicated. Show the distribution of shear stresses in the section with figures similar to those in Example 14–7.

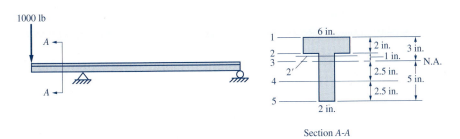

FIGURE P14–29

Section 14–6   Composite Beams

14–30 Rework Example 14–8 by using a transformed section in wood.

14–31 A timber beam reinforced with steel plates has the cross-section shown in Fig. P14–31. Determine the maximum normal stress in each material due to a bending moment of 100 kN · m. The moduli of elasticity are: E_{st} = 210 MPa, and E_{wd} = 10.5 MPa.

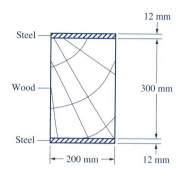

FIGURE P14–31

14–32 A composite beam of steel and aluminum has the cross-section shown in Fig. P14–32. Determine the maximum normal stress in each material due to a positive bending moment of 300 kip · in. The moduli of elasticity are: E_{st} = 30 000 ksi, and E_{al} = 10 000 ksi.

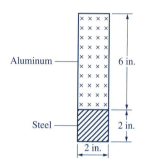

FIGURE P14–32

14–33 A composite beam consists of a wood section reinforced with two steel plates. The cross-section of the beam is shown in Fig. P14–33. Determine the allowable bending moment of the beam about the horizontal neutral axis. The given data are: E_{st} = 210 GPa, E_{wd} = 10.5 GPa, $(\sigma_{st})_{allow}$ = 160 MPa, and $(\sigma_{wd})_{allow}$ = 8.5 MPa.

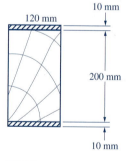

FIGURE P14–33

14–34 Rework Problem 14–33 for the composite beam section of wood and steel plate shown in Fig. P14–34. The other data remain the same.

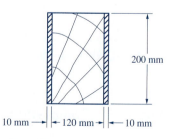

FIGURE P14–34

14–35 A beam is built up with a standard steel I-beam S12 × 31.8 and full size 5 in. × $2\frac{1}{2}$ in. timbers, as shown in Fig. P14–35. The components are adequately fastened so that they act as one unit. Determine the allowable uniform load in lb/ft that the beam can carry over a simple span of 20 ft. The given data are: $E_{st} = 30 \times 10^6$ psi, $E_{wd} = 1.5 \times 10^6$ psi, $(\sigma_{st})_{allow} = 24\,000$ psi, and $(\sigma_{wd})_{allow} = 1200$ psi.

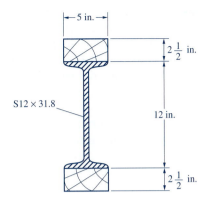

FIGURE P14–35

Section 14–7 Reinforced Concrete Beams

14–36 A rectangular concrete beam is reinforced with three $\frac{3}{4}$-in.-diameter steel bars ($A_{st} = 1.32$ in.^{2}), as shown in Fig. P14–36. Locate the neutral axis of the section. Assume that $n = E_{st}/E_{cn} = 10$.

FIGURE P14–36

14–37 A reinforced concrete beam has the cross-section shown in Fig. P14–37. Determine the maximum normal stress in concrete and in the steel bars due to a positive bending moment of 70 kip · ft. Assume that $n = E_{st}/E_{cn} = 9$.

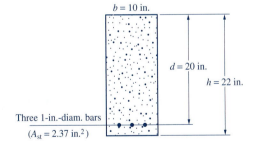

FIGURE P14–37

14–38 A reinforced concrete beam has the cross-section shown in Fig. P14–38. Determine the allowable positive moment that the beam can resist, given that $n = E_{st}/E_{cn} = 8$, $(\sigma_{st})_{allow} = 20\,000$ psi, and $(\sigma_{cn})_{allow} = 1800$ psi.

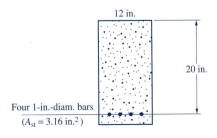

FIGURE P14–38

14–39 A 16-ft simply supported, reinforced concrete beam has the cross-section shown in Fig. P14–39. Knowing that $n = E_{st}/E_{cn} = 10$, determine (a) the location of the neutral axis if both the concrete and the steel are stressed to their maximum allowable values of $(\sigma_{st})_{allow} = 20\,000$ psi and $(\sigma_{cr})_{allow} = 1000$ psi, (b) the required area of the steel bars A_{st}, and (c) the allowable uniform load w lb/ft (including its own weight) over the entire span. [Hints: (a) In the transformed section, the stress on top is $(\sigma_{cn})_{allow}$; on the centerline of steel, it is $(\sigma_{st})_{allow}/n$. The value of x can be determined by proportion. (b) Solve A_{st} from Equation 14–18. (c) Solve M_{max} from Equation 14–20 or Equation 14–21 and then equate it to $wL^2/8$.]

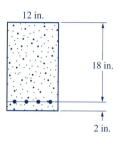

FIGURE P14–39

Section 14–8   Elastic Analysis of Beams

14–40 Determine the maximum shear stress and the maximum flexural stress in the beam with the box section shown in Fig. P14–40.

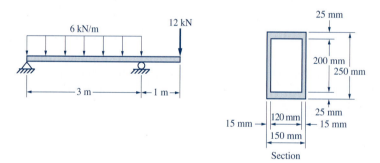

FIGURE P14–40

14–41 A beam on simple supports has a box section with the dimensions indicated in Fig. P14–41. If the moment diagram of the beam is sketched as shown, determine the maximum flexural stress and the maximum shear stress in the beam. (*Hint:* Find shear forces by $V = \Delta M/\Delta x$.)

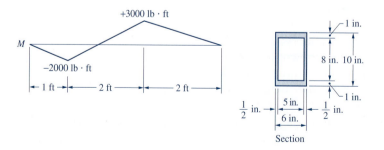

FIGURE P14–41

14–42 The cast-iron beam in Fig. P14–42 has the inverted T-section shown. Determine the maximum shear stress, the maximum tensile stress, and the maximum compressive stress in the beam.

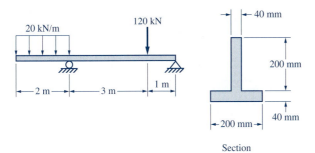

FIGURE P14–42

14–43 The W14 × 38 beam shown in Fig. P14–43 supports a uniform load of 3 kips/ft, including the weight of the beam. Determine the normal and shear stresses acting on the elements at *A* and *B*. Show the directions of these stresses on the elements.

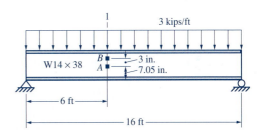

FIGURE P14–43

14–44 Determine the allowable superimposed, uniformly distributed load that may be placed on a simple beam having an 8-ft span and a W18 × 50 section. The allowable flexural stress is 24 ksi and the allowable shear stress is 14.5 ksi.

14–45 Determine the allowable superimposed, uniformly distributed load that may be placed on a simple beam having a 4-m span and a W410 × 0.83 section. The allowable flexural stress is 165 MPa and the allowable shear stress is 100 MPa.

14–46 A simply supported timber beam having a 100 × 250 nominal section is subjected to two symmetrically placed, concentrated loads *P*, as shown in Fig. P14–46. The beam is made of Douglas fir having an allowable flexural stress of 10 MPa and an allowable shear stress parallel to the grain of 660 kPa. Determine the allowable load *P*. Be sure to include the weight of the beam.

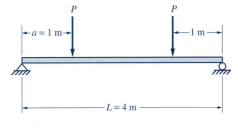

FIGURE P14–46

14–47 Refer to Fig. P14–47. The floor system of a residential building is supported by 3 × 10 Douglas-fir joists spaced 2 ft on centers and having a simple span length of 15 ft. (*a*) Calculate the allowable uniform load *w* for each joist in lb/ft. (*b*) Calculate the allowable life load on the floor in lb/ft² (psf). Assume that the floor system is 2 in. thick at a specific weight of 45 lb/ft³. The allowable flexural stress is 1450 psi and the allowable shear stress is 95 psi.

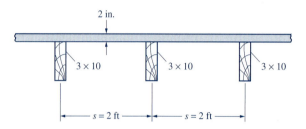

FIGURE P14–47

Section 14–9   Plastic Analysis of Beams

14–48 Calculate the elastic section modulus *S*, the plastic section modulus *Z*, and the shape factor for the beam cross-section in Fig. P14–48.

14–49 For the beam section in Fig. P14–48, calculate (*a*) the maximum elastic moment M_y, (*b*) the plastic moment M_p, and (*c*) the uniformly distributed load that the beam can carry at yield and at full plastification for a simple span of 30 ft. Assume that the beam is made of an elastoplastic material for which $\sigma_y = 36$.

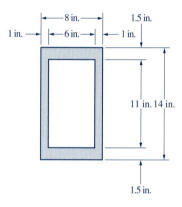

FIGURE P14–48

14–50 Calculate the elastic section modulus *S*, the plastic section modulus *Z*, and the shape factor for the beam cross-section in Fig. P14–50.

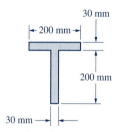

FIGURE P14–50

14–51 For the beam section in Fig. P14–50, calculate (*a*) the maximum elastic moment M_y, (*b*) the plastic moment M_p, and (*c*) the uniformly distributed load that the beam can carry at yield and at full plastification for a simple span of 12 m. Assume that the beam is made of an elastoplastic material for which $\sigma_y = 250$ MPa.

14–52 A simple beam with a 12-ft span and the channel section shown in Fig. P14–52 is made of an elastoplastic material for which $\sigma_y = 36$ ksi. Determine the uniformly distributed load w that can be applied to the beam based on the plastic moment M_p.

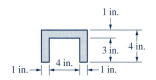

FIGURE P14–52

14–53 The simple beam with the boxed section shown in Fig. P14–53 is made of an elastoplastic material for which $\sigma_y = 250$ MPa. Determine the load P that the beam can carry based on the plastic moment M_p. Neglect the weight of the beam.

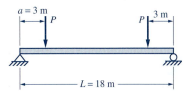

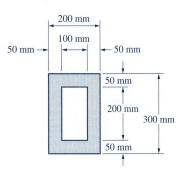

FIGURE P14–53

Computer Program Assignments

For each of the following problems, write a computer program using an appropriate programming language with which you are most familiar. Make the program user friendly by incorporating plenty of comments and input prompts so that the user will understand the input data to be entered and the limitations of their values. The output should include the data entered and the computed results, and they must be well labeled to identify each quantity. If a tabulated format is used, a proper heading must be included at the top of the table. Do not limit the program to any specific unit system. Indicate the consistent U.S. customary or SI units that can be used.

C14–1 Write a computer program that can be used to calculate the maximum flexural and the maximum shear stresses in a simple beam or a cantilever subjected to the loads shown in Fig. C14–1a and b. Let the user have the option of selecting one of three sections: a rectangular section, a circular section, or a W-shape (or an S-shape). The user input should include (1) the span length L; (2) the loads w, P_1, P_2, and a for a simple beam or the loads w, P, and a for a cantilever beam (the downward loads are treated as positive); (3) the section properties: for the rectangular section, the width b and the height h; for the circular section, the diameter d; and for the W-shape or the S-shape, the overall depth d, the web thickness t_w, and the section modulus S_x. To calculate the maximum shear stress, use Equation 14–11 for the rectangular section, Equation 14–12 for the circular section, and Equation 14–13 for the W-shape. Use this program to solve (a) Problem 14–1, (b) Problem 14–5, (c) Problem 14–6, (d) Problem 14–8, (e) Example 14–2, and (f) Problem 14–3.

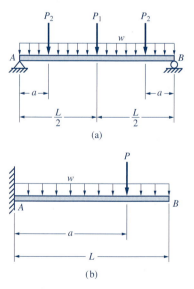

FIGURE C14–1

C14–2 Write a computer program that can be used to calculate the allowable concentrated load at the midspan of a simple beam or at the free end of a cantilever beam, or the allowable uniform load over a simple or cantilever beam span. Let the user have the option of selecting one of the three sections: a rectangular section, a circular section, a W-shape, or an S-shape. The user input should include (1) the type of beam; (2) the span length L; (3) the allowable flexural and shear stresses; and (4) the section properties: for the rectangular section, the width b and height h; for the circular section, the diameter d; for the W-shape or S-shape, the overall depth d, the web thickness t_w, and the section modulus S_x. Use this program to solve (a) Problem 14–19, (b) Problem 14–20, (c) Problem 14–24, and (d) Problem 14–25.

CHAPTER 15

DESIGN OF BEAMS FOR STRENGTH

15–1
INTRODUCTION

As mentioned in Section 14–6, beam problems are of two major types: analysis and design. Beam analysis has been discussed in the preceding chapter. In this chapter, we will deal with the design of beams for strength. We will consider only the design of *prismatic beams,* i.e., straight beams with a uniform cross-section. Particular attention will be given to rolled steel sections of W (wide-flange) shape and timber beams with a rectangular cross-section. Generally the span length, supporting conditions, and loading are given; the shape and size of the beam are to be selected.

The complete design of a beam includes the selection of materials and the type of sections, and consideration to bending and shear strength, deflection, lateral support of the compression flange, web crippling, and support details. Among these items, only the design for bending and shear strength is treated in this chapter. The methods for computing deflections will be presented in the next chapter.

Two approaches are currently employed in structural design for strength: the conventional *allowable stress design* and the newer *ultimate strength design.* The major concern in this chapter will be the allowable stress design. The ultimate strength design will be introduced at the end of the chapter.

The size of a beam must be such that the maximum flexural stress and the maximum shear stress at the critical sections will stay within the allowable limits. One critical section occurs where the absolute value of the bending moment is a maximum; the second critical section occurs where the absolute value of the shear force is a maximum. The shear force and bending moment diagrams of a given beam are useful for determining the locations of these critical sections along the beam. For some typical loadings, however, the construction of shear and moment diagrams may not be necessary. Handbooks provide formulas for the maximum shear force and the maximum moment for many different loading conditions. Table 13–1 listed some simple cases.

The primary concern in structural design is safety: the structure must be able to sustain the given design load without failure. In the allowable stress design, the way to ensure safety is to keep the maximum flexural stress and the maximum shear stress in the beam within the allowable limits. The elastic beam analysis established in Chapter 14 also applies to the design of beams. Our discussion of beam design involves no new principles, so there are only a few techniques and steps that we need to learn.

15–2
BASIC CONSIDERATIONS IN BEAM DESIGN

Framing Layout.   Figure 15–1 illustrates two common methods of framing layout: part a shows center-point concentration and part b shows third-point concentration. When the area enclosed by the columns (called a *bay*) is not a square, the usual custom is to run the beams in the long direction, with the girders on the shorter span. However, this may not always be so. Two or more alternative layouts may be designed and then compared for cost. The area of floor supported by a beam is equal to the span length multiplied by the sum of half of the distances to adjacent beams. The reactions on the beams are applied to the girder as concentrated loads.

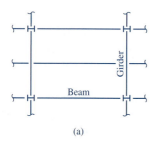

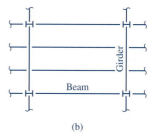

(a)   (b)

FIGURE 15–1

Loads on the floor supported by the beams can be divided into two major categories:

Dead Load.   The dead load consists of the weight of walls, partitions, columns, floors, and roofs. In the design of a beam, the dead load must include an allowance for its own weight.

Live Load.   The live load on a floor represents the effect created by occupancy. It includes the weight of humans, furniture, equipment, stored materials, and so on. Building codes provide minimum live loads to be used in the design of various types of buildings. Table 15–1 lists a few typical values.

Beam Materials.   The most commonly used materials for beams are steel and timber. Generally speaking, for longer span lengths and/or heavier loads, steel beams are selected; for shorter span lengths or lighter loads, timber beams are preferred. Special applications may require the use of

TABLE 15–1 Minimum Live Loads

Description	Uniform Live Load	
	(psf)	(kN/m^2)
Warehouse (heavy)	250	12.0
Warehouse (light)	125	6.0
Heavy manufacturing	125	6.0
Light manufacturing	75	3.6
Wholesale stores	100	4.8
Retail stores	75	3.6
Offices	50	2.4
Residential	40	1.9

other materials. For example, for lightweight framework, as required by aircraft or racing bikes, aluminum alloy or titanium alloy may be a better choice.

Beam Sections.   Since beams are often made of steel and timber, sections made of these materials are discussed below:

Rolled Steel Shapes.   Most manufactured steel beams are produced by rolling a hot ingot of steel until the desired shape is formed. The properties of these structural shapes are tabulated in the American Institute of Steel Construction (AISC) manual. Selected W shapes (wide-flange sections) and S shapes (I-beams) are listed in the appendix, Tables A–1 and A–2. These shapes are designated by their depth and weight per unit length; for example, W21 × 83 indicates a wide-flange section (W shape) having a nominal depth of 21 in. (actual depth is 21.43 in.) and a weight of 83 lb/ft. Other properties of this section can be obtained by referring to the appendix: Table A–1(a) in U.S. customary units and Table A–1(b) in SI units. Tables A–3 to A–5 list properties of other shapes, such as channels, angles, and circular pipes.

Built-Up Steel Sections.   A rolled steel section may not be large enough to resist very large loads. A heavier section may be built up by using roller shapes and/or plates. A few examples of such sections are shown in Fig. 15–2.

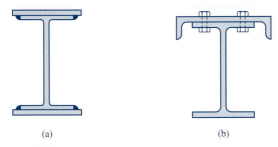

(a)   (b)

FIGURE 15–2

Timber Sections.   Most beams made of timber have rectangular cross-sections because such beams are easy to manufacture, ship, and frame. Lumber is identified by its nominal dimensions, such as 2 × 4 (2 in. by 4 in.). The nominal size indicates the dimensions of rough-sawed lumber.

Dressed or surfaced (after planing) lumber is $\frac{1}{2}$ or $\frac{3}{4}$ in. smaller in dimension than the rough-sawed timber. For example, a 4×8 section is actually $3\frac{1}{2}$ in. $\times 7\frac{1}{4}$ dressed. Properties of representative timber sections are listed in the appendix, Table A–6(a) for U.S. customary units and Table A–6(b) for SI units. All the properties are computed based on the dressed sizes.

Built-Up Timber Sections. Built-up wood beams are usually in the form of a box-section or a T-section, as shown in Fig. 15–3a and b, respectively. For very long spans, *glulam beams* are used. These beams are made from several boards glue-laminated together to form a single unit, as shown in Fig. 15–3c.

The component parts in a built-up steel or timber section must be connected properly by fasteners such as weld, solder, glue, nails, bolts, etc., so the beam can act as a single unit when subjected to loads. The shear strength in the connectors must be investigated, and the techniques for doing so will be discussed in Section 15–5.

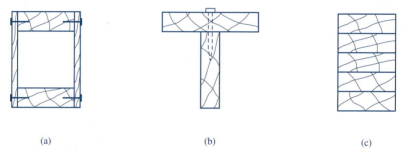

(a)   (b)   (c)

FIGURE 15–3

Lateral Bracing. A beam must be braced properly to prevent unstable sidesway and buckling of the compression flange. The full value of the allowable bending stress can be used only when the compression flange is braced adequately. In buildings of the usual type, all floor beams are laterally supported because the floors they support are fastened to them.

Web Crippling. An excessive end reaction on a beam or an excessive concentrated load may cause crippling or localized yielding of the web of a steel beam. When these concentrated forces are over certain limits, AISC specifications require that the web be reinforced with stiffeners, the length of the bearing be increased, or a beam with a thicker web be selected.

Beam-Bearing Plates. Beams that are supported on walls or piers of masonry or concrete usually rest on steel-bearing plates. The purpose of the plate is to provide an ample bearing area. The plate also helps to seat the beam at its proper elevation. Properly placed bearing plates allow a uniform distribution of the beam reaction over the area of contact with the support material.

15–3
DESIGN OF STEEL BEAMS

The allowable stress design for steel beams will be covered in this section. The allowable stresses used in the design are often specified by the design code. The code frequently used for structural steel design is the American Institute of Steel Construction (AISC) specifications. Under the most favorable conditions, the AISC specifications prescribe an allowable value of $0.66\sigma_y$ for flexural stress, and an allowable value of $0.4\sigma_y$ for shear stress. For the most commonly used structural steel, designated by A36, the yield strength σ_y is 36 ksi (250 MPa). Thus, the allowable flexural stress is

$$\sigma_{allow} = 0.66 \times 36 = 24 \text{ ksi (165 MPa)}$$

and the allowable shear stress is

$$\tau_{allow} = 0.4 \times 36 = 14.5 \text{ ksi (100 MPa)}$$

The following general procedure describes the steps required for the design of steel beams for strength.

Step 1: Determine the beam span, support conditions, allowable stresses, and other design limitations. Identify or compute the loads.

Step 2: Determine the maximum shear force and the maximum bending moment along the beam. For simple loadings, use the formulas from Table 13–1.

Step 3: Using the maximum value of the bending moment, regardless of the sign, compute the minimum required section modulus from the flexure formula:

$$S_{req} = \frac{M_{max}}{\sigma_{allow}} \tag{15–1}$$

Step 4: Scan the tables in the appendix, Table A–1, and list several possible choices of W shape that have a section modulus (about the strong axis) greater than the required value computed in step 3. Select the one with the lightest weight per unit length as a trial section (it may not have the smallest section modulus S). Make sure that this section has a value of S slightly greater than S_{req} so that the beam can resist the additional moment produced by the weight of the beam.

Step 5: The beam selected is checked for shear stress by using the average web shear formula, Equation 14–13:

$$\tau_{avg} = \frac{V_{max}}{dt_w} \tag{15–2}$$

This calculated stress is usually much lower than τ_{allow}, so shear stress rarely dictates the selection of the size of a steel beam. And the effect of the weight of the beam on the shear stress can usually be neglected.

─── **EXAMPLE 15–1** ──────────────────────────────

Select the lightest W shape steel beam for a 15-ft simple span carrying a uniform load of 3 kip/ft. The beam is supported laterally for its entire length. Use A36 steel.

Solution. From case 4 in Table 13–1, the maximum shear and the maximum moment due to the superimposed uniform load are

$$V_{max} = \frac{wL}{2} = \frac{(3 \text{ kips/ft})(15 \text{ ft})}{2} = 22.5 \text{ kips}$$

$$M_{max} = \frac{wL^2}{8} = \frac{(3 \text{ kips/ft})(15 \text{ ft})^2}{8} = 84.4 \text{ kip} \cdot \text{ft}$$

$$= 1013 \text{ lb} \cdot \text{in.}$$

For A36 steel, the allowable flexural stress and the allowable shear stress are

$$\sigma_{allow} = 24 \text{ ksi} \qquad \tau_{allow} = 14.5 \text{ ksi}$$

From Equation 15–1, the minimum required section modulus is

$$S_{req} = \frac{M_{max}}{\sigma_{allow}} = \frac{1013 \text{ kip} \cdot \text{in.}}{24 \text{ kips/in.}^2} = 42.2 \text{ in.}^3$$

From the appendix, Table A –1(a), the following W shapes fulfill the requirement on the minimum required section modulus:

$$\text{W14} \times 34\text{:} \qquad S = 48.6 \text{ in.}^3$$

$$\text{W12} \times 35\text{:} \qquad S = 45.6 \text{ in.}^3$$

$$\text{W10} \times 45\text{:} \qquad S = 49.1 \text{ in.}^3$$

The lightest shape W14 × 34 is selected as a trial section. We find

$$\frac{\text{Beam weight}}{\text{Load}} = \frac{34}{3000} = 0.011 = 1.1\%$$

$$\frac{\text{Extra } S}{S_{req}} = \frac{48.6 - 42.2}{42.2} = 0.152 = 15.2\%$$

Note that the percentage of extra section modulus is greater than the percentage of the weight of the beam to the uniform load. Thus, the beam is satisfactory for bending.

Now we need to check the shear stress. From the appendix, Table A–1(a), for W14 × 34 we find

$$d = 13.98 \text{ in.} \qquad t_w = 0.285 \text{ in.}$$

From Equation 15–2, the average web shear stress in the beam is

$$\tau_{avg} = \frac{V_{max}}{dt_w} = \frac{22.5 \text{ kips}}{(13.98 \text{ in.})(0.285 \text{ in.})} = 5.65 \text{ ksi}$$

which is well within the allowable shear stress of 14.5 ksi. Thus, the section is satisfactory for shear strength.

Use W14 × 34   ⇐ **Ans.**

―――――― **EXAMPLE 15–2** ――――――――――――――――――――――――――――――

A simply supported girder has a span of 5.5 m with a concentrated load of 250 kN at the midspan. The girder is braced laterally throughout its length. Select the lightest W shape to carry the load for A36 steel.

Solution. From case 1 in Table 13–1, the maximum shear and the maximum moment due to the concentrated load are

$$V_{max} = \frac{P}{2} = \frac{250 \text{ kN}}{2} = 125 \text{ kN}$$

$$M_{max} = \frac{PL}{4} = \frac{(250 \text{ kN})(5.5 \text{ m})}{4} = 344 \text{ kN} \cdot \text{m}$$

For A36 steel, the allowable flexural stress and the allowable shear stress are

$$\sigma_{allow} = 165 \text{ MPa} \qquad \tau_{allow} = 100 \text{ MPa}$$

From Equation 15–3, the minimum required section modulus is

$$S_{req} = \frac{M_{max}}{\sigma_{allow}} = \frac{344 \text{ kN} \cdot \text{m}}{165\,000 \text{ kN/m}^2} = 2.08 \times 10^{-3} \text{ m}^3$$

From the appendix, Table A–1(b), the following W shapes fulfill the requirement on the minimum required section modulus:

W460 × 1.42: $S = 3.08 \times 10^{-3} \text{ m}^3$

W410 × 1.30: $S = 2.54 \times 10^{-3} \text{ m}^3$

W360 × 1.31: $S = 2.34 \times 10^{-3} \text{ m}^3$

The lightest shape, W410 × 1.30, is selected as a trial section. We find

$$M_{wt} = \frac{wL^2}{8} = \frac{(1.30 \text{ kN/m})(5.5 \text{ m})^2}{8} = 4.92 \text{ kN} \cdot \text{m}$$

$$\frac{M_{wt}}{M_{max}} = \frac{4.92}{344} = 0.014 = 1.4\%$$

$$\frac{\text{Extra } S}{S_{req}} = \frac{2.54 \times 10^{-3} - 2.08 \times 10^{-3}}{2.08 \times 10^{-3}} = 0.221 = 21.1\%$$

Note that the percentage of extra section modulus is much greater than the percentage of the ratio of the moment caused by the weight of the beam to the moment caused by the uniform load. Thus, the beam is satisfactory for bending.

Now we need to check the shear stress. From the appendix, Table A–1(b), for W 410 × 1.30 we find

$$d = 425 \text{ mm} \qquad t_w = 13.3 \text{ mm}$$

The average web shear stress in the beam is

$$\tau_{avg} = \frac{V_{max}}{dt_w} = \frac{125 \text{ kN}}{(0.425 \text{ m})(0.0133 \text{ m})} = 22\ 100 \text{ kPa}$$

which is equivalent to 22.1 MPa and is well within the allowable limit of 100 MPa. Thus, the shape selected is also satisfactory for shear stress.

Use W410 × 1.30 (or W16 × 89)   ⇐ **Ans.**

EXAMPLE 15–3

A simple beam has a span of 4 m with a uniformly distributed load of 60 kN/m, and a concentrated load of 120 kN applied 1 m from the right support, as shown in Fig. E15–3(1). The beam is braced laterally throughout its length. Select the lightest W shape for A36 steel to carry the load.

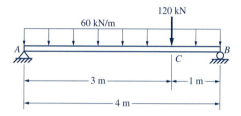

FIGURE E15–3(1)

Solution. To determine the maximum shear and the maximum moment, the shear and moment diagrams must be sketched, as shown in Fig. E15–3(2). From the diagrams, we find

$$V_{max} = 210 \text{ kN}$$

$$M_{max} = 187.5 \text{ kN} \cdot \text{m}$$

For A36 steel, the allowable flexural stress and the allowable shear stress are

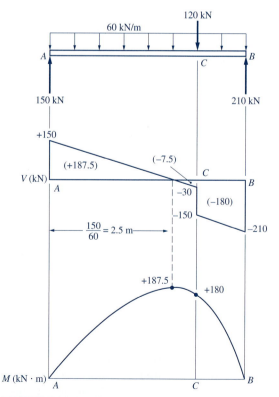

FIGURE E15–3(2)

$$\sigma_{allow} = 165 \text{ MPa} \qquad \tau_{allow} = 100 \text{ MPa}$$

From Equation 15–1, the minimum required section modulus is

$$S_{req} = \frac{M_{max}}{\sigma_{allow}} = \frac{187.5 \text{ kN} \cdot \text{m}}{165\ 000 \text{ kN/m}^2} = 1.14 \times 10^{-3} \text{ m}^3$$

From the appendix, Table A–1(b), the following W shapes fulfill the requirement on the minimum required section modulus:

$$\text{W460} \times 0.67: \qquad S = 1.29 \times 10^{-3} \text{ m}^3$$

$$\text{W410} \times 0.73: \qquad S = 1.33 \times 10^{-3} \text{ m}^3$$

$$\text{W360} \times 0.77: \qquad S = 1.28 \times 10^{-3} \text{ m}^3$$

The lightest shape, W460 $\times$ 0.67, is selected as a trial section. We find

$$M_{wt} = \frac{wL^2}{8} = \frac{(0.67 \text{ kN/m})(4 \text{ m})^2}{8} = 1.34 \text{ kN} \cdot \text{m}$$

$$\frac{M_{wt}}{M_{max}} = \frac{1.34}{187.5} = 0.007 = 0.7\%$$

$$\frac{\text{Extra } S}{S_{req}} = \frac{1.29 \times 10^{-3} - 1.14 \times 10^{-3}}{1.14 \times 10^{-3}} = 0.132 = 13.2\%$$

Note that the percentage of extra section modulus is much greater than the percentage of the ratio of additional moment caused by the weight of the beam to the maximum moment caused by the load. Thus, the beam is satisfactory for bending.

From the appendix, Table A–1(b), for W460 $\times$ 0.67 we find

$$d = 459 \text{ mm} \qquad t_w = 9.14 \text{ mm}$$

The average web shear stress in the beam is

$$\tau_{avg} = \frac{V_{max}}{dt_w} = \frac{210 \text{ kN}}{(0.459 \text{ m})(0.00914 \text{ m})} = 50\ 100 \text{ kPa}$$

$$= 50.1 \text{ MPa}$$

which is well within the allowable limit of 100 MPa. Thus, the shape selected is satisfactory for shear stress.

Use W460 $\times$ 0.67 (or W18 $\times$ 46) ⇐ **Ans.**

EXAMPLE 15–4

The floor plan of an interior bay of a warehouse has the layout sketched in Fig. E15–4. The steel beams run in the horizontal direction, and the steel girders run in the vertical direction. A 6-in.-thick, reinforced concrete floor is supported on the beams. The live load on the floor is 125 psf. Assume that both the beams and the girders are supported laterally for their entire length. Use A36 steel. Select the lightest W shapes for (*a*) the beam and (*b*) the girder.

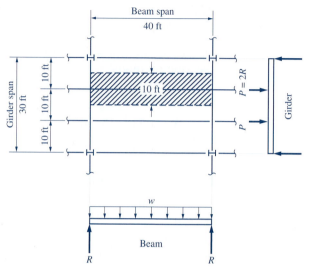

FIGURE E15–4

Solution. Beams must be designed first.

(*a*) *Design of Beams.* Beams are supported on girders or directly on columns. The spans, load, and supporting conditions of all interior beams are the same. Therefore, the design is typical for all interior beams. The beams are assumed to be simply supported. The loading on the beam includes the dead load and the live load. The dead load consists of the weight of a 6-in. concrete floor (reinforced concrete weighs 150 pounds per cubic foot) and the weight of the beam itself, which is not known and hence it will not be included for now. The given live load on the floor is 125 psf.

The width of the floor area supported by an interior beam is the sum of half of the distances to the adjacent beams on centers. Since the spacings are equal, the width of this floor area must be equal to the spacing, which is 10 ft, as indicated by the cross-hatched area in the floor plan. Thus, we find the dead load, the live load, and the total design load per linear foot of beam as:

$$
\begin{aligned}
\text{Dead load} &= (150 \text{ lb/ft}^3)\left(\frac{6}{12} \text{ ft}\right)(10 \text{ ft}) = & 750 \text{ lb/ft} \\
\text{Live load} &= (125 \text{ lb/ft}^2)(10 \text{ ft}) &= 1250 \text{ lb/ft} \\
\hline
\text{Total design load} & &= 2000 \text{ lb/ft} \\
& &= 2 \text{ kips/ft}
\end{aligned}
$$

which is a uniform load on the beam. Thus, the maximum shear and the maximum moment in the beam are

$$V_{max} = \frac{wL}{2} = \frac{(2 \text{ kips/ft})(40 \text{ ft})}{2} = 40 \text{ kips}$$

$$M_{max} = \frac{wL^2}{8} = \frac{(2 \text{ kips/ft})(40 \text{ ft})^2}{8} = 400 \text{ kip} \cdot \text{ft}$$

$$= 4800 \text{ kip} \cdot \text{in.}$$

For A36 steel, the allowable flexural stress and the allowable shear stress are

$$\sigma_{allow} = 24 \text{ ksi} \qquad \tau_{allow} = 14.5 \text{ ksi}$$

The minimum required section modulus is

$$S_{req} = \frac{M_{max}}{\sigma_{allow}} = \frac{4800 \text{ kip} \cdot \text{in.}}{24 \text{ kips/in.}^2} = 200 \text{ in.}^3$$

From the appendix, Table A–1(a), the lightest W shape that fulfills the requirement on the minimum required section modulus is

$$W27 \times 94 \ (d = 26.92 \text{ in.}, \ t_w = 0.49 \text{ in.}, \ S = 243 \text{ in.}^3)$$

We find

$$\frac{\text{Beam weight}}{\text{Load}} = \frac{94}{2000} = 0.047 = 4.7\%$$

$$\frac{\text{Extra } S}{S_{req}} = \frac{243 - 200}{200} = 0.215 = 21.5\%$$

Note that the percentage of extra section modulus is greater than the percentage of the weight of the beam to the uniform load. Thus, the beam is satisfactory for bending.

The average web shear stress in the beam is

$$\tau_{avg} = \frac{V_{max}}{dt_w} = \frac{40 \text{ kips}}{(26.92 \text{ in.})(0.490 \text{ in.})} = 3.03 \text{ ksi}$$

which is well within the allowable limit. Thus, the shape selected is satisfactory for shear strength.

Use W27 × 94 for the interior beams   ⟸ **Ans.**

(b) Design of Girders.   The span length of the girder is the distance from center to center of the supporting columns and is considered a simple beam. The loading on the girder consists of two symmetrically placed, concentrated loads equal to twice the beam reaction (one on each side for the interior girders). The beam reaction R is

$$R = \frac{(2000 \text{ lb/ft} + 94 \text{ lb/ft})(40 \text{ ft})}{2} = 41\,900 \text{ lb}$$

$$= 41.9 \text{ kips}$$

The two concentrated forces at the third points of the girder have a magnitude of

$$P = 2R = 83.8 \text{ kips}$$

From case 3 in Table 13–1, the maximum shear force and bending moment in the beam are

$$V_{max} = P = 83.8 \text{ kips}$$

$$M_{max} = Pa = (83.8 \text{ kips})(10 \text{ ft}) = 838 \text{ kip} \cdot \text{ft}$$

$$= 10\ 060 \text{ kip} \cdot \text{in.}$$

From Equation 15–3, the minimum required section modulus is

$$S_{req} = \frac{M_{max}}{\sigma_{allow}} = \frac{10\ 060 \text{ kip} \cdot \text{in.}}{24 \text{ kips/in.}^2} = 419 \text{ in.}^3$$

From the appendix, Table A–1(a), the lightest W shape that fulfills the requirement on the section modulus is

$$\text{W36} \times 150 \ (d = 35.85 \text{ in.}, \ t_w = 0.625 \text{ in.}, \ S = 504 \text{ in.}^3)$$

We find

$$M_{wt} = \frac{wL^2}{8} = \frac{(0.150 \text{ kip/ft})(30 \text{ ft})^2}{8} = 16.9 \text{ kip} \cdot \text{ft}$$

$$\frac{M_{wt}}{M_{max}} = \frac{16.9 \text{ kip} \cdot \text{ft}}{838 \text{ kip} \cdot \text{ft}} = 0.020 = 2\%$$

$$\frac{\text{Extra } S}{S_{req}} = \frac{504 - 419}{419} = 0.203 = 20.3\%$$

Note that the percentage of extra section modulus is much greater than the percentage of the ratio of the moment caused by the weight of the beam to the maximum moment caused by the loads. Thus, the beam is satisfactory for bending.

The average web shear stress in the beam is

$$\tau_{avg} = \frac{V_{max}}{dt_w} = \frac{83.8 \text{ kips}}{(35.85 \text{ in.})(0.625 \text{ in.})} = 3.74 \text{ ksi}$$

which is well within the allowable limit. Thus, the shape selected is satisfactory for shear strength.

Use W36 × 150 for the interior girders   ⇐ **Ans.**

15–4
DESIGN OF TIMBER BEAMS

Timber beam design is generally limited to the building industry. The allowable bending and allowable shear stresses vary with different wood species, grades, and the design code used. Some typical values are listed in Table 15–2, p. 554. Note that wood is not an isotropic material; i.e., the mechanical properties of a piece of wood are not the same in all directions. When subjected to normal stresses, either tension or compression, wood is stronger if the load is parallel to the grain rather than perpendicular to the grain. Its shear resistance, however, is much weaker in the direction parallel to the grain. Since shear stresses occur in both the direction perpendicular and the direction parallel to the grain, a weak shear resistance parallel to the grain may dictate the selection of the size of a beam.

The following general procedure describes the steps required for the design of a timber beam for strength:

Step 1: Determine the beam span, support conditions, allowable stresses, and other design limitations. Identify or compute the loads.

Step 2: Determine the maximum shear force and the maximum bending moment along the beam. For simple loadings, use the formulas from Table 13–1.

Step 3: Using the largest value of the bending moment, regardless of the sign, compute the minimum required section modulus from the flexure formula:

$$S_{req} = \frac{M_{max}}{\sigma_{allow}} \tag{15–3}$$

Step 4: Timber beams are usually available in rectangular sections for which the maximum shear stress is 1.5 times the average shear stress (Equation 14–11). In view of the weak shear stress resistance parallel to the grain, the minimum rectangular cross-sectional area required must be calculated from

$$A_{req} = \frac{1.5V_{max}}{\tau_{allow}} \tag{15–4}$$

Step 5: Scan Table A–6 in the appendix and select the lightest rectangular timber section that has a section modulus (about the strong axis) slightly greater than the required value computed in step 3 and an area slightly greater than the required area computed in step 4. Compute the percentage of the extra section modulus provided and the percentage of extra cross-sectional area provided, and make sure that these percentages are greater than the percentage of the ratio of the weight of the beam selected to the total design load.

TABLE 15–2   Allowable Stresses for Timber

| Species | Extreme Fiber in Bending psi (kPa) | Tension Parallel to Grain psi (kPa) | Longi- tudinal Shear psi (kPa) | Compression | |
				Perpen- dicular to Grain psi (kPa)	Parallel to Grain psi (kPa)
Douglas fir	1450 (10 000)	625 (4310)	95 (660)	385 (2650)	1050 (7240)
Eastern hemlock	1350 (9310)	925 (6380)	80 (550)	360 (2480)	950 (6550)
Southern pine	1600 (11 000)	825 (5690)	90 (620)	410 (2830)	1250 (8620)
Ponderosa pine	1100 (7580)	725 (5000)	65 (450)	235 (1620)	750 (5170)
California redwood	1350 (9310)	650 (4480)	100 (690)	270 (1860)	1050 (7240)

In general, narrow and deep timber beams are more effective than wide and shallow beams in resisting bending moments. The preferred depth–width ratio, called the *aspect ratio,* of a timber beam is between 1.5 to 3. For closely spaced joists, the aspect ratio may range from 3 to 6.

───── **EXAMPLE 15–5** ─────

Select a solid, rectangular Douglas fir beam section for a 12-ft simple span carrying a superimposed uniform load of 500 lb/ft.

Solution. The maximum shear and the maximum moment due to the superimposed uniform load are

$$V_{max} = \frac{wL}{2} = \frac{(500 \text{ lb/ft})(12 \text{ ft})}{2} = 3000 \text{ lb}$$

$$M_{max} = \frac{wL^2}{8} = \frac{(500 \text{ lb/ft})(12 \text{ ft})^2}{8} = 9000 \text{ lb} \cdot \text{ft}$$

$$= 108\,000 \text{ lb} \cdot \text{in.}$$

From Table 15–2, the allowable bending stress and the allowable shear stress parallel to the grain for Douglas fir are

$$\sigma_{allow} = 1450 \text{ psi} \qquad \tau_{allow} = 95 \text{ psi}$$

From Equations 15–3 and 15–4, the minimum required section modulus and the minimum required cross-sectional area are

$$S_{req} = \frac{M_{max}}{\sigma_{allow}} = \frac{108\,000 \text{ lb} \cdot \text{in.}}{1450 \text{ lb/in.}^2} = 74.5 \text{ in.}^3$$

$$A_{req} = \frac{1.5 V_{max}}{\tau_{allow}} = \frac{1.5(3000 \text{ lb})}{95 \text{ lb/in.}^2} = 47.4 \text{ in.}^2$$

From the appendix, Table A–6(a), the following rectangular timber sections fulfill both requirements on the section modulus and cross-sectional area:

Section	A (in.²)	S (in.³)	wt (lb/ft)
6 × 10	52.3	82.7	14.5
8 × 10	71.3	113	19.8
10 × 10	90.3	143	25.1

The lightest section, 6×10, is selected as a trial section. We find

$$\frac{\text{Beam weight}}{\text{Load}} = \frac{14.5}{500} = 0.029 = 2.9\%$$

$$\frac{\text{Extra } S}{S_{\text{req}}} = \frac{82.7 - 74.5}{74.5} = 0.11 = 11\%$$

$$\frac{\text{Extra } A}{A_{\text{req}}} = \frac{52.3 - 47.4}{47.4} = 0.103 = 10.3\%$$

Note that both the percentage of extra section modulus and the percentage of extra cross-sectional area are greater than the percentage of the weight of the beam to the uniform load. Thus, the beam is satisfactory for both bending and shear.

Use 6×10 Douglas fir section   ⇐ **Ans.**

EXAMPLE 15–6

Refer to Fig. E15–6(1). Select a California redwood beam of rectangular cross-section for the overhanging beam with a superimposed uniform load of 12 kN/m.

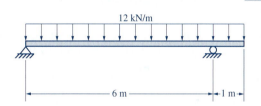

FIGURE E15–6(1)

Solution. To find the maximum shear and the maximum moment due to the superimposed uniform load, the shear and moment diagrams in Fig. E15–6(2) are sketched. From these diagrams, we find

$$V_{\text{max}} = 37.0 \text{ kN}$$

$$M_{\text{max}} = 51.0 \text{ kN} \cdot \text{m}$$

From Table 15–2, the allowable bending stress and the allowable shear stress parallel to the grain for California redwood are

$$\sigma_{\text{allow}} = 9310 \text{ kPa} \qquad \tau_{\text{allow}} = 690 \text{ kPa}$$

From Equations 15–3 and 15–4, the minimum required section modulus and the minimum required cross-sectional area are

$$S_{\text{req}} = \frac{M_{\text{max}}}{\sigma_{\text{allow}}} = \frac{51.0 \text{ kN} \cdot \text{m}}{9130 \text{ kN/m}^2} = 5.59 \times 10^{-3} \text{ m}^3$$

$$A_{\text{req}} = \frac{1.5 V_{\text{max}}}{\tau_{\text{allow}}} = \frac{1.5(37.0 \text{ kN})}{690 \text{ kN/m}^2} = 80.4 \times 10^{-3} \text{ m}^2$$

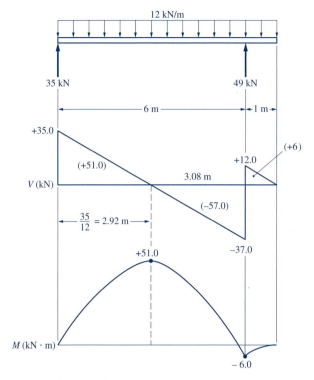

FIGURE E15–6(2)

From the appendix, Table A–6(b), the following rectangular timber sections fulfill both requirements on the section modulus and cross-sectional area:

Section	A ($\times 10^{-3}$ m²)	S ($\times 10^{-3}$ m³)	wt (kN/m)
200 × 460	84.5	6.28	0.533
250 × 410	94.8	6.23	0.597

The lightest section, 200 × 460, is selected as a trial section. We find

$$\frac{\text{Beam weight}}{\text{Load}} = \frac{0.533}{12} = 0.045 = 4.5\%$$

$$\frac{\text{Extra } S}{S_{\text{req}}} = \frac{6.28 \times 10^{-3} - 5.59 \times 10^{-3}}{5.59 \times 10^{-3}} = 0.123 = 12.3\%$$

$$\frac{\text{Extra } A}{A_{\text{req}}} = \frac{84.5 \times 10^{-3} - 80.4 \times 10^{-3}}{80.4 \times 10^{-3}} = 0.051 = 5.1\%$$

Note that both the percentage of extra section modulus and the percentage of extra cross-sectional area are greater than the percentage of the weight of the beam to the uniform load. Thus, the beam is satisfactory for both bending and shear.

Use 200 × 460 Douglas fir section   ⇐ **Ans.**

Note that this is the same section designated 8 × 18 in U.S. customary units.

━━━━━━━ **EXAMPLE 15–7** ━━━━━━━

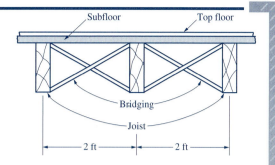

Refer to Fig. E15–7. The joists supporting the floor in a heavy manufacturing building have a span of 16 ft and a spacing of 2 ft on centers, as shown. The floor must support a live load of 125 psf plus 20 psf for the dead weight of the heavy plank subfloor and the top floor. Select the lightest Southern pine joist size.

FIGURE E15–7

Solution. The width of the floor area supported by an interior joist is equal to the 2-ft spacing of the joists. Thus, we find the dead load, the live load, and the total design load per linear foot of joist as:

$$\text{Dead load} = (20 \text{ lb/ft}^2)(2 \text{ ft}) \quad = 40 \text{ lb/ft}$$

$$\text{Live load} = (125 \text{ lb/ft}^2)(2 \text{ ft}) = 250 \text{ lb/ft}$$

$$\overline{\text{Total design load} \qquad\qquad\qquad = 290 \text{ lb/ft}}$$

which is a uniform load on the joist considered simply supported. Thus, the maximum shear and the maximum moment in the beam are

$$V_{\text{max}} = \frac{wL}{2} = \frac{(290 \text{ lb/ft})(16 \text{ ft})}{2} = 2320 \text{ lb}$$

$$M_{\text{max}} = \frac{wL^2}{8} = \frac{(290 \text{ lb/ft})(16 \text{ ft})^2}{8} = 9280 \text{ lb} \cdot \text{ft}$$

$$= 111\,400 \text{ lb} \cdot \text{in.}$$

From Table 15–2, the allowable bending stress and the allowable shear stress parallel to the grain for Southern pine are

$$\sigma_{\text{allow}} = 1600 \text{ psi} \qquad \tau_{\text{allow}} = 90 \text{ psi}$$

From Equations 15–3 and 15–4, the minimum required section modulus and the minimum required cross-sectional area are

$$S_{\text{req}} = \frac{M_{\text{max}}}{\sigma_{\text{allow}}} = \frac{111\,400 \text{ lb} \cdot \text{in.}}{1600 \text{ lb/in.}^2} = 69.6 \text{ in.}^3$$

$$A_{\text{req}} = \frac{1.5V_{\text{max}}}{\tau_{\text{allow}}} = \frac{1.5(2320 \text{ lb})}{90 \text{ lb/in.}^2} = 38.7 \text{ in.}^2$$

From the appendix, Table A–6(a), the following rectangular timber sections fulfill both requirements on the section modulus and cross-sectional area:

Section	A (in.²)	S (in.²)	wt (lb/ft)
4 × 12	39.4	73.8	10.9
4 × 14	46.4	102	12.9
6 × 10	52.3	82.7	14.5

The lightest section, 4 × 12, is selected as a trial section. We find

$$\frac{\text{Beam weight}}{\text{Load}} = \frac{10.9}{290} = 0.038 = 3.8\%$$

$$\frac{\text{Extra } S}{S_{\text{req}}} = \frac{73.8 - 69.6}{69.6} = 0.06 = 6\%$$

$$\frac{\text{Extra } A}{A_{\text{req}}} = \frac{39.4 - 38.7}{38.7} = 0.018 = 1.8\%$$

Note that the percentage of extra cross-sectional area is lower than the ratio of the weight of the beam to the uniform load. Thus, the beam is *not* satisfactory for shear. The next lightest section, 4 × 14, is investigated. We find

$$\frac{\text{Beam weight}}{\text{Load}} = \frac{12.9}{290} = 0.044 = 4.4\%$$

$$\frac{\text{Extra } S}{S_{\text{req}}} = \frac{102 - 69.6}{69.6} = 0.47 = 47\%$$

$$\frac{\text{Extra } A}{A_{\text{req}}} = \frac{46.4 - 38.7}{38.7} = 0.20 = 20\%$$

Note that both the percentage of extra section modulus and the percentage of extra cross-sectional area are greater than the ratio of the weight of the beam to the uniform load. Thus, the beam is satisfactory for both bending and shear.

Use 4 × 14 Southern pine section   ⇐ **Ans.**

15–5
DESIGN OF SHEAR CONNECTORS

As mentioned in Section 15–2, beams are sometimes fabricated by joining several component parts to form a single section. Three typical examples of built-up beams are shown in Fig. 15–4. Figure 15–4a shows a T-beam fabricated by nailing two timber planks together. Figure 15–4b shows plywood and boards nailed together to form a box section. Figure 15–4c shows a fabricated steel beam consisting of a steel channel bolted to the top flange of a W shape.

  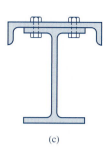

(a) (b) (c)

FIGURE 15–4

 The nails or bolts that connect the components of a built-up beam must be spaced at a proper pitch (the spacing of nails or bolts along the length of the beam) so that the connectors will carry the longitudinal shear force adequately on the contact surface. Consider the T-beam in Fig. 15–5 subjected to a concentrated load at the midspan. Since the shear force is constant throughout the beam, the nails used to connect the two planks are spaced at a constant pitch p. Each nail is required to carry the shear force on the contact surface for each pitch length. The longitudinal shear stress at the contact surface (level 1–1) is

$$\tau_{1-1} = \frac{VQ}{It}$$

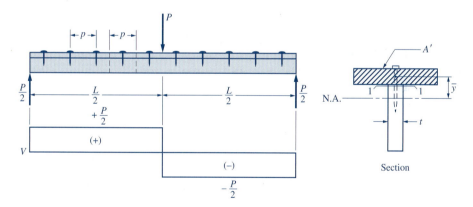

FIGURE 15–5

The quantity $Q = A'\bar{y}'$ represents the first moment of the area A' (area of the plank on the top) about the neutral axis. The shear stress is distributed uniformly in the contact surface, which has an area equal to the product of p and t. Thus, the total shear force in the area is

$$F_s = (p \cdot t)\tau_{1-1} = p\frac{VQ}{I} \tag{15–5}$$

which is the shear force that must be carried by each nail. The quantity VQ/I in Equation 15–5 represents the shear force in a longitudinal section per unit length of beam. This quantity is referred to as the *shear flow* and is denoted by q; that is,

$$q = \frac{VQ}{I} \tag{15-6}$$

If the maximum allowable shear force of each nail is $(F_s)_{\text{allow}}$, then from Equation 15–5, the maximum pitch is

$$p_{\max} = \frac{(F_s)_{\text{allow}}}{q} \tag{15-7}$$

This equation can also be used to determine the pitch of the nails in Fig. 15–4b and c. In Fig. 15–4b, the quantities Q and $(F_s)_{\text{allow}}$ are

Q = the first moment of the area of board A about the neutral axis

$(F_s)_{\text{allow}}$ = allowable shear force of two nails

In Fig. 15–4c, the quantities Q and $(F_s)_{\text{allow}}$ are

Q = the first moment of the area of the steel channel about the neutral axis

$(F_s)_{\text{allow}}$ = allowable shear force of two bolts

EXAMPLE 15–8

A beam is made up of four full-size, 50-mm-thick boards nailed together to form a box section, as shown in Fig. E15–8(1). Each nail is capable of resisting 400 N of shear force. If this beam transmits a constant vertical shear force of 6000 N, determine the pitch of the nails.

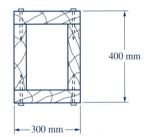

FIGURE E15–8(1)

Solution. The moment of inertia of the section about the neutral axis is calculated first as follows:

$$I = \frac{0.3 \times 0.4^3}{12} - \frac{0.2 \times 0.3^3}{12} = 0.001\ 15 \text{ m}^4$$

See Fig. E15–8(2). The first moment of the shaded area A' about the neutral axis is

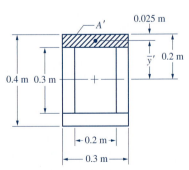

FIGURE E15–8(2)

$$Q = A'\overline{y}' = (0.3 \times 0.05)(0.2 - 0.025)$$
$$= 0.002\ 625\ m^3$$

The shear flow is

$$q = \frac{VQ}{I} = \frac{(6000\ N)(0.002\ 625\ m^3)}{0.001\ 15\ m^4}$$
$$= 13\ 700\ N/m$$

The maximum spacing can now be calculated from Equation 15–7:

$$p_{max} = \frac{(F_s)_{allow}}{q} = \frac{2(400\ N)}{13\ 700\ N/m}$$
$$= 0.0584\ m = 58.4\ mm$$

Use a pitch of 50 mm　　　　　　　⇐ **Ans.**

EXAMPLE 15–9

A girder is subjected to the loads shown in Fig. E15–9(1). It is fabricated with four steel angles riveted to a web plate, as shown in Fig. E15–9(2). Determine the pitch of the rivets if they have a diameter of $\frac{3}{4}$ in. and an allowable shear stress of 15 ksi.

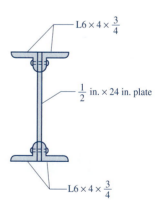

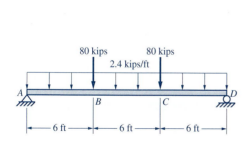

FIGURE E15–9(1)　　　　　　　　　　　FIGURE E15–9(2)

Solution. The neutral axis is located at the mid-depth of the section due to symmetry. The moment of inertia of the entire section about the neutral axis is

$$I = I_{web} + 4I_{angle} = \frac{bh^3}{12} + 4[I + Ad^2]_{angle}$$

$$= \frac{\left(\frac{1}{2}\right)(24)^3}{12} + 4[8.68 + 6.94(12 - 1.08)^2]$$

$$= 3920\ in.^4$$

From the shear diagram plotted in Fig. E15–9(3), the maximum pitch for regions *AB* and *CD* is calculated using the maximum shear force $V_{max} = 101.6$ kips. The value of Q is the first moment of the two angles at the top [shown shaded in the cross section of Fig. E15–9(4)] about the neutral axis. Thus,

$$A' = 2A_{\text{angle}} = 2(6.94) = 13.88 \text{ in.}^2$$

$$Q = A'\overline{y}' = (13.88)(12 - 1.08) = 151.6 \text{ in.}^3$$

$$q = \frac{VQ}{I} = \frac{(101.6 \text{ kips})(151.6 \text{ in.}^3)}{3920 \text{ in.}^4} = 3.93 \text{ kips/in.}$$

The cross-sectional area of each rivet is

$$A = \frac{\pi}{4}(0.75 \text{ in.})^2 = 0.442 \text{ in.}^2$$

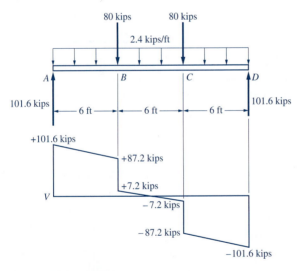

FIGURE E15–9(3)

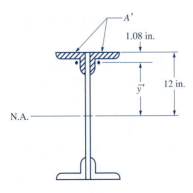

FIGURE E15–9(4)

The rivets are in double shear. Hence, the allowable shear force in each rivet is

$$(F_s)_{\text{allow}} = \tau_{\text{allow}}[2A] = (15 \text{ ksi})[2(0.442 \text{ in.}^2)]$$

$$= 13.3 \text{ kips}$$

From Equation 15–7, the maximum pitch is

$$p_{\text{max}} = \frac{(F_s)_{\text{allow}}}{q} = \frac{13.3 \text{ kips}}{3.93 \text{ kips/in.}} = 3.37 \text{ in.}$$

Use a 3-in. pitch in *AB* and *CD*   ⇐ **Ans.**

To determine the pitch of the rivets in region *BC*, we use $V = 7.2$ kips. Thus,

$$q = \frac{VQ}{I} = \frac{(7.2 \text{ kips})(151.6 \text{ in.}^3)}{3920 \text{ in.}^4} = 0.278 \text{ kip/in.}$$

$$p_{\text{max}} = \frac{(F_s)_{\text{allow}}}{q} = \frac{13.3 \text{ kips}}{0.278 \text{ kip/in.}} = 47.8 \text{ in.}$$

In this case, the maximum pitch permissible by the code will be used for region *BC*. A pitch of 12 in. may be used conveniently, unless the code requires a closer spacing.   ⇐ **Ans.**

15–6
LOAD AND RESISTANCE FACTOR DESIGN

So far, only the allowable stress design based on elastic theory has been discussed. For ductile metal considered to be *elastoplastic* (see Section 11–8), the design approach based on the plastic behavior of such material has been gaining in popularity. The plastic analysis of beams was discussed in Section 14-7. Now we will present the plastic design approach, also known as *ultimate strength design* or *load and resistance factor design*.

Load and resistance factor design (LRFD) for bending is based on the plastic analysis discussed in Section 14–7. The LRFD method was developed in the early 1970s. In 1986, the American Institute of Steel Construction (AISC) published the first LRFD specification. This specification uses load factors and the resistance factors as factors of safety to guard against failure. *Load factors* are used to increase the loads in various types of load combinations. *Resistance factors* are used to decrease the ultimate plastic strength (resistance) of the member. The following terms and criterion must be explained: *design load, required shear and flexural strengths, furnished flexural strength, and the shear strength criterion.*

Design Load.   The *design load* is determined by finding the critical combination of factored loads. A factored load is the product of the load factor and the load. For example, one type of load combination is calculated by

$$P_u \text{ (or } w_u) = 1.2(\text{DL}) + 1.6(\text{LL}) \tag{15–8}$$

where P_u (or w_u) is the design load, 1.2 and 1.6 are the load factors, and DL and LL stand for the dead load and the live load, respectively.

Required Shear and Flexural Strengths.   The design loads calculated above are used to determine the maximum shear V_u and the maximum moment M_u. These are called the *required shear strength* and the *required flexural strength*. For some typical cases, these values can be computed from formulas listed in Table 13–1. In other cases, the shear and moment diagrams may have to be sketched.

Furnished Flexural Strength.   The *furnished flexural strength* is the resistance bending strength of a beam calculated from

$$\phi_b M_n$$

where ϕ_b = the resistance factor for flexure = 0.90

M_n = nominal flexural strength

In this introduction to LRFD, we will assume that the beam has adequate lateral bracing for its compression flange and that localized buckling will not occur. The nominal flexural strength M_n is equal to the plastic moment M_p, which is developed when the entire cross-section has yielded and the plastic hinge is formed at the section (see Section 14–7). From Equation 14–15, the plastic moment can be calculated from

$$M_p = Z\sigma_y$$

where Z is the plastic section modulus with respect to the plastic neutral axis. For sections having a horizontal axis of symmetry (such as the W shapes or S shapes), the plastic neutral axis is located at the horizontal axis of symmetry. The plastic section modulus Z of W shapes and S shapes are listed in the appendix, Tables A–1 and A–2.

For flexural strength, the furnished flexural strength must be greater than the required flexural strength. This criterion can be expressed as

$$\phi_b M_n = 0.90 M_p = 0.90 Z\sigma_y \geq M_u$$

From which the minimum required plastic section modulus can be computed:

$$Z_{\text{req}} = \frac{M_u}{0.90\sigma_y} \tag{15–9}$$

The Shear Strength Criterion.   The furnished or design shear strength that a beam can resist is

$$\phi_v V_n$$

where ϕ_v = the resistance factor for shear = 0.90

V_n = nominal shear strength

The nominal shear strength of a bending member may be obtained from

$$V_n = 0.60\sigma_y A_w \qquad (15\text{--}10)$$

where A_w represents the area of the web of the shape. This equation is valid for all rolled W shapes made of steel with a yield stress of 65 ksi (450 MPa) or less.

The *shear strength criterion* can be expressed as

$$\phi_v V_n = 0.90 V_n \geq V_u \qquad (15\text{--}11)$$

The LRFD design of bending members involves the following steps:

Step 1. Calculate the design load from Equation 15–8.

Step 2. Determine the maximum shear V_u and the maximum moment M_u along the beam produced by the design load.

Step 3. Calculate the required plastic section modulus Z_{req} from Equation 15–9.

Step 4. Scan the list for the W shapes in Table A–1 and select the lightest W section that will provide the required value of Z_{req}.

Step 5. Check to see if the shear stress criterion expressed in Equation 15–11 is satisfied.

———— **EXAMPLE 15–10** ————

The simply supported girder in Fig. E15–10 is subjected to a uniformly distributed dead load of 1.1 kip/ft (which includes the estimated weight of the beam) and two symmetrically placed, concentrated loads of 8 kips each (of which 3 kips is the dead load and 5 kips is the live load). Assume that the compression flange is fully braced and that localized buckling will not occur. Using A36 steel and the LRFD method, select the lightest W shape for the beam.

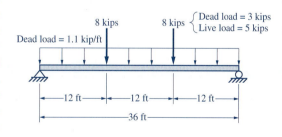

FIGURE E15–10

Solution. For the LRFD method, the following steps are used.

(1) The Design Load. From Equation 15–8, we find

$$w_u = 1.2(1.1 \text{ kip/ft}) = 1.32 \text{ kip/ft}$$

$$P_u = 1.2(3 \text{ kips}) + 1.6(5 \text{ kips}) = 11.6 \text{ kips}$$

(2) The Required Shear and Flexure Strengths. The maximum shear and moment produced by the design loads are

$$V_u = P_u + \frac{w_u L}{2}$$

$$= 11.6 \text{ kips} + \frac{(1.32 \text{ kip/ft})(36 \text{ ft})}{2}$$

$$= 35.4 \text{ kips}$$

$$M_u = P_u a + \frac{w_u L^2}{8}$$

$$= (11.6 \text{ kips})(12 \text{ ft}) + \frac{(1.32 \text{ kip/ft})(36 \text{ ft})^2}{8}$$

$$= 353 \text{ kip} \cdot \text{ft} = 4236 \text{ kip} \cdot \text{in.}$$

(3) The Required Plastic Section Modulus. From Equation 15–9, the minimum plastic section modulus required is

$$Z_{req} = \frac{M_u}{0.90\sigma_y} = \frac{4236 \text{ kip} \cdot \text{in.}}{0.90(36 \text{ kip/in.}^2)} = 131 \text{ in.}^3$$

(4) Select a Section. Scanning Table A–1(a) in the appendix, we select the lightest W shape: W21 × 62, which has a Z_x value of 144 in.^{3}.

(5) Check Shear Strength. From Equation 15–10, the nominal shear strength is

$$V_n = 0.60 \, \sigma_y A_w$$

$$= 0.60(36 \text{ ksi})(20.99 \text{ in.})(0.400 \text{ in.})$$

$$= 181 \text{ kips}$$

The furnished shear strength is

$$\phi_v V_n = 0.90(181 \text{ kips}) = 163 \text{ kips} > V_u = 35.4 \text{ kips}$$

So the shear strength is satisfactory.

Use W21 × 62   ⇐ **Ans.**

EXAMPLE 15–11

The overhanging beam in Fig. E15–11(1) is subjected to the loads shown. The uniform load includes an estimated weight of the beam. Assume that the compression flange is fully braced and that localized buckling will not occur. Using A36 steel and the LRFD method, select the lightest W shape for the beam. For A36 steel, $\sigma_y = 250$ MPa.

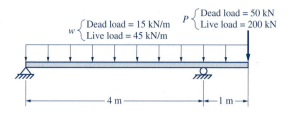

FIGURE E15–11(1)

Solution. For the LRFD method, the following steps are used.

(1) The Design Loads. From Equation 15–8, we find

$$w_u = 1.2(15 \text{ kN/m}) + 1.6(45 \text{ kN/m}) = 90 \text{ kN/m}$$
$$P_u = 1.2(50 \text{ kN}) + 1.6(200 \text{ kN}) = 380 \text{ kN}$$

(2) The Required Shear and Flexure Strengths. Refer to Fig. E15–11(2). Using the design loads, we sketch a new loading diagram. The shear and moment diagrams due to the design loads are sketched as shown. From these diagrams, we identify the maximum shear and the maximum moment as

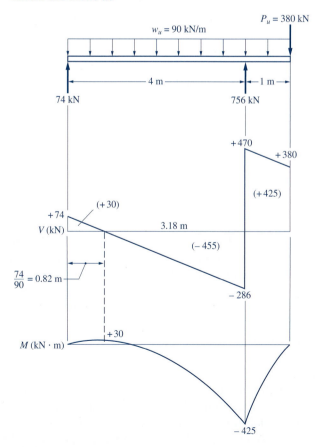

FIGURE E15–11(2)

$$V_u = 470 \text{ kN} \qquad M_u = 425 \text{ kN} \cdot \text{m}$$

(3) The Required Plastic Section Modulus.   From Equation 15–9, the minimum plastic section modulus required is

$$Z_{\text{req}} = \frac{M_u}{0.90\sigma_y} = \frac{425 \text{ kN} \cdot \text{m}}{0.90(250\,000 \text{ kN/m}^2)}$$

$$= 1.89 \times 10^{-3} \text{ m}^3$$

(4) Select a Section.   Scanning Table A–1(b) in the appendix, we select the lightest W shape W460 × 0.88, which provides $Z_x = 2.02 \times 10^{-3}$ m³.

(5) Check Shear Strength.   From Equation 15–10, the nominal shear strength is

$$V_n = 0.60 \ \sigma_y A_w$$

$$= 0.60(250\,000 \text{ kN/m}^2)(0.463 \text{ m} \times 0.0105 \text{ m})$$

$$= 729 \text{ kN}$$

The furnished shear strength is

$$\phi_v V_n = 0.90(729 \text{ kN}) = 656 \text{ kN} > V_u = 470 \text{ kN}$$

So the shear strength is satisfactory.

Use W460 × 0.88   ⇐ Ans.

which is W18 × 60 in the U.S. customary system.

15–7
SUMMARY

Beam Materials and Sections.   The most commonly used materials for beams are steel and timber. The properties of hot-rolled structural shapes are tabulated in the AISC manual. Selected W shapes (wide-flange sections) and S shapes (I-beams) in both U.S. customary and SI units are listed in the appendix, Tables A–1 and A–2. Most beams made of timber have rectangular cross-sections. Properties of representative timber sections are listed in Table A–6 in the appendix for both U.S. customary and SI units.

Lateral Bracing.   A beam must be braced properly to prevent unstable sidesway and the buckling of the compression flange. The full value of the allowable bending stress can be used only when the compression flange is braced adequately. Under the most favorable conditions, including adequate lateral bracings, the AISC specification prescribes an allowable value of $0.66\sigma_y$ for flexural stress and an allowable value of $0.4\sigma_y$ for shear stress.

Steel Beam Design. Steel beams are selected based on the minimum required section modulus, calculated from

$$S_{req} = \frac{M_{max}}{\sigma_{allow}} \tag{15-1}$$

The beam selected is checked for shear stress by requiring that the average web shear stress calculated from the formula below be smaller than the allowable shear stress.

$$\tau_{avg} = \frac{V_{max}}{dt_w} \tag{15-2}$$

Timber Beam Design. The rectangular timber beam is selected based on both the minimum required section modulus and the minimum required cross-sectional area. These values are calculated from the following equations:

$$S_{req} = \frac{M_{max}}{\sigma_{allow}} \tag{15-3}$$

$$A_{req} = \frac{1.5V_{max}}{\tau_{allow}} \tag{15-4}$$

Shear Connector Design. The nails or bolts that connect the components of a built-up beam must be spaced at a proper pitch so that the connectors carry the longitudinal shear force adequately on the contact surface. The maximum pitch can be computed from

$$P_{max} = \frac{(F_s)_{allow}}{q} \tag{15-7}$$

where q is the *shear flow* representing the shear force in a longitudinal section per unit length of beam. The shear flow (q) can be computed from

$$q = \frac{VQ}{I} \tag{15-6}$$

$(F_s)_{allow}$ is the maximum allowable shear force of the nail(s) or bolt(s) in each spacing.

LRFD Design. LRFD is a plastic design approach for ductile metal such as construction grade steel. The design uses load factors and resistance factors as factors of safety to guard against failure. *Load factors* are used to increase the loads in various types of load combinations. *Resistance factors* are used to decrease the ultimate plastic strength (resistance) of the member. For the introductory discussion in this chapter, the compression flange of the beam is assumed to be fully braced. A proper section is selected based on the plastic section modulus, computed from

$$Z_{req} = \frac{M_u}{0.90\sigma_y} \tag{15-9}$$

Shear stress is checked by the criterion that the furnished shear strength of the web be greater than the required shear strength.

PROBLEMS

Section 15–3   Design of Steel Beams

15–1   Select the lightest wide-flange steel section for a simple beam of 20-ft span that will carry a uniform load of 4 kips/ft. Use A36 steel and assume that the beam is supported laterally for its entire length.

15–2   Select the lightest wide-flange steel beam (in SI designation) for a simple beam of 6-m span that will carry a uniform load of 60 kN/m. Use A36 steel and assume that the beam is supported laterally for its entire length.

15–3   Select the lightest wide-flange steel girder for a simple span of 15 ft subjected to a concentrated load of 10 kips at the midspan. Use A36 steel and assume that the beam is supported laterally for its entire length.

15–4   Select the lightest wide-flange steel girder (in SI designation) for a simple span of 5 m that will support a concentrated load of 45 kN at the midspan. Use A36 steel and assume that the beam is supported laterally for its entire length.

15–5   A simply supported beam of 20-ft span is subjected to a uniformly distributed load and a concentrated load, as shown in Fig. P15–5. Select the lightest W shape using A36 steel and assume that the beam is supported laterally for its entire length.

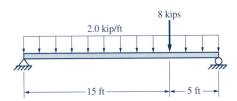

FIGURE P15–5

15–6 to 15–9 Refer to Figs. P15–6 to P15–9. For each of the beams subjected to the loadings shown, the weight of the beam is already included in the uniform load. Select the lightest wide-flange steel shape using A36 steel. Assume that the beam is supported laterally for its entire length.

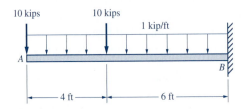

FIGURE P15–6

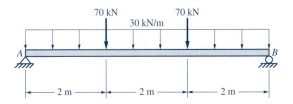

FIGURE P15–7

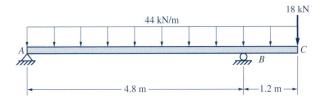

FIGURE P15–8

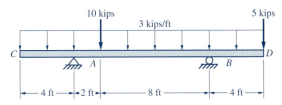

FIGURE P15–9

15–10 The floor plan of an interior bay of a commercial building has the layout sketched in Fig. P15–10. The system is to support a 6-in.-thick, reinforced concrete slab. The live load on the floor is 145 psf. Use A36 steel and assume that both the beams and the girders are supported laterally for their entire length. Select the lightest W shapes for (*a*) the beam and (*b*) the girder.

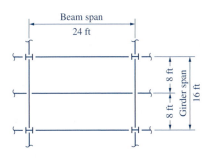

FIGURE P15–10

Section 15–4 Design of Timber Beams

15–11 Select a solid, rectangular, Eastern hemlock beam section for a 16-ft simple span carrying a superimposed uniform load of 800 lb/ft.

15–12 Select a solid, rectangular, Eastern hemlock beam section (in SI designation) for a 5-m simple span carrying a superimposed uniform load of 12 kN/m.

15–13 Select the lightest oak beam of rectangular section (in SI designation) for a simple beam of 4-m span subjected to a concentrated load of 45 kN at the midspan. Assume that the allowable flexural stress is 13 000 kPa and the allowable shear stress parallel to the grain is 1000 kPa.

15–14 Select the lightest, rectangular
California redwood section for
an overhanging beam with the
superimposed uniform load
shown in Fig. P15–14.

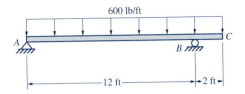

FIGURE P15–14

15–15 Select the lightest, rectangular Douglas fir
section for the cantilever beam subjected
to the loading shown in Fig. P15–15.

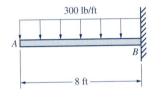

FIGURE P15–15

15–16 Select the lightest, rectangular
Southern pine section for the simply supported girder subjected to
the loading shown in Fig. P15–16.

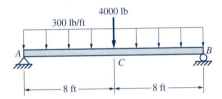

FIGURE P15–16

15–17 Select the lightest, rectangular
California redwood section (in SI
designation) for the overhanging
beam subjected to the loading
shown in Fig. P15–17.

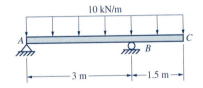

FIGURE P15–17

15–18 Select the lightest, rectangular
Southern pine section for the
overhanging beam subjected to
the loading shown in Fig. P15–18.
The weight of the beam is already
included in the uniform load.

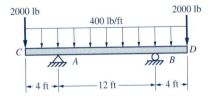

FIGURE P15–18

15–19 Refer to Fig. P15–19. The joists supporting the floor in a warehouse have a
span of 18 ft and a spacing of 1 ft 6 in. on centers, as shown. The floor must
support a live load of 145 psf plus 25 psf for the dead weight of the subfloor
and the top floor. Select the lightest Southern pine joist size.

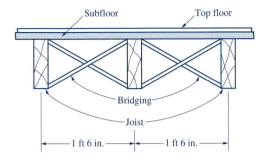

FIGURE P15–19

Section 15–5   Design of Shear Connectors

15–20 Four 25-mm × 150-mm full-size boards are nailed together to form a boxed section, as shown in Fig. P15–20. The beam is subjected to a constant shear force of 4.5 kN. Determine the pitch of the nails if the allowable shear force per nail is 800 N.

FIGURE P15–20

15–21 Two "two by four" Southern pine studs with a dressed size of $1\frac{1}{2}$ in. × $3\frac{1}{2}$ in. are nailed together to make a beam section, as shown in Fig. P15–21. Two 0.192-in.-diameter nails are used in each spacing at a pitch of $2\frac{1}{2}$ in. The allowable shear stress for the nails is 12 000 psi. Determine the maximum uniform load that the beam can carry over a 4-ft span. Neglect the weight of the beam.

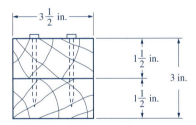

FIGURE P15–21

15–22 Three 50-mm × 100-mm full-size Douglas fir planks are fastened by 5-mm-diameter bolts spaced at a pitch of 40 mm, as shown in Fig. P15–22. The allowable shear stress for the bolts is 100 MPa. Determine the maximum concentrated load that can be applied at the midpoint of a 3-m simple span. Neglect the weight of the beam.

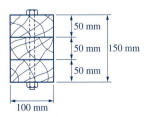

FIGURE P15–22

15–23 Refer to Fig. P15–23. A simple beam of 12-ft span is subjected to a uniform load of 1200 lb/ft. The cross-section of the beam is fabricated by nailing two pieces of full-size, $\frac{1}{2}$-in. × 24-in. plywood to two full-size, 2-in. × 8-in. timber planks, as shown. If the screws are $\frac{1}{8}$ in. in diameter, with an allowable shear stress of 8000 psi, determine the pitch of the screws near the supports.

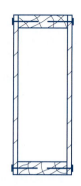

FIGURE P15–23

15–24 The built-up beam in Fig. P15–24 consists of cover plates fastened with $\frac{3}{4}$-in.-diameter rivets to the flange of W21 × 62 section. If the beam is subjected to a constant shear force of 120 kips and the allowable shear stress in the rivets is 15 ksi, determine the pitch of the rivets.

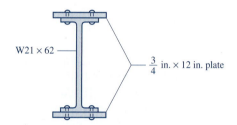

W21 × 62

$\frac{3}{4}$ in. × 12 in. plate

FIGURE P15–24

15–25 Refer to Fig. P15–25. The overhanging beam subjected to a concentrated load is fabricated from two steel channels and two steel cover plates, as shown. If the bolts are $\frac{1}{2}$ in. in diameter, with an allowable shear stress of 15 ksi, determine the pitches of the bolts in segments AB and BC.

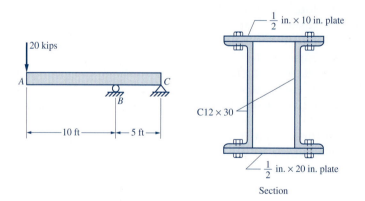

20 kips

A

B

C

10 ft

5 ft

$\frac{1}{2}$ in. × 10 in. plate

C12 × 30

$\frac{1}{2}$ in. × 20 in. plate

Section

FIGURE P15–25

15–26 Refer to Fig. P15–26. The simple beam of 16-ft span is subjected to a vertical load of 50 kips at the midspan, and its own weight is considered to be a uniform load. The cross-section of the beam is fabricated by riveting a

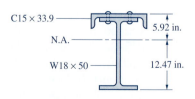

C15 × 33.9

N.A.

W18 × 50

5.92 in.

12.47 in.

FIGURE P15–26

steel channel to a wide-flange section with 1-in.-diameter rivets. The moment of inertia of the section about the neutral axis is 1250 in.[4]. If the allowable shear stress of the rivets is 15 ksi, determine the pitch of the rivets.

Section 15–6   Load and Resistance Factor Design

15–27 A simply supported girder of 20-ft span is subjected to a uniformly distributed dead load of 1 kip/ft (which includes the estimated weight of the beam) and a uniformly distributed live load of 2 kips/ft. Assume that the compression flange is braced fully and that localized buckling will not occur. Using A36 steel and the LRFD method, select the lightest W shape for the beam.

15–28 A simply supported girder of 6-m span is subjected to a uniformly distributed dead load of 20 kN/m (which includes the estimated weight of the beam) and a concentrated load at the midspan of 45 kN (of which 15 kN is the dead load and 30 kN is the live load). Assume that the compression flange is braced fully and that localized buckling will not occur. Using A36 steel and the LRFD method, select the lightest W shape (in SI designation) for the beam.

15–29 to 15–32 See Figs. P15–29 to P15–32. Each of the beams is subjected to the service loads shown. The weight of the beam is already included in the dead load. Select the lightest W shape using A36 steel and the LRFD method. Assume that the compression flange is fully braced and that localized buckling will not occur.

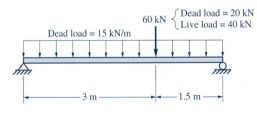

FIGURE P15–29

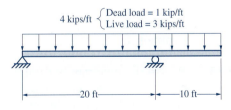

FIGURE P15–30

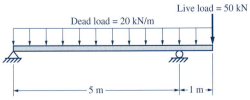

FIGURE P15–31

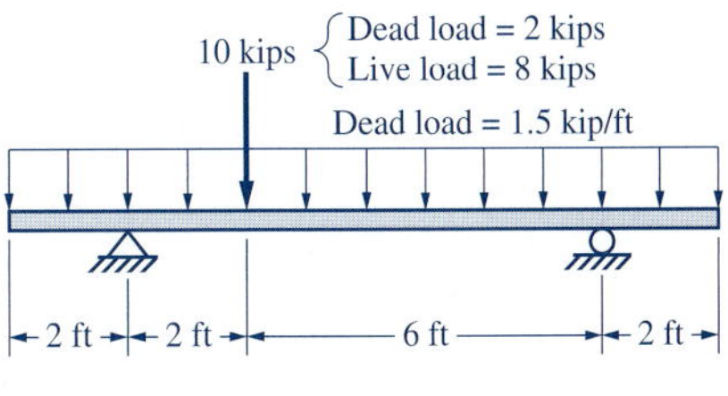

FIGURE P15–32

Computer
Program
Assignments

For each of the following problems, write a computer program using an appropriate programming language with which you are most familiar. Make the program user friendly by incorporating plenty of comments and input prompts so that the user will understand the input data to be entered and the limitations of their values. The output should include the data entered and the computed results, and they must be well labeled to identify each quantity. If a tabulated format is used, a proper heading must be included at the top of the table. Do not limit the program to any specific unit system. Indicate the consistent U.S. customary or SI units that can be used.

C15–1 Write a computer program that can be used to select the required sizes from the appendix, Table A–6, for a simply supported timber beam carrying a uniform load. The user input should include (1) the span length L, (2) the load w, and (3) the allowable flexural and shear stresses. Assume that the dressed sizes are $\frac{1}{2}$ in. less than the nominal size. Use the dressed size to compute the section properties. For the weight of the beam, assume a specific weight of timber to be 40 pcf. Use this program to solve (*a*) Example 15–5, (*b*) Problem 15–11, (*c*) Problem 15–12, and (*d*) Problem 15–19.

C15–2 Revise the program in Computer Program Assignment C15–1 to accommodate the more general loading condition shown in Fig. C15–2. The user input should include (1) the span length L; (2) the uniform load w; (3) the concentrated load P_1 applied at midspan; (4) the symmetrically applied, concentrated force P_2 and the distance a; and (5) the allowable flexural and shear stresses. Use this program to solve (*a*) Problem 15–13 and (*b*) Problem 15–16.

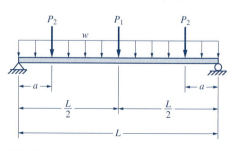

FIGURE C15–2

CHAPTER 16

DEFLECTIONS OF BEAMS

16–1
INTRODUCTION

When a beam is subjected to transverse loads that produce bending moment, the beam will deflect from its unloaded position, as shown in Fig. 16–1. The vertical displacement of a point on a horizontal beam is called the *deflection* of the beam at the point. If the maximum flexural stress in the beam is within the elastic limit of the beam material, the beam undergoes elastic deflection. In this chapter, we will discuss the computation of elastic deflections of beams.

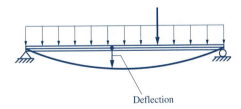

Deflection

FIGURE 16–1

A beam properly designed for strength may not be rigid enough and may produce excessive deflection. Excessive deflections are to be avoided for many reasons. For example, the deflection of beams to which a plastered ceiling is attached must be limited so that the beam will not crack the plaster; power-transmission shafts carrying gears must be rigid enough to ensure proper meshing of the gear teeth. Consideration of deflections of a beam is also needed in solving statically indeterminate beam problems, which will be considered in the next chapter. In addition, a sagging beam visible to the naked eye may arouse suspicion about the safety of a structure or the competency of the designer. Therefore, the maximum deflection of beams must be kept within a certain allowable limit. Allowable deflections are generally established by design specifications or codes, which are based on past acceptable practice and judgment.

Many methods can be used to calculate beam deflections. Two basic methods presented in this chapter are the *formula method* and the *moment-area method*. The formula method is relatively easy to apply. Formulas for deflections in prismatic beams of various loading and supporting conditions are available from manuals or handbooks; a few simple cases are tabulated in this chapter. Numerical values of the quantities involved can be substituted into these formulas to compute beam deflections. The formula method is the most commonly used method in engineering practice; however, it has its limitations. The formulas are available only for beams of uniform cross-section and for limited loading and supporting conditions. The moment-area method is versatile, and it is applicable to more general loading and supporting conditions.

The development of beam deflection methods is based on the following assumptions:

1. The beam is homogeneous and obeys Hooke's law, having an equal modulus of elasticity in tension and compression, and the bending is within the elastic range.
2. The beam has a vertical plane of symmetry on which the loads and reactions act.
3. The deflections are small and are caused by bending only. The deflection due to shear is negligible.

16–2
RELATIONSHIP BETWEEN CURVATURE AND BENDING MOMENT

Consider a beam segment bent into a concave upward curvature due to a positive bending moment, as shown in Fig. 16–2. Two adjacent sections, *ab* and *cd*, remain planar and normal to the axis of the beam. The point of intersection O of lines *ab* and *cd* is called the *center of curvature*, and the distance ρ (the Greek lowercase letter rho) from O to the beam axis is called the *radius of curvature*. The length of fiber *mn* along the axis of the beam remains unchanged, while that of fiber *bd* is elongated. Let the angle $\Delta\theta$ at O be measured in radians. Then the length of *mn* is $\rho\Delta\theta$, and the length of *bd* is $(\rho + c)\Delta\theta$. By definition, the strain ϵ_{max} of the extreme fiber *bd* is

$$\epsilon_{max} = \frac{bd - mn}{mn} = \frac{(\rho + c)\Delta\theta - \rho\Delta\theta}{\rho\Delta\theta} = \frac{c}{\rho}$$

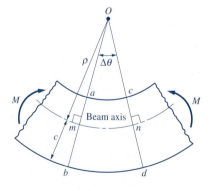

FIGURE 16–2

For elastic bending, the stress is proportional to strain, i.e., $\sigma_{max} = E\epsilon_{max}$. Thus, we have

$$\sigma_{max} = \frac{Ec}{\rho} \tag{16-1}$$

Equating this expression with the flexure formula $\sigma_{max} = Mc/I$ and solving for ρ, we get

$$\rho = \frac{EI}{M} \tag{16-2}$$

In this equation, the product EI is a constant. It states that *the radius of curvature* ρ *of a beam at any section varies inversely with the bending moment M at that section.* This relationship is fundamental to developing methods for determining beam deflections. If M is zero, the radius of curvature becomes infinity and the elastic curve is a straight line. As the value of M increases, the radius of curvature decreases and the smallest value of the radius of curvature occurs where the moment is a maximum. If the bending moment is constant over a segment of a beam, the radius of curvature is also constant and the elastic curve is a circular arc.

────── **EXAMPLE 16–1** ──────

The steel blade for a band saw is 1.0 mm thick and runs on drums that are 300 mm in diameter. Find the flexural stress in the blade when it goes over the drum. Use $E = 210$ GPa for steel.

Solution. As the blade goes over the drum, it deforms into a circular arc of radius $\rho = 150$ mm. From Equation 16–1, the maximum flexural stress in the blade is

$$\sigma_{max} = \frac{Ec}{\rho} = \frac{(210 \times 10^3 \text{ MN/m}^2)\left[\frac{1}{2}(0.001 \text{ m})\right]}{0.150 \text{ m}}$$

$$= 700 \text{ MPa} \qquad \qquad \Leftarrow \textbf{Ans.}$$

────── **EXAMPLE 16–2** ──────

A 3×12 Douglas fir beam is subjected to a constant bending moment of 70 000 lb · in. about its strong axis. Determine the radius of curvature of the beam if $E = 1900$ ksi.

Solution. From the appendix, Table A–6(a), we find the moment of inertia of a 3×12 section as

$$I = 297 \text{ in.}^4$$

From Equation 16–2, the radius of curvature of the beam is

$$\rho = \frac{EI}{M} = \frac{(1\ 900\ 000\ \text{lb/in.}^2)(297\ \text{in.}^4)}{70\ 000\ \text{lb} \cdot \text{in.}} = 8060\ \text{in.}$$

$$= 670\ \text{ft} \qquad \Leftarrow \textbf{Ans.}$$

16–3
THE FORMULA METHOD

Formulas derived by integral calculus are listed in Table 16–1. The table includes formulas for cantilever beams (cases 1 through 4) and simple beams (cases 5 through 8) for several loading conditions. In each case, formulas are provided for the following:

1. The maximum deflection along the beam.
2. The slope at the free end of cantilever beams or at the ends of simple beams.
3. The general deflection equation from which the deflection at any point along the beam can be calculated.

Note that the table lists only the absolute value of deflections or slopes. Generally the deflection is downward due to downward loads, and in most of the cases the direction of deflection can be determined readily by inspection.

From the formulas, we see that deflection and slope are inversely proportional to the product EI, where E is the modulus of elasticity of the material and I is the moment of inertia of the cross-sectional area. The quantity EI is called the *flexural rigidity* of the beam. It is an indication of the resistance of the beam to deflection. A beam with a greater value of EI is stiffer and will deflect less.

The deflections and slopes computed from the formulas are usually very small. A very small angle in radians and its tangent function are nearly the same value. When numerical values are substituted into the formulas, the units must be consistent, as is the case with any physical equations. The resulting unit for deflection is the unit of length and that of slope is in radians, which is unitless.

──── EXAMPLE 16–3 ────

A W14 × 34 steel beam has a 15-ft simple span. The beam carries a uniform load of 3 kips/ft, including its own weight. The allowable deflection is 1/360 of the span length. Determine whether the beam is satisfactory for deflection.

Solution. From case 7 in Table 16–1, the maximum deflection at the midspan for a simple beam subjected to a uniform load can be calculated from the formula

$$\delta_{max} = \frac{5wL^4}{384EI}$$

TABLE 16–1   Beam Deflection Formulas

Beam Loading and Deflection	Maximum Deflection	Slope at End(s)	Deflection Equations
1	$\delta_{max} = \dfrac{PL^3}{3EI}$	$\theta_B = \dfrac{PL^2}{2EI}$	$\delta = \dfrac{Px^2}{6EI}(3L - x)$
2	$\delta_{max} = \dfrac{Pa^2}{6EI}(3L - a)$	$\theta_B = \dfrac{Pa^2}{2EI}$	$\delta_{AC} = \dfrac{Px^2}{6EI}(3a - x)$ $\delta_{CB} = \dfrac{Pa^2}{6EI}(3x - a)$
3	$\delta_{max} = \dfrac{wL^4}{8EI}$	$\theta_B = \dfrac{wL^3}{6EI}$	$\delta = \dfrac{wx^2}{24EI}(x^2 - 4Lx + 6L^2)$
4	$\delta_{max} = \dfrac{ML^2}{2EI}$	$\theta_B = \dfrac{ML}{EI}$	$\delta = \dfrac{Mx^2}{2EI}$
5	$\delta_{max} = \dfrac{PL^3}{48EI}$	$\theta_A = \theta_B = \dfrac{PL^2}{16EI}$	$\delta_{AC} = \dfrac{Px}{48EI}(3L^2 - 4x^2)$
6	For $a > b$: $\delta_{max} = \dfrac{Pb(L^2 - b^2)^{3/2}}{9\sqrt{3}EIL}$ at $x_m = \sqrt{\dfrac{L^2 - b^2}{3}}$	$\theta_A = \dfrac{Pb(L^2 - b^2)}{6EIL}$ $\theta_B = \dfrac{Pa(L^2 - a^2)}{6EIL}$	$\delta_{AC} = \dfrac{Pbx}{6EIL}(L^2 - x^2 - b^2)$ $\delta_{CB} = \dfrac{Pb}{6EIL}\left[\dfrac{L}{b}(x - a)^3 + (L^2 - b^2)x - x^3\right]$
7	$\delta_{max} = \dfrac{5wL^4}{384EI}$	$\theta_A = \theta_B = \dfrac{wL^3}{24EI}$	$\delta = \dfrac{wx}{24EI}(L^3 + x^3 - 2Lx^2)$
8	$\delta_{max} = \dfrac{ML^2}{9\sqrt{3}EI}$ at $x_m = \dfrac{L}{\sqrt{3}}$	$\theta_A = \dfrac{ML}{6EI}$ $\theta_B = \dfrac{ML}{3EI}$	$\delta = \dfrac{Mx}{6EIL}(L^2 - x^2)$

For steel, E = 30 000 ksi [from the appendix, Table A–7(a)] and for W14 × 34, I = 340 in.[4] [also from the appendix, Table A–1(a)]. Thus, the constant value of EI is

$$EI = (30\ 000\ \text{kips/in.}^2)(340\ \text{in.}^4)$$
$$= 10.2 \times 10^6\ \text{kip} \cdot \text{in.}^2 = 70\ 800\ \text{kip} \cdot \text{ft}^2$$

Substituting this value for EI and w = 3 kips/ft and L = 15 ft into the formula for the maximum deflection, we get

$$\delta_{max} = \frac{5wL^4}{384EI} = \frac{5(3\ \text{kips/ft})(15\ \text{ft})^4}{384(70\ 800\ \text{kip} \cdot \text{ft})} = 0.0279\ \text{ft}$$

The allowable deflection is

$$\delta_{allow} = \frac{L}{360} = \frac{15\ \text{ft}}{360} = 0.0417\ \text{ft}$$

The maximum deflection is less than the allowable value; thus, the beam is satisfactory for deflection.   ⇐ **Ans.**

EXAMPLE 16–4

A cantilever beam made of Southern pine has a rectangular 150 × 460 section. The beam carries a concentrated load P = 10 kN, as shown in Fig. E16–4. Determine the deflections and slopes at the free end B and at point C, where the load is applied.

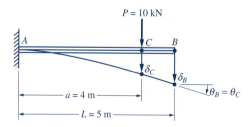

FIGURE E16–4

    Solution. The modulus of elasticity of Southern pine [from Table A–7(b)] is

$$E = 12\ \text{GPa} = 12 \times 10^6\ \text{kN/m}$$

and the moment of inertia [from Table A–6(b)] for a 150 × 460 rectangular section is

$$I = 1022 \times 10^{-6}\ \text{m}^4$$

Thus, the constant value of EI is

$$EI = (12 \times 10^6 \text{ kN/m}^2)(1022 \times 10^{-6} \text{ m}^4)$$

$$= 12\ 260 \text{ kN} \cdot \text{m}^2$$

Substituting this value of EI and $P = 10$ kN, $L = 5$ m, and $a = 4$ m into the formula from case 2 in Table 16–1, we find

$$\delta_B = \delta_{max} = \frac{Pa^2}{6EI}(3L - a)$$

$$= \frac{(10 \text{ kN})(4 \text{ m})^2}{6(12\ 260 \text{ kN} \cdot \text{m}^2)}[3(5 \text{ m}) - 4 \text{ m}]$$

$$= 0.0239 \text{ m} = 23.9 \text{ mm} \qquad\qquad\qquad \Leftarrow \textbf{Ans.}$$

For the deflection at point C, we must use $x = 4$ m in the general deflection equation from case 2 of Table 16–1. Thus,

$$\delta_C = \frac{Px^2}{6EI}(3a - x)$$

$$= \frac{(10 \text{ kN})(4 \text{ m})^2}{6(12\ 260 \text{ kN} \cdot \text{m}^2)}[3(4 \text{ m}) - 4 \text{ m}]$$

$$= 0.0174 \text{ m} = 17.4 \text{ mm} \qquad\qquad\qquad \Leftarrow \textbf{Ans.}$$

Since there is no bending moment along CB, this part must be a straight line. Hence, the slope at B and C must be the same. We have

$$\theta_B = \theta_C = \frac{Pa^2}{2EI} = \frac{(10 \text{ kN})(4 \text{ m})^2}{2(12\ 260 \text{ kN} \cdot \text{m}^2)} = 0.00653 \text{ rad}$$

$$= 0.374° \qquad\qquad\qquad \Leftarrow \textbf{Ans.}$$

Note that, since the angle is small, the angle in radians is nearly the same as the tangent function of the angle; that is,

$$\tan (0.00653 \text{ rad}) = 0.00653$$

16–4
THE METHOD OF SUPERPOSITION

From the formulas in Table 16–1, we see that the load is related linearly to deflection and that deflection is small so that the geometry of the beam does not change significantly. As a result, the deflections due to several loadings may be superimposed; that is, the beam deflection caused by each load can be calculated separately. The algebraic sum of the deflections gives the resultant deflection due to all the loads acting simultaneously. The method consists of using the deflection formulas tabulated in Table 16–1 for each loading condition. Various beam deflection problems can be solved this way, as illustrated in the following examples.

━━━ **EXAMPLE 16–5** ━━━

Find the maximum deflection at the free end C of the cantilever beam in Fig. E16–5. The beam is made of white oak having a 100×200 nominal size rectangular section.

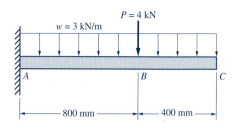

FIGURE E16–5

Solution. From Table A–7(b), the modulus of elasticity for white oak is $E = 12$ GPa. From Table A–6(b), the moment of inertia of a 100×200 section is $I = 46.2 \times 10^{-6}$ m^4. Thus, the flexural rigidity is

$$EI = (12 \times 10^6 \text{ kN/m}^2)(46.2 \times 10^{-6} \text{ m}^4)$$

$$= 554 \text{ kN} \cdot \text{m}^2$$

The cantilever is subjected to a concentrated load and a uniform load; this loading can be considered as the superposition of cases 2 and 3 from Table 16–1. Therefore, the maximum deflection at the free end C of the beam is the sum of the maximum deflections of cases 2 and 3. Thus,

$$\delta_{max} = [\delta_{max}]_P + [\delta_{max}]_w$$

$$= \frac{Pa^2}{6EI}(3L - a) + \frac{wL^4}{8EI}$$

$$= \frac{(4 \text{ kN})(0.8 \text{ m})^2}{6(554 \text{ kN} \cdot \text{m}^2)}[3(1.2 \text{ m}) - 0.8 \text{ m}] + \frac{(3 \text{ kN/m})(1.2 \text{ m})^4}{8(554 \text{ kN} \cdot \text{m}^2)}$$

$$= 0.00216 \text{ m} + 0.00140 \text{ m} = +0.00356 \text{ m}$$

$$= 3.56 \text{ mm} \downarrow \qquad\qquad\qquad\qquad\qquad \Leftarrow \textbf{Ans.}$$

━━━ **EXAMPLE 16–6** ━━━

Refer to Fig. E16–6(1). Use the method of superposition to find the maximum deflection at the free end C of the cantilever beam of wide-flange W12 × 40 steel section due to the uniform load shown.

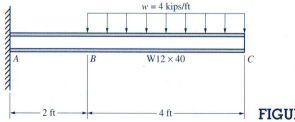

FIGURE E16–6(1)

Solution. From Table A–1(a), the moment of inertia for W12 × 40 is $I = 310$ in.^{4}. The modulus of elasticity for steel is $E = 30 \times 10^3$ ksi. The flexural rigidity is

$$EI = (30 \times 10^3 \text{ kips/in.}^2)(310 \text{ in.}^4)$$
$$= 9.30 \times 10^6 \text{ kip} \cdot \text{in.}^2 = 6.46 \times 10^4 \text{ kip} \cdot \text{ft}^2$$

The uniform load over part of the span may be considered as the superposition of the two uniform loads, one acting downward over the entire length [see Fig. E16–6(2)], the other one acting upward over length AB, as shown in Fig. E16–6(3).

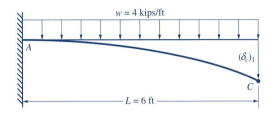

FIGURE E16–6(2)

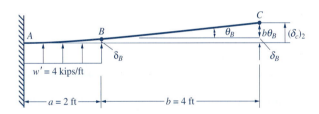

FIGURE E16–6(3)

Deflection Due to w over AC. The deflection at the free end of the beam due to a downward uniform load w over the entire length can be calculated from the formula in case 3:

$$(\delta_C)_1 = \frac{wL^4}{8EI} \quad \downarrow$$

Deflection Due to w over AB. Due to the upward uniform load w over length AB, the upward deflection and the slope at B from case 3 in Table 16–1 are

$$\delta_B = \frac{wa^4}{8EI} \quad \uparrow$$

$$\theta_B = \frac{wa^3}{6EI}$$

Since θ_B in radians is very small, $\theta_B \approx \tan\theta_B$. The BC part of the beam is a straight line. The additional upward deflection at the free end C due to the slope θ_B is

$$b\tan\theta_B = b\theta_B = b\frac{wa^3}{6EI}$$

Thus, the deflection at the free end C due to w over AB is

$$(\delta_C)_2 = \left(\frac{wa^4}{8EI} + b\frac{wa^3}{6EI}\right) \uparrow$$

Deflection Due to the Combined Loading. The given loading is the superposition of the two loadings above. Therefore, the deflection at the free end C is the superposition of the deflection at C due to the above loadings:

$$\delta_C = (\delta_C)_1 - (\delta_C)_2 = \frac{wL^4}{8EI} - \left(\frac{wa^4}{8EI} + b\frac{wa^3}{6EI}\right)$$

$$= \frac{w}{EI}\left(\frac{L^4}{8} - \frac{a^4}{8} - \frac{ba^3}{6}\right)$$

$$= \frac{4}{6.46\times10^4}\left[\frac{(6)^4}{8} - \frac{(2)^4}{8} - \frac{(4)(2)^3}{6}\right]$$

$$= +0.009\,58 \text{ ft} = +0.115 \text{ in.} \downarrow \qquad\qquad \Leftarrow \textbf{Ans.}$$

EXAMPLE 16–7

Use the method of superposition to find the deflection at the midspan C and point D of the simply supported beam in Fig. E16–7. The beam is made of Southern pine having a 100×200 nominal size rectangular section.

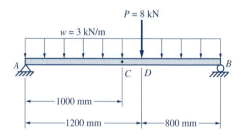

FIGURE E16–7

Solution. From Table A–7(b), the modulus of elasticity for southern pine is $E = 12$ GPa. From Table A–6(b), the moment of inertia of 100×200 section is $I = 46.2 \times 10^{-6}$ m^4. The flexural rigidity is

$$EI = (12 \times 10^6 \text{ kN/m}^2)(46.2 \times 10^{-6} \text{ m}^4)$$

$$= 554 \text{ kN} \cdot \text{m}^2$$

Deflection at C and D Due to the Uniform Load. Using the formulas in case 7 of Table 16–1, we find

$$(\delta_C)_1 = \frac{5wL^4}{384EI} = \frac{5(3)(2.0)^4}{384(554)} = 0.00113 \text{ m}$$

$$(\delta_D)_1 = \frac{wx}{24EI}(L^3 + x^3 - 2Lx^2)$$

$$= \frac{(3)(1.2)}{24(554)}[(2.0)^3 + (1.2)^3 - 2(2.0)(1.2)^2]$$

$$= 0.00107 \text{ m}$$

Deflection at C and D Due to the Concentrated Load. Using the formulas in case 6 of Table 16–1, we find

$$(\delta_C)_2 = \frac{Pbx}{6EIL}(L^2 - x^2 - b^2)$$

$$= \frac{(8)(0.8)(1.0)}{6(554)(2.0)}[(2.0)^2 - (1.0)^2 - (0.8)^2]$$

$$= 0.00227 \text{ m}$$

$$(\delta_D)_2 = \frac{Pbx}{6EIL}(L^2 - x^2 - b^2)$$

$$= \frac{(8)(0.8)(1.2)}{6(554)(2.0)}[(2.0)^2 - (1.2)^2 - (0.8)^2]$$

$$= 0.00222 \text{ m}$$

Deflection at C and D Due to the Combined Loading. The deflections at C and D found above are added. We find

$$\delta_C = (\delta_C)_1 + (\delta_C)_2$$

$$= 0.00113 \text{ m} + 0.00227 \text{ m} = 0.00340 \text{ m}$$

$$= 3.40 \text{ mm} \downarrow \qquad \qquad \Leftarrow \textbf{Ans.}$$

$$\delta_D = (\delta_D)_1 + (\delta_D)_2$$

$$= 0.00107 \text{ m} + 0.00222 \text{ m} = 0.00329 \text{ m}$$

$$= 3.29 \text{ mm} \downarrow \qquad \qquad \Leftarrow \textbf{Ans.}$$

EXAMPLE 16–8

Refer to Fig. E16–8(1). Use the method of superposition to determine the deflections at the midspan C and the free end D of the overhanging beam due to the loads shown. The beam has a W10 × 22 steel section.

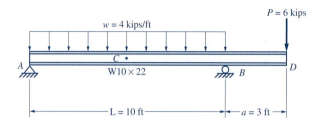

FIGURE E16–8(1)

Solution. From Table A–1(a) in the appendix, for a W10 × 22 steel section, the moment of inertia is $I = 118$ in.^{4}. The flexural rigidity EI is

$$EI = (30 \times 10^3 \text{ kips/in.}^2)(118 \text{ in.}^4)$$

$$= 3.54 \times 10^6 \text{ kip} \cdot \text{in.}^2 = 2.46 \times 10^4 \text{ kip} \cdot \text{ft}^2$$

The deflections at C and D include the following three parts.

(1) Due to the Uniform Load on the Simple Span. Refer to Fig. E16–8(2). The deflections at the midspan C and the free end D due to the uniform load w in segment AB are

$$(\delta_C)_w = \frac{5wL^4}{384EI} = \frac{5(4 \text{ kips/ft})(10 \text{ ft})^4}{384(2.46 \times 10^4 \text{ kip} \cdot \text{ft}^2)}$$

$$= 0.0212 \text{ ft} \downarrow$$

$$(\delta_D)_w = a(\theta_B)_w = a\frac{wL^3}{24EI}$$

$$= (3 \text{ ft})\frac{(4 \text{ kips/ft})(10 \text{ ft})^3}{24(2.46 \times 10^4 \text{ kip} \cdot \text{ft}^2)}$$

$$= 0.0203 \text{ ft} \uparrow$$

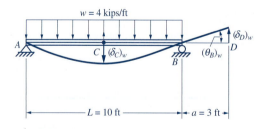

FIGURE E16–8(2)

(2) Due to the Couple M. The load P on the overhang produces a couple $M = Pa = 18$ kip · ft at the support B, as shown in Fig. E16–8(3). The deflections at C and D due to the couple M are

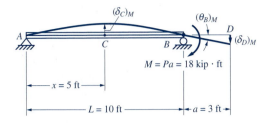

FIGURE E16–8(3)

$$(\delta_C)_M = \frac{Mx}{6EIL}[L^2 - x^2]$$

$$= \frac{(18 \text{ kip} \cdot \text{ft})(5 \text{ft})}{6(2.46 \times 10^4 \text{ kip} \cdot \text{ft}^2)(10 \text{ft})} [(10 \text{ft})^2 - (5 \text{ft})^2]$$

$$= 0.0046 \text{ ft} \uparrow$$

$$(\delta_D)_M = a(\theta_B)_M = a\frac{ML}{3EI} = (3 \text{ ft}) \frac{(18 \text{ kip} \cdot \text{ft})(10 \text{ft})}{3(2.46 \times 10^4 \text{ kip} \cdot \text{ft}^2)}$$

$$= 0.0073 \text{ ft} \downarrow$$

(3) Due to the Load P on the Overhang. The overhang is considered a cantilever, as shown in Fig. E16–8(4). The deflection at D due to P is

$$(\delta_D)_P = \frac{Pa^3}{3EI} = \frac{(6 \text{ kips})(3 \text{ ft})^3}{3(2.46 \times 10^4 \text{ kip} \cdot \text{ft}^2)}$$

$$= 0.0022 \text{ ft} \downarrow$$

FIGURE E16–8(4)

Due to the Combined Loading. The deflections at C and D due to the combined loading are the algebraic sum of the deflections due to each load acting separately. We find

$$\delta_C\left(+\downarrow\right) = (\delta_C)_w + (\delta_C)_M$$

$$= 0.0212 \text{ ft} - 0.0046 \text{ ft}$$

$$= +0.0166 \text{ ft} = 0.199 \text{ in.} \downarrow \qquad\qquad \Leftarrow \textbf{Ans.}$$

$$\delta_D\left(+\uparrow\right) = (\delta_D)_w + (\delta_D)_M + (\delta_D)_P$$

$$= 0.0203 \text{ ft} - 0.0073 \text{ ft} - 0.0022 \text{ ft}$$

$$= +0.0108 \text{ ft} = 0.130 \text{ in.} \uparrow \qquad\qquad \Leftarrow \textbf{Ans.}$$

16–5
THE MOMENT-AREA METHOD

The moment-area method provides a semigraphical technique for finding the slope and deflection of a beam. To develop the moment-area
method, we will consider a simple beam subjected to an arbitrary loading,
as shown in Fig. 16–3a. The moment diagram and the deflection curve of the
beam are sketched as shown in Fig. 16–3b and c. Two points, p and q, on the
deflection curve are at an incremental distance Δx apart. The radius of curvature of the deflection curve at these points is denoted by ρ. The incremental angle $\Delta \theta$, measured in radians between the radii of curvature at
points p and q, is equal to

$$\Delta \theta = \frac{\Delta x}{\rho}$$

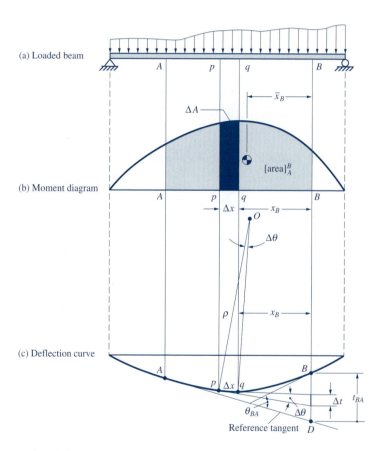

FIGURE 16–3

Substituting $\rho = EI/M$ from Equation 16–2 into the above equation, we get

$$\Delta \theta = \frac{1}{EI} M \, \Delta x = \frac{1}{EI} \, \Delta A$$

where $\Delta A = M \Delta x$ is the area of the moment diagram between p and q. In
Fig. 16–3c, tangents drawn to the deflection curve at p and q also make an

angle $\Delta\theta$. The angle in radians, measured between tangents at any two points A and B on the deflection curve (designated by θ_{BA} in Fig. 16–3c, can be obtained by summing all the incremental angles $\Delta\theta$ from A to B:

$$\theta_{BA} = \Sigma \, \Delta\theta = \Sigma \, \frac{1}{EI} \, \Delta A$$

If the flexural rigidity EI is a constant, then the constant factor $1/EI$ can be factored from each term and brought out of the summation symbol. Thus, we have

$$\theta_{BA} = \frac{1}{EI} \Sigma \, \Delta A = \frac{1}{EI} A_{AB} \qquad (16\text{–}3)$$

where A_{AB} represents the area of the moment diagram between points A and B, shown shaded in Fig. 16–3b. This equation forms the basis for the first moment-area theorem, stated below:

*   **Theorem 1.**   *The angle θ_{AB} between the tangents at two points A and B on the deflection curve is equal to the area of the moment diagram between the two points divided by EI.*

The second moment-area theorem is based on the tangential deviation. The incremental tangential deviation Δt, corresponding to the incremental length Δx, is shown in Fig. 16–3c. Since the slopes and deflections are assumed to be small, the incremental value Δt is nearly equal to the incremental angle $\Delta\theta$ multiplied by the horizontal distance x_B of the element to point B:

$$\Delta t = x_B \, \Delta\theta$$

Summing up the incremental values Δt of all the elements from A to B, we obtain the *tangential deviation* t_{BA}, the vertical distance from point B to the reference tangent at A (the distance BD in Fig. 16–3c). Thus,

$$t_{BA} = \Sigma \, \Delta t = \Sigma \, x_B \, \Delta\theta$$

Substituting $\Delta\theta = \Delta A / EI$ into the above equation and factoring the constant factor EI, we obtain

$$t_{BA} = \frac{1}{EI} \Sigma \, x_B \, \Delta A$$

Since the centroid of an area can be determined from $A\bar{x} = \Sigma x \, \Delta A$, we write

$$t_{BA} = \frac{1}{EI} A_{AB} \, \bar{x}_B \qquad (16\text{–}4)$$

where A_{AB} is the area of the moment diagram between points A and B and $\bar{x}_B$ is the horizontal distance from the centroid of this area to point B. This equation leads to the second moment-area theorem, stated as follows:

*   **Theorem 2.**   *The tangential deviation t_{BA}, measured from point B on the deflection curve vertically to the reference tangent drawn at point A, is equal to the first moment of the area of the moment diagram between A and B with respect to point B divided by EI.*

Although θ_{AB} is equal to θ_{BA}, t_{AB} may *not* be equal to t_{BA} in general. The tangential deviation t_{AB} measured from point A on the deflection curve vertically to the reference tangent drawn at point B, as shown in Fig. 16–4, is

$$t_{AB} = \frac{1}{EI} A_{AB} \, \overline{x}_A \qquad (16\text{–}5)$$

where A_{AB} is the area of the moment diagram between A and B, and $\overline{x}_A$ is the horizontal distance from the centroid of the area to point A. Since $\overline{x}_B$ and $\overline{x}_A$ are not equal in general, the deviations t_{BA} and t_{AB} in Equations 16–4 and 16–5 may not be equal.

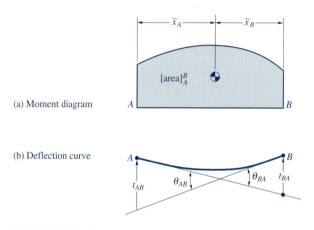

(a) Moment diagram

(b) Deflection curve

FIGURE 16–4

If a beam has a known horizontal tangent, the horizontal tangent can be used as a reference tangent and the angles or deviations from the horizontal tangent will usually yield the simplest solution. In the case of a cantilever beam, the tangent at the fixed end is always horizontal. In the case of a simply supported beam subjected to symmetric loading, the tangent to the deflection curve at the midspan is horizontal.

Application of the moment-area method requires computing areas of the moment diagram and the location of its centroid. If the moment diagram consists of simple geometric shapes or can be divided into several simple areas, then the moment-area method is convenient to apply. Properties of simple areas found in Table 7–2 (printed also inside the front cover) are useful in calculating the area and the location of the centroid, which are needed in the application of the moment-area method. The area of the moment diagram above the baseline (where the moment is positive) is treated as positive, and the area of the moment diagram below the baseline (where the moment is negative) is treated as negative.

In the case of combined or nonsymmetrical loadings, the moment diagram of a beam can become more complex and difficult to compute. In such cases, an alternative method of constructing the moment diagram, called *moment diagram by parts*, is preferable. This method will be presented in the next section. Applications of the moment-area method in the general loading condition will be discussed in Section 16–7.

─────── **EXAMPLE 16–9** ───────

Find the expressions for the maximum slope and the maximum deflection of the cantilever beam due to a concentrated load applied at the free end, as shown in Fig. E16–9(1). Express the results in terms of the load P, span length L, and flexural rigidity EI.

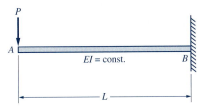

FIGURE E16–9(1)

Solution. The maximum slope and the maximum deflection both occur at the free end A. Since the tangent at the fixed end B is in the horizontal direction, the maximum slope θ_{max} is equal to θ_{AB}, and the maximum deflection δ_{max} is equal to the tangential deviation t_{AB}, as shown in the deflection curve of Fig. E16–9(2).

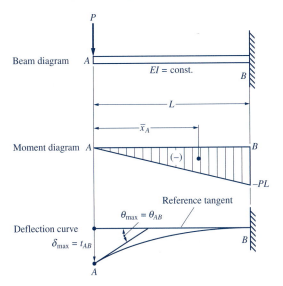

FIGURE E16–9(2)

The moment diagram is a triangle; thus, the area of the moment diagram between A and B is

$$A_{AB} = \frac{bh}{2} = \frac{(L)(-PL)}{2} = -\frac{PL^2}{2}$$

and the centroid of the area from A is

$$\bar{x}_A = \frac{2L}{3}$$

Using the first moment-area theorem, we write

$$\theta_{AB} = \frac{1}{EI} A_{AB} = \frac{1}{EI}\left[-\frac{PL^2}{2}\right] = -\frac{PL^2}{2EI}$$

$$\theta_{max} = \frac{PL^2}{2EI} \qquad\qquad \Leftarrow \textbf{Ans.}$$

Using the second moment-area theorem, we write

$$t_{AB} = \frac{1}{EI} A_{AB}\,\bar{x}_A = \frac{1}{EI}\left[-\frac{PL^2}{2}\right]\left(\frac{2L}{3}\right) = -\frac{PL^3}{3EI}$$

$$\delta_{max} = \frac{PL^3}{3EI} \; \Big\downarrow \qquad\qquad \Leftarrow \textbf{Ans.}$$

These deflection formulas are the same as those listed in Table 16–1. The negative sign in the expression for t_{AB} means that point A on the elastic curve is *below* the tangent at B.

EXAMPLE 16–10

Refer to Fig. E16–10(1). Find the maximum slope and maximum deflection of the simple beam due to the uniform load w over the entire span length L. The beam has a constant flexural rigidity EI.

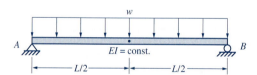

FIGURE E16–10(1)

Solution. The moment diagram and the deflection curve are sketched in Fig. E16–10(2). Due to symmetry, the deflection curve has a horizontal tangent at the midspan. Hence, we see that

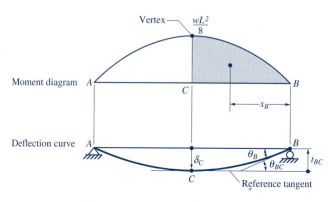

FIGURE E16–10(2)

$$\theta_B = \theta_{BC} \qquad \delta_C = t_{BC}$$

The moment diagram between C and B is a semiparabolic area. Thus, from Table 7–2,

$$A_{BC} = \frac{2}{3}bh = \frac{2}{3}\left(\frac{L}{2}\right)\left(\frac{wL^2}{8}\right) = \frac{wL^3}{24}$$

$$\bar{x}_B = \frac{5}{8}b = \frac{5}{8}\left(\frac{L}{2}\right) = \frac{5L}{16}$$

Using the first moment-area theorem, we write

$$\theta_{BC} = \frac{1}{EI}A_{BC} = +\frac{wL^3}{24EI}$$

$$\theta_{\max} = \frac{wL^3}{24EI} \qquad\qquad \Leftarrow \textbf{Ans.}$$

Using the second moment-area theorem, we write

$$t_{BC} = \frac{1}{EI}A_{BC}\bar{x}_B = \frac{1}{EI}\left(\frac{wL^3}{24}\right)\left(\frac{5L}{16}\right) = +\frac{5wL^4}{384EI}$$

$$\delta_{\max} = \frac{5wL^4}{384EI} \; \Big\downarrow \qquad\qquad \Leftarrow \textbf{Ans.}$$

These deflection formulas are the same as those listed in Table 16–1. The positive sign in the expression for t_{BC} indicates that point B on the elastic curve is *above* the tangent at C.

─────────────────────────────

EXAMPLE 16–11

The overhanging beam shown in Fig. E16–11(1) is a standard weight steel pipe of 4-in. nominal diameter. Find (a) the slope at the free ends, (b) the deflection at the midspan, and (c) the deflection at the free ends due to the uniform loads symmetrically placed as shown.

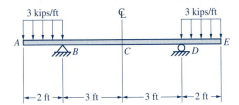

FIGURE E16–11(1)

Solution. From the free-body diagram of the beam in Fig. E16–11(2), we determine the reactions and then draw the shear and moment diagrams. Since the beam and its loading are symmetrical with respect to the midspan C, the tangent at C is horizontal and is used as the reference tangent, as shown on the deflection curve of the beam.

For a 4-in. standard weight steel pipe, $E = 30\ 000$ ksi [from Table A–7(a)], and $I = 7.23$ in.^{3} [from Table A–5(a)]. Thus, the flexural rigidity of the beam is

$$EI = (30\ 000\ \text{kips/ft}^2)(7.23\ \text{in.}^3) = 216\ 900\ \text{kip} \cdot \text{in.}^2$$

$$= 1506\ \text{kip} \cdot \text{ft}^2$$

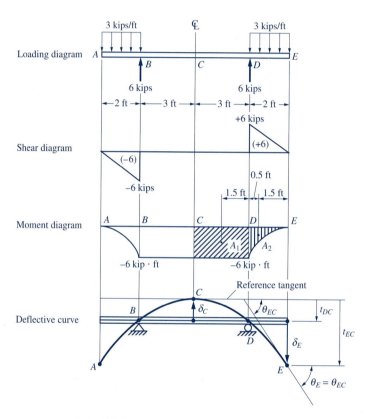

FIGURE E16–11(2)

(*a*) *Slope at E.* Due to symmetry, the absolute value of the slope at the free ends is the same. From the deflection curve shown, the slope θ_E is equal to the angle θ_{EC} between the tangent at E and the reference tangent at C. Using the first moment-area theorem, we find

$$A_1 = (3)(-6) = -18\ \text{kip} \cdot \text{ft}^2$$

$$A_2 = \frac{1}{3}(2)(-6) = -4\ \text{kip} \cdot \text{ft}^2$$

$$\theta_{EC} = \frac{1}{EI} A_{EC} = \frac{1}{EI}(A_1 + A_2)$$

$$= \frac{1}{1506}(-18 - 4) = -0.0146\ \text{rad}$$

$$\theta_E = |\theta_{EC}| = 0.0146\ \text{rad} = 0.837° \qquad \Leftarrow \textbf{Ans.}$$

(b) *Deflection at C.* From the elastic curve, we see that the deflection δ_C at the midspan is equal to the tangential deviation t_{DC} measured from D to the reference tangent. Using the second moment-area theorem, we find

$$t_{DC} = \frac{1}{EI}A_1(\bar{x}_D)_1 = \frac{1}{1506}(-18)(1.5)$$

$$= -0.0179 \text{ ft} = -0.215 \text{ in.}$$

$$\delta_C = |t_{DC}| = 0.215 \text{ in.} \uparrow \qquad\qquad \Leftarrow \textbf{Ans.}$$

(c) *Deflection at E.* Using the second moment-area theorem, the tangential deviation t_{EC} measured from E to the reference tangent is

$$t_{EC} = \frac{1}{EI}[A_1(\bar{x}_E)_1 + A_2(\bar{x}_E)_2]$$

$$= \frac{1}{1506}[(-18)(3.5) + (-4)(1.5)]$$

$$= -0.0458 \text{ ft} = -0.550 \text{ in.}$$

From the elastic curve, we see that

$$\delta_E = |t_{EC}| - |t_{DC}| = 0.335 \text{ in.} \downarrow \qquad\qquad \Leftarrow \textbf{Ans.}$$

16–6
MOMENT DIAGRAM BY PARTS

To simplify the computations involved in the moment-area method, it is better to draw the moment diagrams *by parts*; that is, draw the moment diagram of each load and each reaction separately. Then, by the method of superposition, the algebraic sum of all the moment diagrams drawn separately will be equivalent to the moment diagram drawn in the usual manner.

Three fundamental cantilever loadings and their respective moment diagrams are shown in Fig. 16–5. The areas of the moment diagram produced by each loading are in the shape of a rectangle, a triangle, or a parabolic spandrel. Properties of these areas are listed in Table 7–2.

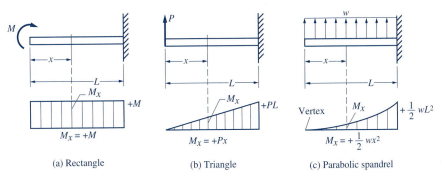

(a) Rectangle $\qquad\qquad$ (b) Triangle $\qquad\qquad$ (c) Parabolic spandrel

FIGURE 16–5

The method of plotting moment diagrams by parts can be applied to beams of any support conditions. After the reactions of a beam are determined, they are treated as applied loads, and an imaginary fixed support can be placed at any section along the beam. The moment diagram can be drawn by parts toward that section assumed to be fixed using the formulas in Fig. 16–5. The following examples illustrate the method.

───── **EXAMPLE 16–12** ─────

Draw the moment diagram by parts for the cantilever beam shown in Fig. E16–12(1).

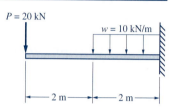

FIGURE E16–12(1)

Solution. Since the given beam is a cantilever beam, the formulas in Fig. 16–5 can be applied directly by considering each load separately, as shown in Fig. E16–12(2). The combined moment diagram is the superposition of the two moment diagrams drawn separately, as shown.

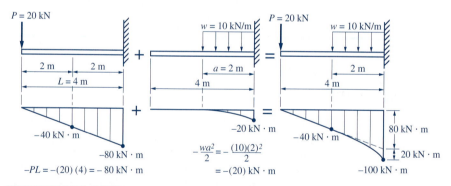

FIGURE E16–12(2)

───── **EXAMPLE 16–13** ─────

Draw the moment diagram by parts for the uniformly loaded simple beam in Fig. E16–13(1).

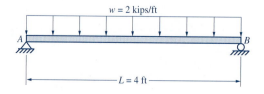

FIGURE E16–13(1)

Solution. Due to symmetry, the reaction at each support is equal to one-half of the total load; thus

$$R_A = R_B = \frac{1}{2}wL = \frac{1}{2}(2 \text{ kips/ft})(4 \text{ ft}) = 4 \text{ kips}$$

The moment diagram can be drawn by parts toward any section desired. If we choose to draw the moment diagram toward end B, the beam is treated as if it were a cantilever beam with a fixed end at B. The moment diagram drawn by parts is shown in Fig. E16–13(2).

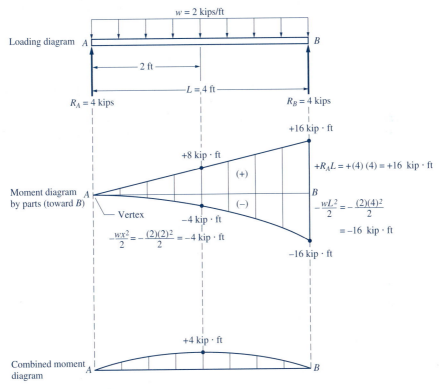

FIGURE E16–13(2)

Now we see why it is preferable to draw the moment diagram by parts. If we want to find the area and the centroid location of the entire moment diagram, then the moment diagram either by the usual method or by parts can be used. But if we need to find the areas and the locations of the centroids of part of the moment diagram between A and some intermediate section, then the moment diagram drawn by parts must be used.

EXAMPLE 16–14

Refer to Fig. E16–14(1). Draw the moment diagram by parts for the over-hanging beam subjected to the loading shown.

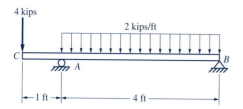

FIGURE E16–14(1)

Solution. The reactions must be determined first by considering the equilibrium condition of the entire beam. The results are indicated in the loading diagram in Fig. E16–14(2). The moment diagram can be drawn toward any section desired. If we choose to draw toward section A, we consider section A as if it were fixed and draw the moment diagrams by parts toward section A as shown. These diagrams can be superimposed into the combined moment diagram shown, which is the same as that obtained with the usual method. We see that the areas in the combined moment diagram are very difficult to compute. The areas and their centroids in the moment diagrams by parts can be computed readily by using the formulas in Table 7–2.

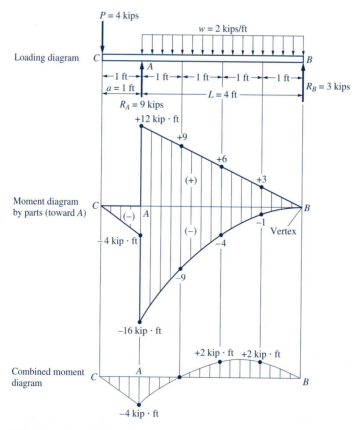

FIGURE E16–14(2)

16–7
APPLICATIONS OF THE MOMENT-AREA METHOD

With the moment diagram drawn by parts, we can now use the moment-area method to find the deflections in beams with a general loading condition. The method of superposition will be used to sum up the areas of the moment diagram and the first moments of the areas about a point. When performing the summation, treat the areas corresponding to positive moments as positive, and treat the areas corresponding to the negative moments as negative.

With the moment-area method, a carefully sketched deflection curve is always necessary and a convenient reference tangent must be chosen. The angles between the tangents and the tangential deviations can be obtained by applying the moment-area theorems. Deflections of points along the beam can be obtained by additional consideration of the geometry of the deflection curve, as illustrated in the following examples.

—————— **EXAMPLE 16–15** ——————

Find the maximum deflection of the cantilever beam of wide-flange W12 × 40 steel section due to the uniform load shown in Fig. E16–15(1).

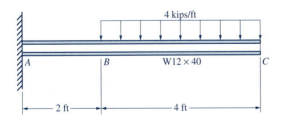

FIGURE E16–15(1)

Solution. From Table A–1(a), for a W12 × 40 steel section, $I = 310$ in.^{4}, and for steel, $E = 30 \times 10^3$ ksi. Then we have

$$EI = (30 \times 10^3 \text{ kips/in.}^2)(310 \text{ in.}^4)$$
$$= 9.30 \times 10^6 \text{ kip} \cdot \text{in.}^2 = 6.46 \times 10^4 \text{ kip} \cdot \text{ft}^2$$

The given load may be considered as the superposition of two uniform loads, as shown in Fig. E16–15(2). The moment diagram is drawn by parts toward the fixed end A, as shown. The elastic curve is drawn. The horizontal tangent at the fixed end will be used as the reference tangent.

Using the second moment-area theorem and units of kip and ft, we write

$$t_{CA} = \frac{1}{EI} A_{CA} \bar{x}_C = \frac{1}{EI}[A_1 \bar{x}_1 + A_2 \bar{x}_2]$$

$$= \frac{1}{6.46 \times 10^4}\left[\frac{1}{3}(2)(+8)(5.5) + \frac{1}{3}(6)(-72)(4.5)\right]$$

$$= -9.58 \times 10^{-3} \text{ ft} = -0.115 \text{ in.}$$

$$\delta_{max} = t_{CA} = 0.115 \text{ in.} \downarrow \qquad\qquad \Leftarrow \textbf{Ans.}$$

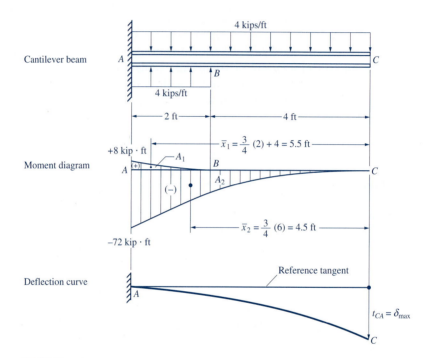

FIGURE E16–15(2)

EXAMPLE 16–16

Refer to Fig. E16–16(1). Determine the deflections at points C, D, and E of the beam due to a uniform load over part of the span, as shown. The beam is a solid, rectangular section of Southern pine of 100×200 nominal size.

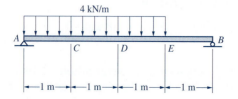

FIGURE E16–16(1)

Solution. For Southern pine, the modulus of elasticity [from Table A–7(b)] is $E = 12$ GPa. From Table A–6(b), the moment of inertia of 100×200 section is $I = 46.2 \times 10^{-6}$ m⁴. Then the flexural rigidity is

$$EI = (12 \times 10^6 \text{ kN/m}^2)(46.2 \times 10^{-6} \text{ m}^4)$$

$$= 554 \text{ kN} \cdot \text{m}^2$$

The reactions of the beam are determined first by considering the equilibrium of the entire beam. The results are shown in the loading diagram in Fig. E16–16(2). Then the moment diagram of the beam is drawn by parts toward A, as shown. The deflection curve of the beam is sketched next. The tangent to the deflection curve at B is chosen as the reference tangent.

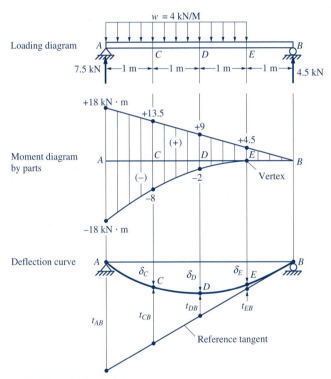

FIGURE E16–16(2)

Using the second moment-area theorem and units of kN and m, the tangential deviations of points A, C, D, and E from the reference tangent are calculated in the following:

$$t_{AB} = \frac{1}{EI} A_{AB} \bar{x}_A = \frac{1}{554}\left[\frac{1}{2}(4)(+18)\left(\frac{4}{3}\right) + \frac{1}{3}(3)(-18)\left(\frac{3}{4}\right)\right]$$

$$= 0.0623 \text{ m}$$

$$t_{CB} = \frac{1}{EI} A_{CB} \bar{x}_C = \frac{1}{554}\left[\frac{1}{2}(3)(+13.5)\left(\frac{3}{3}\right) + \frac{1}{3}(2)(-8)\left(\frac{2}{4}\right)\right]$$

$$= 0.0317 \text{ m}$$

$$t_{DB} = \frac{1}{EI}A_{DB}\bar{x}_D = \frac{1}{554}\left[\frac{1}{2}(2)(+9)\left(\frac{2}{3}\right) + \frac{1}{3}(1)(-2)\left(\frac{1}{4}\right)\right]$$

$$= 0.0105 \text{ m}$$

$$t_{EB} = \frac{1}{EI}A_{EB}\bar{x}_E = \frac{1}{554}\left[\frac{1}{2}(1)(+4.5)\left(\frac{1}{3}\right)\right]$$

$$= 0.0014 \text{ m}$$

From the similar triangles in the elastic curve, we can set up the following proportion:

$$\frac{\delta_C + t_{CB}}{3} = \frac{t_{AB}}{4}$$

From which we get

$$\delta_C = \frac{3}{4}t_{AB} - t_{CB} = \frac{3}{4}(0.0623) - 0.0317 = 0.0150 \text{ m}$$

$$= 15.0 \text{ mm} \;\downarrow \qquad\qquad\qquad \Leftarrow \textbf{Ans.}$$

Similarly,

$$\delta_D = \frac{1}{2}t_{AB} - t_{DB} = \frac{1}{2}(0.0623) - 0.0105 = 0.0207 \text{ m}$$

$$= 20.7 \text{ mm} \;\downarrow \qquad\qquad\qquad \Leftarrow \textbf{Ans.}$$

$$\delta_E = \frac{1}{4}t_{AB} - t_{EB} = \frac{1}{4}(0.0623) - 0.0014 = 0.0142 \text{ m}$$

$$= 14.2 \text{ mm} \;\downarrow \qquad\qquad\qquad \Leftarrow \textbf{Ans.}$$

EXAMPLE 16–17

Refer to Fig. E16–17(1). Determine the deflection at points C and D of the overhanging beam of wide flange W10 × 22 steel section due to the loads shown.

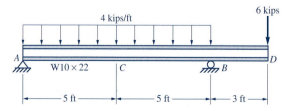

FIGURE E16–17(1)

Solution. From Table A–1(a), for a W10 × 22 steel section, $I = 118$ in.^{4}. The flexural rigidity EI is

$$EI = (30 \times 10^3 \text{ kips/in.}^2)(118 \text{ in.}^4)$$
$$= 3.54 \times 10^6 \text{ kip} \cdot \text{in.}^2 = 2.46 \times 10^4 \text{ kip} \cdot \text{ft}^2$$

The reactions of the beam are first determined by considering the equilibrium of the entire beam. The results are indicated in the loading diagram of the beam in Fig. E16–17(2). Then the moment diagram of the beam is drawn by parts toward B as shown. The deflection curve of the beam is also sketched. The tangent to the deflection curve at A is chosen as the reference tangent.

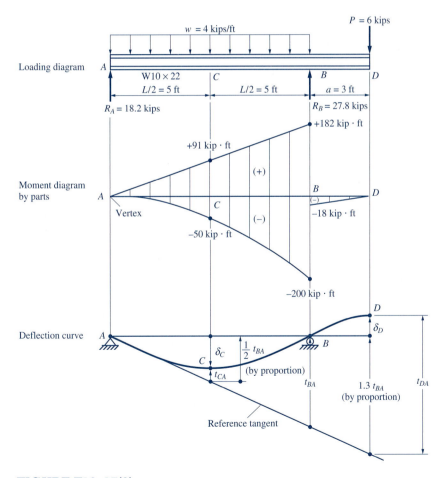

FIGURE E16–17(2)

Using the second moment-area theorem and the units kips and feet, we obtain the following tangential deviations of points C, B, and D from the reference tangent:

$$t_{CA} = \frac{1}{EI} A_{CA} \bar{x}_C$$

$$= \frac{1}{2.46 \times 10^4} \left[\frac{1}{2}(5)(91)\left(\frac{5}{3}\right) + \frac{1}{3}(5)(-50)\left(\frac{5}{4}\right) \right]$$

$$= +0.0112 \text{ ft}$$

$$t_{BA} = \frac{1}{EI} A_{BA} \bar{x}_B$$

$$= \frac{1}{2.46 \times 10^4} \left[\frac{1}{2}(10)(182)\left(\frac{10}{3}\right) + \frac{1}{3}(10)(-200)\left(\frac{10}{4}\right) \right]$$

$$= +0.055\ 56 \text{ ft}$$

$$t_{DA} = \frac{1}{EI} A_{DA} \bar{x}_D$$

$$= \frac{1}{2.46 \times 10^4} \left[\frac{1}{2}(10)(182)\left(3 + \frac{10}{3}\right) + \frac{1}{3}(10)(-200)\left(3 + \frac{10}{4}\right) + \frac{1}{2}(3)(-18)\left(\frac{2}{3} \times 3\right) \right]$$

$$= +0.0830 \text{ ft}$$

From the geometry of the elastic curve, the deflections of points C and D are

$$\delta_C = \frac{1}{2} t_{BA} - t_{CA} = \frac{1}{2}(+0.055\ 56 \text{ ft}) - (+0.0112 \text{ ft})$$

$$= +0.0166 \text{ ft}$$

$$= +0.199 \text{ in.} \downarrow \qquad\qquad \Leftarrow \textbf{Ans.}$$

$$\delta_D = t_{DA} - 1.3 t_{BA} = (+0.0830 \text{ ft}) - 1.3(+0.055\ 56 \text{ ft})$$

$$= +0.0108 \text{ ft}$$

$$= +0.130 \text{ in.} \uparrow \qquad\qquad \Leftarrow \textbf{Ans.}$$

The positive signs for δ_C and δ_D indicate that points C and D are deflected in the assumed direction shown in the deflection curve. It is hard to tell whether the deflection of point D is upward or downward in the beginning. The deflection δ_D has been assumed to be upward. If the computed value is positive, δ_D is upward as assumed; if the computed value is negative, δ_D must be downward.

16–8
SUMMARY

Beam Deflection. The vertical displacement of a point on a horizontal beam is called the *deflection* of the beam at the point. The maximum deflection of beams must be kept within a certain allowable limit. Allowable deflections are generally established by design codes.

Relationship Between Curvature and Bending Moment. In a beam subjected to pure bending, the maximum flexural stress in the extreme fibers is inversely proportional to the radius of curvature of the elastic curve, as expressed in the following equation:

$$\sigma_{max} = \frac{Ec}{\rho} \qquad (16\text{–}1)$$

The radius of curvature ρ of a beam at any section is directly proportional to the *elastic rigidity EI* and inversely proportional to the bending moment M, as given by the equation:

$$\rho = \frac{EI}{M} \qquad (16\text{–}2)$$

The Formula Method. This method is based on the formulas available from manuals or handbooks. Table 16–1 lists the deflection formulas for some simple types of loading. Using the *principle of superposition*, the formula method may be used to compute beam deflections due to combined loading conditions or deflections of overhanging beams.

The Moment-Area Method. This method provides a semigraphical technique for finding the slope and deflection of a beam subjected to a general loading condition. It involves the area and the first moment of the area of the moment diagram, as stated in the following two theorems:

Theorem 1. *The angle θ_{AB} between the tangents at two points A and B on the deflection curve is equal to the area of the moment diagram between the two points divided by EI.*

Theorem 2. *The tangential deviation t_{BA}, measured from point B vertically to the reference tangent drawn at point A, is equal to the first moment of the area of the moment diagram between A and B with respect to point B divided by EI.*

Moment Diagram by Parts. To simplify the computations involved in the moment-area method, an alternative method of drawing a moment diagram can be used. The reactions of the beam are determined first. Any section along the beam can be regarded as fixed. The moment diagram of each load can be drawn toward that section. Each part of the moment diagram is a simple area and can be computed easily.

PROBLEMS

Section 16–2   Relationship Between Curvature and Bending Moment

16–1   A steel rod of $\frac{1}{8}$-in. diameter is bent into a loop of 10-ft diameter. Compute the maximum flexural stress in the rod.

16–2   A copper rod of 3-mm diameter is bent into a loop of 3-m diameter. Compute the maximum flexural stress in the rod.

16–3   A $\frac{1}{32}$-in.-thick steel blade is wrapped around a drum of 3-ft diameter. Compute the maximum flexural stress in the blade.

16–4   A 2-mm diameter copper wire is wound into a coil. Determine the minimum diameter of the coil that the wire can be wound around if the allowable flexural stress is 60 MPa.

16–5   A $\frac{1}{16}$-in.-diameter steel wire is wound into a coil. Determine the minimum diameter of the coil if the allowable flexural stress is 24 ksi.

16–6   A 10-mm-diameter steel rod is bent by a bending moment of 20 N · m. Compute the radius of curvature of the rod.

Section 16–3   The Formula Method

16–7   A W16 × 36 steel section is used in a 30-ft simple span. Compute the maximum deflection due to a concentrated load of 12 kips at the midspan.

16–8   Rework Problem 16–7. Assume that a uniform load of 0.4 kip/ft is applied to the entire span.

16–9   A 250 × 360 rectangular California redwood section is used in a 5-m simple span. Compute the maximum deflection due to a uniform load of 4 kN/m.

16–10   Rework Problem 16–9. Assume that a concentrated load of 20 kN is applied to the midspan.

16–11   A 150 × 360 rectangular Southern pine section is used in a 3-m cantilever span. Compute the maximum deflection and the maximum slope due to a uniform load of 15 kN/m.

16–12   A 4 × 10 rectangular Southern pine section is used in a 10-ft cantilever span. Compute the deflections at the quarter points due to a uniform load of 300 lb/ft.

16–13   A W18 × 60 steel section is used in a 25-ft simple span. Determine the maximum allowable uniform load w that the beam can carry if the allowable flexural stress is 24 ksi, the allowable shear stress is 15 ksi, and the allowable deflection is $\frac{1}{360}$ of the span length.

16–14 A W410 × 0.83 steel section is used in a 10-m simple span. Determine the maximum allowable uniform load w that the beam can carry if the allowable flexural stress is 165 MPa, the allowable shear stress is 100 MPa, and the allowable deflection is $\frac{1}{360}$ of the span length.

Section 16–4 The Method of Superposition

16–15 See Fig. P16–15. A 3 × 6 rectangular Southern pine section is used in an 8-ft cantilever span subjected to the loads shown. Compute the deflections at point A.

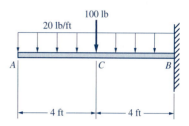

FIGURE P16–15

16–16 See Fig. P16–16. A W410 × 1.30 steel section is used in a 3-m cantilever span subjected to the loads shown. Compute the deflections at points B and C.

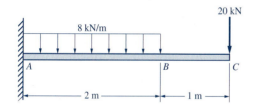

FIGURE P16–16

16–17 See Fig. P16–17. An 80 × 250 rectangular white oak section is used in a 4-m cantilever span subjected to the uniform load shown. Compute the deflections at points B and C due to the load.

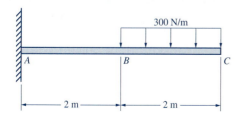

FIGURE P16–17

16–18 A W16 × 36 steel section is used in a 21-ft simple span. Determine the maximum deflection in the beam due to two concentrated loads of 15 kips each and its own weight. The concentrated loads are applied at the third points of the span.

16–19 A 100×360 rectangular Douglas fir beam is used in a 6-m simple span. Determine the maximum deflection in the beam due to two concentrated loads of 5 kN each and its own weight. The concentrated loads are applied at the third points of the span.

16–20 to 16–25 See Figs. P16–20 to P16–25. Determine the deflection at the midspan C of each simply supported beam subjected to the loads shown. The flexural rigidity of each beam is indicated.

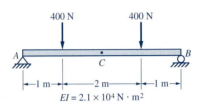

400 N   400 N

$EI = 2.1 \times 10^4$ N · m²

FIGURE P16–20

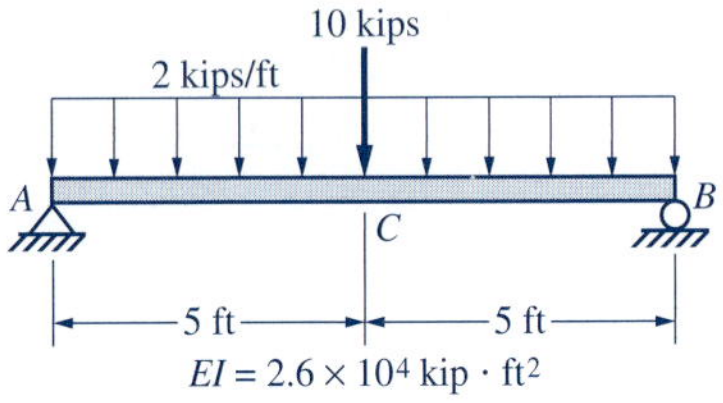

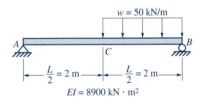

10 kips

2 kips/ft

5 ft   5 ft

$EI = 2.6 \times 10^4$ kip · ft²

FIGURE P16–21

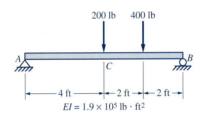

200 lb   400 lb

4 ft   2 ft   2 ft

$EI = 1.9 \times 10^5$ lb · ft²

FIGURE P16–22

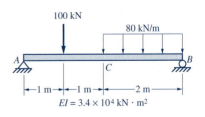

$w = 50$ kN/m

$\dfrac{L}{2} = 2$ m $\dfrac{L}{2} = 2$ m

$EI = 8900$ kN · m²

FIGURE P16–23

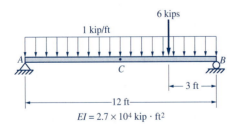

6 kips

1 kip/ft

3 ft

12 ft

$EI = 2.7 \times 10^4$ kip · ft²

FIGURE P16–24

100 kN

80 kN/m

1 m   1 m   2 m

$EI = 3.4 \times 10^4$ kN · m²

FIGURE P16–25

16–26 Find the deflection at the free end of an overhanging beam (with overhang on one side only) due to a uniform load of 60 kN/m over the entire length of the beam. The beam has a steel section W300 × 0.51. The length between the supports is 3 m and the overhanging length is 1 m.

16–27 to 16–30 Refer to Figs. P16–27 to P16–30. Determine the deflections at the midspan C and the free end D of each overhanging beam subjected to the loads shown. The flexural rigidity of each beam is indicated.

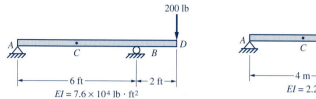

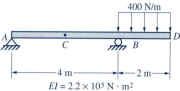

FIGURE P16–27   **FIGURE P16–28**

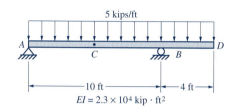

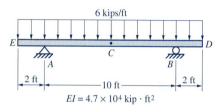

FIGURE P16–29   **FIGURE P16–30**

Section 16–5   The Moment-Area Method

16–31 and 16–32 See Figs. P16–31 and P16–32. Use the moment-area method to find the expressions for the maximum slope and the maximum deflection of each cantilever beam due to the load shown.

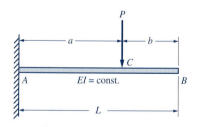

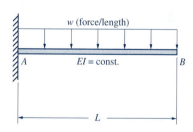

FIGURE P16–31   **FIGURE P16–32**

16–33 Use the moment-area method to find the expression for the maximum slope and the maximum deflection of a simple beam of span length L due to a concentrated load P at the midspan. The beam has a constant value of EI.

16–34 A W16 × 36 steel section is used in a 30-ft simple span. Use the moment-area method to compute the maximum deflection due to a concentrated load of 12 kips at the midspan.

16–35 Rework Problem 16–34 and assume that a uniform load of 0.4 kip/ft is applied to the entire span.

16–36 A 250 × 360 rectangular California redwood section is used in a 5-m simple span. Use the moment-area method to compute the maximum deflection due to a uniform load of 4 kN/m.

16–37 Rework Problem 16–36. Assume that a concentrated load of 20 kN is applied to the midspan.

16–38 See Fig. P16–38. Use the moment-area method to find the deflection at the midspan C of the simple beam due to the loads shown.

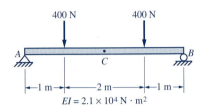

FIGURE P16–38

16–39 Refer to Fig. P16–39. Use the moment-area method to find the deflection at the midspan C and the free end D of the overhanging beam due to the loads shown.

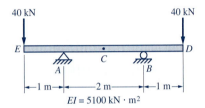

FIGURE P16–39

Section 16–6   Moment Diagram by Parts

16–40 to 16–47 Refer to Figs. P16–40 to P16–47. Draw the moment diagram by parts toward the section indicated for each beam subjected to the loading shown.

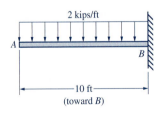

FIGURE P16–40

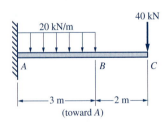

FIGURE P16–41

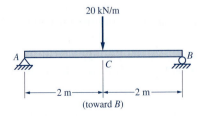

FIGURE P16–42

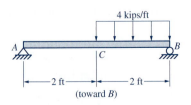

FIGURE P16–43

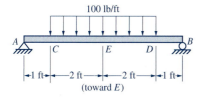

FIGURE P16–44

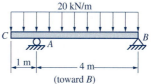

FIGURE P16–45

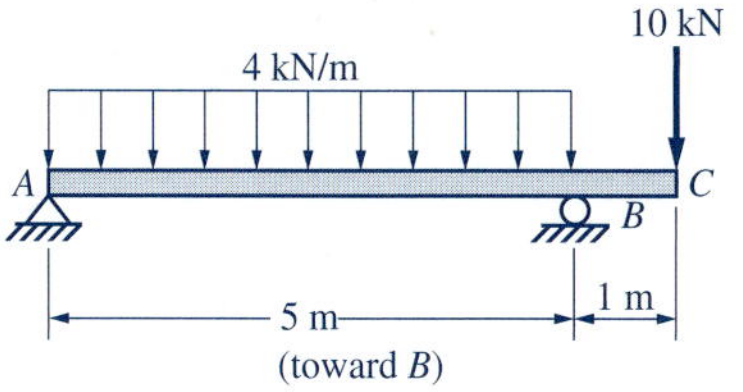

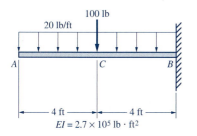

FIGURE P16–46

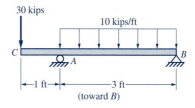

FIGURE P16–47

Section 16–7 Applications of the Moment-Area Method

16–48 to **16–51** Refer to Figs. P16–48 to P16–51. Use the moment-area method to find the maximum deflection of each cantilever beam due to the loading shown. Use the *EI* value indicated in each figure.

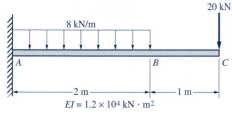

FIGURE P16–48

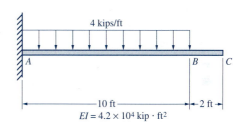

FIGURE P16–49

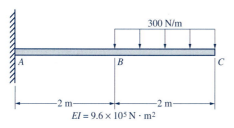

FIGURE P16–50

FIGURE P16–51

16–52 Rework Example 16–10. Use the moment diagram plotted by parts toward the midspan C.

16–53 to 16–58 Refer to Figs. P16–53 to P16–58. Use the moment-area method to find the deflection at the midspan C of each simple beam due to the loading shown. Use the EI value indicated in each figure.

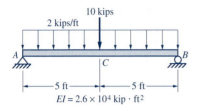

FIGURE P16–53

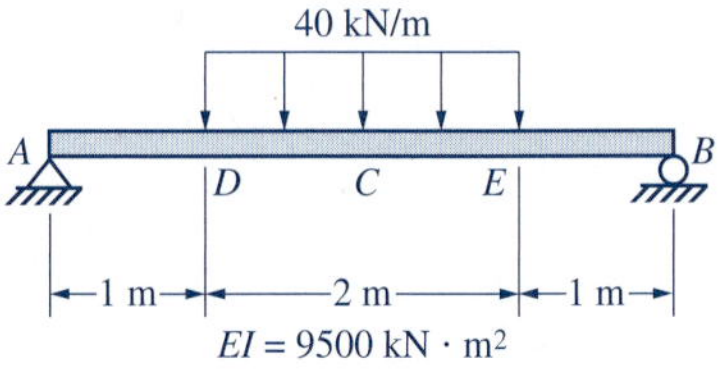

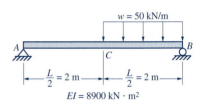

FIGURE P16–54

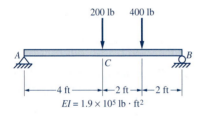

FIGURE P16–55

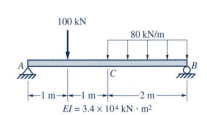

FIGURE P16–56

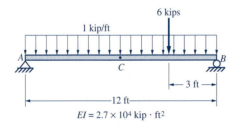

FIGURE P16–57

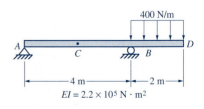

FIGURE P16–58

16–59 to 16–62 Refer to Figs. P16–59 to P16–62. Use the moment-area method to find the deflection at the midspan C and the free end D of each overhanging beam due to the loading shown. Use the EI value indicated in each figure.

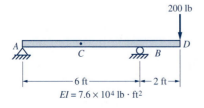

FIGURE P16–59

FIGURE P16–60

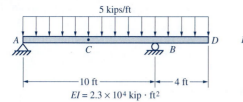

FIGURE P16–61

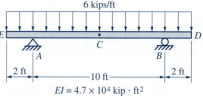

FIGURE P16–62

Computer Program Assignments

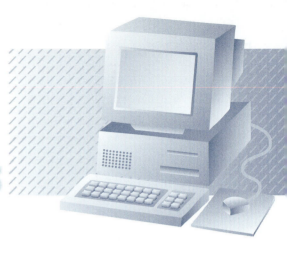

For each of the following problems, write a computer program using an appropriate programming language with which you are most familiar. Make the program user friendly by incorporating plenty of comments and input prompts so that the user will understand the input data to be entered and the limitations of their values. The output should include the data entered and the computed results, and they must be well labeled to identify each quantity. If a tabulated format is used, a proper heading must be included at the top of the table. Do not limit the program to any specific unit system. Indicate the consistent U.S. customary or SI units that can be used.

C16–1 Refer to Fig. C16–1. Write a computer program that can be used to compute the deflection of the cantilever beam subjected to a uniform load and a concentrated load, as shown. The user input should include (1) the span length L; (2) the flexural stiffness EI; (3) the intensity and location of the uniform load, w, b, and c; and (4) the magnitude and location of the concentrated force, P, and d. The downward loads are considered positive. The user should be able to choose two options for the output: (1) the deflection at a specified point along the beam, or (2) the deflection at every tenth point along the beam. Use this program to solve (*a*) Example 16–5, (*b*) Problem 16–15, (*c*) Problem 16–16, and (*d*) Problem 16–17.

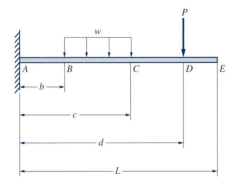

FIGURE C16–1

C16–2 Refer to Fig. C16–2. Write a computer program that can be used to calculate the deflection of the simple beam subjected to the loading shown. The user input should include (1) the span length L; (2) the flexural stiffness EI; (3) the intensity of the uniform load w; (4) the total number of concentrated loads n; and (5) the magnitude and location of each concentrated load, P_i and a_i. The downward loads are considered positive. The user should be able to choose two options for the output: (1) the deflection at a specified point along the beam, and (2) the deflection at every tenth point along the beam. Use this program to solve (a) Example 16–7, (b) Problem 16–19, (c) Problem 16–21, and (d) Problem 16–22.

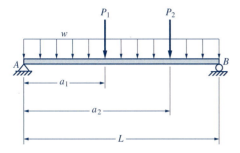

FIGURE C16–2

CHAPTER 17

STATICALLY INDETERMINATE BEAMS

17–1
INTRODUCTION

Three types of beams have been considered previously: simple beams, overhanging beams, and cantilever beams. These are all *statically determinate* beams, for which the reactions at the supports can be determined from the equilibrium conditions. Beams for which the equilibrium equations alone are insufficient to solve for reactions at the supports are *statically indeterminate*. Additional equations based on the deflection conditions must be introduced to solve for the external reactions on a statically indeterminate beam.

In a statically indeterminate beam, there are more than three reaction components, which are more than are needed to maintain the equilibrium of the beam. The extra constraints are called *redundant constraints*. The number of the redundant constraints of a beam is referred to as the *degree of indeterminacy* of the beam. Figure 17–1 shows three

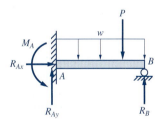

(a) Propped cantilever beam

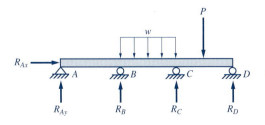

(b) Continuous beam

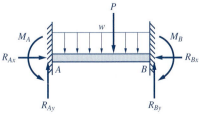

(c) Fixed beam

FIGURE 17–1

examples of statically indeterminate beams. The *propped cantilever beam* shown in Fig. 17–1a has one redundant constraint, so it is statically indeterminate to the first degree. The *continuous beam* shown in Fig. 17–1b has two redundant constraints; it is statically indeterminate to the second degree. The *fixed beam* in Fig. 17–1c has three redundant constraints; it is statically indeterminate to the third degree. A continuous beam over a larger number of spans may have a higher degree of statical indeterminacy.

In this chapter, we shall discuss methods for determining reactions on a statically indeterminate beam. Only horizontal beams with homogeneous material having a constant cross-section will be considered.

17–2
METHOD OF SUPERPOSITION

The method of superposition used in Chapter 16 for finding beam deflections may also be used conveniently to solve problems involving statically indeterminate beams. If a beam is statically indeterminate to the first degree, one of the reactions is designated as a redundant constraint. Which support element should be regarded as redundant is primarily a matter of convenience. After the redundant constraint is removed, the beam will be reduced to either a cantilever beam or a simple beam. The reaction at the redundant constraint is then treated as an unknown load. The slope or deflection at the redundant support is set to a value compatible with the original supporting conditions, which gives rise to an equation that can be used to solve for the redundant reaction. The remaining reaction components can be determined by the equilibrium equations. Once the reactions at the supports have been found, the effects of the external forces on quantities such as shear forces, bending moments, flexural and shear stresses, and the deflections may be determined in the usual way.

If a beam is statically indeterminate to the second degree, we need to designate two of the support elements as redundant constraints and the reactions at the redundant constraints are treated as unknown loads. The slope or deflection conditions at the redundant supports compatible with the original supports will give rise to two equations that can be used to solve for the unknown reactions at the redundant supports. Similarly, if a beam is statically indeterminate to the third degree, we need to designate three support elements as redundant constraints and use the slope or deflection conditions at the redundant support elements to set up three equations for solving the three unknown reactions at the redundant supports.

17–3
PROPPED CANTILEVER BEAMS

A *propped cantilever beam* has a fixed support on one end and a roller support on the other end, as shown in Fig. 17–2a, or a roller support at some intermediate point, as shown in Fig. 17–2b. A propped cantilever beam is statically indeterminate to the first degree. Usually the roller support is designated as redundant and is removed. Then the beam is reduced to a cantilever beam. The deflection at the redundant roller support must be set to zero. The resulting equation can be solved for the reaction at the redundant roller support. The remaining reaction components at the fixed support can be determined from the equilibrium equations.

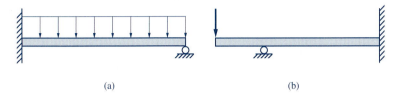

                      (a)   (b)

FIGURE 17–2

—————— **EXAMPLE 17–1** ——————————————————————————————————

For a uniformly loaded, propped cantilever beam with a constant EI, as shown in Fig. E17–1(1), find the reactions and sketch the shear force and bending moment diagrams of the beam. Express the results in terms of the load intensity w and the span length L.

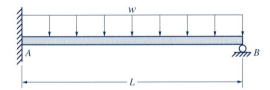

FIGURE E17–1(1)

    Solution. We designate the roller support at B as redundant. The beam is reduced to a cantilever beam after the redundant roller support is removed. The reaction R_B is treated as an external load applied on the cantilever beam, as shown in Fig. E17–1(2).

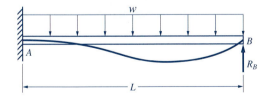

FIGURE E17–1(2)

    By the method of superposition, the deflection at B of the cantilever beam due to R_B and w can be obtained by using the formulas from case 1 and case 3 in Table 16–1. Thus,

$$\delta_B(+\uparrow) = (\delta_B)_{R_B} - (\delta_B)_w = \frac{R_B L^3}{3EI} - \frac{wL^4}{8EI}$$

To be compatible with the original supporting conditions, the deflection at B must be zero. Thus, we write

$$\frac{R_B L^3}{3EI} - \frac{wL^4}{8EI} = 0$$

From which we get

$$R_B = \frac{3}{8}wL \uparrow \qquad\qquad \Leftarrow \textbf{Ans.}$$

With R_B determined, the other reaction components can be calculated by applying the equilibrium equations to the free-body diagram in Fig. E17–1(3). We write

$$\Sigma F_y = R_B - wL + \frac{3}{8}wL = 0$$

$$R_B = \frac{5}{8}wL \uparrow \qquad\qquad \Leftarrow \textbf{Ans.}$$

$$\Sigma M_A = M_A - wL\left(\frac{L}{2}\right) + \frac{3}{8}wL(L) = 0$$

$$M_A = \frac{1}{8}wL^2 \;\circlearrowleft \qquad\qquad \Leftarrow \textbf{Ans.}$$

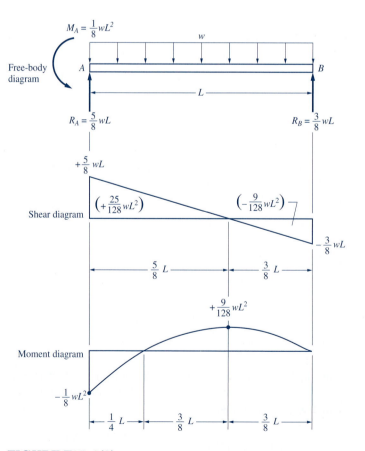

FIGURE E17–1(3)

After all the reactions are determined, the loading diagram, shear diagram, and moment diagram are sketched in the usual way, as shown in Fig. E17–1(3). The results may be used as formulas for the propped beam subjected to a uniform load.

EXAMPLE 17–2

The propped cantilever beam in Fig. E17–2(1) supports the concentrated load shown. Select the lightest W shape for the beam if the allowable flexural stress is 24 ksi and the allowable shear stress is 14.5 ksi. Neglect the weight of the beam.

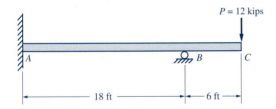

FIGURE E17–2(1)

Solution. Refer to Fig. E17–2(2). Designate the roller support as redundant. The given beam is reduced to a cantilever beam after the redundant roller support is removed. Show R_B at B. By the method of superposition, the deflection at B due to R_B and P can be obtained by using the formulas from case 1 in Table 16–1. Thus,

$$(\delta_B)_{R_B} = \frac{R_B a^3}{3EI} = \frac{R_B(18)^3}{3EI} = \frac{1944R_B}{EI} \uparrow$$

$$(\delta_B)_P = \frac{Px^2}{6EI}(3L - x) = \frac{(12)(18)^2}{6EI}(3 \times 24 - 18)$$

$$= \frac{34\,990}{EI} \downarrow$$

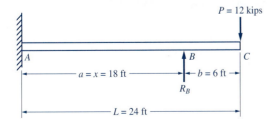

FIGURE E17–2(2)

The deflection of the given beam at B must be zero; thus, we write

$$\delta_B(+\uparrow) = (\delta_B)_{R_B} + (\delta_B)_P = \frac{1944R_B}{EI} - \frac{34\,990}{EI} = 0$$

From which we get

$$R_B = 18.0 \text{ kips} \uparrow$$

With R_B determined, the other reaction components can be calculated by applying the equilibrium equations to the free-body diagram in Fig. E17–2(3). We write

$$\Sigma F_y = R_A + 18 - 12 = 0$$
$$R_A = -6 \text{ kips} \downarrow$$

$$\Sigma M_A = M_A + 18(18) - 12(18 + 6) = 0$$
$$M_A = -36 \text{ kip} \cdot \text{ft} \circlearrowright$$

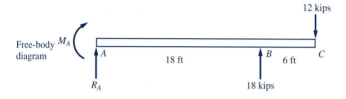

FIGURE E17–2(3)

After all the reactions are determined, the loading diagram, shear diagram, and moment diagram can be sketched in the usual way, as shown in Fig. E17–2(4). From these diagrams, we identify:

$$V_{\max} = 12 \text{ kips} \qquad M_{\max} = 72 \text{ kip} \cdot \text{ft}$$

The size of the beam is based on the required section modulus:

$$S_{req} = \frac{M_{\max}}{\sigma_{allow}} = \frac{(72 \times 12) \text{ kip} \cdot \text{in.}}{24 \text{ kips/in.}^2} = 36 \text{ in.}^3$$

From the appendix, Table A–1(a), the lightest W shape, W12 × 30, is selected. The section provides $S = 38.6 \text{ in.}^3$. The average web shear stress is

$$\tau_{avg} = \frac{V_{\max}}{dt_w} = \frac{12 \text{ kips}}{(12.34 \text{ in.})(0.260 \text{ in.})} = 3.74 \text{ ksi}$$

which is well within the allowable shear stress, so the section selected is satisfactory for both bending and shear strength.

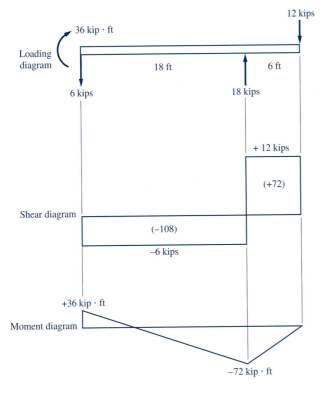

FIGURE E17–2(4)

Use W12 × 30   ⇐ **Ans.**

17–4
FIXED BEAMS

A *fixed beam* has fixed supports on both ends of the beam. A *fixed support* is assumed to have complete fixity, which means that at such a support, the slope of the tangent to the deflection curve is zero. In general, a fixed beam is statically indeterminate to the third degree. If we assume that the axial forces are absent, as indicated by the loading conditions shown in Fig. 17–3a and b, there are four unknown reaction elements and two static equilibrium equations available (the equation $\Sigma F_x = 0$ is a trivial equation). Therefore, a fixed beam without axial loads has two degrees of statical indeterminacy. If we regard the moment reaction elements on the fixed ends as redundant and release the moment constraints on the fixed ends, the beam becomes simply supported. The loads on the simple beam consist of the redundant moment reactions M_A and M_B and the given loads. The slopes at the ends of the beam must be set to zero to be compatible with the fixed-ends condition. The resulting equations can be solved for the redundant moment reactions M_A and M_B. These moments at the fixed supports are called the *fixed-end moments.* The remaining reaction components can be determined from the equilibrium equations.

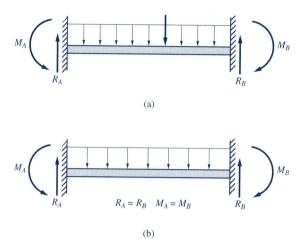

(a)

(b)

FIGURE 17–3

For a fixed beam subjected to symmetrical loading, as shown in Fig. 17–3b, we can take advantage of symmetry by equating the two redundant moment reactions. Then only one equation based on the slope at either end is needed for solving the redundant moment reactions.

EXAMPLE 17–3

The fixed beam in Fig. E17–3(1) is subjected to the uniform load shown. Find the reactions at the fixed supports and plot the shear and moment diagrams. Express the results in terms of the load intensity w and span length L. The flexural rigidity EI of the beam is a constant.

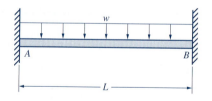

FIGURE E17–3(1)

Solution. Without a horizontal load, the horizontal components of the reactions are zero. There are four unknown reaction components, but only two equilibrium equations are available ($\Sigma F_x = 0$ is a trivial equation). The beam is therefore statically indeterminate to the second degree. Treat M_A and M_B as redundant, and the beam is reduced to the simple beam in Fig. E17–3(2). Due to symmetry, we have $M_A = M_B$. To be compatible with the fixed-ends condition, we set the slope at end A to zero and write

$$\theta_A(\measuredangle \ +) = [\theta_A]_{M_A} + [\theta_A]_{M_B} - [\theta_A]_w = 0$$

Using the formulas from Table 16–1, we write

$$\frac{M_A L}{3EI} + \frac{M_B L}{6EI} - \frac{wL^3}{24EI} = 0$$

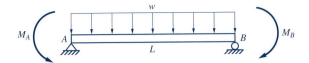

FIGURE E17–3(2)

Multiplied by $6EI/L$, the above equation is reduced to

$$2M_A + M_B - \frac{wL^2}{4} = 0$$

From which the fixed-end moment is found to be

$$M_A = M_B = \frac{wL^2}{12}$$ ⇐ **Ans.**

Now the reactions R_A and R_B can be determined from the equilibrium conditions. We get

$$R_A = R_B = \frac{wL}{2} \uparrow$$ ⇐ **Ans.**

The loading diagram, shear diagram, and moment diagram can now be sketched as shown in Fig. E17–3(3).

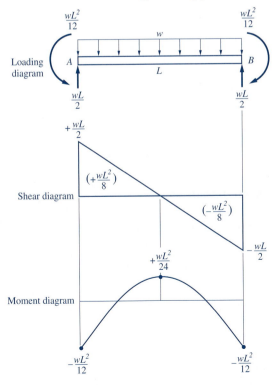

FIGURE E17–3(3)

━━━━ **EXAMPLE 17–4** ━━━━

A fixed beam of 10-m span supports a uniform load of 50 kN/m. Select the most economical W shape for the beam if the allowable flexural stress is 165 MPa and the allowable shear stress is 100 MPa.

   Solution. The maximum shear force and the maximum bending moment may be calculated from the formulas developed in Example 17–3. We find

$$V_{max} = \frac{wL}{2} = \frac{(50 \text{ kN/m})(10 \text{ m})}{2} = 250 \text{ kN}$$

$$M_{max} = \frac{wL^2}{12} = \frac{(50 \text{ kN/m})(10 \text{ m})^2}{12} = 417 \text{ kN} \cdot \text{m}$$

The size of the beam is based on the required section modulus:

$$S_{req} = \frac{M_{max}}{\sigma_{allow}} = \frac{417 \text{ kN} \cdot \text{m}}{165 \text{ 000 kN/m}^2} = 2.53 \times 10^{-3} \text{ m}^3$$

From the appendix, Table A–1(b), the lightest W shape, W530 × 1.21, is selected. The section provides $S = 2.80 \times 10^{-3} \text{ m}^3$. The extra section modulus provided is more than enough to compensate for the additional bending produced by the weight of the beam. The average web shear stress is

$$\tau_{avg} = \frac{V_{max}}{dt_w} = \frac{250 \text{ kN}}{(0.544 \text{ m})(0.0131 \text{ m})}$$

$$= 35 \text{ 100 kPa} = 35.1 \text{ MPa}$$

which is well within the allowable shear stress, so the section selected is satisfactory for both bending and shear strength.

Use W530 × 1.21 (W21 × 83)   ⇐ **Ans.**

━━━━ **EXAMPLE 17–5** ━━━━

The fixed beam in Fig. E17–5(1) is subjected to the concentrated load P shown. Express the fixed-end moments at the fixed supports in terms of P, a, and b.

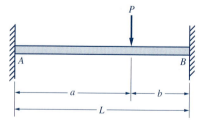

   Solution. The beam is statically indeterminate to the second degree. Treat the fixed-end moments M_A and M_B as redundant; the beam is thus reduced to the simple beam shown in Fig. E17–5(2). These moments must be such that the slopes at A and B are equal to zero to satisfy the required conditions of zero slope at the fixed ends. Thus,

FIGURE E17–5(1)

$$\theta_A\left(\underline{\triangle} \ +\right) = [\theta_A]_{M_A} + [\theta_A]_{M_B} - [\theta_A]_P = 0$$

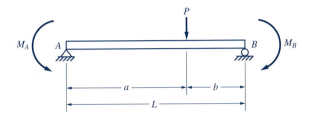

FIGURE E17–5(2)

Using the formulas from Table 16–1, we write

$$\frac{M_A L}{3EI} + \frac{M_B L}{6EI} - \frac{Pb(L^2 - b^2)}{6EIL} = 0$$

Multiplying each term of this equation by $6EI/L$, we get

$$2M_A + M_B = \frac{Pb(L^2 - b^2)}{L^2} \qquad \text{(a)}$$

Similarly, the slope at the fixed end B must be set to zero. We require

$$\theta_B\left(+ \ \underline{\triangle} \ \right) = [\theta_B]_{M_A} + [\theta_B]_{M_B} - [\theta_B]_P = 0$$

Using the formulas from Table 16–1, we write

$$\frac{M_A L}{6EI} + \frac{M_B L}{3EI} - \frac{Pa(L^2 - a^2)}{6EIL} = 0$$

Multiplying each term of this equation by $6EI/L$, we get

$$M_A + 2M_B = \frac{Pa(L^2 - a^2)}{L^2} \qquad \text{(b)}$$

Solving Equations (a) and (b) simultaneously gives

$$M_A = \frac{P}{3L^2}[2b(L^2 - b^2) - a(L^2 - a^2)]$$

$$M_B = \frac{P}{3L^2}[2a(L^2 - a^2) - b(L^2 - b^2)]$$

Substituting $L = a + b$ for L inside the bracket and simplifying, we get

$$M_A = \frac{Pab^2}{L^2} \ \circlearrowright \qquad\qquad \Leftarrow \textbf{Ans.}$$

$$M_B = \frac{Pa^2 b}{L^2} \ \circlearrowleft \qquad\qquad \Leftarrow \textbf{Ans.}$$

If the load P is applied at the midspan, the fixed-end moments can be determined by substituting $a = b = L/2$ into the above expressions. Thus,

$$M_A = M_B = \frac{1}{8}PL$$

For this case, the loading diagram, shear diagram, and moment diagram are sketched as shown in Fig. E17–5(3).

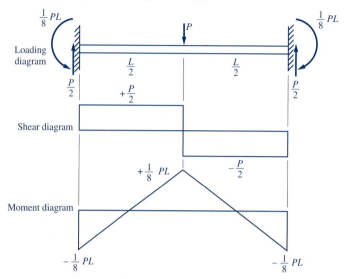

FIGURE E17–5(3)

17–5
CONTINUOUS BEAMS

A *continuous beam* continues over more than one span, as shown in Fig. 17–4. The continuity effectively reduces the maximum moment in a continuous beam compared to that of a simple beam, thereby reducing the size and resulting in a more economical design. In general, the method of superposition cannot be used conveniently to analyze continuous beams. It can be applied only to a few simple cases, as illustrated in the following examples. The more general cases can be solved by other methods. One of the methods, the theorem of three moments, will be presented in the next section.

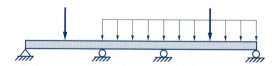

FIGURE 17–4

EXAMPLE 17–6

Determine the reactions for a two-span continuous beam subjected to a concentrated load on the second span, as shown in Fig. E17–6(1).

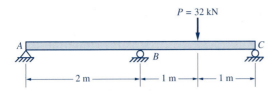

FIGURE E17–6(1)

Solution. Treat the roller support B as redundant. The continuous beam is reduced to a simple beam, as shown in Fig. E17–6(2). The magnitude of the reaction R_B at the redundant support should be such that it will make the deflection at B equal to zero to satisfy the support condition of the original beam. We require that

$$\delta_B(+\uparrow) = [\delta_B]_{R_B} - [\delta_B]_P = 0$$

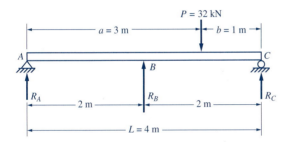

FIGURE E17–6(2)

Using the formulas from Table 16–1, we have

$$\frac{R_B L^3}{48EI} - \frac{Pb}{48EI}(3L^2 - 4b^2) = 0$$

$$\frac{R_B(4 \text{ m})^3}{48EI} - \frac{(32 \text{ kN})(1 \text{ m})}{48EI}[3(4 \text{ m})^2 - 4(1 \text{ m})^2] = 0$$

From which we get

$$R_B = +22 \text{ kN} \uparrow \qquad\qquad \Leftarrow \textbf{Ans.}$$

The equilibrium conditions of the free-body diagram of the beam in Fig. E17–6(2) require that

$$\Sigma M_C = -R_A(4 \text{ m}) - (22 \text{ kN})(2 \text{ m}) + (32 \text{ kN})(1 \text{ m}) = 0$$

From which we get

$$R_A = -3 \text{ kN} \downarrow \qquad\qquad \Leftarrow \textbf{Ans.}$$

$$\Sigma M_A = R_C(4 \text{ m}) + (22 \text{ kN})(2 \text{ m}) - (32 \text{ kN})(3 \text{ m}) = 0$$

From which we get

$$R_C = +13 \text{ kN} \uparrow \qquad\qquad \Leftarrow \textbf{Ans.}$$

Check. The equilibrium condition along the y direction serves as a check:

$$\Sigma F_y = -3 + 22 - 32 + 13 = 0 \qquad\qquad \text{(Checks)}$$

EXAMPLE 17–7

A two-span continuous beam supports the uniform load shown in Fig. E17–7(1). Select the lightest W shape for the beam if the allowable flexural stress is 24 ksi and the allowable shear stress is 14.5 ksi.

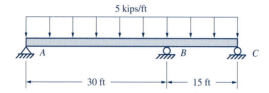

FIGURE E17–7(1)

Solution. Treat the roller support B as redundant. The continuous beam is reduced to a simple beam, with the reaction R_B exerted at B as a load, as shown in Fig. E17–7(2). To be compatible with the given beam, we require that

$$\delta_B \left(+ \uparrow \right) = [\delta_B]_{R_B} - [\delta_B]_w = 0$$

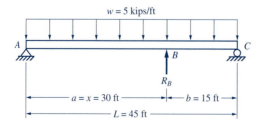

FIGURE E17–7(2)

Using kip and ft units and the formulas from Table 16–1, we have

$$\frac{R_B bx}{6EIL}(L^2 - x^2 - b^2) - \frac{wx}{24EI}(L^3 + x^3 - 2Lx^2) = 0$$

$$\frac{R_B(15)(30)}{6EI(45)}(45^2 - 30^2 - 15^2) - \frac{(5)(30)}{24EI}(45^3 + 30^3 - 2 \times 45 \times 30^2) = 0$$

Multiplied by EI, the above equation is reduced to

$$1500\,R_B - 232\,000 = 0$$

From which we get

$$R_B = +155 \text{ kips } \uparrow$$

The equilibrium conditions of the free-body diagram of the beam in Fig. E17–7(3) require that

$$\Sigma M_C = -R_A(45 \text{ kips}) + (5 \text{ kip/ft})(45 \text{ ft})\left(\frac{45 \text{ ft}}{2}\right) - (155 \text{ kips})(15 \text{ ft}) = 0$$

From which we get

$$R_A = +60.8 \text{ kips } \uparrow$$

$$\Sigma M_A = R_C(45 \text{ ft}) - (5 \text{ kips/ft})(45 \text{ ft})\left(\frac{45 \text{ ft}}{2}\right) + (155 \text{ kips})(30 \text{ ft}) = 0$$

From which we get

$$R_A = +9.2 \text{ kips } \uparrow$$

Check. The equilibrium condition along the y direction serves as a check:

$$\Sigma F_y = 60.8 + 155 + 9.2 - 5 \times 45 = 0 \qquad \text{(Checks)}$$

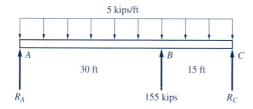

FIGURE E17–7(3)

Now the loading diagram, shear diagram, and moment diagram are sketched as shown in Fig. E17–7(4). From these diagrams, we find

$$V_{\max} = 89.2 \text{ kips} \qquad M_{\max} = 423 \text{ kip} \cdot \text{ft}$$

The size of the beam is based on the required section modulus:

$$S_{\text{req}} = \frac{M_{\max}}{\sigma_{\text{allow}}} = \frac{(423 \times 12) \text{ kip} \cdot \text{in.}}{24 \text{ kip/in.}^2} = 212 \text{ in.}^3$$

From the appendix, Table A–1(a), the lightest W shape, W21 × 111, is selected. The section provides $S = 249$ in.^{3}. The extra section modulus is enough to resist the additional bending moment caused by the weight of the beam. The average web shear stress is

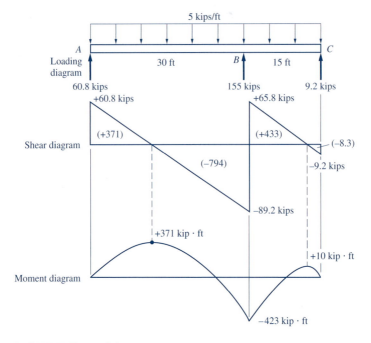

FIGURE E17–7(4)

$$\tau_{avg} = \frac{V_{max}}{dt_w} = \frac{89.2 \text{ kips}}{(21.51 \text{ in.})(0.550 \text{ in.})} = 7.54 \text{ ksi}$$

which is well within the allowable shear stress, so the section selected is satisfactory for both bending and shear strength.

Use W21 × 111   ⇐ **Ans.**

17–6
THE THREE-MOMENT THEOREM

The theorem of three moments presents the relationship among the internal moments at the sections over three consecutive supports in a continuous beam. This relationship is expressed as an equation in terms of the three moments. Three-moment equations may be written for the moments in any three consecutive supports. In this way, we will get enough equations to solve for the internal moments at all the supports. After the moments at the supports have been computed, the vertical reactions at all the supports can be determined from the static equilibrium conditions. Then the shear and moment diagrams can be sketched in the usual manner.

To establish the three-moment equation, consider two adjacent spans of a continuous beam subjected to the uniform loads w_1 and w_2, and the concentrated loads P_1 and P_2, as shown in Fig. 17–5a. Using the method of sections, separate the two spans into two simple beams, with end moments M_A, M_B, and M_C applied at the appropriate ends of the simple beams, as shown in Fig. 17–5b and c (where the moments are shown in the positive direction according to the beam sign convention). In the first span shown in Fig. 17–5b, the slope $(\theta_B)_1$ at B due to the end moments and the applied loads on the first span are

$$(\theta_B)_1 \; (+ \; \overline{\searurrow}) = (\theta_B)_{M_A} + (\theta_B)_{M_B} + (\theta_B)_{w_1} + (\theta_B)_{P_1}$$

$$= \frac{M_A L_1}{6EI} + \frac{M_B L_1}{3EI} + \frac{w_1 L_1^3}{24EI} + \frac{P_1 a_1 (L_1^2 - a_1^2)}{6EIL_1}$$

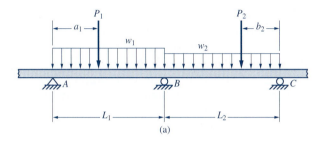

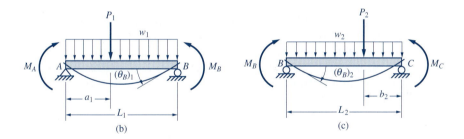

FIGURE 17–5

In the second span shown in Fig. 17–5c, the slope $(\theta_B)_2$ at B due to the end moments and the loads on the second span are

$$(\theta_B)_2 \; (\overline{\swarrow} \; +) = (\theta_B)_{M_B} + (\theta_B)_{M_C} + (\theta_B)_{w_2} + (\theta_B)_{P_2}$$

$$= \frac{M_B L_2}{3EI} + \frac{M_C L_2}{6EI} + \frac{w_2 L_2^3}{24EI} + \frac{P_2 b_2 (L_2^2 - b_2^2)}{6EIL_2}$$

For a continuous beam, the deflection curve must have the same slope on either side over support B. Thus, we require $(\theta_B)_1 = -(\theta_B)_2$, where the negative sign is introduced because the positive direction of $(\theta_B)_1$ and $(\theta_B)_2$ are chosen in the opposite directions. Thus, we write

$$\frac{M_AL_1}{6EI} + \frac{M_BL_1}{3EI} + \frac{w_1L_1^3}{24EI} + \frac{P_1a_1(L_1^2 - a_1^2)}{6EIL_1}$$

$$= -\left[\frac{M_BL_2}{3EI} + \frac{M_CL_2}{6EI} + \frac{w_2L_2^3}{24EI} + \frac{P_2b_2(L_2^2 - b_2^2)}{6EIL_2}\right]$$

Multiplying by $6EI$ and rearranging the terms, we get the *three-moment equation*:

$$M_AL_1 + 2M_B(L_1 + L_2) + M_CL_2$$

$$= -\frac{w_1L_1^3}{4} - \frac{w_2L_2^3}{4} - \frac{P_1a_1(L_1^2 - a_1^2)}{L_1} - \frac{P_2b_2(L_2^2 - b_2^2)}{L_2} \qquad (17\text{–}1)$$

The three-moment equation is applicable to any number of multispan continuous beams. For any three successive supports, one three-moment equation can be written. The number of three-moment equations available is two less than the number of supports. If the end support is simple with or without an overhang the moment at the end support is either zero if there is no overhang or equal to the moment produced by the load on the overhang about the end support. In this case, there will be enough three-moment equations to solve for the support moments.

If either one or both supports at the ends is fixed, the moment at the fixed support will be unknown. It would appear that there are not enough equations for the unknown support reactions. Additional equations, however, may be introduced by creating an auxiliary span of zero span length, thereby creating a new three-moment equation at each fixed support. Then we will have enough equations to solve for the support moments. This approach is illustrated in Example 17–10. After the support moments are determined, the vertical reactions may be determined from the static equilibrium conditions.

—— **EXAMPLE 17–8** ——

Determine the reactions for the two-span continuous beam loaded as shown in Fig. E17–8(1). The flexural rigidity EI for the beam is a constant.

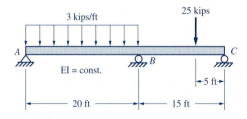

FIGURE E17–8(1)

Solution. Note that the ends of the beam are simply supported; therefore,

$$M_A = M_C = 0$$

Substituting these values and the other given quantities shown in Fig. E17–8(2) into Equation 17–1, we get

$$0 + 2M_B(20 + 15) + 0 = -\frac{(3)(20)^3}{4} - \frac{(25)(5)(15^2 - 5^2)}{15}$$

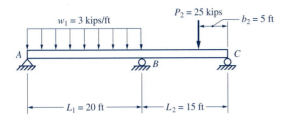

FIGURE E17–8(2)

From which we get

$$M_B = -110 \text{ kip} \cdot \text{ft}$$

To determine the vertical reactions, we cut the beam at B and separate it into two simple spans. The free-body diagrams of the two spans are sketched as shown in Fig. E17–8(3). Note that section B has been cut. The internal moment M_B becomes external and it must be shown on the free-body diagrams. The equilibrium equations for the first span give

$$\Sigma M_B = -R_A(20) + (3 \times 20)(10) - 110 = 0$$
$$R_A = +24.5 \text{ kips} \uparrow \qquad \Leftarrow \textbf{Ans.}$$

$$\Sigma M_A = (R_B)_1(20) - (3 \times 20)(10) - 110 = 0$$
$$(R_B)_1 = +35.5 \text{ kips}$$

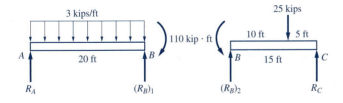

FIGURE E17–8(3)

The equilibrium equations for the second span give

$$\Sigma M_B = +R_C(15) - (25)(10) + 110 = 0$$
$$R_C = +9.3 \text{ kips} \uparrow \qquad \Leftarrow \textbf{Ans.}$$

$$\Sigma M_C = -(R_B)_2(15) + (25)(5) + 110 = 0$$
$$(R_B)_2 = +15.7 \text{ kips}$$

The reaction at B is the sum of $(R_B)_1$ and $(R_B)_2$ found above. Thus,

$$R_B = 35.5 + 15.7 = 51.2 \text{ kips} \uparrow \qquad \Leftarrow \textbf{Ans.}$$

EXAMPLE 17–9

Refer to Fig. E17–9(1). Determine the reactions and plot the shear force and bending moment diagrams for the two-span continuous beam loaded as shown. The flexural rigidity EI for the beam is a constant.

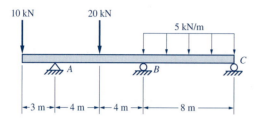

FIGURE E17–9(1)

Solution. Note that the right end is simply supported; therefore,

$$M_C = 0$$

The left end of the beam has an overhang. The internal moment at section A over the support is

$$M_A = -(10 \text{ kN})(3 \text{ m}) = -30 \text{ kN} \cdot \text{m}$$

Note that the moment M_A is negative according to the beam sign convention. Substituting these values and the other given quantities shown in Fig. E17–9(2) into Equation 17–1, we get

$$M_A = -30 \text{ kN} \cdot \text{m}$$

FIGURE E17–9(2)

$$(-30)(8) + 2M_B(8 + 8) + 0 = -\frac{(20)(4)(8^2 - 4^2)}{8} - \frac{(5)(8)^3}{4}$$

From which we get

$$M_B = -27.5 \text{ kN} \cdot \text{m}$$

To determine the vertical reactions, we cut the beam at B and separate the given beam into two simple spans. The free-body diagrams of the two spans are sketched in Fig. E17–9(3). The equilibrium equations for the first span give

$$\Sigma M_B = -R_A(8 \text{ m}) + (10 \text{ kN})(11 \text{ m}) + (20 \text{ kN})(4 \text{ m}) - 27.5 \text{ kN} \cdot \text{m} = 0$$

$$R_A = +20.3 \text{ kN} \uparrow \qquad\qquad \Leftarrow \textbf{Ans.}$$

$$\Sigma M_A = (R_B)_1(8 \text{ m}) + (10 \text{ kN})(3 \text{ m}) - (20 \text{ kN})(4 \text{ m}) - 27.5 \text{ kN} \cdot \text{m} = 0$$

$$(R_B)_1 = +9.7 \text{ kN}$$

The equilibrium equations for the second span give

$$\Sigma M_B = R_C(8 \text{ m}) - (5 \text{ kN/m} \times 8 \text{ m})(4 \text{ m}) + 27.5 \text{ kN} \cdot \text{m} = 0$$

$$R_C = +16.6 \text{ kN} \uparrow \qquad\qquad \Leftarrow \textbf{Ans.}$$

$$\Sigma M_C = -(R_B)_2(8 \text{ m}) + (5 \text{ kN/m} \times 8 \text{ m})(4 \text{ m}) + 27.5 \text{ kN} \cdot \text{m} = 0$$

$$(R_B)_2 = +23.4 \text{ kN}$$

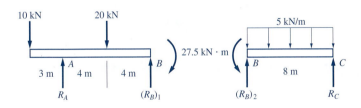

FIGURE E17–9(3)

The reaction at B is the sum of $(R_B)_1$ and $(R_B)_2$ found above. Thus,

$$R_B = 9.7 + 23.4 = 33.1 \text{ kN} \uparrow \qquad\qquad \Leftarrow \textbf{Ans.}$$

Now the loading diagram, shear diagram, and moment diagram can be sketched as shown in Fig. E17–9(4).

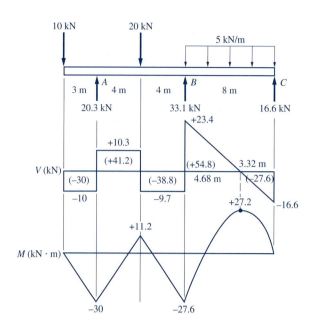

FIGURE E17–9(4)

EXAMPLE 17–10

Refer to Fig. E17–10(1). Determine the reactions and plot the shear force and bending moment diagrams for the two-span continuous beam loaded as shown. The left support at A is fixed. The flexural rigidity EI for the beam is a constant.

FIGURE E17–10(1)

Solution. Note that the support at the right end is a simple support; therefore,

$$M_C = 0$$

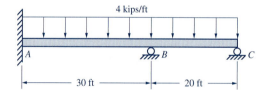

FIGURE E17–10(2)

See Fig. E17–10(2). At the fixed support, we introduce an auxiliary span of zero span length. The three-moment equation for the auxiliary span and the first span is

$$M_0(0) + 2M_A(0 + 30) + M_B(30) = -\frac{(4)(30)^3}{4}$$

which reduces to

$$60M_A + 30M_B = -27\ 000 \tag{a}$$

The three-moment equation for the two spans of the beam is

$$M_A(30) + 2M_B(30 + 20) + 0 = -\frac{(4)(30)^3}{4} - \frac{(4)(20)^3}{4}$$

which reduces to

$$30M_A + 100M_B = -35\ 000 \tag{b}$$

Solving Equations (a) and (b) for M_A and M_B, we get

$$M_A = -324 \text{ kip} \cdot \text{ft} \qquad M_B = -253 \text{ kip} \cdot \text{ft}$$

From the free-body diagrams of the isolated spans (not shown), we find

$$R_A = 62.4 \text{ kips} \uparrow \qquad\qquad \Leftarrow \textbf{Ans.}$$

$$R_B = 110.3 \text{ kips} \uparrow \qquad\qquad \Leftarrow \textbf{Ans.}$$

$$R_C = 27.3 \text{ kips} \uparrow \qquad\qquad \Leftarrow \textbf{Ans.}$$

Now the loading diagram, shear diagram, and moment diagram can be sketched as shown in Fig. E17–10(3).

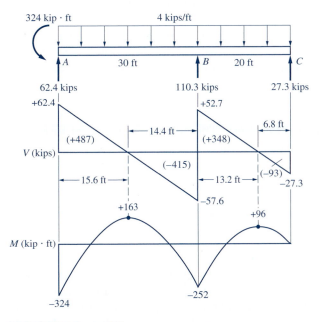

FIGURE E17–10(3)

17-7
SUMMARY

Statically Indeterminate Beams. Beams for which the equilibrium equations alone are insufficient to solve for the reactions at the supports are *statically indeterminate*. Additional equations based on deflections or end rotations must be introduced to solve for the external reactions on a statically indeterminate beam.

Degree of Indeterminacy. The number of the *redundant constraints,* which are the extra constraints that are unnecessary for the equilibrium of a beam, is referred to as the *degree of indeterminacy* of the beam.

Propped Cantilever Beam. A propped cantilever beam has a fixed support on one end and a roller support on the other end or at some intermediate point. It is statically indeterminate to the first degree. The roller support can be designated as redundant and reduces the beam to a cantilever beam. The deflection at the redundant roller support must be set to zero to be compatible with the original supporting conditions. The resulting equation can be solved for the redundant reaction.

Fixed Beam. A *fixed beam* has fixed supports on both ends of the beam. In general, a fixed beam is statically indeterminate to the third degree. In the absence of the axial forces, it is statically indeterminate to the second degree. The moment reactions on the fixed end can be designated as redundant. With the moment constraints released, the beam becomes simply supported. The slopes at the ends of the beam must be set to zero to be compatible with the fixed-ends condition. The resulting equations can be solved for the redundant moment reactions.

Continuous Beam. A *continuous beam* continues over more than one span. The method of superposition can be applied only to continuous beams of a degree of statical indeterminacy no higher than two. The more general cases can be solved conveniently by using the theorem of three moments.

Three-Moment Theorem. The internal moments at three consecutive supports are related by the three-moment equation:

$$M_A L_1 + 2M_B(L_1 + L_2) + M_C L_2$$

$$= -\frac{w_1 L_1^3}{4} - \frac{w_2 L_2^3}{4} - \frac{P_1 a_1 (L_1^2 - a_1^2)}{L_1} - \frac{P_2 b_2 (L_2^2 - b_2^2)}{L_2} \qquad (17\text{--}1)$$

The three-moment equation is applicable to any number of multispan continuous beams. For any three successive supports, one three-moment equation can be written. It is always possible to write enough three-moment equations to solve for the unknown support moments. After the support moments are determined, the vertical reactions may be determined from the static equilibrium conditions, and the shear and moment diagrams can be constructed in the usual way.

PROBLEMS

Section 17–3   Propped Cantilever Beams

17–1 to 17–4   See Figs. P17–1 to P17–4. The flexural rigidity EI of each propped cantilever beam loaded as shown is a constant. Use the method of superposition to determine the reactions of each beam.

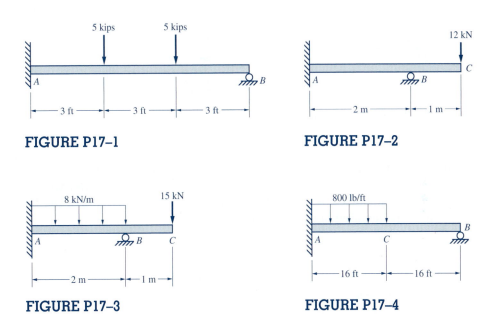

FIGURE P17–1

FIGURE P17–2

FIGURE P17–3

FIGURE P17–4

17–5 to 17–8   See Figs. P17–5 to P17–8. The flexural rigidity EI of each propped cantilever beam loaded as shown is a constant. Use the method of superposition to determine the reactions and plot the shear and moment diagrams.

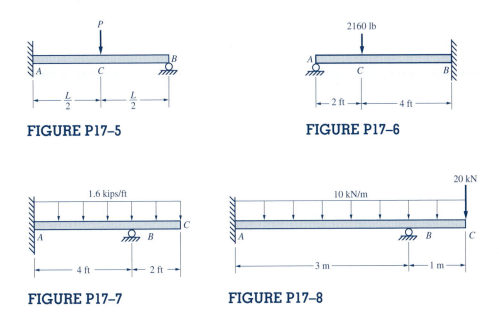

FIGURE P17–5

FIGURE P17–6

FIGURE P17–7

FIGURE P17–8

17–9 Select the lightest W shape for the propped beam subjected to the uniform load shown in Fig. P17–9. The allowable flexural stress is 24 ksi and the allowable shear stress is 14.5 ksi. Assume that the weight of the beam is already included in the uniform load. Use the expressions for the maximum shear and maximum moment obtained in Example 17–1.

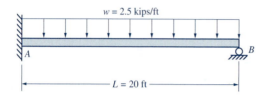

FIGURE P17–9

17–10 Select the lightest W shape for the propped beam subjected to the uniform load shown in Fig. P17–10. The allowable flexural stress is 165 MPa and the allowable shear stress is 100 MPa. Assume that the weight of the beam is already included in the uniform load.

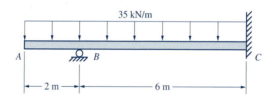

FIGURE P17–10

Section 17–4   Fixed Beams

17–11 to 17–14   See Figs. P17–11 to P17–14. The flexural rigidity EI of each fixed beam shown is a constant. Use the method of superposition to determine the fixed-end moments from the expressions obtained in Examples 17–3 and 17–5. Then find the other reaction components and plot the shear and moment diagrams.

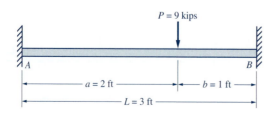

FIGURE P17–11

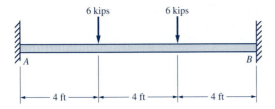

FIGURE P17–12

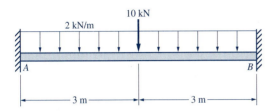

FIGURE P17–13

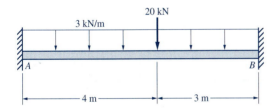

FIGURE P17–14

17–15 Refer to Fig. P17–15. Select the lightest W shape for the fixed beam subjected to the centrally placed, concentrated load shown. The allowable flexural stress is 165 MPa and the allowable shear stress is 100 MPa. Neglect the weight of the beam. Use the expressions for the maximum shear and maximum moment obtained in Example 17–5.

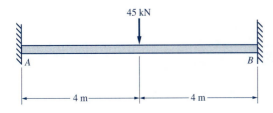

FIGURE P17–15

17-16   See Fig. P17-16. Select the lightest W shape for the fixed girder subjected to the loads shown. The allowable flexural stress is 24 ksi and the allowable shear stress is 14.5 ksi. Assume that the weight of the girder is already included in the uniform load.

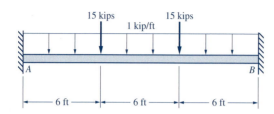

FIGURE P17-16

Section 17-5   Continuous Beams

17-17 and **17-18**   Refer to Figs. P17-17 and P17-18. The flexural rigidity EI of each continuous beam loaded as shown is a constant. Use the method of superposition to determine the reactions of each beam.

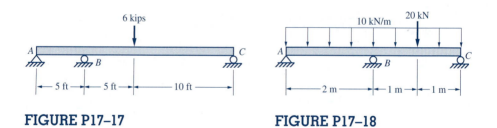

FIGURE P17-17   **FIGURE P17-18**

17-19 and **17-20**   See Figs. P17-19 and P17-20. The flexural rigidity EI of each continuous beam loaded as shown is a constant. Use the method of superposition to determine the reactions and plot the shear and moment diagrams.

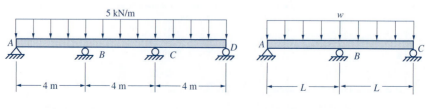

FIGURE P17-19   **FIGURE P17-20**

17–21 Refer to Fig. P17–21. Select the lightest W shape for the two-span continuous beam subjected to the uniform load shown. The allowable flexural
stress is 24 ksi and the allowable shear stress is 14.5 ksi. Assume that the
weight of the beam is already included in the uniform load. Use the expressions for the maximum shear and maximum moment obtained in Problem
17–20.

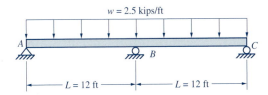

FIGURE P17–21

17–22 Refer to Fig. P17–22. Select the lightest W shape for the three-span continuous beam subjected to the uniform load shown. The allowable flexural
stress is 165 MPa and the allowable shear stress is 100 MPa. Assume that
the weight of the beam is already included in the uniform load.

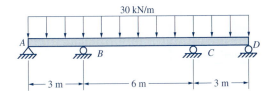

FIGURE P17–22

Section 17–6   The Three-Moment Theorem

17–23 to 17–26   See Figs. P17–23 to P17–26. The flexural rigidity EI of each continuous beam loaded as shown is a constant. Use the three-moment theorem
to determine the reactions and plot the shear and moment diagrams.

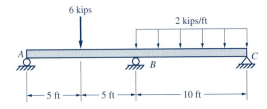

FIGURE P17–23

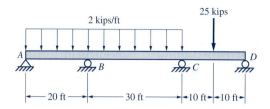

FIGURE P17–24

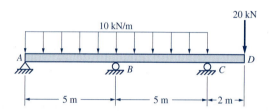

FIGURE P17–25

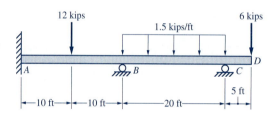

FIGURE P17–26

17–27   Refer to Fig. P17–27. Select the lightest W shape for the continuous beam
subjected to the uniform load shown. The allowable flexural stress is 165
MPa and the allowable shear stress is 100 MPa. Assume that the weight of
the beam is already included in the uniform load.

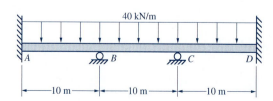

FIGURE P17–27

17–28 Refer to Fig. P17–28. Select the lightest W shape for the continuous beam subjected to the loads shown. The allowable flexural stress is 24 ksi and the allowable shear stress is 14.5 ksi. Neglect the weight of the beam.

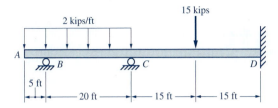

FIGURE P17–28

Computer Program Assignments

For each of the following problems, write a computer program using an appropriate programming language with which you are most familiar. Make the program user friendly by incorporating plenty of comments and input prompts so that the user will understand the input data to be entered and the limitations of their values. The output should include the data entered and the computed results, and they must be well labeled to identify each quantity. If a tabulated format is used, a proper heading must be included at the top of the table. Do not limit the program to any specific unit system. Indicate the consistent U.S. customary or SI units that can be used.

C17–1 Refer to Fig. C17–1. Write a computer program that can be used to compute the reactions of the fixed beam subjected to the loads shown. To find the fixed-end moments, use the method of superposition and the expressions in Examples 17–3 and 17–5. The user input should include (1) the span length L; (2) the intensity of the uniform load w; and (3) the magnitude and location of the concentrated loads P_1, a_1, P_2, and a_2. The downward loads are treated as positive. The output should include (1) the fixed-end moments M_A and M_B and (2) the reactions R_A and R_B. Use this program to solve (a) Problem 17–12, (b) Problem 17–13, and (c) Problem 17–14.

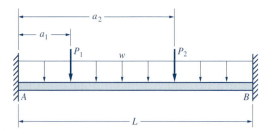

FIGURE C17–1

C17–2 Refer to Fig. C17–2. Write a computer program that can be used to compute the reactions of the two-span continuous beam with overhangs subjected to the loads shown by using the three-moment equation. The user input should include (1) the span lengths L_0, L_1, L_2, and L_3; (2) the loads on the first span w_1, P_1, and a_1; (3) the loads on the second span w_2, P_2, and a_2; (4)

if $L_0 \neq 0$, the loads on the left overhang w_0, P_0, and a_0; and (5) if $L_3 \neq 0$, the loads on the right overhang w_3, P_3, and a_3. The downward loads are treated as positive. The output should include (1) the moments M_A, M_B, and M_C and (2) the reactions R_A, R_B, and R_C. Use this program to solve (a) Example 17–9, (b) Problem 17–23, and (c) Problem 17–25.

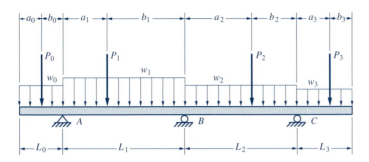

FIGURE C17–2

CHAPTER 18

COMBINED STRESSES

18–1
INTRODUCTION

The fundamental formulas for calculating the stresses in a member subjected to only one type of loading have been developed in the previous chapters. These formulas are listed for reference in Table 18–1. The formulas were derived based on the assumption that the stresses were caused by only one type of loading, and the maximum stress in the member was within the elastic limit of the material, within which stress was proportional to strain.

In many engineering applications, more than one type of loading may be applied to a member. Therefore, a technique is needed for finding the combined stress in a member due to several types of loading. The *method of superposition* is used to determine the combined stresses caused by two or more types of loading. Using this method and the fundamental formulas in Table 18–1, the same type of stresses caused by each loading are determined separately. The algebraic sum of these stresses gives the combined stresses caused by all the loadings acting simultaneously. The method of superposition is valid only if the maximum stress is within the elastic limit of the material and if the deformations are small.

In this chapter, the normal stresses caused by simultaneous action of the axial force and bending moment are discussed first, followed by discussion of bending about two perpendicular axes and the effects of eccentric loading. We will discuss also the general state of plane stress at a point, as expressed on a small rectangular element enclosing the point, that consists of biaxial normal stresses and shear stresses acting simultaneously on the element. We will derive equations for calculating the normal and shear stresses on any inclined plane in an element where the general state of plane stress in two perpendicular directions is given. The graphical representation of the equations, called Mohr's circle, will also be developed. From Mohr's circle, the maximum normal and shear stresses can be readily determined. These maximum stresses are important factors to consider in structural design.

TABLE 18–1 List of the Fundamental Formulas

Type of Load	Type of Stress	Formula	Equation Number
Axial load	Direct normal stress	$\sigma = \dfrac{P}{A}$	(9–1)
Internal pressure in thin-walled vessels	Circumferential stress	$\sigma_c = \dfrac{Pr_i}{t}$	(9–16)
	Longitudinal stress	$\sigma_l = \dfrac{Pr_i}{2t}$	(9–17)
Beam bending load	Flexural stress	$\sigma = \dfrac{My}{I}$	(14–3)
		$\sigma_{max} = \dfrac{Mc}{I}$	(14–2)
		$\sigma_{max} = \dfrac{M}{S}$	(14–7)
Direct shear load	Direct shear stress	$\tau_{avg} = \dfrac{P}{A}$	(9–4)
Torque in circular shaft	Torsional shear stress	$\tau = \dfrac{T\rho}{J}$	(12–2)
		$\tau_{max} = \dfrac{Tc}{J}$	(12–1)
Beam shear force	Beam shear stress	$\tau = \dfrac{VQ}{It}$	(14–10)
	Maximum shear stress in rectangular section	$\tau_{max} = 1.5\dfrac{V}{A}$	(14–11)
	Maximum shear stress in circular section	$\tau_{max} = \dfrac{4V}{3A}$	(14–12)

18–2
COMBINED AXIAL AND BENDING STRESSES

Many structural and machine members are subjected to axial forces and bending moments exerted simultaneously. Both produce normal stresses along the longitudinal directions. The normal stresses due to each load can be calculated separately and added algebraically to find the combined stresses, as illustrated in the following two examples.

──── **EXAMPLE 18–1** ────

Refer to Fig. E18–1(1). The wide-flange shape W360 × 0.99 is used as a simple beam of 3-m span. The beam is subjected to a uniform load w of 100 kN/m (including the weight of the beam) and an axial tensile force P of 500 kN. Determine the normal stresses at points A and B, and plot the normal stress variation between A and B.

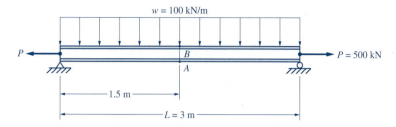

FIGURE E18–1(1)

Solution. From the appendix, Table A–1(b), the properties of a W360 × 0.99 shape are

$$A = 12.9 \times 10^{-3} \text{ m}^2$$

$$I = 301 \times 10^{-6} \text{ m}^4$$

$$S = 1.69 \times 10^{-3} \text{ m}^3$$

The beam is subjected to two types of loadings: the axial force and the bending moment, as shown in Fig. E18–1(2). Due to the axial force, the direct normal stress is a constant tension P/A throughout the beam. The flexural stresses at A and B can be computed by $\pm M/S$. For a simple span with a uniform load, the moment at the midspan is a maximum equal to

$$M_{\text{max}} = \frac{wL^2}{8} = \frac{(100 \text{ kN/m})(3 \text{ m})^2}{8} = 112.5 \text{ kN} \cdot \text{m}$$

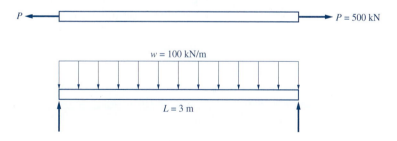

FIGURE E18–1(2)

This moment is positive. It will cause tension at A and compression at B. The normal stresses at A and B are

$$\sigma_A = +\frac{P}{A} + \frac{M}{S} = \frac{500 \text{ kN}}{12.9 \times 10^{-3} \text{ m}^2} + \frac{112.5 \text{ kN} \cdot \text{m}}{1.69 \times 10^{-3} \text{ m} 10^3}$$

$$= +38\ 800 \text{ kPa} + 66\ 600 \text{ kPa}$$

$$= +105\ 400 \text{ kPa} = +105.4 \text{ MPa (T)} \qquad \Leftarrow \textbf{Ans.}$$

$$\sigma_B = +\frac{P}{A} - \frac{M}{S}$$

$$= +38\ 800\ \text{kPa} - 66\ 600\ \text{kPa}$$

$$= -27\ 800\ \text{kPa} = -27.8\ \text{MPa (C)} \qquad \Leftarrow \textbf{Ans.}$$

The distributions of axial stress, flexural stresses, and the combined stresses in the midspan are plotted as shown in Fig. E18–1(3). Note that when the beam is subjected only to flexural stresses, the neutral axis passes through the centroid of the section. When the beam is subjected to combined stresses due to both the axial force and bending moment, the line of zero stress shifts upward.

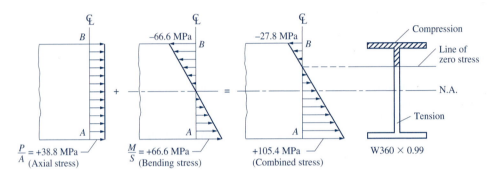

FIGURE E18–1(3)

Remark. Since the beam deflects due to bending, the axial compressive force produces an additional bending moment equal to $P\delta$. However, since deflection is very small, this moment is usually negligible. For the beam in this example, the maximum deflection at the midspan due to the uniform load is

$$\delta_{max} = \frac{5wL^4}{384EI} = \frac{5(100)(3)^4}{384(210 \times 10^6)(301 \times 10^{-6})}$$

$$= 0.00167\ \text{m}$$

The additional moment due to axial compressive force is

$$P\delta = (500\ \text{kN})(0.00167\ \text{m}) = 0.83\ \text{kN} \cdot \text{m}$$

This is less than 1% of the maximum moment and it can therefore be neglected without any appreciable error. However, a compression member must always be investigated for the possibility of buckling. (Refer to Chapter 19 for the determination of a safe buckling load.)

─────── **EXAMPLE 18–2** ───────────────────────────────────

Refer to Fig. E18–2(1). A crane with a swinging arm is designed to hoist a maximum weight of 2 kips. If the allowable compressive stress is 13 ksi, select a W shape for the arm AB.

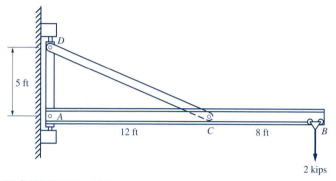

FIGURE E18–2(1)

Solution. The free-body diagram of the arm AB is constructed as shown in Fig. E18–2(2), where T is the tension in rod CD, and T_x and T_y are the components of T.

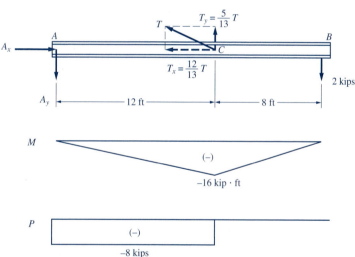

FIGURE E18–2(2)

The equilibrium equations of the free body give

$$\Sigma M_A = \frac{5}{13}T(12 \text{ ft}) - (2 \text{ kips})(20 \text{ ft}) = 0$$
$$T = 8.67 \text{ kips}$$

$$\Sigma F_x = A_x - \frac{12}{13}(8.67 \text{ kips}) = 0$$
$$A_x = 8.00 \text{ kips}$$

$$\Sigma F_y = -A_y + \frac{5}{13}(8.67 \text{ kips}) - 2 \text{ kips} = 0$$
$$A_y = 1.33 \text{ kips}$$

The bending moment and axial force diagrams of the arm AB are sketched as shown in Fig. E18–2(2). From these diagrams, we see that the critical section occurs just to the left of C, where the maximum negative moment is 16 kip · ft (or 192 kip · in.) and the compressive axial force is 8 kips.

For a tentative selection of a W shape, consider the bending moment only, which requires

$$S_{req} = \frac{M}{\sigma_{allow}^{(C)}} = \frac{192 \text{ kip} \cdot \text{in.}}{13 \text{ kips/in.}^2} = 14.8 \text{ in.}^3$$

From Table A–1(a), we tentatively select W8 × 18 ($A = 5.26$ in.^{2}, $S = 15.2$ in.^{3}). The maximum compressive stress at the critical section is

$$\left| \sigma_{max}^{(C)} \right| = \frac{P}{A} + \frac{M}{S} = \frac{8 \text{ kips}}{5.26 \text{ in.}^2} + \frac{192 \text{ kip} \cdot \text{in.}}{15.2 \text{ in.}^3}$$

$$= 14.2 \text{ ksi} > \sigma_{allow}^{(C)} = 13 \text{ ksi}$$

The W8 × 18 is too small. Try the next larger size: W8 × 21 ($A = 6.16$ in.^{2}, $S = 18.2$ in.^{3}).

$$\left| \sigma_{max}^{(C)} \right| = \frac{P}{A} + \frac{M}{S} = \frac{8 \text{ kips}}{6.16 \text{ in.}^2} + \frac{192 \text{ kip} \cdot \text{in.}}{18.2 \text{ in.}^3}$$

$$= 11.8 \text{ ksi} < \sigma_{allow}^{(C)} = 13 \text{ ksi}$$

$$\text{Use a W8} \times 21 \text{ shape} \qquad \qquad \Leftarrow \textbf{Ans.}$$

18–3
BIAXIAL BENDING

Problems can arise when a load is inclined at an angle with respect to the vertical plane of symmetry of the beam, as shown in Fig. 18–1. The load P can be resolved into two components, P_x and P_y, in the directions of the two axes of symmetry at the section on the free end. The vertical component P_y causes bending about the horizontal axis, and the horizontal component P_x causes bending about the vertical axis; thus, *biaxial bending* results. The stresses caused by either type of bending are normal stresses along the longitudinal direction; hence, the method of superposition can be applied. The bending about each axis is analyzed separately and the results are added algebraically.

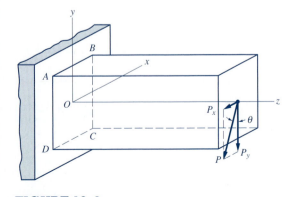

FIGURE 18–1

─────── **EXAMPLE 18–3** ───────

A simply supported timber beam of 3-m span has a rectangular cross-section with a nominal size of 150 mm × 200 mm. The beam carries a uniform load w of 4 kN/m and is supported at the ends in the tilted position shown in Fig. E18–3. Determine the maximum flexural stresses in the beam.

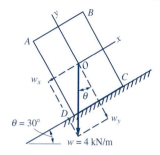

FIGURE E18–3

Solution. The uniform load w can be resolved into x and y components:

$$w_x = w \sin \theta = (4 \text{ kN/m}) \sin 30° = 2.00 \text{ kN/m}$$

$$w_y = w \cos \theta = (4 \text{ kN/m}) \cos 30° = 3.46 \text{ kN/m}$$

The component w_y produces bending about the x axis; the component w_x produces bending about the y axis. The maximum bending moments at the midspan about the x and y axes are, respectively,

$$M_x = \frac{W_y L^2}{8} = \frac{(3.46 \text{ kN/m})(3 \text{ m})^2}{8} = 3.89 \text{ kN} \cdot \text{m}$$

$$M_y = \frac{W_x L^2}{8} = \frac{(2.00 \text{ kN/m})(3 \text{ m})^2}{8} = 2.25 \text{ kN} \cdot \text{m}$$

From Table A–6(b), we find that the section modulus of a 150 × 200 section about the x axis is

$$S_x = 0.851 \times 10^{-3} \text{ m}^3$$

The section modulus about the y axis is not listed in Table A–6(b). It can be computed from the dressed size as follows:

$$S_y = \frac{(0.191 \text{ m})(0.140 \text{ m})^2}{6} = 0.624 \times 10^{-3} \text{ m}^3$$

Bending About the x Axis. The maximum stress due to bending about the x axis is

$$\sigma_1 = \frac{M_x}{S_x} = 3.89 \text{ kN} \cdot \frac{\text{m}}{0.851 \times 10^{-3} \text{ m}^3} = 4570 \text{ kPa} = 4.57 \text{ MPa}$$

Due to M_x, the flexural stresses of points along CD are in tension and those of points along AB are in compression.

Bending About the y Axis. The maximum stress due to bending about the y axis is

$$\sigma_2 = \frac{M_y}{S_y} = \frac{2.25 \text{ kN} \cdot \text{m}}{0.624 \times 10^{-3} \text{ m}^3} = 3610 \text{ kPa} = 3.61 \text{ MPa}$$

Due to M_y, the flexural stresses of points along AD are in tension and those of points along BC are in compression.

Combined Stresses. Using the method of superposition, the stresses at corners A, B, C, and D at the midspan are

$$\sigma_A = -\sigma_1 + \sigma_2 = -4.57 + 3.61 = -0.96 \text{ MPa (C)}$$

$$\sigma_B = -\sigma_1 - \sigma_2 = -4.57 - 3.61 = -8.18 \text{ MPa (C)}$$

$$\sigma_C = +\sigma_1 - \sigma_2 = +4.57 - 3.61 = +0.96 \text{ MPa (T)}$$

$$\sigma_D = +\sigma_1 + \sigma_2 = +4.57 + 3.61 = +8.18 \text{ MPa (T)}$$

Thus,

$$\sigma_{max}^{(T)} = \sigma_D = +8.18 \text{ MPa} \qquad\qquad \Leftarrow \textbf{Ans.}$$

$$\sigma_{max}^{(C)} = \sigma_B = -8.18 \text{ MPa} \qquad\qquad \Leftarrow \textbf{Ans.}$$

EXAMPLE 18–4

A bridge-type crane consists of a beam that can roll on rails fastened to supports. The load lifted by the crane is attached to a cart that can slide along the beam, as shown in Fig. E18–4(1). The W12 × 30 beam has a 12-ft simple span and carries a maximum weight, including the weight of the movable cart, of $P = 8$ kips. When the beam moves on the rail, the load P makes an angle θ of 15° with the longitudinal vertical plane, as shown in Fig. E18–4(2), the cross-section of the beam. Determine the maximum tensile stress in the beam.

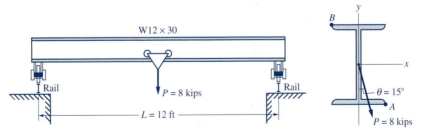

FIGURE E18–4(1)   FIGURE E18–4(2)

Solution. The maximum moment occurs when the movable cart is located at the midspan of the beam and the critical section is at the midspan. The load P can be resolved into components along the x and y axes:

$$P_x = P \sin \theta = 8 \sin 15° = 2.07 \text{ kips}$$

$$P_y = P \cos \theta = 8 \cos 15° = 7.73 \text{ kips}$$

The weight of the beam acts along the y axis; thus,

$$w_y = 30 \text{ lb/ft} = 0.03 \text{ kip/ft}$$

Bending About the x Axis. The maximum moment at the midspan about the x axis is

$$M_x = \frac{P_y L}{4} + \frac{w_y L^2}{8}$$

$$= \frac{(7.73 \text{ kips})(12 \text{ ft})}{4} + \frac{(0.03 \text{ kip/ft})(12 \text{ ft})^2}{8}$$

$$= 23.7 \text{ kip} \cdot \text{ft} = 285 \text{ kip} \cdot \text{in.}$$

From Table A–1(a), the section moduli of a W12 × 30 shape are

$$S_x = 38.6 \text{ in.}^3 \qquad S_y = 6.24 \text{ in.}^3$$

The maximum tensile stress at the bottom of the midspan is

$$\sigma_1 = \frac{M_x}{S_x} = \frac{285 \text{ kip} \cdot \text{in.}}{38.6 \text{ in.}^3} = 7.38 \text{ ksi}$$

Bending About the y Axis. The maximum moment at the midspan about the y axis is

$$M_y = \frac{P_x L}{4} = \frac{(2.07 \text{ kips})(12 \text{ ft})}{4}$$

$$= 6.21 \text{ kip} \cdot \text{ft} = 74.5 \text{ kip} \cdot \text{in.}$$

The maximum tensile stress at the outer fibers on the right-hand side of the midspan is

$$\sigma_2 = \frac{M_y}{S_y} = \frac{74.5 \text{ kip} \cdot \text{in.}}{6.24 \text{ in.}^3} = 11.9 \text{ ksi}$$

The Combined Stresses. Due to bending about both axes, the combined stresses can be determined by superposition. The maximum tensile stress due to biaxial bending at point A of the midspan is

$$\sigma_A = +\sigma_1 + \sigma_2 = +7.38 + 11.9 = +19.3 \text{ ksi}$$

$$\sigma_{max}^{(T)} = \sigma_A = +19.3 \text{ ksi} \qquad\qquad \Leftarrow \textbf{Ans.}$$

Remark. To see the effect of the inclined load, consider a load P applied in the vertical direction with $\theta = 0°$. The maximum moment at the midspan is

$$M_{max} = \frac{PL}{4} + \frac{wL^2}{8}$$

$$= \frac{(8 \text{ kips})(12 \text{ ft})}{4} + \frac{(0.03 \text{ kip/ft})(12 \text{ ft})^2}{8}$$

$$= 24.5 \text{ kip} \cdot \text{ft} = 294 \text{ kip} \cdot \text{in.}$$

The maximum bending stress in the beam is

$$\sigma_{max} = \frac{M_{max}}{S_x} = \frac{294 \text{ kip} \cdot \text{in.}}{38.6 \text{ in.}^2} = 7.62 \text{ ksi}$$

Note that when the load is inclined at 15° from the longitudinal vertical plane, the maximum flexural stress is 19.3 ksi, which is 2.5 times the stress caused by a vertical load only. The bending resistance of a wide-flange section about the y axis is much weaker than its strength about the x axis. Therefore, the adverse effect of bending about the y axis of the section must not be overlooked.

18–4
ECCENTRICALLY LOADED MEMBERS

Eccentric loading is a special case of combined axial and flexural stresses. The method of superposition can be applied to a short compression member that has small deflections and will not buckle under compressive loads. When the axial load is applied through the centroid of the cross-sections (i.e., without eccentricity), the normal stress produced is distributed uniformly over the cross-section of the member. If the axial load is applied at E with an eccentricity e from the centroid C, as shown in Fig. 18–2a, the normal stresses will no longer be distributed uniformly in a cross-section. An eccentric axial load can be replaced by a concentric force and a couple. To determine the equivalent force-couple system, we place two equal and opposite forces P at D, as shown in Fig. 18–2b.

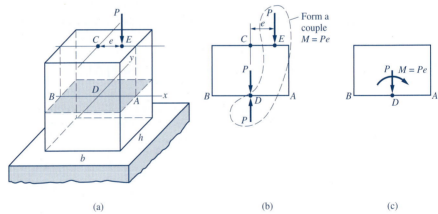

(a)   (b)   (c)

FIGURE 18–2

The original downward force at E and the upward force at D form a couple $M = Pe$. The system is thus reduced to a concentric force P at the centroid D and a couple M, as shown in Fig. 18–2c. Now we have an axial force and a bending moment, and the combined stresses can be determined in the same way as in Section 18–2. The axial force produces a uniform compressive stress throughout the section. The bending moment produces maximum compressive stress at A and maximum tensile stress at B. Using the method of superposition, the normal stresses at points A and B are, respectively,

$$\sigma_A = -\frac{P}{A} - \frac{M}{S} \qquad (18\text{–}1)$$

$$\sigma_B = -\frac{P}{A} + \frac{M}{S} \qquad (18\text{–}2)$$

Maximum Eccentricity. From Equations 18–1 and 18–2, we see that the normal stress at A is always compressive, while the normal stress at B could be compressive, tensile, or zero, depending on the eccentricity of the load. Some materials, such as concrete, are very weak in resisting tension. For these materials, it is important to keep the eccentricity of a compressive load to a certain maximum limit so that no tensile stress will develop anywhere in the member. The maximum eccentricity may be obtained by equating Equation 18–2 to zero and solving for e. If the width and height of the cross-section are b and h, respectively, we write

$$\sigma_B = -\frac{P}{A} + \frac{M}{S} = -\frac{P}{bh} + \frac{Pe}{hb^2/6} = 0$$

From which we get

$$e = \frac{b}{6} \qquad (18\text{–}3)$$

This is the maximum eccentricity for which tensile stress will not occur anywhere in the member. At an eccentricity greater than $h/6$, tensile stresses will develop in some fibers in the member. The eccentricity on the other side has the same limiting value. Hence, we can say that in general the compressive load applied along either centroidal axis of a rectangular section must be within the *middle third* of the section if no tensile stress is to occur. Since a gravitational dam must not be subjected to tensile stress at any point at its base, the resultant force must be acting within the middle third of the base.

EXAMPLE 18–5

Refer to Fig. E18–5(1). A full-size, 6 in. × 10 in. rectangular short timber post carries an eccentrically placed axial load $P = 12\ 000$ lb as shown. Determine the normal stresses at points A and B.

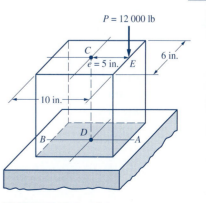

$P = 12\ 000$ lb

$e = 5$ in.

6 in.

10 in.

FIGURE E18–5(1)

Solution. Refer to Fig. E18–5(2). The given load at E is equivalent to a concentric axial force at D and a couple

$$M = Pe = (12\ 000\ \text{lb})(5\ \text{in.}) = 60\ 000\ \text{lb} \cdot \text{in.}$$

The normal stress at point A is

FIGURE E18–5(2)

$$\sigma_A = -\frac{P}{A} - \frac{M}{S}$$

$$= -\frac{12\ 000\ \text{lb}}{(6\ \text{in.})(10\ \text{in.})} - \frac{60\ 000\ \text{lb} \cdot \text{in.}}{(6\ \text{in.})(10\ \text{in.})^2/6}$$

$$= -200\ \text{psi} - 600\ \text{psi} = -800\ \text{psi (C)} \qquad \Leftarrow \textbf{Ans.}$$

The normal stress at point B is

$$\sigma_B = -\frac{P}{A} + \frac{M}{S}$$

$$= -200\ \text{psi} + 600\ \text{psi} = +400\ \text{psi (T)} \qquad \Leftarrow \textbf{Ans.}$$

EXAMPLE 18–6

A punch press has a cast-iron frame with the cross-section shown in Fig. E18–6(1). Determine the maximum tensile and compressive normal stresses in section 1–1 due to a load $P = 45$ kN on the press.

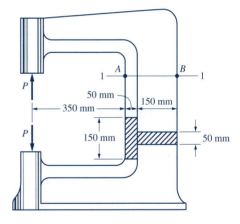

FIGURE E18–6(1)

Solution. See Fig. E18–6(2). The properties of the cross-section must be computed first. The cross-sectional area is

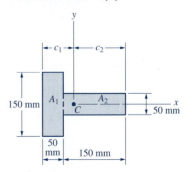

FIGURE E18–6(2)

$$A = A_1 + A_2$$

$$= (0.05 \text{ m})(0.15 \text{ m}) + (0.15 \text{ m})(0.05 \text{ m})$$

$$= 0.015 \text{ m}^2$$

Using the left side of the cross-section as a reference axis, the centroid C is located at

$$\overline{x} = \frac{A_1 x_1 + A_2 x_2}{A}$$

$$= \frac{(0.05 \times 0.15)(0.025) + (0.15 \times 0.05)(0.125)}{0.015}$$

$$= 0.075 \text{ m}$$

Thus,

$$c_1 = \overline{x} = 0.075 \text{ m}$$

$$c_2 = 0.20 \text{ m} - 0.075 \text{ m} = 0.125 \text{ m}$$

The moment of inertia of the section with respect to the centroidal y axis is

$$I = \Sigma[\overline{I} + A(\overline{x} - x)^2]$$

$$= \left[\frac{0.15(0.05)^3}{12} + 0.0075(0.075 - 0.025)^2\right]$$

$$+ \left[\frac{0.05(0.15)^3}{12} + 0.0075(0.075 - 0.125)^2\right]$$

$$= 5.31 \times 10^{-5} \text{ m}^4$$

Relative to section 1–1 of the frame, the load P has a large eccentricity e from the centroid of the section, as shown in Fig. E18–6(3), the free-body diagram of the upper part of the frame. The eccentric load results in a tensile force P at the centroid C and a bending moment $M = Pe$. The eccentricity is

$$e = 0.35 \text{ m} + 0.075 \text{ m} = 0.425 \text{ m}$$

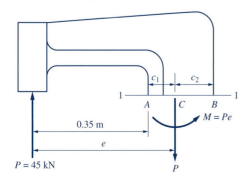

FIGURE E18–6(3)

The moment at the section is

$$M = Pe = (45 \text{ kN})(0.425 \text{ m}) = 19.1 \text{ kN} \cdot \text{m}$$

Note that the tensile stress due to the direct tensile force is distributed uniformly over the cross-section. The bending moment will cause a maximum tension at A and a maximum compression at B. Hence, by the method of superposition, we get

$$\sigma_A = +\frac{P}{A} + \frac{Mc_1}{I} = \frac{45}{0.015} + \frac{(19.1)(0.075)}{5.31 \times 10^{-5}}$$

$$= +3000 + 27\,000 = +30\,000 \text{ kPa}$$

$$\sigma_{max}^{(T)} = 30 \text{ MPa} \qquad \Leftarrow \textbf{Ans.}$$

$$\sigma_B = +\frac{P}{A} - \frac{Mc_2}{I} = +\frac{45}{0.015} - \frac{(19.1)(0.125)}{5.31 \times 10^{-5}}$$

$$= +3000 - 45\,000 = -42\,000 \text{ kPa}$$

$$\sigma_{max}^{(C)} = 42 \text{ MPa} \qquad \Leftarrow \textbf{Ans.}$$

18–5
DOUBLE ECCENTRICITY

The eccentric load discussed in the preceding section is on one of the centroidal axes. In this section, we will consider the stresses produced by an eccentric load that is not on either one of the centroidal axes, as shown in Fig. 18–3a. This case is referred to as *double eccentricity.* The eccentric load can be replaced by its equivalent centroidal force and moments about both axes, as shown in Fig. 18–3b. Using the method of superposition, the combined normal stress at any point is the algebraic sum of the normal stresses caused by the direct axial force, as well as the bending moments about both centroidal axes. The stresses in the fibers at the corner of the member can be computed from

$$\sigma = -\frac{P}{A} \pm \frac{M_x}{S_x} \pm \frac{M_y}{S_y} \qquad (18\text{-}4)$$

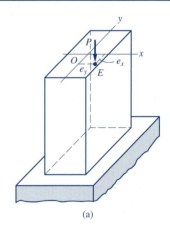

(a)

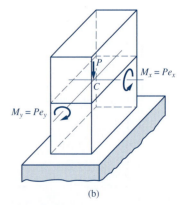

(b)

FIGURE 18–3

───────── **EXAMPLE 18–7** ─────────

The rectangular block in Fig. E18–7(1) is subjected to the axial force $P = 20$ kips applied at E as shown. The member has a full size 4-in. × 8-in. section. Determine the normal stresses at points A, B, C, and D, and show the normal stress distribution in the section.

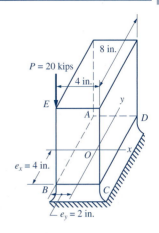

FIGURE E18–7(1)

Solution. See Fig. E18–7(2). The equivalent loading for the section $ABCD$ consists of an axial compressive force $P = 20$ kips and bending moments M_x and M_y equal to, respectively,

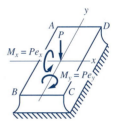

FIGURE E18–7(2)

$$M_x = Pe_x = (20 \text{ kips})(4 \text{ in.}) = 80 \text{ kip} \cdot \text{in.}$$

$$M_y = Pe_y = (20 \text{ kips})(2 \text{ in.}) = 40 \text{ kip} \cdot \text{in.}$$

Using the method of superposition, the normal stresses at points A, B, C, and D due to the axial compression and the bending moments can be computed in the following:

$$\sigma_A = -\frac{P}{A} + \frac{M_x}{S_x} - \frac{M_y}{S_y}$$

$$= -\frac{20}{(4)(8)} + \frac{80}{(4)(8)^2/6} - \frac{40}{(8)(4)^2/6}$$

$$= -0.625 + 1.875 - 1.875$$

$$= -0.625 \text{ ksi (C)} \qquad\qquad ⇐ \textbf{Ans.}$$

$$\sigma_B = -\frac{P}{A} - \frac{M_x}{S_x} - \frac{M_y}{S_y}$$

$$= -0.625 - 1.875 - 1.875$$

$$= -4.375 \text{ ksi (C)} \qquad\qquad ⇐ \textbf{Ans.}$$

$$\sigma_C = -\frac{P}{A} - \frac{M_x}{S_x} + \frac{M_y}{S_y}$$

$$= -0.625 - 1.875 + 1.875$$

$$= -0.625 \text{ ksi (C)} \qquad\qquad \Leftarrow \textbf{Ans.}$$

$$\sigma_D = -\frac{P}{A} + \frac{M_x}{S_x} + \frac{M_y}{S_y}$$

$$= -0.625 + 1.875 + 1.875$$

$$= +3.125 \text{ ksi (T)} \qquad\qquad \Leftarrow \textbf{Ans.}$$

The normal stress distribution is plotted as shown in Fig. E18–7(3). Note that there is no normal stress along the line of zero stress.

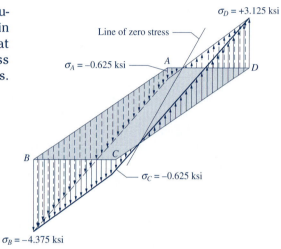

FIGURE E18–7(3)

EXAMPLE 18–8

Find the zone over which a vertical downward force may be applied without causing tensile stresses at any point in the rectangular block shown in Fig. E18–8(1). Neglect the weight of the block.

Solution. Let the force P be placed at point $E(x, y)$ in the first quadrant of the x–y coordinate system shown. If tensile stresses exist at any point in section $ABCD$, the maximum tensile stress must occur at point B and be equal to

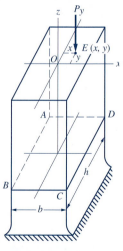

FIGURE E18–8(1)

$$\sigma_B = -\frac{P}{A} + \frac{M_x}{S_x} + \frac{M_y}{S_y} = -\frac{P}{A} + \frac{Py}{bh^2/6} + \frac{Px}{hb^2/6}$$

Since $A = bh$, the factor P/A can be factored. Thus,

$$\sigma_B = \frac{P}{A}\left(-1 + \frac{6y}{h} + \frac{6x}{b}\right)$$

Setting the stress at B to zero would fulfill the limiting condition of the problem.

$$\sigma_B = \frac{P}{A}\left(-1 + \frac{6y}{h} + \frac{6x}{b}\right) = 0$$

From which we get

$$\frac{6x}{b} + \frac{6y}{h} = 1$$

which is the equation of a straight line. From this equation, it follows that when $x = 0$, $y = h/6$; and when $y = 0$, $x = b/6$. These results locate two points $G(0, h/6)$ and $H(b/6, 0)$, which define the line L shown in Fig. E18–8(2). A vertical force applied to the block at a point along the line L between G and H will cause the stress at B to be equal to zero, and compressive stresses will occur at all the other points in the section. If the force P acts anywhere to the upper right of this line, tensile stresses will occur in the member. As far as the first quadrant is concerned, the point of application of the force P should be within the shaded area OGH so that no tensile stresses occur anywhere in the section.

A similar situation occurs in other quadrants. The diamond-shaped area [shown shaded in Fig. E18–8(3)] is called the *kern* of the section. A vertical compressive force applied at a point within the kern will not cause tensile stresses at any point in the member. Note that if a compressive load is applied along an axis of symmetry, the maximum limit within the kern is consistent with the result obtained in Section 18–4.

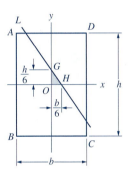

FIGURE E18–8(2)

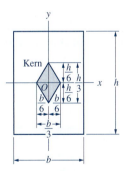

FIGURE E18–8(3)

18–6
STRESSES ON AN INCLINED PLANE

Plane Stress. When an element is subjected to normal and shear stresses in only two directions, as shown in Fig. 18–4a, the element is said to be in a *state of plane stress*. A plane stress can be represented by a two-dimensional figure, as shown in Fig. 18–4b, where σ_x and τ_x are the normal and shear stresses acting on the vertical planes whose normals are along the x axis, and σ_y and τ_y are the normal and shear stresses acting on the horizontal planes whose normals are along the y axis. When all the stresses, σ_x, σ_y, τ_x, and τ_y, co-exist, the element is in a state of *general plane stress*.

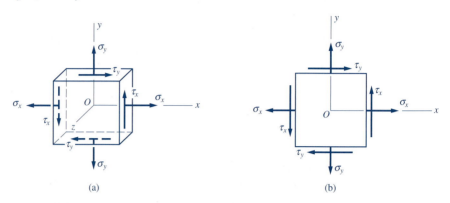

(a)   (b)

FIGURE 18–4

The state of general plane stress usually occurs in the combined loading condition. For example, when a thin-walled cylindrical vessel is subjected simultaneously to internal pressure and external torque, a rectangular element on the wall is in a state of general plane stress, as shown later in Fig. 18–5. The following are special cases of plane stress:

Uniaxial Stress.   An element is said to be in the *uniaxial stress* condition if the normal stress occurs only along one direction and there is no shear stress.

Biaxial Stress.   An element is said to be in the *biaxial stress* condition if the normal stresses occur along two perpendicular directions and there is no shear stress.

Pure Shear.   An element is said to be in *pure shear* if it is subjected to shear stresses only.

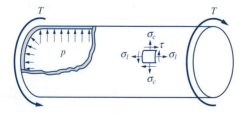

FIGURE 18–5

Stresses on an Inclined Plane.   In Section 9–6, stresses on inclined planes in an axially loaded member (a uniaxial stress condition) were discussed. In this section, formulas for stresses on an inclined plane for the general state of plane stress will be developed. Consider an element in a state of general plane stress, as shown in Fig. 18–6a. The formulas for the normal and shear stresses on the inclined plane BC, denoted by σ_θ and τ_θ, respectively, are to be derived. Before doing so, it is important to establish the sign conventions for the quantities involved.

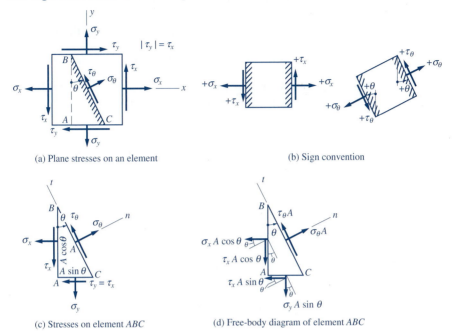

(a) Plane stresses on an element

(b) Sign convention

(c) Stresses on element ABC

(d) Free-body diagram of element ABC

FIGURE 18–6

1. ***Sign for the Normal Stress.*** Tensile stresses are considered positive, and compressive stresses are considered negative.
2. ***Sign for the Shear Stress.*** Shear stresses are considered positive when the pair of shear stresses acting on the opposite sides of the element form a counterclockwise couple, as shown in Fig. 18–6b.
3. ***Sign for the Angle of Inclination.*** The angle of inclination θ is measured from the vertical plane toward the inclined plane. The counterclockwise measurement is considered positive, and the clockwise measurement is considered negative, as shown in Fig. 18–6b.

From these sign conventions, we see that all the stresses are shown in the positive direction in the element of Fig. 18–6a, except τ_y, which is negative. Because the shear stresses on perpendicular planes must be equal (Section 12–4), the absolute value of the shear stress τ_y must be equal to τ_x.

To derive the formula for σ_θ and τ_θ, a triangular element ABC is isolated from the element in Fig. 18–6a. The inclined plane BC makes an angle θ with the vertical direction. If the area of the inclined plane is A, the area of the horizontal plane AC is $A \sin \theta$, and the area of the vertical plane AB is $A \cos \theta$, as shown in Fig. 18–6c. A free-body diagram of the wedge is shown in Fig. 18–6d, where the normal and shear forces acting on an area are obtained by multiplying the normal and shear stresses by the area. Writing the equilibrium equation along the direction normal to the incline, we obtain:

$$\Sigma F_n = \sigma_\theta A - (\sigma_x A \cos \theta)(\cos \theta) - (\sigma_y A \sin \theta)(\sin \theta)$$
$$- (\tau_x A \cos \theta)(\sin \theta) - (\tau_x A \sin \theta)(\cos \theta) = 0$$

From which we get

$$\sigma_\theta = \sigma_x \cos^2 \theta + \sigma_y \sin^2 \theta + 2\tau_x \sin \theta \cos \theta \tag{a}$$

Similarly, the equilibrium equation along the tangential direction will lead to

$$\tau_\theta = -(\sigma_x - \sigma_y) \sin \theta \cos \theta + \tau_x (\cos^2 \theta - \sin^2 \theta) \tag{b}$$

The above equations can be expressed in a more convenient form if the following trigonometric identities are introduced:

$$\cos^2 \theta = \frac{1}{2}(1 + \cos 2\theta)$$

$$\sin^2 \theta = \frac{1}{2}(1 - \cos 2\theta)$$

$$\sin \theta \cos \theta = \frac{1}{2}(\sin 2\theta)$$

Substituting these identities into Equations (a) and (b), we obtain

$$\sigma_\theta = \frac{\sigma_x + \sigma_y}{2} + \frac{\sigma_x - \sigma_y}{2} \cos 2\theta + \tau_x \sin 2\theta \tag{18–5}$$

$$\tau_\theta = -\frac{\sigma_x - \sigma_y}{2} \sin 2\theta + \tau_x \cos 2\theta \tag{18–6}$$

These formulas may be used to compute the normal and shear stresses on an inclined plane. It is important to note that the sign conventions presented earlier in this section must be followed exactly.

EXAMPLE 18–9

For a small element with the state of plane stress as shown in Fig. E18–9(1), determine the stresses acting on the inclined plane m–m.

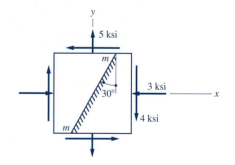

FIGURE E18–9(1)

Solution. According to the sign convention established in this section, the state of plane stress of the element and the angle of inclination of the inclined plane are

$$\sigma_x = -3 \text{ ksi}$$

$$\sigma_y = +5 \text{ ksi}$$

$$\tau_x = -4 \text{ ksi}$$

$$\theta = -30° \qquad 2\theta = -60°$$

Substituting these values into Equations 18–5 and 18–6, we find

$$\sigma_\theta = \frac{-3+5}{2} + \frac{-3-5}{2} \cos{(-60°)} + (-4) \sin{(-60°)}$$

From which we get

$$\sigma_\theta = +2.46 \text{ ksi} \qquad \Leftarrow \textbf{Ans.}$$

$$\tau_\theta = \frac{-3-5}{2} \sin{(-60°)} + (-4) \cos{(-60°)}$$

From which we get

$$\tau_\theta = -5.46 \text{ ksi} \qquad \Leftarrow \textbf{Ans.}$$

These stresses on the inclined plane are shown in Fig. E18–9(2). Note that a positive normal stress indicates that the plane is in tension. The negative sign for the shear stress indicates that the pair of shear stresses on the incline form a clockwise couple.

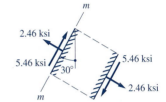

FIGURE E18–9(2)

18–7
MOHR'S CIRCLE

When the parameter θ is eliminated from Equations 18–5 and 18–6, a functional relationship between σ_θ and τ_θ is obtained. To do this, the two equations are rewritten as

$$\sigma_\theta - \frac{\sigma_x + \sigma_y}{2} = \frac{\sigma_x - \sigma_y}{2} \cos 2\theta + \tau_x \sin 2\theta$$

$$\tau_\theta = -\frac{\sigma_x - \sigma_y}{2} \sin 2\theta + \tau_x \cos 2\theta$$

Squaring both sides of each equation gives

$$\left(\sigma_\theta - \frac{\sigma_x + \sigma_y}{2}\right)^2 = \left(\frac{\sigma_x - \sigma_y}{2}\right)^2 \cos^2 2\theta + (\sigma_x - \sigma_y)\tau_x \sin 2\theta \cos 2\theta + \tau_x^2 \sin^2 2\theta \quad \text{(a)}$$

$$\tau_\theta^2 = \left(\frac{\sigma_x - \sigma_y}{2}\right)^2 \sin^2 2\theta - (\sigma_x - \sigma_y)\tau_x \sin 2\theta \cos 2\theta + \tau_x^2 \cos^2 2\theta \quad \text{(b)}$$

Adding Equations (a) and (b) and using the trigonometric identity $\sin^2 2\theta + \cos^2 2\theta = 1$, we get

$$\left(\sigma_\theta - \frac{\sigma_x + \sigma_y}{2}\right)^2 + \tau_\theta^2 = \left(\frac{\sigma_x - \sigma_y}{2}\right)^2 + \tau_x^2 \qquad \text{(18–7)}$$

If we let

$$a = \frac{\sigma_x + \sigma_y}{2}$$

and

$$r = \sqrt{\left(\frac{\sigma_x - \sigma_y}{2}\right)^2 + \tau_x^{\,2}} \qquad (18\text{–}8)$$

then Equation 18–7 becomes

$$(\sigma_\theta - a)^2 + (\tau_\theta - 0)^2 = r^2 \qquad (18\text{–}9)$$

This is the equation of a circle with its center at $(a, 0)$ and radius r. When this circle is plotted on σ–τ coordinate axes, it is called *Mohr's circle of stress*. The coordinates of a point on Mohr's circle represent the normal and shear stresses on a certain inclined plane.

To draw Mohr's circle, the following steps are recommended.

1. Sketch a rectangular element and indicate on the element the given state of plane stress (Fig. 18–7a). The sign conventions established in Section 18–6 must be followed strictly.

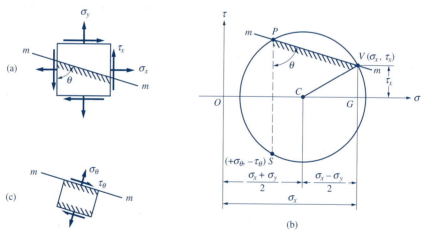

FIGURE 18–7

2. Set up a rectangular coordinate system with the origin O located at a proper point. The axis for normal stress is in the horizontal direction, to the right of O as positive; the axis for shear stress is in the vertical direction, above O as positive. See Fig. 18–7b. Select a proper scale for the axes.

3. To draw Mohr's circle (Fig. 18–7b), first locate the center C along the horizontal σ axis at $(\sigma_x + \sigma_y)/2$ from the origin O. Then plot point V at (σ_x, τ_x) corresponding to the stress condition on the vertical plane. With C as the center and CV as the radius, draw a circle passing through V. The circle constructed is Mohr's circle for the element with the given state of plane stress.

Mohr's circle is useful for problems involving plane stresses. In this section, we will use Mohr's circle for the following two purposes: to find the stresses on an inclined plane, and to find an inclined plane with the stresses given. Another important application will be presented in the next section.

To Find the Stresses on an Inclined Plane.   From point V on Mohr's circle (Fig. 18-7b), draw line VP parallel to the inclined plane and locate point P on the circle. The coordinates of S, a point on the circle vertically opposite from P, give the stresses acting on the inclined plane. The coordinates of the point S are identified as $(+\sigma_\theta, -\tau_\theta)$. A positive σ_θ indicates a tensile stress, and a negative value of τ_θ indicates that the shear stresses on the opposite sides of an element form a clockwise couple. The stresses on the inclined plane m–m are shown in Fig. 18-7c.

To Find an Inclined Plane with the Stresses Given.   To find the plane whose stresses are indicated by a point S on Mohr's circle, we can proceed by reversing the steps outlined in the preceding paragraph. Locate the point P on the circle vertically opposite from S; then the desired inclined plane is parallel to the line PV.

EXAMPLE 18-10 ————————————————————————————————

Rework Example 18-9 by constructing a Mohr's circle for the element. The element and the given state of plane stress are sketched as shown in Fig. E18-10(1).

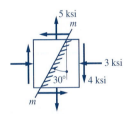

5 ksi
m
3 ksi
4 ksi
30°
m

FIGURE E18-10(1)

Solution. The center C of Mohr's circle is located at

$$\frac{(\sigma_x + \sigma_y)}{2} = \frac{(-3 \text{ ksi} + 5 \text{ ksi})}{2} = +1 \text{ ksi}$$

on the σ axis. The stresses on the vertical plane of the element are $(-3, -4)$, which are the coordinates of point V on the circle. These points are plotted to scale, and Mohr's circle is drawn by using C as the center and CV as the radius, as shown in Fig. E18-10(2).

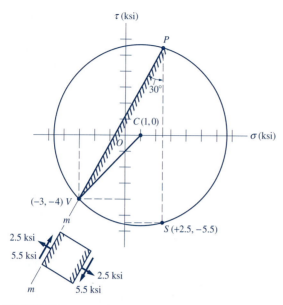

FIGURE E18–10(2)

A line VP drawn parallel to the plane $m-m$ locates point P. The coordinates of point S vertically opposite from P give the stresses acting on the inclined plane $m–m$. The coordinates of point S are

$$S(+2.5, -5.5)$$

Therefore,

$$\sigma_\theta = +2.5 \text{ ksi} \qquad \tau_\theta = -5.5 \text{ ksi} \qquad \Leftarrow \textbf{Ans.}$$

These stresses are indicated on the incline as shown in Fig. E18–10(2).

18–8
PRINCIPAL STRESSES AND MAXIMUM SHEAR STRESSES

The maximum and minimum normal stresses in an element are called the *principal stresses*. The planes where the principal stresses occur are called the *principal planes*. Mohr's circle can be used readily to determine the principal stresses and the principal planes.

The Principal Stresses. Consider a stressed element and its corresponding Mohr's circle shown in Fig. 18–8a and b. The maximum normal stress (denoted by σ_1) is given by the abscissa of point A, and the minimum normal stress (denoted by σ_2) is given by the abscissa of point B. The abscissa of A is $OC + r$ and the abscissa of B is $OC - r$. Therefore, the principal stresses are

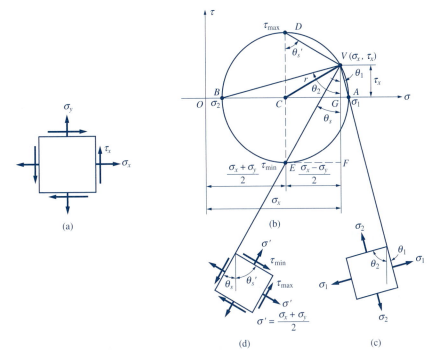

FIGURE 18–8

$$\sigma_1 = a + r \tag{18–10a}$$

$$\sigma_2 = a - r \tag{18–10b}$$

where

$$a = \frac{\sigma_x + \sigma_y}{2} \qquad b = \frac{\sigma_x - \sigma_y}{2} \qquad r = \sqrt{b^2 + \tau_x^2} \tag{18–11}$$

The Principal Planes. Recall from Section 18–7 that the plane whose stresses are equal to the coordinates of a point S on Mohr's circle is parallel to line VP, where P is vertically opposite to S. In the Mohr's circle in Fig. 18–8b, the point vertically opposite to A is point A itself. Therefore, the plane for the maximum normal stress is parallel to line VA, and the plane for the minimum normal stress is parallel to line VB. Since angle AVB is 90°, the principal planes are perpendicular to each other. Figure 18–8c shows the principal planes and the principal stresses. Note that no shear stress acts on the principal planes.

From the Mohr's circle in Fig. 18–8b, we can get a formula for calculating θ_1 for the plane of maximum normal stress σ_1. In triangle AVG, we find

$$\theta_1 = \angle AVG = \tan^{-1}\frac{AG}{VG} = \tan^{-1}\left[\frac{r - \frac{1}{2}(\sigma_x - \sigma_y)}{\tau_x}\right]$$

Hence, we obtain

$$\theta_1 = \tan^{-1}\left[\frac{r - b}{\tau_x}\right] \tag{18–12}$$

where b and r are defined in Equation 18–11.

Since the principal planes are $90°$ apart, the angle θ_2 for the minimum normal stress σ_2 is

$$\theta_2 = \theta_1 \pm 90° \qquad (18\text{–}13)$$

which means that the angle θ_2 for the minimum normal stress is θ_1 plus or minus $90°$. Usually the sign that will give a positive or negative acute angle of θ_2 is chosen.

The Maximum and Minimum Shear Stresses. In Mohr's circle, the maximum shear stress is given by the ordinate of point D, and the minimum shear stress is given by the ordinate of point E. At these points, the absolute value of shear stress is equal to the radius r of Mohr's circle. Thus,

$$\tau_{max} = +r \qquad (18\text{–}14a)$$

$$\tau_{max} = -r \qquad (18\text{–}14b)$$

where b and r are defined in Equation 18–11.

Planes for Maximum and Minimum Shear Stresses. The plane on which the maximum shear stress acts is parallel to line VE, where E is vertically opposite to D. The plane on which the minimum shear stress acts is parallel to line VD, where D is vertically opposite to E. The two planes are $90°$ apart. Note that the angle between lines VE and VA is $45°$; therefore, the plane for maximum shear stress and the principal planes are $45°$ apart. Figure 18–8d shows the maximum and minimum shear stresses and the planes on which they act. The normal stresses on the planes of maximum and minimum shear stresses, called the *associated normal stresses,* are given by the expression

$$\sigma' = \sigma_{avg} = a \qquad (18\text{–}15)$$

where a is defined in Equation 18–11.

From the Mohr's circle in Fig. 18–8b, we can get a formula for calculating θ_s for the plane with the maximum shear stress τ_{max}. In triangle VEF, we have

$$\theta_s = -\angle VEF = -\tan^{-1}\left[\frac{EF}{FV}\right] = -\tan^{-1}\left[\frac{\frac{1}{2}(\sigma_x - \sigma_y)}{r + \tau_x}\right]$$

Hence we obtain

$$\theta_s = -\tan^{-1}\left[\frac{b}{r + \tau_x}\right] \qquad (18\text{–}16)$$

where b and r are defined in Equation 18–11.

Since the planes for the maximum and minimum shear stresses are 90° apart, the angle θ_s' for the plane with the minimum shear stress is

$$\theta_s' = \theta_s \pm 90° \qquad (18\text{–}17)$$

which means that the angle θ_s' for the plane of the minimum shear stress is θ_s plus or minus 90°. Usually the sign that will give a positive or negative acute angle of θ_s' is chosen.

Two special cases of plane stress merit particular attention: all-around tension and pure shear.

All-Around Tension.   In Section 9–7, it was pointed out that the normal stress in the wall of a spherical vessel subjected to internal pressure is the same along any direction. Thus, the state of stress on an element isolated from the wall is as shown in Fig. 18–9a, where σ_x and σ_y are equal. The Mohr's circle of the element is a point circle, as shown in Fig. 18–9b. Therefore, the normal stress on any inclined plane is equal to σ_x and shear stresses do not exist on any plane. This state of stress is called *all-around tension*.

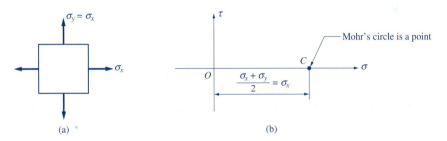

(a)   (b)

FIGURE 18–9

Pure Shear.   A condition of *pure shear* occurs when only shear stresses exist in two mutually perpendicular directions, as shown in Fig. 18–10a. The Mohr's circle of the element is shown in Fig. 18–10b. The principal stresses and principal planes are shown in Fig. 18–10c. Note that the principal planes form angles of +45° and −45° with the vertical direction. The principal stresses and the maximum shear stress are equal to

$$\sigma_1 = -\sigma_2 = \tau_{max} = \tau_x$$

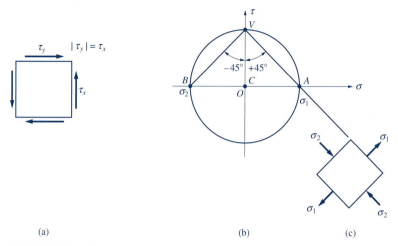

(a)   (b)   (c)

FIGURE 18–10

The condition of pure shear occurs at an element on the surface of a torsion bar, as shown in Fig. 18–11a. The maximum shear stress and the principal stresses are shown acting on the respective elements. From the torsion formula,

$$\tau_x = \frac{Tc}{J} = \frac{\pi \dfrac{d}{2}}{\dfrac{\pi d^4}{32}} = \frac{16T}{\pi d^3}$$

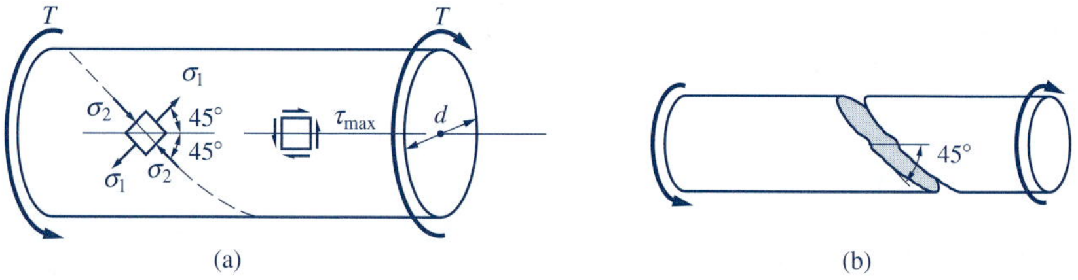

(a)   (b)

FIGURE 18–11

Most brittle materials are weak in tension. When subjected to torsion, brittle materials fail by tearing along a line perpendicular to the direction of σ_1. This torsional failure can be demonstrated in the classroom by twisting a piece of chalk until it fails, as shown in Fig. 18–11b. The failure takes place along a helix at 45° from the longitudinal direction. Shafts made from brittle materials such as cast iron, sandstone, or concrete fail a torsion test in this manner.

EXAMPLE 18–11

An element of a machine member is subjected to the stresses shown in Fig. 18–11(1). Determine (a) the principal stresses and the principal planes, and (b) the maximum and minimum shear stresses and the planes on which they occur. Show the results for both parts on properly oriented elements.

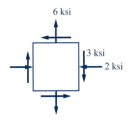

FIGURE E18–11(1)

Solution. Mohr's circle of stress can be constructed based on the following data:

1. Center C along the σ axis is

$$a = \frac{\sigma_x + \sigma_y}{2} = \frac{-2 + 6}{2} = +2$$

2. From the values of normal and shear stresses on the vertical plane, the coordinates of V are $(-2, -3)$ ksi.

Mohr's circle is drawn with the center at $C(2, 0)$ ksi passing through point $V(-2, -3)$ ksi, as shown in Fig. E18–11(2).

(a) The Principal Stresses. The parameter $a = 2$ ksi has already been computed above. The parameter b and the radius r of the circle are

$$b = \frac{\sigma_x - \sigma_y}{2} = \frac{-2 - 6}{2} = -4 \text{ ksi}$$

$$r = \sqrt{b^2 + \tau_x^2} = \sqrt{(-4)^2 + (-3)^2} = 5 \text{ ksi}$$

Thus,

$$\sigma_1 = a + r = 2 + 5 = +7 \text{ ksi}$$

$$\sigma_2 = a - r = 2 - 5 = -3 \text{ ksi}$$

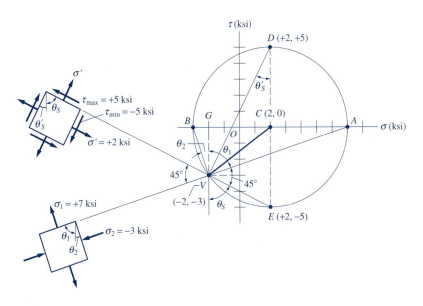

FIGURE E18–11(2)

The plane for the maximum normal stress is along line *VA*, and the plane for the minimum normal stress is along line *VB*. The principal planes and the principal stresses are sketched as shown in Fig. E18–11(2). The orientation of the principal planes can be determined either by measuring the angle with a protractor or computing from Equations 18–12 and 18–13. By direct measurement, the angle θ_1 is found to be approximately 71.5°. Using Equations 18–12 and 18–13, we find

$$\theta_1 = \tan^{-1}\left[\frac{r-b}{\tau_x}\right]$$

$$= \tan^{-1}\left[\frac{5-(-4)}{-3}\right] = -71.57° \qquad \Leftarrow \textbf{Ans.}$$

$$\theta_2 = \theta_1 + 90° = -71.57° + 90°$$

$$= 18.43° \qquad \Leftarrow \textbf{Ans.}$$

(b) The Maximum and Minimum Shear Stresses. The maximum shear stress τ_{max} and the associated normal stress σ' are given by the coordinates of point *D* (+2, +5). Thus,

$$\tau_{max} = +5 \text{ ksi} \qquad \sigma' = +2 \text{ ksi} \qquad \Leftarrow \textbf{Ans.}$$

These stresses act on the plane parallel to line *VE*. The angle θ_s for the plane of maximum shear stress and θ_s' for the plane of minimum shear stress can be calculated from Equations 18–16 and 18–17 as

$$\theta_s = -\tan^{-1}\left[\frac{b}{r+\tau_x}\right]$$

$$= -\tan^{-1}\left[\frac{-4}{5+(-3)}\right] = +63.43° \qquad \Leftarrow \textbf{Ans.}$$

$$\theta_s' = \theta_s - 90° = 63.43° - 90°$$

$$= -26.57° \qquad \Leftarrow \textbf{Ans.}$$

EXAMPLE 18–12

In the bracket subjected to the 50-kN load shown in Fig. E18–12(1), determine the principal stresses and the principal planes at point *A*, which is on the web just below the top flange of the cross-section.

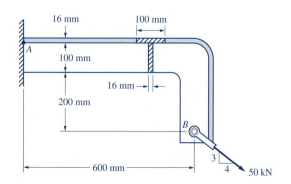

FIGURE E18–12(1)

Solution. Refer to Fig. E18–12(2). The area, centroid, and moment of inertia of the cross-section must be determined first.

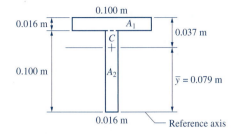

FIGURE E18–12(2)

$$A = A_1 + A_2 = 0.100(0.016) + 0.016(0.100)$$
$$= 0.0016 + 0.0016 = 0.0032 \text{ m}^2$$

$$\bar{y} = \frac{A_1 y_1 + A_2 y_2}{A}$$

$$= \frac{0.0016(0.108) + 0.0016(0.050)}{0.0032} = 0.079 \text{ m}$$

$$I = \Sigma[\bar{I} + A(\bar{y} - y)^2]$$

$$= \left[\frac{0.100(0.016)^3}{12} + 0.0016(0.079 - 0.108)^2\right]$$

$$+ \left[\frac{0.016(0.100)^3}{12} + 0.0016(0.079 - 0.050)^2\right]$$

$$= 4.06 \times 10^{-6} \text{ m}^4$$

The 50-kN load is resolved into horizontal and vertical components, as shown in Fig. E18–12(3), the free-body diagram of the bracket. The internal forces, including the axial force through the centroid C, the shear force, and a bending moment at the fixed section, are also shown. From the equilibrium conditions, we find

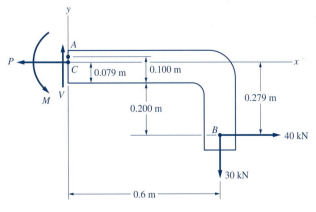

FIGURE E18–12(3)

The axial force: $P = 40$ kN (T)

The shear force: $V = 30$ kN

The bending moment: $M = 30(0.6) - 40(0.279)$

$= 6.84$ kN $\cdot$ m

These forces will cause a combined stress at point A. We have

$$\sigma_x = +\frac{P}{A} + \frac{My}{I}$$

$$= +\frac{40 \text{ kN}}{0.0032 \text{ m}^2} + \frac{(6.84 \text{ kM} \cdot \text{m})(0.100 \text{ m} - 0.079 \text{ m})}{4.06 \times 10^{-6} \text{ m}^4}$$

$$= 12\ 500 \text{ kPa} + 35\ 400 \text{ kPa}$$

$$= 47\ 900 \text{ kPa} = 47.9 \text{ MPa}$$

$$\sigma_y = 0$$

$$|\tau| = \frac{VQ}{It} = \frac{VA\bar{y}'}{It}$$

$$= \frac{(30 \text{ kN})(0.0016 \text{ m})(0.037 \text{ m} - 0.008 \text{ m})}{(4.06 \times 10^{-6} \text{ m}^4)(0.016 \text{ m})}$$

$$= 21\ 400 \text{ kPa} = 21.4 \text{ MPa}$$

Thus, the element at A is subjected to the stresses shown in Fig. E18–12(4). According to the sign convention presented in Section 18–6, the given stresses are

$$\sigma_x = 47.9 \text{ MPa}$$

$$\sigma_y = 0$$

$$\tau_x = -21.4 \text{ MPa}$$

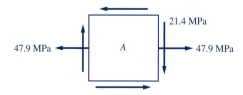

FIGURE E18–12(4)

The parameters, a and b, and the radius r are

$$a = \frac{\sigma_x + \sigma_y}{2} = \frac{47.9 + 0}{2} = 24.0 \text{ MPa}$$

$$b = \frac{\sigma_x - \sigma_y}{2} = \frac{47.9 - 0}{2} = 24.0 \text{ MPa}$$

$$r = \sqrt{b^2 + \tau_x^2} = \sqrt{(24.0)^2 + (-21.4)^2} = 32.1 \text{ MPa}$$

Thus,

$$\sigma_1 = a + r = 24.0 + 32.1 = +56.1 \text{ MPa} \qquad \Leftarrow \textbf{Ans.}$$

$$\sigma_2 = a - r = 24.0 - 32.1 = -8.1 \text{ MPa} \qquad \Leftarrow \textbf{Ans.}$$

The principal planes are oriented by the angles

$$\theta_1 = \tan^{-1}\left[\frac{r - b}{\tau_x}\right]$$

$$= \tan^{-1}\left[\frac{32.1 - 24.0}{-21.4}\right] = -20.7° \qquad \Leftarrow \textbf{Ans.}$$

$$\theta_2 = \theta_1 + 90° = -20.7° + 90° = 69.3° \qquad \Leftarrow \textbf{Ans.}$$

The principal stresses and the principal planes are sketched in Fig. E18–12(5).

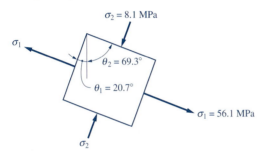

FIGURE E18–12(5)

Remark. Since the principal stresses and the orientation of the principal planes can be determined by computations, as shown above, Mohr's circle can be omitted.

18–9
SUMMARY

Combined Stresses. Stresses due to several loadings applied simultaneously are called *combined stresses*. The *method of superposition* is used to determine the combined stresses caused by two or more loadings.

Combined Axial and Bending Stresses. When a member is subjected to axial forces and a bending moment, the normal stresses due to each load can be calculated separately and added algebraically to find the combined stresses.

Biaxial Bending. Loads on a beam may cause bending about both centroidal axes of the cross-section of the beam. Normal stresses caused by bending about both axes are along the longitudinal direction; hence, the method of superposition can be applied. The bending about each direction is analyzed separately and the results are added algebraically.

Eccentric Loading. When a short member is subjected to a compressive load with eccentricity from one of the centroidal axes, the load will produce an axial compressive force and a bending moment. The axial force produces a uniform compressive stress throughout the section. The bending moment produces maximum compressive stress on one side of the section and maximum tension on the other side of the section. Using the method of superposition, the normal stresses on the two sides of the section are

$$\sigma = -\frac{P}{A} \pm \frac{M}{S} \qquad\qquad (18\text{--}1, 18\text{--}2)$$

where $M = Fe$. The maximum eccentricity of the compressive load, for which no tensile stresses will occur anywhere in the member, is

$$e = \frac{b}{6} \qquad\qquad (18\text{--}3)$$

Double Eccentricity. When a load is applied with eccentricity from both centroidal axes, the load is said to have *double eccentricity*. The eccentric load can be replaced by its equivalent centroidal force and moments about both axes. The combined normal stresses in the fibers at the corner of the member can be computed from

$$\sigma = -\frac{P}{A} \pm \frac{M_x}{S_x} \pm \frac{M_y}{S_y} \qquad\qquad (18\text{--}4)$$

Plane Stress. When an element is subjected to normal and shear stresses in only two directions, the element is said to be in a state of *plane stress*.

Stresses on an Inclined Plane. The following formulas may be used to compute the normal and shear stresses on an inclined plane of an element subjected to plane stresses:

$$\sigma_\theta = \frac{\sigma_x + \sigma_y}{2} + \frac{\sigma_x - \sigma_y}{2}\cos 2\theta + \tau_x \sin 2\theta \qquad\qquad (18\text{--}5)$$

$$\tau_\theta = -\frac{\sigma_x - \sigma_y}{2}\sin 2\theta + \tau_x \cos 2\theta \qquad\qquad (18\text{--}6)$$

The following sign conventions must be followed exactly:

1. *Sign for the Normal Stress.* Tensile stresses are considered positive.
2. *Sign for the Shear Stress.* Shear stresses are considered positive when the pair of shear stresses acting on the opposite sides of the element form a counterclockwise couple.
3. *Sign for the Angle of Inclination.* The angle of inclination θ is measured from the vertical plane toward the inclined plane. The counterclockwise measurement is considered positive.

Mohr's Circle. Mohr's circle is a graphical method for finding the stresses on any inclined plane of an element subjected to plane stresses. Use the following steps to draw a Mohr's circle:

1. Indicate the state of plane stress on the given element.
2. Set up a rectangular coordinate system with the normal stress axis in the horizontal direction and the shear stress axis in the vertical direction.
3. Locate the center C along the horizontal σ axis at $(\sigma_x + \sigma_y)/2$ from the origin O. With C as the center, draw the Mohr's circle passing through point $V (\sigma_x, \tau_x)$ corresponding to the stress condition on the vertical plane.

Mohr's circle is especially useful for finding the principal stresses and the maximum and minimum shear stresses.

Principal Stresses. The maximum and minimum normal stresses may be computed from

$$\sigma_1 = a + r \tag{18–10a}$$

$$\sigma_2 = a - r \tag{18–10b}$$

where

$$a = \frac{\sigma_x + \sigma_y}{2} \qquad b = \frac{\sigma_x - \sigma_y}{2} \qquad r = \sqrt{b^2 + \tau_x^2} \tag{18–11}$$

The principal planes may be oriented by the angles

$$\theta_1 = \tan^{-1}\left[\frac{r - b}{\tau_x}\right] \tag{18–12}$$

$$\theta_2 = \theta_1 \pm 90° \tag{18–13}$$

The Maximum and Minimum Shear Stresses. The maximum and minimum shear stresses and the associated normal stresses are

$$\tau_{max} = +r \tag{18–14a}$$

$$\tau_{min} = -r \tag{18–14b}$$

$$\tau' = a \tag{18–15}$$

The plane on which the maximum and minimum shear stresses act may be oriented by the angles

$$\theta_s = -\tan^{-1}\left[\frac{b}{r + \tau_x}\right] \tag{18–16}$$

$$\theta_s' = \theta_s \pm 90° \tag{18–17}$$

PROBLEMS

Section 18–2   Combined Axial and Bending Stresses

18–1   A beam of W16 × 50 section has a simple span of 20 ft. The beam is subjected to a uniform load of 2 kips/ft and an axial tensile force of 10 kips.
Determine the maximum tensile and compressive stresses in the beam.

18–2   Refer to Fig. P18–2. The timber beam of full-size 50-mm × 100-mm rectangular section supports a load applied at the free end, as shown. Determine
the normal stresses at points A and B.

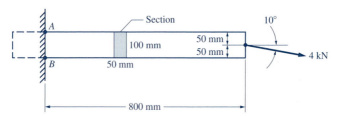

FIGURE P18–2

18–3   A concrete block of full-size 75-mm × 150-mm section is subjected to the
loads shown in Fig. P18–3. Determine (*a*) the magnitude of the load F such
that the normal stress at A is equal to zero and (*b*) the corresponding normal stress at B.

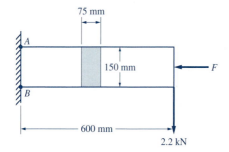

FIGURE P18–3

18–4   A beam consisting of two standard C9 × 15 steel channels, arranged back
to back as shown in Fig. P18–4, is subjected to a load of 6 kips. Determine
the maximum tensile and compressive stresses along the beam.

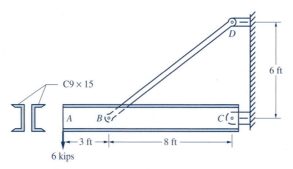

FIGURE P18–4

18–5 The horizontal beam of the jib crane in Fig. P18–5 is made of two standard steel channels. The maximum load, including the weight of the moving cart that the crane is designed to carry, is 8 kips. If the allowable compressive stress is 15 ksi, select a proper size for the pair of channels.

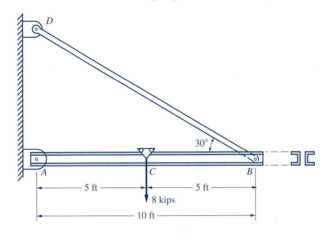

FIGURE P18–5

18–6 A wide-flange steel beam W250 × 1.46 is lifted by the crane shown in Fig. P18–6. Determine the maximum tensile and compressive stresses in the beam.

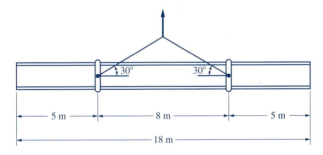

FIGURE P18–6

18–7 A gravity dam has a rectangular cross-section 3 ft by 6 ft, as shown in Fig. P18–7. The water pressure varies linearly from zero at the free surface to γh at the bottom, where γ is the weight of water per unit volume, equal to 62.4 lb/ft^3. If the weight of concrete per unit volume is 150 lb/ft^3, determine the height of the water level h at which the foundation pressure at A is just equal to zero. (*Hint:* For the purpose of calculation, consider one linear foot of length along the longitudinal direction of the dam.)

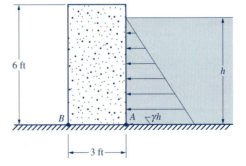

FIGURE P18–7

Section 18–3 Biaxial Bending

18–8   Refer to Fig. P18–8. The 10-ft-long, simply supported timber beam of full-size, 6-in. × 8-in. section is supported in such a way that the vertical concentrated load $P = 2$ kips applied at the centroid of the midspan passes through the diagonal AC, as shown. Find the normal stresses at points A, B, C, and D in the midspan due to the load.

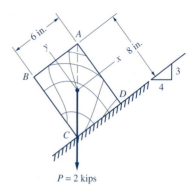

FIGURE P18–8

18–9   Refer to Fig. P18–9. The cantilever beam has a 2-m horizontal span and is built into a concrete pier on one end at the tilted position shown. The beam has a full-size, 50-mm × 100-mm timber section. A vertical load $P = 270$ N is applied at the free end through the centroid. Find the maximum tensile and compressive stresses in the beam.

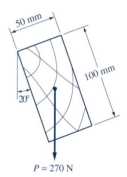

FIGURE P18–9

18–10   The cantilever timber beam shown in Fig. P18–10 has a span length $L = 4$ ft and a full-size rectangular section with $a = 3$ in. It carries a horizontal load $P_x = 120$ lb and a vertical load $P_y = 300$ lb. Determine the normal stresses at points A, B, C, and D.

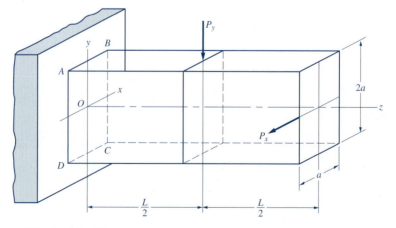

FIGURE P18–10

18–11 In Fig. P18–10, assume that $L = 3$ m, $P_x = 600$ N, and $P_y = 1000$ N. Determine the required dimension a of the section if the allowable flexural stress for timber is 10 MPa. Neglect the weight of the beam.

18-12 Refer to Fig. P18–12. The simply supported, standard S6 × 17.3 steel I-beam has a span length $L = 12$ ft and carries a load $P = 1.5$ kips applied at the midspan. The load passes through the centroid of the section and makes an angle $\theta = 20°$ with the longitudinal vertical plane, as shown. Determine the maximum tensile and compressive stresses in the beam.

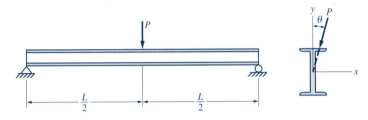

FIGURE P18–12

18–13 In Fig. P18–12, assume that the span length is $L = 16$ ft and the inclined load is $P = 6$ kips at an angle of inclination of $\theta = 10°$. Select a proper W shape for the beam if the allowable flexural stress is 20 ksi.

Section 18–4 Eccentrically Loaded Members

18–14 Refer to Fig. P18–14. Find the normal stresses at points A and B due to the eccentrically applied, axial compressive load $P = 600$ kN acting on the circular post as shown.

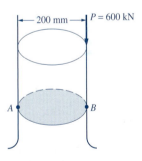

FIGURE P18–14

18–15 Refer to Fig. P18–15. The timber bar has a full-size, rectangular section 3 in. × 4 in. At section 1–1, the 4-in. width is reduced to 2 in., as shown. Determine the normal stresses at points A and B due to the axial load $P = 1800$ lb.

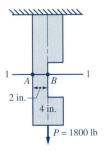

FIGURE P18–15

18–16   See Fig. P18–16. Determine the maximum eccentricity e at which the vertical compressive load
P can be applied to the wide-flange W14 × 90
steel section without causing tensile stress anywhere in the section. Neglect the weight of the
section.

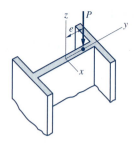

FIGURE P18–16

18–17   The steel bracket in Fig. P18–17 is loaded as shown. Determine the normal
stresses at points A and B, and plot the normal stress distribution along AB.

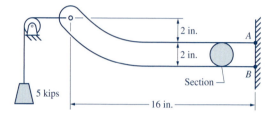

FIGURE P18–17

18–18   See Fig. P18–18. The short steel
post of wide-flange section W14 × 34
is subjected to the load shown.
Determine the normal stresses at A
and B, and plot the normal stress distribution along AB.

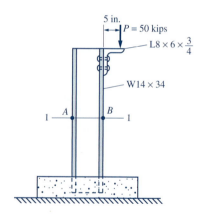

FIGURE P18–18

18–19   See Fig. P18–19. The machine part is subjected to an eccentric pull of 100 kN,
as shown. Determine the normal stresses at A and B.

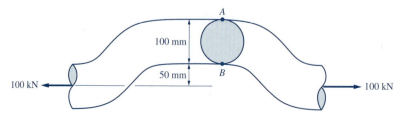

FIGURE P18–19

18–20 The frame of the hydraulic press has the dimensions and the properties of section 1–1 shown in Fig. P18–20. If $P = 1600$ kN, determine the normal stresses at points A and B.

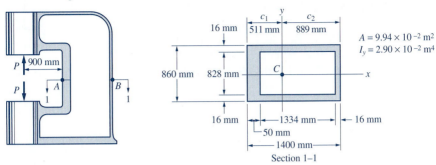

FIGURE P18–20

18–21 A cast-iron frame for a punch press has the dimensions and the properties of the cross-section 1–1 shown in Fig. P18–21. If the allowable stresses are 4000 psi in tension and 12 000 psi in compression, determine the maximum allowable force P that can be applied.

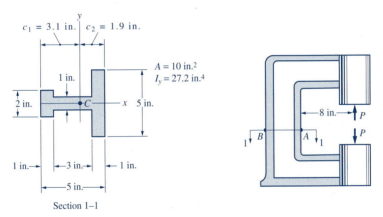

FIGURE P18–21

18–22 A load $P = 70$ kN is applied to the bracket in Fig. P18–22, which is welded to a wide-flange W410 $\times$ 0.53 steel section as shown. Determine the normal stresses at points A and B.

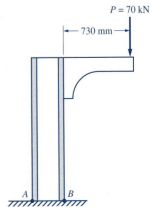

FIGURE P18–22

Section 18–5 Double Eccentricity

18–23 Refer to Fig. P18–23. The short timber block having a full-size, 0.2-m × 0.3-m rectangular section supports an eccentric load $P = 80$ kN, as shown. Determine the normal stresses at points A, B, C, and D due to the load.

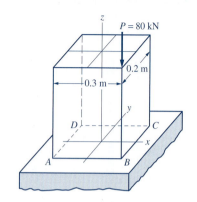

FIGURE P18–23

18–24 See Fig. P18–24. A short post with a full-size, 150-mm × 100-mm section is subjected to the loads $P = 30$ kN and $Q = 4$ kN, as shown. Determine the normal stresses at points A, B, C, and D. Neglect the weight of the post.

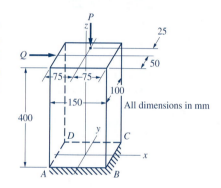

FIGURE P18–24

18–25 An aluminum-alloy block is subjected to an eccentric axial compressive load P as shown in Fig. P18–25. The dimensions are $a = 6$ in., $b = 3$ in., and $c = 1$ in. The linear strain produced by the load at point A in the vertical direction is 7.2×10^{-4} in./in. If the modulus of elasticity of the aluminum alloy is 10×10^6 psi, determine the magnitude of the load P.

18–26 In Fig. P18–25, assume that the aluminum-alloy block is subjected to an eccentric axial compressive load $P = 250$ kN, and the dimensions are $a = 150$ mm, $b = 75$ mm, and $c = 25$ mm. Determine the linear strain produced by the load at point A in the vertical direction if the modulus of elasticity of the aluminum alloy is 70 GPa.

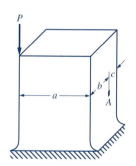

FIGURE P18–25

18–27 See Fig. P18–27. Prove that the kern of a circular section is a concentric circular area having a diameter equal to one-fourth of the diameter of the section.

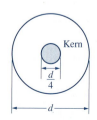

FIGURE P18–27

Section 18–6   Stresses on an Inclined Plane

18–28 to **18–33** Refer to Figs. P18–28 to P18–33. The state of stress on each element is as shown. Determine the normal and shear stresses on the inclined plane *m–m*, and sketch the stresses acting on the inclined plane.

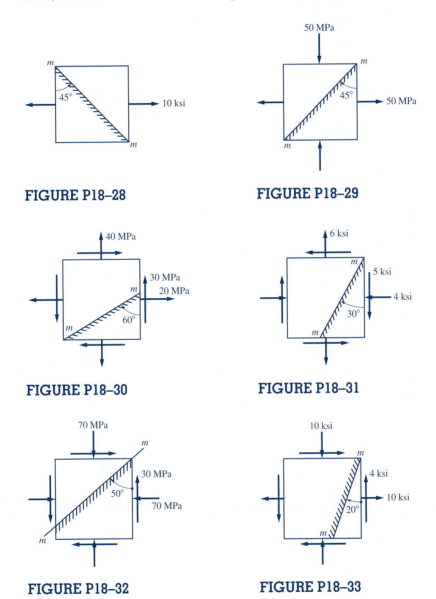

FIGURE P18–28

FIGURE P18–29

FIGURE P18–30

FIGURE P18–31

FIGURE P18–32

FIGURE P18–33

Section 18–7   Mohr's Circle

18–34 to **18–39** Refer to Figs. P18–34 to P18–39. The state of stress on each element is as shown. Drawing the Mohr's circle for each element and use the circle to determine the normal and shear stresses on the inclined plane *m–m*. Show the stresses acting on each inclined plane.

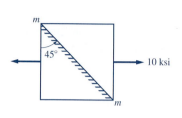

FIGURE P18–34

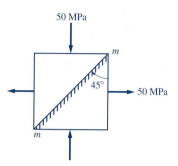

FIGURE P18–35

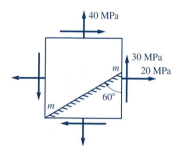

FIGURE P18–36

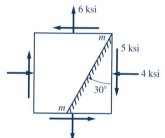

FIGURE P18–37

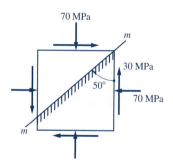

FIGURE P18–38

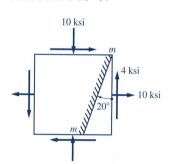

FIGURE P18–39

18–40   See Fig. P18–40. The boiler, of inside diameter 1 m and wall thickness 20 mm, is subjected to an internal pressure of 6 MPa. Determine (*a*) the state of stress in the rectangular element shown, and (*b*) the normal and shear stresses along the inclined plane *m*–*m*.

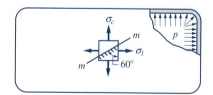

FIGURE P18–40

18–41   Refer to Fig. P18–41. The simply supported timber beam with a full-size, 2-in. × 4-in. rectangular section supports a concentrated load at the midspan, as shown. Determine (*a*) the state of stress of point *C* at section 1–1 just to the left of the concentrated load, and (*b*) the normal and shear stresses on the inclined plane *m*–*m* passing through point *C* as shown.

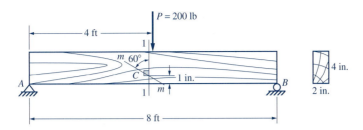

FIGURE P18–41

Section 18–8   Principal Stresses and Maximum Shear Stresses

18–42 to **18–45**   Refer to Figs. P18–42 to P18–45. Draw the Mohr's circle for the state of stress indicated in each figure and find (*a*) the principal stresses and the principal planes, and (*b*) the maximum and minimum shear stresses, the associated normal stresses, and the planes for these stresses. In both cases, show the results in properly oriented elements.

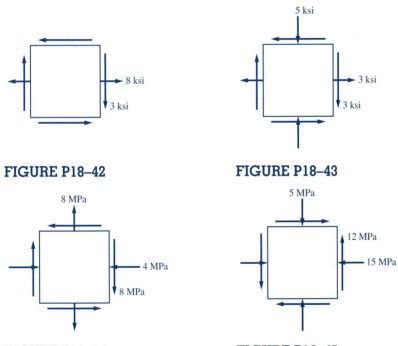

FIGURE P18–42   **FIGURE P18–43**

FIGURE P18–44   **FIGURE P18–45**

In Problems 18–46 to 18–49, show the given state of stress on a rectangular element and draw the Mohr's circle for the element. Find (a) the principal stresses and the principal planes, and (b) the maximum and minimum shear stresses, the associated normal stresses, and the planes for these stresses. In both cases, show the results in properly oriented elements.

18–46 $\sigma_x = +4$ ksi
$\qquad\quad$ $\sigma_y = +16$ ksi
$\qquad\quad$ $\tau_x = +8$ ksi

18–47 $\sigma_x = -20$ MPa

$\sigma_y = +40$ MPa

$\tau_x = -30$ MPa

18–48 $\sigma_x = +400$ psi

$\sigma_y = -800$ psi

$\tau_x = -800$ psi

18–49 $\sigma_x = +18$ ksi

$\sigma_y = 0$

$\tau_x = -12$ ksi

18–50 Refer to Fig. P18–50. The timber beam with a full-size, 4-in. × 6-in. rectangular section has a simple span of 6 ft and supports a uniform load of 500 lb/ft, including its own weight. Determine the principal stresses and the directions of the principal planes at points A, B, and C. Show the results on a properly oriented element.

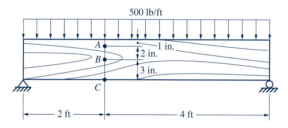

500 lb/ft

FIGURE P18–50

18–51 The cantilever cast-iron beam in Fig. P18–51 supports a uniform load of 6 kN/m, including its own weight. Determine the principal stresses at points A, B, and C at the fixed support. The cross-section and the section properties are as shown.

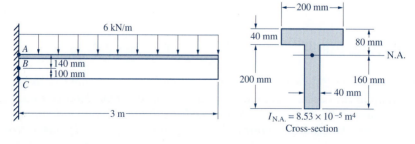

FIGURE P18–51

18–52 See Fig. P18–52. The maximum shear stress at point A in the simple beam of full-size, rectangular section is 100 psi. Determine the magnitude of the force P. Neglect the weight of the beam.

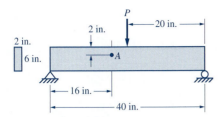

FIGURE P18–52

18–53 A clevis transmits a force $P = 10$ kN to the bracket in Fig. P18–53. Determine (*a*) the state of stress of the element at point A, (*b*) its principal stresses, and (*c*) its maximum shear stress and the associated normal stresses. Show the results on properly oriented elements.

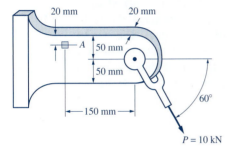

FIGURE P18–53

18–54 A cylindrical pressure vessel of 500-mm inside diameter and 15-mm wall thickness is subjected simultaneously to an internal pressure $p = 6$ MPa and an external torque T. If the maximum shear stress in the wall is limited to 100 MPa, determine the maximum permissible torque T.

Computer Program Assignments

For each of the following problems, write a computer program using an appropriate programming language with which you are most familiar. Make the program user friendly by incorporating plenty of comments and input prompts so that the user will understand the input data to be entered and the limitations of their values. The output should include the data entered and the computed results, and they must be well labeled to identify each quantity. If a tabulated format is used, a proper heading must be included at the top of the table. Do not limit the program to any specific unit system. Indicate the consistent U.S. customary or SI units that can be used.

C18–1 Write a computer program that can be used to calculate the stresses on the inclined plane *m–m* for the element in Fig. C18–1 subjected to a state of plane stress. The user input should include (1) the values of the stresses σ_x, σ_y, and τ_x, and (2) the angle θ measured from the vertical to the incline. The output results are to be the normal stress σ_θ and the shear stress τ_θ on the incline. Follow the sign convention described in Section 18–6. Use this program to solve (*a*) Example 18–9, (*b*) Problem 18–29, (*c*) Problem 18–31, and (*d*) Problem 18–32.

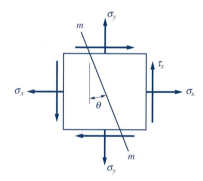

FIGURE C18–1

C18–2 Modify the program written in Computer Program Assignment C18–1 to compute the principal stresses, principal planes, maximum and minimum shear stresses, and the corresponding shear planes for the element in Fig. C18–1. The user input is to be the values of stresses σ_x, σ_y, and τ_x. The output should include (1) the principal stresses and the angles θ_1 and θ_2 in degrees measured from the vertical to the principal planes, and (2) the maximum and minimum shear stresses, the associated normal stresses, and the angles θ_s and θ_s' in degrees measured from the vertical to the planes of maximum and minimum shear stresses. Use this program to solve (*a*) Example 18–11, (*b*) Problem 18–44, and (*c*) Problem 18–45.

CHAPTER 19

COLUMNS

19–1
INTRODUCTION

Short bars compressed by axial forces were discussed in Chapter 9. Short compression members subjected to eccentric axial loads were considered in Chapter 18. In both cases, the members were assumed to be short so that the load-carrying capacity of the member depended solely on the strength of the materials. Failure of these members occurs only when the normal or shear stresses become excessive. When the length of a compression member is large compared to the transverse dimensions, however, the member tends to buckle before high stress levels are reached. When buckling occurs, the member tends to deflect laterally and to lose load-carrying stability. A small additional axial load will cause the member to collapse suddenly without warning. Such long compression members are called *columns*. Stability considerations of columns are the primary concern of this chapter. Formulas will be established for computing the load-carrying capacity of columns of different lengths and end conditions.

19–2
EULER FORMULA FOR PIN-ENDED COLUMNS

The long column shown in Fig. 19–1 has pin supports at both ends and is subjected to an axial compressive load. When the load P is small, the column remains straight and in stable equilibrium. When the compressive load is gradually increased so that it equals or exceeds a critical value P_{cr}, the column becomes highly unstable and any small disturbance or imperfection in the column material could trigger the buckling of the column, followed by a sudden collapse.

FIGURE 19–1

In 1757, Leonard Euler, a famous Swiss mathematician, developed the following formula, now called the *Euler formula*, for the critical (buckling) load on a long column with pinned ends made of homogeneous material:

$$P_{cr} = \frac{\pi^2 EI}{L^2} \tag{19-1}$$

where

P_{cr} = the critical (buckling) load, or the largest axial compressive load that a long column can carry before failure due to buckling

E = the modulus of elasticity of the column material

I = the least moment of inertia of the cross-sectional area of the column (buckling usually occurs about the axis with respect to which the moment of inertia is the smallest)

L = the length of the column between the pins

The Euler formula was derived based on the elastic behavior of the materials. Columns that buckle elastically are slender columns. The critical load calculated from the Euler formula is called the elastic buckling load. Note that the critical load is independent of the strength of the column material. The only material property involved is the elastic modulus E, which represents the stiffness characteristic of the material. Hence, according to the Euler formula, a column made of high-strength alloy steel will have the same buckling load as a column made of ordinary structural steel because the elastic moduli of the two kinds of steel are the same.

19-3
EULER FORMULA FOR OTHER SUPPORTING CONDITIONS

The Euler formula presented in the preceding section applies to columns with pinned ends. This condition is referred to as the *fundamental case*. The Euler column formula can be modified for other end-supporting conditions through the use of an *effective length* L_e in place of the actual length. Figure 19–2 shows the deflection curves for several end-supporting conditions. In each case, the effective length is the distance between points of inflection (points where the curve concavity changes). The points of inflection occur where the moment is zero, and thus these points may be considered analogous to pinned ends. For the fundamental case shown in Fig. 19–2a, L_e is equal to L. As we see from Fig. 19–2b, c, and d, effective length varies as the end-supporting condition changes.

The end-supporting condition can be accounted for through the use of an effective length factor k, as in the expression $L_e = kL$. The values of k for the four different end-supporting conditions shown in Fig. 19–2 are:

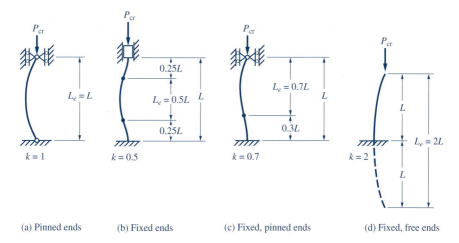

(a) Pinned ends   (b) Fixed ends   (c) Fixed, pinned ends   (d) Fixed, free ends

FIGURE 19–2

For pinned ends (fundamental case): $k = 1$

For fixed ends: $k = 0.5$

For fixed, pinned ends: $k = 0.7$

For fixed, free ends: $k = 2$

Using the effective length in place of the actual length, the Euler column formula becomes

$$P_{cr} = \frac{\pi^2 EI}{L_e^{\,2}} = \frac{\pi^2 EI}{(kL)^2} \tag{19–2}$$

Consider the fixed-ends condition shown in Fig. 19–2b, which has an effective length factor $k = 0.5$. From Equation 19–2, the critical load is

$$P_{cr} = \frac{\pi^2 EI}{(0.5L)^2} = \frac{4\pi^2 EI}{L^2}$$

This indicates that the elastic buckling load of a slender column with fixed ends is four times that of the same column with pinned ends.

Note that for the "flagpole" case shown in Fig. 19–2d, the lower end is fixed and the top of the column is free. In this case, we consider the free end as an inflection point and an imaginary point of inflection at a distance L below the base of the column. Therefore, the theoretical effective length factor k is 2.0. Such a column has an elastic buckling strength only one-quarter of that of the same column with pinned ends. We may conclude that in general, *the stiffer the end constraints, the higher the buckling strength of the column.*

When a column with pinned ends is braced at the midpoint, as shown in Fig. 19–3a, the effective length of the column is $L/2$. This column may be regarded as two columns, one on top of the other. Similarly, when a pin-ended column is braced at the third points, as shown in Fig. 19–3b, the effective length of the column is $L/3$. To increase the load-carrying capacity, a column is sometimes braced at its intermediate points in the weaker direction. In this case, buckling about both axes must be investigated.

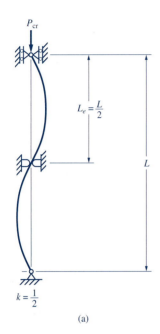

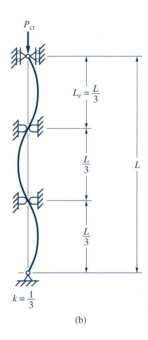

FIGURE 19–3

19–4
LIMITATION OF THE EULER FORMULA

As mentioned before, the Euler formula was derived based on the elastic behavior of the material. Therefore, the Euler formula is applicable only when the critical stress (the stress corresponding to the critical load) in the column is within the proportional limit of the material.

Critical Stress. To find the expression for the critical stress, first we recall that $I = Ar^2$, where A is the cross-sectional area and r is the radius of gyration of the cross-section of the column. Substituting this into Equation 19–2 gives

$$P_{cr} = \frac{\pi^2 EI}{(kL)^2} = \frac{\pi^2 EAr^2}{(kL)^2}$$

Thus, the critical stress is

$$\sigma_{cr} = \frac{P_{cr}}{A} = \frac{\pi^2 E}{(kL/r)^2} \tag{19–3}$$

where

σ_{cr} = the critical compressive stress in the column corresponding to the critical load P_{cr}

r = the *least radius of gyration* of the cross-section of the column corresponding to the minimum value of I

kL/r = the *slenderness ratio* of the column, defined as the ratio of the effective length of the column to the least radius of gyration

Limiting Slenderness Ratio. The Euler formula applies when σ_{cr} is less than the stress at the proportional limit σ_p, that is, when

$$\sigma_{cr} = \frac{\pi^2 E}{(kL/r)^2} \leq \sigma_p$$

Therefore, the Euler formula is applicable only when the column has a slenderness ratio greater than or equal to

$$\left(\frac{kL}{r}\right)_{\min} = \sqrt{\frac{\pi^2 E}{\sigma_p}} \qquad\qquad (19\text{–}4)$$

For example, for structural steel, with $\sigma_p = 30$ ksi and $E = 30 \times 10^3$ ksi, the minimum slenderness ratio is

$$\left(\frac{kL}{r}\right)_{\min} = \sqrt{\frac{\pi^2 \times 30 \times 10^3 \text{ ksi}}{30 \text{ ksi}}} \approx 100$$

A graphical representation of Equation 19–3 for structural steel is shown in Fig. 19–4 by plotting the slenderness ratio kL/r as the abscissa and the critical stress σ_{cr} as the ordinate. Point A is the upper limit of applicability of the Euler formula. The curve for the Euler formula is valid for long columns with $kL/r > 100$ (the part of the curve to the right of point A). The Euler formula is not valid for the AC part of the curve where $kL/r < 100$ since in this region the compressive stress is greater than σ_p, and the material no longer behaves elastically. The buckling in this region is inelastic buckling.

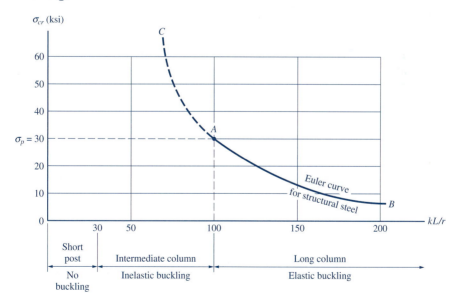

FIGURE 19–4

Many columns encountered in machine and building design buckle inelastically and thus the Euler formula does not apply. For these cases, many semi-empirical formulas have been developed. The most frequently used formula in machine design and structural steel design is the J. B. Johnson formula, to be discussed in the next section.

When a compression member is short, with its slenderness ratio less than a certain value, say, 30 in Fig. 19–4, it will not buckle and its load-carrying capacity depends only on its strength.

Allowable Load. A column is designed to carry an allowable load P_{allow} equal to the critical load divided by the factor of safety, that is,

$$P_{allow} = \frac{P_{cr}}{\text{F.S.}} \tag{19–5}$$

The factor of safety to be used depends on many factors and is usually specified by the design code. For structural steel design, a factor of safety of 1.92 is used for long columns. For conditions of greater uncertainty, a factor of safety of 3 or more may be used.

EXAMPLE 19–1

A 1.5-m-long, pin-ended Douglas fir column has a full-size, rectangular cross-section 50 mm × 100 mm. Determine the maximum compressive axial load that the column can carry before buckling. For Douglas fir, $E = 13$ GPa and $\sigma_p = 28$ MPa.

Solution. The minimum moment of inertia of the rectangular cross-section is

$$I_{min} = \frac{(0.100 \text{ m}) \ (0.050 \text{ m})^3}{12} = 1.04 \times 10^{-6} \text{ m}^4$$

Hence, by definition,

$$r = r_{min} = \sqrt{\frac{I_{min}}{A}} = \sqrt{\frac{1.04 \times 10^{-6}}{0.100 \times 0.050}} = 0.0144 \text{ m}$$

Or from the formula in Table 8–1, we find

$$r = r_{min} = \frac{b}{\sqrt{12}} = \frac{0.050 \text{ m}}{\sqrt{12}} = 0.0144 \text{ m}$$

The slenderness ratio is

$$\frac{kL}{r} = \frac{1(1.5 \text{ m})}{0.0144 \text{ m}} = 104$$

From Equation 19–4, the minimum slenderness ratio for which the Euler formula applies is

$$\left(\frac{kL}{r}\right)_{min} = \sqrt{\frac{\pi^2 E}{\sigma_p}} = \sqrt{\frac{\pi^2 (13 \times 10^3 \text{ MPa})}{28 \text{ MPa}}} = 68$$

Since the actual slenderness ratio of the given column is greater than the minimum slenderness ratio, the Euler formula applies. From Equation 19–3, we find

$$\sigma_{cr} = \frac{\pi^2 E}{(kL/r)^2} = \frac{\pi^2 (13 \times 10^6 \text{ kN/m}^2)}{(104)^2}$$

$$= 11\ 900 \text{ kN/m}^2 = 11.9 \text{ MPa}$$

which is less than the proportional limit $\sigma_p = 28$ MPa, as is expected for a long column.

$$P_{cr} = \sigma_{cr} A = (11\ 900\ \text{kN/m}^2)(0.100\ \text{m} \times 0.050\ \text{m})$$

$$= 59.5\ \text{kN} \qquad\qquad \Leftarrow \textbf{Ans.}$$

EXAMPLE 19–2

Determine the allowable load of a 36-ft-long, W10 × 33 steel column with pinned ends. The column is braced at the midpoint in the weaker direction. Use $E = 29 \times 10^3$ ksi, $\sigma_p = 34$ ksi, and F.S. = 1.92.

Solution. From Equation 19–4, the minimum slenderness ratio for which the Euler equation applies is

$$\left(\frac{kL}{r}\right)_{\text{min}} = \sqrt{\frac{\pi^2 E}{\sigma_p}} = \sqrt{\frac{\pi^2(29\ 000\ \text{ksi})}{34\ \text{ksi}}} = 92$$

From the appendix, Table A–1(a), the radii of gyration for the W10 × 33 steel section are $r_x = 4.19$ in. and $r_y = 1.94$ in., and the cross-sectional area of the column is $A = 9.71$ in.^{2}. The slenderness ratios about the two axes are

$$\left(\frac{kL_x}{r_x}\right) = \frac{1(36 \times 12\ \text{in.})}{4.19\ \text{in.}} = 103$$

$$\left(\frac{kL_y}{r_y}\right) = \frac{1(18 \times 12\ \text{in.})}{1.94\ \text{in.}} = 111$$

Thus, buckling will occur about the y axis. Since $kL_y/r_y = 111$ is greater than $(kL/r)_{\text{min}} = 92$, the column is a long column and the Euler formula applies. From Equation 19–3, we find

$$\sigma_{cr} = \frac{\pi^2 E}{(kL/r)^2} = \frac{\pi^2(29\ 000\ \text{ksi})}{(111)^2} = 23.2\ \text{ksi}$$

which is less than $\sigma_p = 34$ ksi, as is expected for a long column. Thus,

$$P_{cr} = \sigma_{cr} A = (23.2\ \text{kips/in.}^2)\ (9.71\ \text{in.}^2) = 225\ \text{kips}$$

$$P_{\text{allow}} = \frac{P_{cr}}{\text{F.S.}} = \frac{225\ \text{kips}}{1.92} = 117.3\ \text{kips} \qquad \Leftarrow \textbf{Ans.}$$

19–5
J. B. JOHNSON FORMULA

Since the Euler formula does not apply for the intermediate columns, many semi-empirical formulas have been developed. One of these formulas, the J. B. Johnson formula, is used extensively in steel structure design and machine design. As indicated in Fig. 19–5, the J. B. Johnson formula is the equation of a parabola having its vertex at the point on the vertical axis with ordinate equal to σ_y. The parabola is tangent to the Euler curve at the transition slenderness ratio $kL/r = C_c$, corresponding to one-half of the yield stress σ_y of the steel. The value of the transition slenderness ratio C_c can be determined from Equation 19–3. We write

$$\sigma_{cr} = \frac{1}{2}\sigma_y = \frac{\pi^2 E}{(kL/r)^2} = \frac{\pi^2 E}{C_c^2}$$

From which we get

$$C_c = \sqrt{\frac{2\pi^2 E}{\sigma_y}} \qquad (19\text{–}6)$$

The J. B. Johnson formula is

$$\sigma_{cr} = \frac{P_{cr}}{A} = \left[1 - \frac{(kL/r)^2}{2C_c^2}\right]\sigma_y \qquad (19\text{–}7)$$

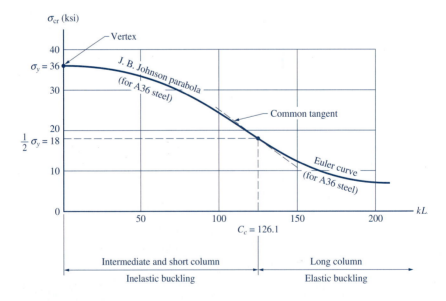

FIGURE 19–5

The Euler formula applies when kL/r is greater than C_c, and the J. B. Johnson formula applies when kL/r is less than C_c. For $kL/r = C_c$, both formulas give the same result. Note that the Euler formula applies to all materials, whereas the J. B. Johnson formula applies mainly to ductile steel.

——————— **EXAMPLE 19–3** ———————————————————————

Determine the allowable compressive load of a 4-in., standard weight steel pipe that is 25 ft long. The column is made of A36 steel with $\sigma_y = 36$ ksi and is welded to fixed supports at both ends. Use F.S. = 2 and $E = 29 \times 10^3$ ksi.

Solution. From the appendix, Table A–5(a), for a 4-in., standard weight steel pipe, $A = 3.17$ in.^{2} and $r = 1.51$ in. The slenderness ratio is

$$\frac{kL}{r} = \frac{0.5(25 \times 12 \text{ in.})}{1.51 \text{ in.}} = 99.3$$

From Equation 19–7, the value of the transition slenderness ratio C_c is

$$C_c = \sqrt{\frac{2\pi^2 E}{\sigma_y}} = \sqrt{\frac{2\pi^2(29\ 000 \text{ ksi})}{36 \text{ ksi}}} = 126.1$$

Since $kL/r < C_c$, the J. B. Johnson formula applies. From Equation 19–6, we find

$$\sigma_{cr} = \left[1 - \frac{(kL/r)^2}{2C_c^{\,2}}\right]\sigma_y$$

$$= \left[1 - \frac{(99.3)^2}{2(126.1)^2}\right](36 \text{ ksi})$$

$$= 24.8 \text{ ksi}$$

Thus,

$$P_{cr} = \sigma_{cr} A = (24.8 \text{ kips/in.}^2)(3.17 \text{ in.}^2) = 78.6 \text{ kips}$$

$$P_{\text{allow}} = \frac{P_{cr}}{\text{F.S.}} = \frac{78.6 \text{ kips}}{2} = 39.3 \text{ kips} \qquad \Leftarrow \textbf{Ans.}$$

19–6
THE AISC COLUMN FORMULAS

The American Institute of Steel Construction (AISC) manual gives formulas for computing the allowable compressive stresses to be used in steel column design. The AISC column formulas are essentially the critical buckling stresses from the Euler and J. B. Johnson formulas divided by the factor of safety. The AISC formulas are:

1. For long columns: $kL/r \geq C_c = \sqrt{2\pi^2 E/\sigma_y}$

$$\sigma_{\text{allow}} = \frac{\pi^2 E/(kL/r)^2}{\text{F.S.}} = \frac{\pi^2 E/(kL/r)^2}{1.92} \qquad (19\text{–}8)$$

where $E = 29 \times 10^3$ ksi for steel.

2. For intermediate and short columns: $kL/r \le C_c = \sqrt{2\pi^2 E/\sigma_y}$

$$\sigma_{\text{allow}} = \frac{\left[1 - \dfrac{(kL/r)^2}{2C_c^2}\right]\sigma_y}{\text{F.S.}} \tag{19-9}$$

where the factor of safety, F.S., is computed from the equation

$$\text{F.S.} = \frac{5}{3} + \frac{3(kL/r)}{8C_c} - \frac{(kL/r)^3}{8C_c^3} \tag{19-10}$$

Note that the factor of safety varies from $\frac{5}{3}$ (or 1.67) when $kL/r = 0$ to $\frac{23}{12}$ (or 1.92) when $kL/r = C_c$. We see that the factor of safety is more conservative for longer columns. From Equation 19-9, we find $\sigma_{\text{allow}} = \sigma_y / \left(\frac{5}{3}\right) = 0.6\,\sigma_y$ when $kL/r = 0$. This means that when the column is very short, the compressive load capacity of a column is a function of its strength only.

Table 19-1 shows the value of the AISC recommended effective length factor k for steel column design when the end-supporting conditions are approximated. As a design aid, values of the allowable compressive stress computed from the AISC formulas corresponding to $\sigma_y = 36$ ksi and $\sigma_y = 50$ ksi are tabulated for kL/r values from 1 to 200 in Tables 19-2 and 19-3. Slenderness ratios in excess of 200 are too sensitive to the imperfection of the column. For example, a very slight crookedness of the column may trigger buckling before the theoretical critical load is reached. So the AISC manual does not allow slenderness ratios to exceed 200.

TABLE 19-1   AISC Recommended k Values

End Conditions	Pinned Ends	Fixed Ends	Fixed, Pinned Ends	Fixed, Free Ends
Theoretical k value	1.0	0.5	0.7	2.0
AISC recommended k value	1.0	0.65	0.8	2.10

———— **EXAMPLE 19-4** ————

Determine the allowable axial compressive load for a 10-ft-long standard L6 × 4 × $\frac{1}{2}$ steel angle of A36 steel if the supporting conditions are (*a*) pinned at both ends or (*b*) fixed at both ends. Use the AISC formulas and the recommended k values.

Solution. From the appendix, Table A-4(a), for an L6 × 4 × $\frac{1}{2}$ steel angle, $A = 4.75$ in.^{2} and the least radius of gyration is $r_z = 0.870$ in.

(*a*) *For Pinned Ends.* The effective length factor is $k = 1.0$. Thus, the slenderness ratio is

$$\frac{kL}{r} = \frac{(1)(10 \times 12 \text{ in.})}{0.870 \text{ in.}} = 137.9$$

TABLE 19–2   AISC Allowable Compressive Stress for Steel Columns for $\sigma_y = 36$ ksi (250 MPa)

$\frac{kL}{r}$	σ_{allow} (ksi)	$\frac{kL}{r}$	σ_{allow} (ksi)	$\frac{kL}{r}$	σ_{allow} (ksi)	$\frac{kL}{r}$	σ_{allow} (ksi)	$\frac{kL}{r}$	σ_{allow} (ksi)
1	21.56	41	19.11	81	15.24	121	10.14	161	5.76
2	21.52	42	19.03	82	15.13	122	9.99	162	5.69
3	21.48	43	18.95	83	15.02	123	9.85	163	5.62
4	21.44	44	18.86	84	14.90	124	9.70	164	5.55
5	21.39	45	18.78	85	14.79	125	9.55	165	5.49
6	21.35	46	18.70	86	14.67	126	9.41	166	5.42
7	21.30	47	18.61	87	14.56	127	9.26	167	5.35
8	21.25	48	18.53	88	14.44	128	9.11	168	5.29
9	21.21	49	18.44	89	14.32	129	8.97	169	5.23
10	21.16	50	18.35	90	14.20	130	8.84	170	5.17
11	21.10	51	18.26	91	14.09	131	8.70	171	5.11
12	21.05	52	18.17	92	13.97	132	8.57	172	5.05
13	21.00	53	18.08	93	13.84	133	8.44	173	4.99
14	20.95	54	17.99	94	13.72	134	8.32	174	4.93
15	20.89	55	17.90	95	13.60	135	8.19	175	4.88
16	20.83	56	17.81	96	13.48	136	8.07	176	4.82
17	20.78	57	17.71	97	13.35	137	7.96	177	4.77
18	20.72	58	17.62	98	13.23	138	7.84	178	4.71
19	20.66	59	17.53	99	13.10	139	7.73	179	4.66
20	20.60	60	17.43	100	12.98	140	7.62	180	4.61
21	20.54	61	17.33	101	12.85	141	7.51	181	4.56
22	20.48	62	17.24	102	12.72	142	7.41	182	4.51
23	20.41	63	17.14	103	12.59	143	7.30	183	4.46
24	20.35	64	17.04	104	12.47	144	7.20	184	4.41
25	20.28	65	16.94	105	12.33	145	7.10	185	4.36
26	20.22	66	16.84	106	12.20	146	7.01	186	4.32
27	20.15	67	16.74	107	12.07	147	6.91	187	4.27
28	20.08	68	16.64	108	11.94	148	6.82	188	4.23
29	20.01	69	16.53	109	11.81	149	6.73	189	4.18
30	19.94	70	16.43	110	11.67	150	6.64	190	4.14
31	19.87	71	16.33	111	11.54	151	6.55	191	4.09
32	19.80	72	16.22	112	11.40	152	6.46	192	4.05
33	19.73	73	16.12	113	11.26	153	6.38	193	4.01
34	19.65	74	16.01	114	11.13	154	6.30	194	3.97
35	19.58	75	15.90	115	10.99	155	6.22	195	3.93
36	19.50	76	15.79	116	10.85	156	6.14	196	3.89
37	19.42	77	15.69	117	10.71	157	6.06	197	3.85
38	19.35	78	15.58	118	10.57	158	5.98	198	3.81
39	19.27	79	15.47	119	10.43	159	5.91	199	3.77
40	19.19	80	15.36	120	10.28	160	5.83	200	3.73

Notes: 1. $C_c = 126.1$.

2. To obtain the allowable stress in MPa, multiply the tabulated value by 6.895.

TABLE 19–3   AISC Allowable Compressive Stress for Steel Columns for $\sigma_y = 50$ ksi (345 MPa)

$\dfrac{kL}{r}$	σ_{allow} (ksi)	$\dfrac{kL}{r}$	σ_{allow} (ksi)	$\dfrac{kL}{r}$	σ_{allow} (ksi)	$\dfrac{kL}{r}$	σ_{allow} (ksi)	$\dfrac{kL}{r}$	σ_{allow} (ksi)
1	29.94	41	25.69	81	18.81	121	10.20	161	5.76
2	29.87	42	25.55	82	18.61	122	10.03	162	5.69
3	29.80	43	25.40	83	18.41	123	9.87	163	5.62
4	29.73	44	25.26	84	18.20	124	9.71	164	5.55
5	29.66	45	25.11	85	17.99	125	9.56	165	5.49
6	29.58	46	24.96	86	17.79	126	9.41	166	5.42
7	29.50	47	24.81	87	17.58	127	9.26	167	5.35
8	29.42	48	24.66	88	17.37	128	9.11	168	5.29
9	29.34	49	24.51	89	17.15	129	8.97	169	5.23
10	29.26	50	24.35	90	16.94	130	8.84	170	5.17
11	29.17	51	24.19	91	16.72	131	8.70	171	5.11
12	29.08	52	24.04	92	16.50	132	8.57	172	5.05
13	28.99	53	23.88	93	16.29	133	8.44	173	4.99
14	28.90	54	23.72	94	16.06	134	8.32	174	4.93
15	28.80	55	23.55	95	15.84	135	8.19	175	4.88
16	28.71	56	23.39	96	15.62	136	8.07	176	4.82
17	28.61	57	23.22	97	15.39	137	7.96	177	4.77
18	28.51	58	23.06	98	15.17	138	7.84	178	4.71
19	28.40	59	22.89	99	14.94	139	7.73	179	4.66
20	28.30	60	22.72	100	14.71	140	7.62	180	4.61
21	28.19	61	22.55	101	14.47	141	7.51	181	4.56
22	28.08	62	22.37	102	14.24	142	7.41	182	4.51
23	27.97	63	22.20	103	14.00	143	7.30	183	4.46
24	27.86	64	22.02	104	13.77	144	7.20	184	4.41
25	27.75	65	21.85	105	13.53	145	7.10	185	4.36
26	27.63	66	21.67	106	13.29	146	7.01	186	4.32
27	27.52	67	21.49	107	13.04	147	6.91	187	4.27
28	27.40	68	21.31	108	12.80	148	6.82	188	4.23
29	27.28	69	21.12	109	12.57	149	6.73	189	4.18
30	27.15	70	20.94	110	12.34	150	6.64	190	4.14
31	27.03	71	20.75	111	12.12	151	6.55	191	4.09
32	26.90	72	20.56	112	11.90	152	6.46	192	4.05
33	26.77	73	20.38	113	11.69	153	6.38	193	4.01
34	26.64	74	20.19	114	11.49	154	6.30	194	3.97
35	26.51	75	19.99	115	11.29	155	6.22	195	3.93
36	26.38	76	19.80	116	11.10	156	6.14	196	3.89
37	26.25	77	19.61	117	10.91	157	6.06	197	3.85
38	26.11	78	19.41	118	10.72	158	5.98	198	3.81
39	25.97	79	19.21	119	10.55	159	5.91	199	3.77
40	25.83	80	19.01	120	10.37	160	5.83	200	3.73

Notes: 1. $C_c = 107.0$.

2. To obtain the allowable stress in MPa, multiply the tabulated value by 6.895.

For A36 steel, $\sigma_y = 36$ ksi, then

$$C_c = \sqrt{\frac{2\pi^2 E}{\sigma_y}} = \sqrt{\frac{2\pi^2\,(29\,000)}{36}} = 126.1$$

Since $kL/r > C_c$, Equation 19–8 applies. Thus,

$$\sigma_{allow} = \frac{\pi^2 E/(kL/r)^2}{1.92}$$

$$= \frac{\pi^2(29\,000\ \text{ksi})/(137.9)^2}{1.92}$$

$$= 7.84\ \text{ksi}$$

Or from Table 19–2, for $\sigma_y = 36$ ksi and $kL/r = 138$ (rounded to the nearest whole number for use in the table; interpolation is not necessary), the allowable compressive stress is $\sigma_{allow} = 7.84$ ksi, the same as calculated above. Thus,

$$P_{allow} = \sigma_{allow}\, A = (7.84\ \text{kips/in.}^2)(4.75\ \text{in.}^2)$$

$$= 37.3\ \text{kips} \qquad\qquad \Leftarrow \textbf{Ans.}$$

(b) For Fixed Ends. From Table 19–1, the AISC recommended effective length factor is $k = 0.65$. The slenderness ratio is

$$\frac{kL}{r} = \frac{(0.65)(10 \times 12\ \text{in.})}{0.870\ \text{in.}} = 89.7$$

which is less than $C_c = 126.1$; thus, Equation 19–9 applies. The factor of safety is calculated first from Equation 19–10. We obtain

$$\text{F.S.} = \frac{5}{3} + \frac{3(kL/r)}{8C_c} - \frac{(kL/r)^3}{8C_c^{\,3}}$$

$$= \frac{5}{3} + \frac{3(89.7)}{8(126.1)} - \frac{(89.7)^3}{8(126.1)^3} = 1.89$$

Substituting into Equation 19–9 gives

$$\sigma_{allow} = \frac{\left[1 - \dfrac{(kL/r)^2}{2C_c^{\,2}}\right]\sigma_y}{\text{F.S.}}$$

$$= \frac{\left[1 - \dfrac{(89.7)^2}{2(126.1)^2}\right](36\ \text{ksi})}{1.89}$$

$$= 14.23\ \text{ksi}$$

Or from Table 19–2, for $\sigma_y = 36$ ksi and $kL/r = 90$, the allowable compressive stress is $\sigma_{allow} = 14.20$ ksi. Thus,

$$P_{allow} = \sigma_{allow}\, A = (14.23\ \text{kips/in.}^2)(4.75\ \text{in.}^2)$$

$$= 67.6\ \text{kips} \qquad\qquad \Leftarrow \textbf{Ans.}$$

Hence, we see that the allowable compressive load of a steel column is substantially increased by imposing stiffer constraints on the ends of the column.

EXAMPLE 19–5

Two C300 × 0.438 steel channels of A441 steel with $\sigma_y = 345$ MPa form a 15-m-long compression member, as shown in Fig. E19–5(1). The channels are tied together by end tie plates and lacing bars to make the two channels act as one unit. The tie plates and the lacings are not effective in resisting compression. If the ends of the column are fixed, determine the allowable axial force of the member according to the AISC specification.

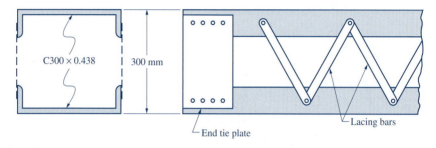

FIGURE E19–5(1)

Solution. Refer to Fig. E19–5(2). From the appendix, Table A–3(b), for a single C300 × 0.438 channel section,

$$A = 5.69 \times 10^{-3} \text{ m}^2 \qquad I_x = 67.4 \times 10^{-6} \text{ m}^4$$

$$\bar{x} = 17.1 \text{ mm} \qquad I_y = 2.14 \times 10^{-6} \text{ m}^4$$

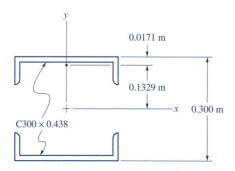

FIGURE E19–5(2)

Since it is not clear for which axis the radius of gyration is the least, the moments of inertia about both centroidal axes must be determined:

$$I_x = \Sigma \, [I + Ad^2]$$
$$= 2[2.14 \times 10^{-6} + (5.69 \times 10^{-3})(0.1329)^2]$$
$$= 2.05 \times 10^{-4} \text{ m}^4$$

$$I_y = 2(67.4 \times 10^{-6}) = 1.35 \times 10^{-4} \text{ m}^4$$

Thus, the least radius of gyration about the y axis is

$$r_y = \sqrt{\frac{I_y}{A}} = \sqrt{\frac{1.35 \times 10^{-4}}{2(5.69 \times 10^{-3})}} = 0.109 \text{ m}$$

For fixed ends, the AISC recommended k value from Table 19–1 is $k = 0.65$. Thus, the slenderness ratio is

$$\frac{kL}{r} = \frac{(0.65)(15 \text{ m})}{0.109 \text{ m}} = 89$$

From Table 19–3, for $\sigma_y = 345$ MPa (50 ksi) and $kL/r = 89$, the allowable compressive stress is $\sigma_{\text{allow}} = 17.15$ ksi. Converting to SI units, we get

$$\sigma_{\text{allow}} = 17.15(6.895) = 118.2 \text{ MPa}$$

The allowable load is

$$P_{\text{allow}} = \sigma_{\text{allow}} A$$
$$= (118\ 200 \text{ kN/m}^2)(2 \times 5.69 \times 10^{-3} \text{ m}^2)$$
$$= 1350 \text{ kN} \qquad\qquad \Leftarrow \textbf{Ans.}$$

19–7
STEEL COLUMN DESIGN

The AISC specification is widely used for the design of steel-framed buildings. It is recognized nationally and it is incorporated into most state and municipal building codes. The AISC column formulas are used for the design of almost all structural steel columns.

The most commonly used structural steel compression members are W shapes, steel pipes, and structural tubing. The AISC manual contains allowable axial load tables for these members, which may be used to simplify the design process. In the absence of these tables, columns can be designed by using AISC column formulas or the allowable compressive stress tables (Tables 19–2 and 19–3).

Before the size of a column is selected, the least radius of gyration is not known, so the slenderness ratio cannot be computed. Therefore, the design of a column is not a straightforward process; a trial-and-error procedure must be employed. The following procedure is recommended:

1. Decide what type of steel section should be used.
2. Assume a reasonable value for the radius of gyration. Calculate the slenderness ratio using the assumed radius of gyration.
3. Find the allowable compressive stress from Table 19–2 or Table 19–3. Calculate the approximate value of the required cross-sectional area from

$$A_{req} = \frac{P}{\sigma_{allow}} \tag{19-11}$$

where P is the axial compressive load on the column.
4. From the appendix, Table A–1 or Table A–5, list several sections of W shape or steel pipe that have an area approximately equal to the required area calculated above. Use the lightest section as the first trial section.
5. Find the allowable axial compressive load of the trial section and compare it with the design load. The section is satisfactory if the allowable load is greater than the design load. Try several other sections if necessary to be sure that the section selected is the most economical (which is usually the one with the least weight).

───── **EXAMPLE 19–6** ─────

Select the lightest W shape for a column subjected to an axial compressive load of 490 kips. The unbraced length of the column is 15 ft and the ends are fixed. Use A441 steel ($\sigma_y = 50$ ksi) and the AISC specification.

Solution. Since the size of the W shape is unknown, a trial-and-error procedure is necessary. Assume the least radius of gyration to be $r = 3$ in. From Table 19–1, for fixed ends, the AISC recommended k value is $k = 0.65$. Then the slenderness ratio is

$$\frac{kL}{r} = \frac{(0.65)(15 \times 12 \text{ in.})}{3 \text{ in.}} = 39$$

From Table 19–3, for $\sigma_y = 50$ ksi and $kL/r = 39$, the allowable compressive stress is $\sigma_{allow} = 25.97$ ksi. Thus, the required cross-sectional area of the column is

$$A_{req} = \frac{P}{\sigma_{allow}} = \frac{490 \text{ kips}}{25.97 \text{ kips/in.}^2} = 18.9 \text{ in.}^2$$

From Table A–1(a), we find the following W shapes that have a cross-sectional area approximately equal to the required area calculated above:

Section	A (in.²)	r_y (in.)
W16 × 89	26.2	2.49
W14 × 68	20.0	2.46
W12 × 65	19.1	3.02
W10 × 77	22.6	2.60

Of the sections listed above, the W12 × 65 shape is the lightest and yet it has the greatest value of r_y. Thus, we use W12 × 65 for the trial section. The slenderness ratio of the section is

$$\frac{kL}{r} = \frac{(0.65)(15 \times 12 \text{ in.})}{3.02 \text{ in.}} = 39$$

From Table 19–3, $\sigma_{allow} = 25.97$ ksi; thus,

$$\begin{aligned} P_{allow} &= \sigma_{allow} A \\ &= (25.97 \text{ kips/in.}^2)(19.1 \text{ in.}^2) \\ &= 496 \text{ kips} \end{aligned}$$

which is more than the design load of 490 kips. Hence, the section is satisfactory.

Use a W12 × 65 section   ⇐ **Ans.**

EXAMPLE 19–7

Select the lightest steel pipe section for a column subjected to an axial compressive load of 400 kN. The column has pinned ends and an unbraced length of 4 m. Use A36 steel ($\sigma_y = 250$ MPa) and the AISC specification.

Solution. Since the size of the steel pipe is unknown, a trial-and-error procedure is necessary. Assume the radius of gyration to be $r = 0.05$ m. For pinned ends, $k = 1.0$. The slenderness ratio is

$$\frac{kL}{r} = \frac{(1.0)(4 \text{ m})}{0.05 \text{ m}} = 80$$

From Table 19–2, for $\sigma_y = 250$ MPa and $kL/r = 80$, the allowable compressive stress is

$$\sigma_{allow} = 15.36 \text{ ksi} = 15.36 \times 6.895 = 105.9 \text{ MPa}$$

where 6.895 is the conversion factor from ksi to MPa. Thus, the required cross-sectional area of the column is

$$A_{req} = \frac{P}{\sigma_{allow}} = \frac{400 \text{ kN}}{105\,900 \text{ kN/m}^2} = 3.78 \times 10^{-3} \text{ m}^2$$

From the appendix, Table A–5(b), we find the following steel pipes that have a cross-sectional area approximately equal to the required area calculated above:

Pipe Section	$A\ (\times 10^{-3}\ m^2)$	$r\ (m)$
150 mm, standard weight	3.60	0.0572
200 mm, standard weight	5.42	0.0747
150 mm, extra weight	5.42	0.0556

Of the pipes listed above, the 150-mm-diameter, standard weight steel pipe is the lightest because it has the smallest cross-sectional area. Thus, we use this pipe for the trial section. The slenderness ratio of the section is

$$\frac{kL}{r} = \frac{(1.0)(4\text{ m})}{0.0572\text{ m}} = 70$$

From Table 19–2,

$$\sigma_{\text{allow}} = 16.43\text{ ksi} = 16.43 \times 6.895 = 113.3\text{ MPa}$$

Thus,

$$\begin{aligned} P_{\text{allow}} &= \sigma_{\text{allow}} A \\ &= (113\ 300\text{ kN/m}^2)(3.60 \times 10^{-3}\text{ m}^2) \\ &= 408\text{ kN} \end{aligned}$$

which is more than the design load of 400 kN. Hence, the pipe section is satisfactory.

Use a 150-mm-diameter, standard weight steel pipe   ⇐ **Ans.**

19–8
SUMMARY

Buckling of Columns. Columns are members subjected to compressive loads that tend to buckle the column laterally. Once buckling occurs, a column will loose its load-carrying stability. The tendency of buckling increases with slenderness.

Euler Formula. The *Euler formula* for the critical load of a long column with pinned ends made of homogeneous material is

$$P_{cr} = \frac{\pi^2 EI}{L^2} \tag{19–1}$$

The Euler formula was derived based on the elastic behavior of the materials. The critical load calculated from the Euler formula is called the *elastic buckling load.*

Euler Formulas for Other Supporting Conditions. The Euler formula for a pin-ended column, given in Equation 19–1, is referred to as the *fundamental case*. The formulas can be modified for other end-supporting conditions through the use of an *effective length* L_e equal to kL, where the effective length factor k takes the following values, depending on the end-supporting conditions of the column:

Pinned ends (fundamental case):	$k = 1$
Fixed ends:	$k = 0.5$
Fixed, pinned ends:	$k = 0.7$
Fixed, free ends:	$k = 2$

The Euler column formula becomes

$$P_{cr} = \frac{\pi^2 EI}{L_e^{\,2}} = \frac{\pi^2 EI}{(kL)^2} \tag{19-2}$$

Limitation of the Euler Formula. The Euler formula can be written in the form:

$$\sigma_{cr} = \frac{P_{cr}}{A} = \frac{\pi^2 E}{(kL/r)^2} \tag{19-3}$$

where kL/r = the *slenderness ratio* of the column. The Euler formula is valid only if the critical stress is less than the proportional limit of the material. Hence, the Euler formula is applicable only when the slenderness ratio is greater than or equal to

$$\left(\frac{kL}{r}\right)_{\min} = \sqrt{\frac{\pi^2 E}{\sigma_p}} \tag{19-4}$$

A column is designed to carry an allowable load P_{allow} equal to the critical load divided by the factor of safety; that is,

$$P_{\text{allow}} = \frac{P_{cr}}{\text{F.S.}} \tag{19-5}$$

The factor of safety to be used depends on many factors and is usually specified by the design code. For structural steel design, a factor of safety of 1.92 is used for long columns.

J. B. Johnson Formula. For inelastic buckling of steel columns of short and intermediate length, where the slenderness ratio of the column is less than or equal to

$$C_c = \sqrt{\frac{2\pi^2 E}{\sigma_y}} \tag{19-6}$$

the following J. B. Johnson formula must be used:

$$\sigma_{cr} = \frac{P_{cr}}{A} = \left[1 - \frac{(kL/r)^2}{2C_c^{\,2}}\right]\sigma_y \tag{19-7}$$

The AISC Column Formulas. The American Institute of Steel Construction (AISC) manual gives the following formulas for steel column design.

1. For long columns: $kL/r \geq C_c = \sqrt{2\pi^2 E/\sigma_y}$

$$\sigma_{\text{allow}} = \frac{\pi^2 E/(kL/r)^2}{\text{F.S.}} = \frac{\pi^2 E/(kL/r)^2}{1.92} \tag{19-8}$$

2. For intermediate and short columns: $kL/r \le C_c = \sqrt{2\pi^2 E / \sigma_y}$

$$\sigma_{\text{allow}} = \frac{\left[1 - \dfrac{(kL/r)^2}{2C_c^2}\right]\sigma_y}{\text{F.S.}} \qquad (19\text{--}9)$$

where the factor of safety (F.S.) is computed from the equation

$$\text{F.S.} = \frac{5}{3} + \frac{3(kL/r)}{8C_c} - \frac{(kL/r)^3}{8C_c^3} \qquad (19\text{--}10)$$

When the end-supporting conditions are approximated, the AISC recommended effective length factor k listed in Table 19–1 must be used.

Steel Column Design. The AISC specification is widely used for the design of structural steel columns. The most commonly used structural steel compression members are W shapes, steel pipes, and structural tubing. Steel columns can be designed by using AISC column formulas or the allowable compressive stress tables (Tables 19–2 and 19–3).

PROBLEMS

Section 19–2   Euler Formula for Pin-Ended Columns

Section 19–3   Euler Formula for Other Supporting Conditions

Section 19–4   Limitation of the Euler Formula

19–1   Determine the slenderness ratio of an 8-ft column of full-size, rectangular section 2 in. × 4 in. with pinned ends.

19–2   Determine the slenderness ratio of a 3-m column of 200-mm standard weight steel pipe section with fixed, free ends.

19–3   If a square section and a circular section have the same cross-sectional area A, which one is a better column section? (*Hint*: Express their radii of gyration in terms of the cross-sectional area A and compare the values.)

19–4   For structural steel with the stress–strain diagram shown in Fig. P19–4, determine the lowest limit of the slenderness ratio for which the elastic Euler formula applies.

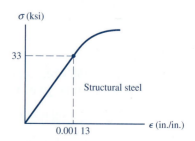

FIGURE P19–4

19–5   Determine the critical buckling load for a 2.5-m timber column of full-size, square section 100 mm × 100 mm. The column has pinned ends. Assume that the modulus of elasticity is 12 GPa and that the Euler formula applies.

19–6   Determine the critical buckling load of an aluminum column 4 m long of 80-mm-diameter circular section. The column has fixed ends. Assume that the modulus of elasticity is 70 GPa and that the proportional limit is 230 MPa.

19–7   Determine the critical buckling load of an L4 × 4 × $\frac{1}{2}$ steel angle 18 ft long with fixed, pinned ends. Assume that the modulus of elasticity is 29×10^3 ksi and that the proportional limit is 34 ksi.

19–8   Determine the critical buckling load of a 2-in. standard weight steel pipe section 4 ft long with fixed, free ends. Assume that the modulus of elasticity is 29×10^3 ksi and that the proportional limit is 34 ksi.

19–9   A W12 × 87 structural steel section of A36 steel ($E = 29 \times 10^3$ ksi, $\sigma_y = 36$ ksi, and $\sigma_p = 34$ ksi) is used as an axially loaded compression member. Determine the allowable axial compressive load using a factor of safety of 2.0. The column is 25 ft long and is pin connected at both ends.

19–10   A W300 × 0.58 structural steel section of A36 steel ($E = 200$ GPa, $\sigma_y = 250$ MPa, and $\sigma_p = 200$ MPa) is used as an axially loaded compression member. Determine the allowable axial compressive load using a factor of safety of 2.0. The column is 15 m long and is fixed at both ends.

19–11   Find the minimum required dimension b of the square section of a 1.2-m long steel strut with pinned ends that must support an axial compressive load of 20 kN. Use the Euler formula with F.S. = 2 and $E = 200$ GPa.

19–12   Find the minimum required diameter d of the circular section of a 20-ft-long steel strut with fixed ends that must support an axial compressive load of 5 kips. Use the Euler formula with F.S. = 2 and $E = 29 \times 10^3$ ksi.

19–13   A 200-lb worker climbs a flagpole made of $\frac{3}{4}$-in. standard weight steel pipe. If the pole is 10 ft tall, can the worker get to the top before the pole buckles? If not, how high can the worker climb before it does? The pole is fixed at the bottom and is free at the top. Neglect the weight of the pole and assume that the worker's center of gravity is always along the axis of the pole. Use the Euler formula and $E = 29 \times 10^6$ psi.

19–14   The jib crane in Fig. P19–14 has a steel boom of square section 60 mm × 60 mm. Determine the maximum weight W in kN that the crane can lift based on the allowable axial compressive load of the boom AB. Use the Euler formula with F.S. = 3 and $E = 200$ GPa. Neglect the weight of the member.

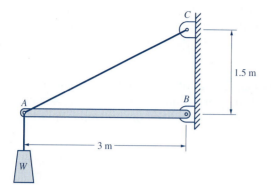

FIGURE P19–14

19–15 For the jib crane in Fig. P19–15, determine the smallest size of the standard steel pipe that can be used for member AB if the crane has a capacity of $2\frac{1}{2}$ tons. Use the Euler formula with F.S. = 3 and $E = 29 \times 10^3$ ksi. Neglect the weight of the member.

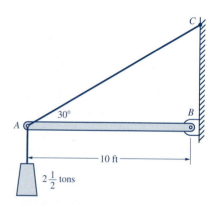

FIGURE P19–15

19–16 The pin-ended timber member AB in Fig. P19–16, has a length of 2 m and a full-size rectangular cross-section of 50 mm × 100 mm. Determine the maximum weight W that can be supported by the structure based on the allowable axial compressive load of member AB. Use the Euler formula with F.S. = 2 and $E = 12$ GPa.

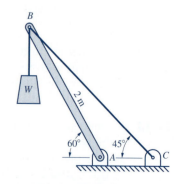

FIGURE P19–16

19–17 A toggle press is a mechanism that causes a large compressive force to be exerted on the block D in Fig. P19–17. If the weight is $W = 1\frac{1}{2}$ tons, determine the minimum required diameter of the circular steel rod for the two arms, AB and AC. Use the Euler formula with F.S. = 3 and $E = 29 \times 10^3$ ksi.

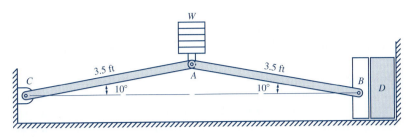

FIGURE P19–17

19–18 The steel column AB in Fig. P19–18 has a full-size, 1-in. × $2\frac{1}{4}$-in. rectangular section and rounded ends, as shown. The column is braced at the midpoint in the weak direction. Determine the allowable force F that can be applied at D based on the allowable axial compressive load of the column. Use the Euler formula with $E = 30 \times 10^3$ ksi and F.S. = 2.

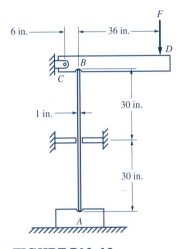

FIGURE P19–18

Section 19–5 J. B. Johnson Formula

19–19 Determine the critical buckling load of a column with a full-size, 100-mm × 200-mm rectangular section having a length of 2.5 m and fixed ends. Assume that the modulus of elasticity is 200 GPa and the yield stress is 250 MPa.

19–20 Determine the critical buckling load of a column with a circular section of 100-mm diameter and is 3.5 m long with fixed, pinned ends. Assume that the modulus of elasticity is 200 GPa and the yield stress is 250 MPa.

19–21 Determine the critical buckling load of a 6-ft long standard weight steel pipe section of 5-in. nominal diameter with fixed, free ends. Assume that the column is made of A441 steel with $E = 29 \times 10^3$ ksi and $\sigma_y = 50$ ksi.

19–22 Determine the critical buckling load of a W14 × 74 section 40 ft long with fixed ends. Assume that the column is made of A242 steel with $E = 29 \times 10^3$ ksi and $\sigma_y = 50$ ksi.

19–23 Determine the allowable axial compressive load that a 10-m column of W360 × 1.31 section with fixed, pinned ends can support. Use a factor of safety of 2.0. Assume that the column is made of A441 steel with $E = 200$ GPa and $\sigma_y = 345$ MPa.

19–24 Determine the allowable axial compressive load of W12 × 65 section having a length of 25 ft and fixed ends. Use a factor of safety of 2.0. Assume that the column is made of A441 steel with $E = 29 \times 10^3$ ksi and $\sigma_y = 50$ ksi.

19–25 The compression member *BD* in Fig. P19–25 has a rectangular section 1 in. × 2 in. and is made of A441 steel with $\sigma_y = 50$ ksi. Determine the maximum weight *W* that can be supported by the assembly. Assume that the beam *AC* and the connections are all properly designed. Use a factor of safety of 2.5.

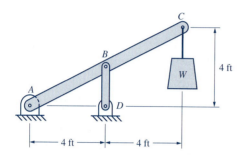

FIGURE P19–25

19–26 Compression member *AB* acts as a spreader bar between the cables shown in Fig. P19–26. The bar has a circular section of 100-mm diameter and is made of A36 steel with $\sigma_y = 250$ MPa and $E = 200$ GPa. Determine the maximum pulling force *F* that can be applied to the assembly based on the allowable axial compressive load of the bar for a factor of safety of 3.

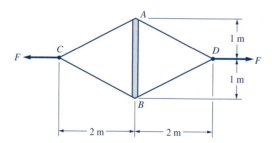

FIGURE P19–26

Section 19–6 The AISC Column Formulas

19–27 A 5-ft column having a 3-in., standard weight steel pipe section is made of A36 steel with $E = 29 \times 10^3$ ksi and $\sigma_y = 36$ ksi. The column has fixed, free ends. Calculate the allowable axial compressive load using the AISC column formulas and the recommended *k* values. Use the allowable stress listed in Table 19–2 to verify the computations.

19–28 Rework Problem 19–27. Assume that the length of the column is 8 ft.

19–29 A 20-ft column having a W8 × 40 section is made of A441 steel with $E = 29 \times 10^3$ ksi and $\sigma_y = 50$ ksi. Both ends of the column are fixed. Calculate the allowable axial compressive load using the AISC column formulas and the recommended k values. Use the allowable stress listed in Table 19–3 to verify the computations.

19–30 Rework Problem 19–29. Assume that the length of the column is 30 ft.

19–31 A 9-m column having a W250 × 1.63 section is made of A36 steel with $E = 200$ GPa and $\sigma_y = 250$ MPa. The column has fixed, pinned ends. Calculate the allowable axial compressive load using the AISC column formulas and the recommended k values. Use the allowable stress listed in Table 19–2 to verify the computations.

19–32 Rework Problem 19–31. Assume that the length of the column is 12 m.

19–33 A 3-m column having an L127 × 127 × 12.7 angle section is made of A242 steel with $E = 200$ GPa and $\sigma_y = 345$ MPa. The column is fixed at both ends. Calculate the allowable axial compressive load using the AISC formulas and the recommended k values. Use the allowable stress listed in Table 19–3 to verify the computations.

19–34 Rework Problem 19–33. Assume that the length of the column is 4.5 m.

19–35 Two standard steel C10 × 20 channels form a 25-ft-long square compression member, as shown in Fig. P19–35. The channels are tied together by end tie plates and lacing bars to make the two channels act as one unit. The channels are made of A36 steel with $\sigma_y = 36$ ksi, and the ends of the columns are considered hinged. Determine the allowable axial force of the member according to the AISC specification.

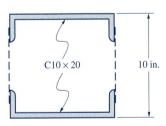

FIGURE P19–35

19–36 A compression member made of two C300 × 0.438 channels is arranged as shown in Fig. P19–36. The unbraced length of the column is 12 m. The channels are properly laced together to act as one unit. They are made of A441 steel with $E = 200$ GPa and $\sigma_y = 345$ MPa. The ends of the column are considered hinged. Determine (*a*) the value of the distance *b* so that the section will have equal moments of inertia about the *x* and *y* axes, and (*b*) the allowable axial compressive load of the member according to the AISC specification.

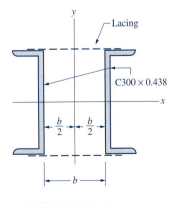

FIGURE P19–36

Section 19–7 Steel Column Design

19–37 Select the lightest W shape for a column subjected to an axial compressive load of 400 kips. The column is pin-connected at both ends and has an unbraced length of 20 ft. Assume that the column is made of A36 steel with $\sigma_y = 36$ ksi. Use the AISC specification and the allowable stress listed in Table 19–2.

19–38 Select the lightest W shape for a column subjected to an axial compressive load of 150 kips. The column has an unbraced length of 35 ft. It is pin-connected at one end and fixed at the other. Assume that the column is made of A36 steel with $\sigma_y = 36$ ksi. Use the AISC specification and the allowable stress listed in Table 19–2.

19–39 Select the lightest W shape (in SI designation) for a column subjected to an axial compressive load of 600 kN. The column has an unbraced length of 10 m and is pin-connected at both ends. Assume that the column is made of A36 steel with $\sigma_y = 250$ MPa. Use the AISC specification and the allowable stress listed in Table 19–2.

19–40 Select the lightest W shape (in SI designation) for a column subjected to an axial compressive load of 1500 kN. The column has an unbraced length of 12 m and is fixed at both ends. Assume that the column is made of A441 steel with $\sigma_y = 345$ MPa. Use the AISC specification and the allowable stress listed in Table 19–3.

19–41 Select the lightest W shape for a column subjected to an axial compressive load of 200 kips. The column has an unbraced length of 30 ft and is fixed at both ends. Assume that the column is made of A441 steel with $\sigma_y = 50$ ksi. Use the AISC specification and the allowable stress listed in Table 19–3.

19–42 Select the lightest W shape for a column subjected to an axial compressive load of 500 kips. The column has an unbraced length of 40 ft and is fixed at both ends. Assume that the column is made of A242 steel with $\sigma_y = 50$ ksi. Use the AISC specification and the allowable stress listed in Table 19–3.

19–43 Select the lightest standard weight steel pipe section (in SI designation) to support an axial compressive load of 700 kN. The column has an unbraced length of 10 m and fixed, pinned ends. Assume that the column is made of A36 steel with $\sigma_y = 250$ MPa. Use the AISC specification and the allowable stress listed in Table 19–2.

19–44 Select the lightest standard weight steel pipe section to support an axial compressive load of 80 kips. The column has an unbraced length of 20 ft and is pinned at both ends. Assume that the column is made of A242 steel with $\sigma_y = 50$ ksi. Use the AISC specification and the allowable stress listed in Table 19–3.

Computer Program Assignments

For each of the following problems, write a computer program using an appropriate programming language with which you are most familiar. Make the program user friendly by incorporating plenty of comments and input prompts so that the user will understand the input data to be entered and the limitations of their values. The output should include the data entered and the computed results, and they must be well labeled to identify each quantity. If a tabulated format is used, a proper heading must be included at the top of the table. Do not limit the program to any specific unit system. Indicate the consistent U.S. customary or SI units that can be used.

C19–1 Write a computer program that can be used to produce a table similar to Tables 19–2 and 19–3 for a specified value of σ_y from the AISC column formulas. The user input is to be the yield strength σ_y of the column. Use this program to produce two tables listing the values of the allowable stresses corresponding to the slenderness ratio varying from 1 to 200 at steps of 1 for σ_y equals (*a*) 60 ksi and (*b*) 345 MPa.

C19–2 Write a computer program that can be used to compute the allowable compressive load of a column using the AISC column formulas. The user input should include (1) the length, the cross-sectional area, and the least radius of gyration of the column; (2) the yield strength of the material; and (3) the supporting condition of the column. The output should include the AISC allowable compressive stress and load. Use this program to solve (*a*) Example 19–4, (*b*) Example 19–5, (*c*) Problem 19–31, and (*d*) Problem 19–35.

CHAPTER 20

CONNECTIONS

20–1
INTRODUCTION

Steel structures, such as building frames, bridge trusses, cranes, machine components, etc., are assemblies of beams and columns and other types of members joined together. These members must be adequately connected so that loads can be transmitted from one member to the other.

The major connectors used in steel structures are rivets, bolts, and welding. Riveted connections were used extensively for many years. The use of large rivets requiring hot forming has declined sharply in recent decades due to competition from welding and high-strength bolts. However, the introduction of cold heading, softer rivet materials and the recent advent of blind rivets (which require access to only one side of the workpiece) have made riveting more competitive, especially in connecting some minor machine components.

This chapter presents the analysis and design of several types of connections for structural members. Riveted connections are discussed first. High-strength steel bolts are presented next, followed by the discussion on welded connections.

20–2
RIVETED CONNECTIONS

A rivet is a short metal pin with a preformed head and a shank that can be worked into a second head following assembly. Rivets are used to connect plates, structural steel shapes, sheet metals, and other relatively thin components. In general, the design of a connection is concerned with the transfer of forces from one component to another through the connection. For a riveted connection, the forces are transmitted through shear forces in the rivets and the bearing force between the rivets and the connected plates. A rivet can be in single or double shear, depending on whether one or two sections of the rivet are subjected to shear forces, as discussed in

Section 9–4. The bearing stress between a rivet and a plate was discussed in Section 9–5. The actual distribution of bearing stress is rather complicated (see Fig. 9–7). In practice, the bearing stress distribution is approximated on the basis of an average bearing stress acting over the projected area of the rivet's shank onto the cross-section of a plate, that is, of a rectangular area td, as shown in Fig. 9–7c. The following assumptions are made for rivet connections:

1. Holes $\frac{1}{16}$ in. (1.5 mm) larger than the rivet diameter are punched or drilled for the insertion of rivets. The rivets completely fill the holes.
2. The friction forces between the connected plates are ignored.
3. A load applied to the member without any eccentricity with respect to the centroid of the rivets is assumed to be shared equally by all the rivets.
4. The shear stress is assumed to be distributed uniformly over a section (or two sections in double shear) of a rivet. The bearing stress is assumed to be distributed uniformly over the projected area td, where t is the thickness of the plate and d is the diameter of the rivet. Stress concentrations (see Section 11–7) at rivet holes in the plate are ignored, and tensile stress is assumed to be distributed uniformly across the net section of a plate.

Rivets are available in standard sizes from $\frac{1}{2}$ to $1\frac{1}{2}$ in., at increments of $\frac{1}{8}$ in. In structural applications, rivets of $\frac{3}{4}$ in. and $\frac{7}{8}$ in. diameters are two of the most commonly used sizes. It is preferable that all rivets on the same structure be of one size. Steel rivets used for structural purposes are classified as ASTM[*] A502, grades 1 and 2; these are designated as A502-1 and A502-2, respectively. Several typical arrangements of riveted connections are shown in Fig. 20–1.

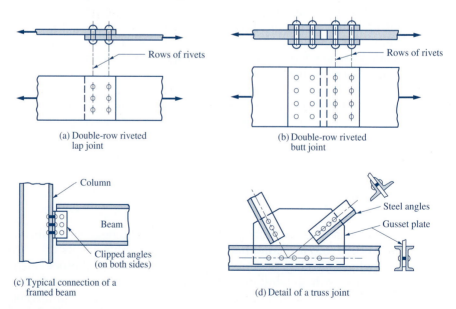

(a) Double-row riveted lap joint

(b) Double-row riveted butt joint

(c) Typical connection of a framed beam

(d) Detail of a truss joint

FIGURE 20–1

[*] American Society for Testing and Materials.

20–3
STRENGTH OF RIVETED CONNECTIONS

Riveted joints may fail in one of the following three modes:

1.   Failure in shear of rivets, as shown in Fig. 20–2a and b.
2.   Failure in bearing when the rivets crush the material of the plate against which the rivets bear, as shown in Fig. 20–2c.
3.   Failure in tension when the connected plate is torn apart at the critical section, which is weakened by the rivet holes, as shown in Fig. 20–2d.

It is presumed that the joint will ultimately fail in one of the three ways listed above. The strength of a connection based on these three methods of failure is outlined below.

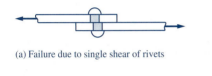

(a) Failure due to single shear of rivets

(b) Failure due to double shear of rivets

(c) Failure due to crushing of plate

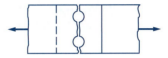

(d) Failure due to tension through the net section

FIGURE 20–2

Shear Strength. The strength of a joint based on the shear of the rivets may be determined from

$$P_s = nA_s\tau_{\text{allow}} \tag{20–1}$$

where

P_s = the strength of the joint based on the shear stress of the rivets

n = total number of shear planes of the rivets in the joint

A_s = the cross-sectional area of a rivet = $\pi d^2/4$

d = the diameter of the rivets

τ_{allow} = the allowable shear stress of the rivet material

The allowable shear stresses for rivets, specified in the AISC manual for both single shear and double shear, are

$$\tau_{\text{allow}} = 15 \text{ ksi (103 MPa) for A502-1}$$

$$\tau_{\text{allow}} = 20 \text{ ksi (138 MPa) for A502-2}$$

Bearing Strength. The strength of a joint based on the bearing of rivets on plates may be determined from

$$P_b = n(td)(\sigma_b)_{\text{allow}} \tag{20-2}$$

where

P_b = the strength of the joint base on the bearing of rivets on the plate

n = total number of bearing surfaces on the plate

t = the thickness of the plate

d = the diameter of the rivets

$(\sigma_b)_{\text{allow}}$ = the allowable bearing stress of the plate

The AISC specification gives the allowable bearing stress on the projected area as

$$(\sigma_b)_{\text{allow}} = 1.35\sigma_y$$

where σ_y is the yield strength of the connected plate.

Tensile Strength.   The strength of a joint based on the allowable tensile load of the plate through the critical section may be determined from

$$P_t = (b_{\text{net}}t)(\sigma_t)_{\text{allow}} \tag{20-3}$$

where

P_t = the strength of the joint based on the allowable tension of the plate

b_{net} = the net width of the plate through the critical section: $b_{\text{net}} = b - n\,(d + c)$

b = width of plate

n = number of rivet holes in the critical section

d = rivet diameter

c = a constant factor equal to $\frac{1}{8}$ in. or 3 mm, to be added to the rivet diameter for the size of the hole

t = thickness of plate

$(\sigma_t)_{\text{allow}}$ = allowable tensile stress of plate

Note that the factor c is added to the rivet diameter to account for the fact that the holes are $\frac{1}{16}$ in. (1.5 mm) larger than the rivet diameters, plus an additional $\frac{1}{16}$ in. (1.5 mm) for possible damage to the rim of the hole when it is punched. According to the AISC manual, the allowable tensile stress in the net section of the plate is

$$(\sigma_t)_{\text{allow}} = 0.60\sigma_y$$

where σ_y is the yield strength of the connected plate.

Joint Strength and Joint Efficiency. The smallest of the three allowable loads is the *strength of the joint*. The ratio of this strength divided by the strength of a solid plate or member without holes, expressed as a percentage, is called the *joint efficiency*. We write

$$\text{Joint efficiency} = \frac{\text{Strength of the joint}}{\text{Strength of the solid plate}} \times 100\% \qquad (20\text{–}4)$$

EXAMPLE 20–1

Determine the strength and efficiency of the lap joint in Fig. E20–1. The $\frac{7}{8}$-in.-diameter rivets are made of A502-1 steel, and the plates are made of A36 steel with $\sigma_y = 36$ ksi.

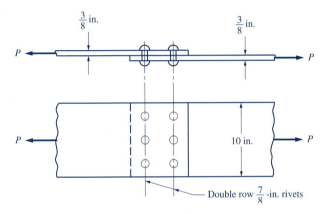

FIGURE E20–1

Solution. The shearing, bearing, and tensile strengths are determined as follows.

Shear Strength. The cross-sectional area of the rivet is

$$A = \frac{1}{4}\pi d^2 = \frac{1}{4}\pi\left(\frac{7}{8}\text{ in.}\right)^2 = 0.601\text{ in.}^2$$

For A502-1 rivets, the allowable shear stress is $\tau_{\text{allow}} = 15$ ksi. With six rivets in single shear, the shear strength, from Equation 20–1, is

$$P_s = nA_s\tau_{\text{allow}}$$

$$= 6(0.601\text{ in.}^2)(15\text{ kips/in.}^2)$$

$$= 54.1\text{ kips}$$

Bearing Strength. With six bearing surfaces on each plate, the bearing strength, from Equation 20–2, is

$$P_b = n(td)(\sigma_b)_{\text{allow}}$$

$$= 6\left(\frac{3}{8} \times \frac{7}{8}\text{ in.}\right)(1.35 \times 36\text{ kips/in.}^2)$$

$$= 95.7\text{ kips}$$

Tensile Strength.   The net width through the critical section is

$$b_{\text{net}} = b - n(d + c) = 10 - 3\left(\frac{7}{8} + \frac{1}{8}\right) = 7.00 \text{ in.}$$

The tensile strength of the plate, from Equation 20–3, is

$$P_t = b_{\text{net}}t\,(\sigma_t)_{\text{allow}}$$

$$= (7.00 \text{ in.})\left(\frac{3}{8} \text{ in.}\right)(0.60 \times 36 \text{ kips/in.}^2)$$

$$= 56.7 \text{ kips}$$

Joint Strength and Efficiency.   The strength of the joint equals the shear strength, which is the smallest of the three:

$$P = P_s = 54.1 \text{ kips} \qquad\qquad \Leftarrow \textbf{Ans.}$$

The tensile strength of a solid plate is

$$P_g = A_g(0.6\sigma_y)$$

$$= \left(10 \times \frac{3}{8} \text{ in.}^2\right)(0.60 \times 36 \text{ kips/in.}^2) = 81.0 \text{ ksi}$$

The joint efficiency, from Equation 20–4, is

$$\text{Joint efficiency} = \frac{P}{P_g} \times 100\% = \frac{54.1 \text{ kips}}{81.0 \text{ kips}} \times 100\%$$

$$= 66.8\% \qquad\qquad \Leftarrow \textbf{Ans.}$$

EXAMPLE 20–2

Determine the strength and efficiency of the butt joint in Fig. E20–2(1). The 20-mm-diameter rivets are made of A502-2 steel, and the plates are made of A441 steel with $\sigma_y = 345$ MPa.

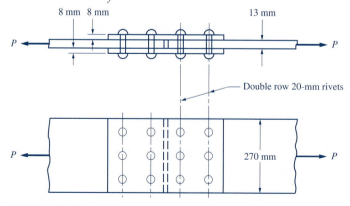

FIGURE E20–2(1)

   Solution. The shear, bearing, and tensile strengths are determined as follows.

Shear Strength. The load is transmitted from the main plate at each side by six rivets to the cover plates, as shown in Fig. E20–2(2), the free-body diagram of one main plate and the cover plates.

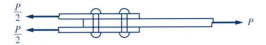

FIGURE E20–2(2)

The six rivets are in double shear; this makes the total number of rivet sections under shear stress equal to 12. The cross-sectional area of the rivet is

$$A = \frac{1}{4}\pi d^2 = \frac{1}{4}\pi (0.020 \text{ m})^2 = 3.14 \times 10^{-4} \text{ m}^2$$

For A502-2 rivets, the allowable shear stress is τ_{allow} = 138 MPa. Thus, the shear strength is

$$P_s = nA_s\tau_{allow}$$

$$= 12(3.14 \times 10^{-4} \text{ m}^2)(138\,000 \text{ kN/m}^2)$$

$$= 520 \text{ kN}$$

Bearing Strength. The bearing on the main plate is more critical than the bearing on the cover plates since the combined thickness of the cover plate is greater than that of the main plate.

$$P_b = n(td)(\sigma_b)_{allow}$$

$$= 6(0.013 \text{ m})(0.020 \text{ m})(1.35 \times 345\,000 \text{ kN/m}^2)$$

$$= 727 \text{ kN}$$

Tensile Strength. The net width at the critical section is

$$b_{net} = b - n\,(d + c) = 270 - 3(20 + 3) = 201 \text{ mm}$$

The tensile strength at the critical section is

$$P_t = b_{net}t\,(\sigma_t)_{allow}$$

$$= (0.201 \text{ m})(0.013 \text{ m})(0.6 \times 345\,000 \text{ kN/m}^2)$$

$$= 541 \text{ kN}$$

Joint Strength and Efficiency. The strength of the joint equals the shear strength, which is the smallest of the three:

$$P = P_s = 520 \text{ kN} \qquad\qquad \Leftarrow \textbf{Ans.}$$

The tensile strength of a solid plate is

$$P_g = A_g(0.6\sigma_y)$$
$$= (0.270 \text{ m})(0.013 \text{ m})(0.6 \times 345\,000 \text{ kN/m}^2)$$
$$= 727 \text{ kN}$$

The joint efficiency, from Equation 20–4, is

$$\text{Joint efficiency} = \frac{P}{P_g} \times 100\% = \frac{520 \text{ kN}}{727 \text{ kN}} \times 100\%$$

$$= 71.5\% \qquad\qquad \Leftarrow \textbf{Ans.}$$

EXAMPLE 20–3

Refer to Fig. E20–3. A tension member A composed of a pair of L4 $\times$ 3 $\times$ $\frac{5}{16}$ angles, arranged back to back, is connected to the $\frac{1}{2}$-in.-thick gusset plate at the joint. Both the angle and the gusset plate are made of A36 steel. The four $\frac{3}{4}$-in.-diameter rivets are made of A502-2 steel. Determine the strength and efficiency of the connection.

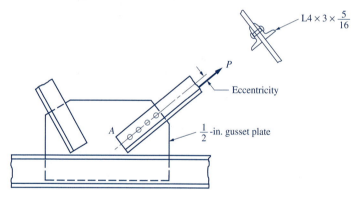

FIGURE E20–3

Solution. The shear, bearing, and tensile strengths are computed as follows.

Shear Strength. The cross-sectional area of the rivet is

$$A_s = \frac{1}{4}\pi d^2 = \frac{1}{4}\pi\left(\frac{3}{4} \text{ in.}\right)^2 = 0.442 \text{ in.}^2$$

We have four rivets in double shear. For A502-2 rivets, the allowable shear stress is $\tau_{\text{allow}} = 20$ ksi. Thus, the shear strength of the joint is

$$P_s = nA_s\tau_{\text{allow}}$$
$$= 8(0.442 \text{ in.}^2)(20 \text{ kips/in.}^2)$$
$$= 70.7 \text{ kips}$$

Bearing Strength. Each angle has four bearing surfaces. There are eight bearing surfaces in two angles; thus, the strength of the joint based on the bearing on the angles is

$$P_b = n(td)(\sigma_b)_{\text{allow}}$$

$$= 8\left(\frac{5}{16} \times \frac{3}{4} \text{ in.}^2\right)(1.35 \times 36 \text{ kips/in.}^2)$$

$$= 91.1 \text{ kips}$$

The rivets bear on the gusset plate at four surfaces. Thus, the strength of the joint based on the bearing on the gusset plate is

$$P_b = n(td)(\sigma_b)_{\text{allow}}$$

$$= 4\left(\frac{1}{2} \times \frac{3}{4} \text{ in.}^2\right)(1.35 \times 36 \text{ kips/in.}^2)$$

$$= 72.9 \text{ kips}$$

Tensile Strength. The net cross-sectional area of the angles is obtained by deducting the area of the rivet holes from the gross area of the angles. The net cross-sectional area for two angles is

$$A_{\text{net}} = 2[A_{\text{angle}} - t(d + c)]$$

$$= 2\left[2.09 - \frac{5}{16}\left(\frac{3}{4} + \frac{1}{8}\right)\right] = 3.63 \text{ in.}^2$$

The tensile strength of the angles is

$$P_t = A_{\text{net}}(\sigma_t)_{\text{allow}}$$

$$= (3.63 \text{ in.}^2)(0.60 \times 36 \text{ kips/in.}^2)$$

$$= 78.5 \text{ kips}$$

Joint Strength and Efficiency. We see that the strength of the connection is governed by the shear of the rivets:

$$P = P_s = 70.7 \text{ kips} \qquad \Leftarrow \textbf{Ans.}$$

The tensile strength of two solid angles is

$$P_g = 2A_{\text{angle}}(0.6\sigma_y)$$

$$= 2(2.09 \text{ in.}^2)(0.6 \times 36 \text{ kips/in.}^2)$$

$$= 90.3 \text{ kips}$$

The joint efficiency, from Equation 20–4, is

$$\text{Joint efficiency} = \frac{P}{P_g} \times 100\% = \frac{70.7 \text{ kips}}{90.3 \text{ kips}} \times 100\%$$

$$= 78.3\% \qquad \Leftarrow \textbf{Ans.}$$

Remark. Note that there is a small eccentricity of the line of action of tensile force P (which acts through the centroid of the angles) from the centerline of the rivets. This eccentricity is small, however, and is usually ignored.

──────── **EXAMPLE 20–4** ────────────────────────────────────

Refer to Fig. E20–4. Find the strength of the connection joining the W16 × 57 beam to the W12 × 87 column. The connection consists of two 9-in.-long clipped L$3\frac{1}{2}$ × $3\frac{1}{2}$ × $\frac{1}{4}$ angles connected to the web of the beam and the flange of the column by $\frac{7}{8}$-in. rivets made of A502-2 steel. The beam, the column, and the clipped angles are all made of A441 steel with σ_y = 50 ksi.

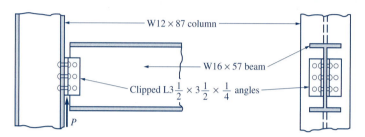

FIGURE E20–4

Solution. This connection between a beam and a column is considered to be a simple support. The loads on the beam are transmitted to the column via shear load on the rivets and bearing load on the bearing surfaces.

Shear Strength. The cross-sectional area of the rivet is

$$A_s = \tfrac{1}{4}\pi d^2 = \tfrac{1}{4}\pi (\tfrac{7}{8} \text{ in.})^2 = 0.601 \text{ in.}^2$$

Where the angles are connected to the beam, we have three rivets in double shear. Where the angles are connected to the column, we have six rivets in single shear. In both cases, we have six shear planes. For A502-2 rivets, the allowable shear stress is τ_{allow} = 20 ksi. Thus,

$$P_s = nA_s\tau_{allow}$$

$$= 6(0.601 \text{ in.}^2)(20 \text{ kips/in.}^2)$$

$$= 72.1 \text{ kips}$$

Bearing Strength. Bearing stress occurs between three rivets and the web of the W16 × 57 beam. The projected bearing area is

$$A_b = 3(t_w d) = 3\left(0.430 \times \frac{7}{8}\right) = 1.13 \text{ in.}^2$$

Bearing stress also occurs between six rivets and the flange of the W12 × 87 column, and between six rivets and the clipped L$3\frac{1}{2}$ × $3\frac{1}{2}$ × $\frac{1}{4}$ angles. Since the $\frac{1}{4}$ thickness of the angle is less than the thickness of the flange of the column, which is 0.810 in., the bearing on the angle is more critical and its projected area is

$$A_b = 6(td) = 6\left(\frac{1}{4} \times \frac{7}{8}\right) = 1.31 \text{ in.}^2$$

The projected bearing area on the web of the beam is smaller, so the bearing strength is based on the bearing on the web of the beam. We have

$$P_b = A_b(\sigma_b)_{\text{allow}}$$

$$= (1.13 \text{ in.}^2)(1.35 \times 50 \text{ kips/in.}^2)$$

$$= 76.3 \text{ kips}$$

Strength of the Connection. We see that the strength of the connection is governed by the shear of the rivets:

$$P = P_s = 72.1 \text{ kips} \qquad \Leftarrow \textbf{Ans.}$$

20–4
HIGH-STRENGTH BOLTED CONNECTIONS

Although the use of high-strength steel bolts is a relatively new development, they have already become the leading fastener for connections done in the field. Generally speaking, any joint that can be connected by rivets can be connected by bolts. The design considerations of a bolted joint are similar to those of a riveted joint.

The two basic types of high-strength steel bolts used in structural joints are the ASTM A325 and A490. Both are threaded structural bolts used with heavy hex nuts. Bolts are available in standard sizes from $\frac{1}{2}$ to $1\frac{1}{2}$ in., in increments of $\frac{1}{8}$ in. For structural applications, bolts of $\frac{3}{4}$-in. and $\frac{7}{8}$-in. diameters are two of the most commonly used sizes. They are inserted in holes having a diameter $\frac{1}{16}$ in. larger than the diameter of the shank of the bolt. High-strength steel bolts have tensile strengths several times those of ordinary bolts, so the high-strength bolts can be tightened to a specified minimum initial tension.

Joints connected by high-strength steel bolts are of two types: the *friction type* and the *bearing type*. In a friction-type joint, the bolt is tightened to a specified minimum initial tension equal to 70% of the bolt tensile strength. The resulting tension in the bolt develops a reliable clamping force, and the load is transmitted by friction between the surfaces of the connected plates, which are pressed tightly together. This type of connection is used for structures subjected to impact and vibration, where a high factor of safety against slippage is necessary.

In a bearing-type joint, the load is transmitted by the bearing of the bolts against the joined parts. This type of connection may be installed by tightening the bolts until all plies in a joint are in firm contact. Friction resistance probably shares the load but is not considered in the design analysis. This type of connection is usually used for structures subjected primarily to a static load.

For a friction-type connection, no bearing stresses need to be considered since the joint is not supposed to slip. For a bearing-type connection, the allowable bearing stress on the projected area of the bolts is the same as that for rivets:

$$(\sigma_b)_{\text{allow}} = 1.35\sigma_y$$

where σ_y is the yield strength of the connected part.

The allowable shear stresses specified by the AISC code for A325 and A490 bolts are given in Table 20–1. The allowable shear stress for a bearing-type connection is considerably higher because a smaller factor of safety against slippage is used. Thus, a smaller factor of safety is used for the allowable shear stress.

TABLE 20–1 Allowable Shear Stresses in High-Strength Bolts

Bolt Material	Friction Type	Bearing Type
A325	15 ksi (103 MPa)	22 ksi (152 MPa)
A490	20 ksi (138 MPa)	32 ksi (221 MPa)

— **EXAMPLE 20–5** —

Rework Example 20–1. Assume that A325 bearing-type, high-strength steel bolts, $\frac{7}{8}$ in. in diameter, are used instead of rivets.

Solution. The cross-sectional area of the $\frac{7}{8}$-in.-diameter bolt is

$$A = \frac{1}{4}\pi d^2 = \frac{1}{4}\pi\left(\frac{7}{8}\text{ in.}\right)^2 = 0.601\text{ in.}^2$$

From Table 20–1, the allowable shear stress for the A325 bearing-type bolts is 22 ksi. Thus, the shear strength of the joint is

$$P_s = nA_s\tau_{\text{allow}}$$
$$= 6(0.601\text{ in.}^2)(22\text{ kips/in.}^2)$$
$$= 79.4\text{ kips}$$

The bearing strength, tensile strength of the joint, and tensile strength of the solid plate are the same as those calculated in Example 20–1, which gives

$$P_b = 95.7\text{ kips}$$
$$P_t = 56.7\text{ kips}$$
$$P_g = 81.0\text{ kips}$$

Thus, the capacity of the joint is governed by the tensile strength at the critical section:

$$P = P_t = 56.7\text{ kips} \qquad \Leftarrow \textbf{Ans.}$$

The joint efficiency is

$$\text{Joint efficiency} = \frac{P}{P_g}\times 100\% = \frac{56.7\text{ kips}}{81.0\text{ kips}}\times 100\%$$
$$= 70.0\% \qquad \Leftarrow \textbf{Ans.}$$

———— **EXAMPLE 20–6** ————

Rework Example 20–2. Assume that A490 bearing-type, high-strength steel bolts, 18 mm in diameter, are used instead of rivets.

Solution. The cross-sectional area of the bolt is

$$A = \frac{1}{4}\pi d^2 = \frac{1}{4}\pi (0.020)^2 = 3.14 \times 10^{-4}\ \text{m}^2$$

From Table 20–1, the allowable shear stress for the A490 bearing type is 221 MPa. Thus, the shear strength of the joint is

$$P_s = nA_s\tau_{\text{allow}}$$
$$= 12(3.14 \times 10^{-4}\ \text{m}^2)(221\ 000\ \text{kN/m}^2)$$
$$= 833\ \text{kN}$$

The bearing strength, tensile strength of the joint, and tensile strength of the solid plate are the same as those calculated in Example 20–2, which gives

$$P_b = 727\ \text{kN}$$
$$P_t = 541\ \text{kN}$$
$$P_g = 727\ \text{kN}$$

The capacity of the joint is governed by the tensile strength. Thus, the strength of the joint is

$$P = P_t = 541\ \text{kN} \qquad\qquad \Leftarrow \textbf{Ans.}$$

The joint efficiency is

$$\text{Joint efficiency} = \frac{P}{P_g} \times 100\% = \frac{541\ \text{kN}}{727\ \text{kN}} \times 100\%$$
$$= 74.4\% \qquad\qquad \Leftarrow \textbf{Ans.}$$

———— **EXAMPLE 20–7** ————

Rework Example 20–3. Assume that A490 friction-type, high-strength steel bolts, $\frac{3}{4}$ in. in diameter, are used instead of rivets.

Solution. From Table 20–1, the allowable shear stress for A490 friction-type steel is 20 ksi, which happens to be the same as that for the rivet connection in Example 20–3. Thus, the shear strength of the bolt is the same as that calculated in Example 20–3 which is

$$P_s = 70.7\ \text{kips}$$

The tensile strength is also the same as that calculated in Example 20–3, which gives

$$(P_t)_{\text{allow}} = 78.5\ \text{kips}$$

Since the joint is not supposed to slip for the friction-type connection, failure due to bearing will not occur, so the bearing strength need not be considered. Thus, the strength of the joint is

$$P = 70.7 \text{ kips} \qquad \Leftarrow \textbf{Ans.}$$

which is the same as that of the riveted connection in Example 20–3. The joint efficiency should also be the same; that is,

$$\text{Joint efficiency} = 78.3\% \qquad \Leftarrow \textbf{Ans.}$$

Remark. Although the strength and efficiency of the bolt connection in this example is the same as that of the riveted connection in Example 20–3, the two types of connections are significantly different. Because the friction-type connection has a high factor of safety against slippage, the connection by high-strength bolts in this example will prove to have a higher fatigue strength and a better resistance to the dynamic loads to which a bridge truss is usually subjected.

20–5
ECCENTRICALLY LOADED RIVETED OR BOLTED CONNECTIONS

In the previous sections, the line of action of the applied load passes through the centroid of the connectors; the load is assumed to be shared equally by all the connectors. When the load is applied with eccentricity from the centroid of the connectors, the eccentric load produces direct shear as well as torsion. The rivets are no longer subjected to equal forces. Figure 20–3a shows a connection subjected to a load P applied at an eccentricity e from the centroid of the connectors. The eccentric load is equivalent to a direct shear force P and a torque Pe, as shown in Fig. 20–3b. The direct shear force P is resisted equally by the four connectors; each carries a load $P/4$, as shown in Fig. 20–3c. To resist the torque Pe, the resisting force on each connector is proportional to its distance to the centroid of the connectors, and acts in the direction perpendicular to the line joining the centroid and the connector, as shown in Fig. 20–3d. We have

$$\frac{F_1}{d_1} = \frac{F_2}{d_2} = \frac{F_3}{d_3} = \frac{F_4}{d_4}$$

or

$$F_2 = \frac{F_1 d_2}{d_1} \qquad F_3 = \frac{F_1 d_3}{d_1} \qquad F_4 = \frac{F_1 d_4}{d_1} \qquad (20\text{–}5)$$

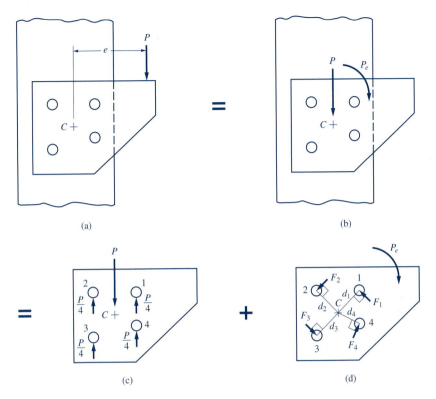

FIGURE 20–3

The sum of the moments of the resisting forces F_1, F_2, F_3, and F_4 must balance the couple Pe. Thus,

$$F_1 d_1 + F_2 d_2 + F_3 d_3 + F_4 d_4 = Pe \qquad (20\text{–}6)$$

Substituting Equation 20–5 into Equation 20–6, we get

$$F_1 d_1 + \frac{F_1 d_2^{\,2}}{d_1} + \frac{F_1 d_3^{\,2}}{d_1} + \frac{F_1 d_4^{\,2}}{d_1} = Pe$$

or

$$\frac{F_1}{d_1}(d_1^{\,2} + d_2^{\,2} + d_3^{\,2} + d_4^{\,2}) = \frac{F_1}{d_1}\Sigma d^2 = Pe$$

From which we get

$$F_1 = \frac{Pe}{\Sigma d^2} d_1 = K d_1$$

Substituting into Equation 20–5, we have

$$F_2 = K d_2 \qquad F_3 = K d_3 \qquad F_4 = K d_4$$

or in general, we have

$$F_i = K d_i \qquad (20\text{–}7)$$

where

$$K = \frac{Pe}{\Sigma d^2} \qquad (20\text{--}8)$$

Let the horizontal and vertical components of the distance d_i be represented by x_i and y_i, respectively. Then

$$d_i^2 = x_i^2 + y_i^2$$

$$\Sigma d^2 = \Sigma x^2 + \Sigma y^2$$

$$K = \frac{Pe}{\Sigma x^2 + \Sigma y^2} \qquad (20\text{--}9)$$

Let the horizontal and vertical components of the force F_i on the ith connector be represented by H_i and V_i, as shown in Fig. 20–4. The force F_i is perpendicular to the line d_i. If the angle between d_i and x_i is θ, the angle between F_i and V_i is also θ. We have

$$H_i = F_i \sin \theta = K d_i \sin \theta$$
$$V_i = F_i \cos \theta = K d_i \cos \theta$$

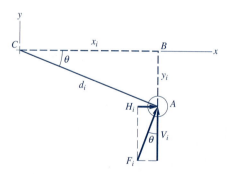

FIGURE 20–4

From triangle ABC, we have

$$d_i \sin \theta = y_i \qquad d_i \cos \theta = x_i$$

Therefore,

$$H_i = K y_i \qquad V_i = K x_i \qquad (20\text{--}10)$$

────── **EXAMPLE 20–8** ──────────────────────────────────

Determine the required size of the A502-1 rivets and the required thickness of the gusset plate in the connection between a beam seat and a W14 × 90 column, as shown in Fig. E20–8(1). The column and the gusset plate are of A36 steel.

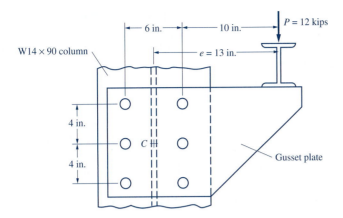

FIGURE E20–8(1)

Solution. The eccentric load P is equivalent to a direct shear force $P = 12$ kips and a torque $Pe = (12 \text{ kips})(13 \text{ in.}) = 156 \text{ kip} \cdot \text{in.}$

Critical Load. Due to the direct shear load, the resisting force on each rivet is

$$\frac{P}{6} = \frac{12 \text{ kips}}{6} = 2 \text{ kips}$$

The force is acting upward on each rivet, as shown in Fig. E20–8(2).

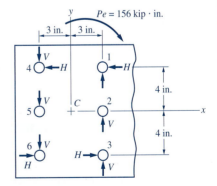

FIGURE E20–8(2)

Due to the torque Pe, the horizontal and vertical components of the resisting force on each rivet are as shown in Fig. E20–8(3). To find these components, we first compute

FIGURE E20–8(3)

$$\Sigma x^2 + \Sigma y^2 = 6(3 \text{ in.})^2 + 4(4 \text{ in.})^2 = 118 \text{ in.}^2$$

$$K = \frac{Pe}{\Sigma x^2 + \Sigma y^2} = \frac{156 \text{ kip} \cdot \text{in.}}{118 \text{ in.}^2} = 1.322 \text{ kips/in.}$$

The y coordinates of rivets 2 and 5 are zero; hence, the horizontal components of the two rivets due to the torque are zero. The horizontal components of the other rivets are equal because these rivets have the same absolute value of y. The magnitude of the horizontal component H is

$$H = Ky = (1.322 \text{ kips/in.})(4 \text{ in.}) = 5.29 \text{ kips}$$

The vertical force components V of all the rivets are equal because they all have the same absolute value of x. The magnitude of the vertical component V is

$$V = Kx = (1.322 \text{ kips/in.})(3 \text{ in.}) = 3.97 \text{ kips}$$

The vertical components of rivets 1 and 3 due to direct shear force and torque are both upward, as shown in Fig. E20–8(4). Hence, these two rivets are most critically loaded. The critical load on either of the two rivets is

$$F = \sqrt{(5.29 \text{ kips})^2 + (2 \text{ kips} + 3.97 \text{ kips})^2}$$
$$= 7.98 \text{ kips}$$

 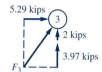

FIGURE E20–8(4)

Diameter of the Rivet.   The allowable shear stress for an A501-1 rivet is 15 ksi. Thus, the required cross-sectional area of the rivet is

$$A_s = \frac{F}{\sigma_{\text{allow}}} = \frac{7.98 \text{ kips}}{15 \text{ kips/in.}^2} = 0.532 \text{ in.}^2$$

Since $A_s = (\pi/4)d^2 = 0.7854d^2$, the required diameter of the rivet is

$$d = \sqrt{\frac{0.532 \text{ in.}^2}{0.7854}} = 0.823 \text{ in.}$$

Use a $\frac{7}{8}$-in. (0.875-in.) diameter rivet   ⇐ **Ans.**

Thickness of the Gusset Plate.   The allowable bearing stress of the gusset plate is

$$(\sigma_b)_{\text{allow}} = 1.35\sigma_y = 1.35(36 \text{ ksi}) = 48.6 \text{ ksi}$$

The required bearing area is

$$A_b = \frac{F}{(\sigma_b)_{\text{allow}}} = \frac{7.98 \text{ kips}}{48.6 \text{ kips/in.}^2} = 0.164 \text{ in.}^2$$

Since the projected bearing area is dt, the required thickness of the gusset plate is

$$t = \frac{A_b}{d} = \frac{0.164 \text{ in.}^2}{0.875 \text{ in.}} = 0.187 \text{ in.}$$

Use a $\frac{1}{4}$-in. gusset plate   ⇐ **Ans.**

20–6
WELDED CONNECTIONS

Welding is the process of connecting metallic parts by heating the surfaces to a plastic or fluid state and allowing the melted parts to join together. A welded connection requires no holes for fasteners; therefore, the gross cross-sectional area is effective in resisting tension, and 100% efficiency is possible for a welded joint.

Welding is a widely used method for joining metallic parts. It is often used in combination with bolting in "shop-welded and field-bolted construction." In this practice, clipped angles with holes in the outstanding legs may be welded to a beam in the shop and then bolted to a girder or column in the field.

Arc Welding. Welding is accomplished by heating the designated surfaces until they melt and flow together. Heat may be supplied by an electric arc or a gas flame. *Electric arc welding* is used in almost all steel building and steel bridge construction, in which the heat required for fusion is generated by an electric arc between the workpiece and an *electrode.* Most electrodes contain a filler metal that is consumed as welding progresses. Electrodes are designated by the letter E (for "electrode") followed by a number, such as E70, E90, E100, etc. The number denotes the ultimate tensile strength in ksi of the electrode. Thus, an E70 electrode has an ultimate strength of 70 ksi.

The two main types of welds are butt welds and fillet welds, as shown in Fig. 20–5. Most structural connections are made with fillet welds. The main problem in a butt weld is that it is difficult to get the pieces to fit together in the field. For the fillet weld, the amount of overlap can be adjusted freely.

Butt weld   Fillet weld

FIGURE 20–5

Strength of the Fillet Weld. Fillet welds are designated by the *leg size,* as shown in Fig. 20–6. Although the weld surface is usually curved, the smallest inscribed triangle, shown by dashed lines, is considered to be the theoretical dimensions of the fillet weld. The corner of the two legs is called the *root.* The smallest distance from the root to the opposite side of the triangle is called the *throat* of a fillet weld. For fillet welds with equal legs, the throat is equal to

$$(\text{size}) \sin 45° = 0.707 \ (\text{size})$$

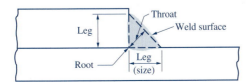

FIGURE 20–6

Tests have shown that failure of a fillet weld commonly occurs through the throat due to shear stress. Therefore, the strength of a fillet weld, regardless of the direction of the applied load, is equal to the cross-sectional area at the throat multiplied by the allowable shear stress for the weld metal. According to the AISC specification, the allowable shear stress is 0.3 times the tensile strength of the electrode; that is,

$$\tau_{\text{allow}} = 0.3\sigma_u \qquad\qquad (20\text{--}11)$$

where σ_u is the ultimate tensile strength of the electrode. The allowable shear force q per inch of length of the fillet weld is

$$q = \tau_{\text{allow}}\, 0.707\,(\text{size}) \qquad\qquad (20\text{--}12)$$

Substituting the expression from Equation 20–11 into Equation 20–12, we get

$$q = 0.212\sigma_u\,(\text{size}) \qquad\qquad (20\text{--}13)$$

The allowable load of a fillet weld of length L (regardless of the direction of the weld with respect to the applied load) is

$$P = qL \qquad\qquad (20\text{--}14)$$

For example, the allowable shear force q per unit length of a $\frac{5}{16}$-in. fillet weld with an E70 electrode is

$$q = 0.212(70\text{ ksi})\left(\frac{5}{16}\text{ in.}\right) = 4.64\text{ kips/in.}$$

If the length of the $\frac{5}{16}$-in. fillet weld in the joint is 10 in., the allowable load of the joint would be (4.64 kips/in.)(10 in.) = 46.4 kips.

Size of the Fillet Weld. Depending on the thickness of the base plate to be welded, the AISC specification gives the minimum sizes of fillet welds as listed in Table 20–2. The maximum sizes of fillet welds are specified as follows:

1. Along edges of material less than $\frac{1}{4}$ in. thick, the maximum size may be equal to the thickness of the material.
2. Along edges of material equal to or greater than $\frac{1}{4}$ in. thick, the maximum size should be $\frac{1}{16}$ in. less than the thickness of the material.

Table 20–2 Minimum Sizes of Fillet Welds

Thickness of Material	Minimum Size of Fillet Weld
To $\frac{1}{4}$ in., inclusive	$\frac{1}{8}$ in.
Over $\frac{1}{4}$ in. to $\frac{1}{2}$ in.	$\frac{3}{16}$ in.
Over $\frac{1}{2}$ in. to $\frac{3}{4}$ in.	$\frac{1}{4}$ in.
Over $\frac{3}{4}$ in.	$\frac{5}{16}$ in.

━━━━━ **EXAMPLE 20–9** ━━━━━━━━━━━━━━━━━━━━━━━━━━━━━━━━━━━━━

The $\frac{5}{16}$-in. fillet weld of E70 electrode is used to connect the two plates shown in Fig. E20–9(1). Determine the total length of fillet weld required if the full strength of the 6-in. $\times \frac{3}{8}$-in. plate made of A36 steel ($\sigma_y = 36$ ksi) is developed.

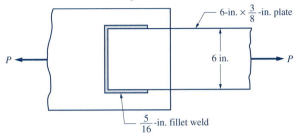

FIGURE E20–9(1)

Solution. The full capacity of a 6-in. $\times \frac{3}{8}$-in. plate is

$$P = A(\sigma_t)_{\text{allow}} = (6 \times \tfrac{3}{8} \text{ in.}^2)(0.6 \times 36 \text{ kips/in.}^2)$$
$$= 48.6 \text{ kips}$$

For an E70 electrode, $\sigma_u = 70$ ksi. From Equation 20–13, the allowable load per inch of a $\frac{5}{16}$-in. fillet weld is

$$q = 0.212(70 \text{ ksi})\left(\frac{5}{16} \text{ in.}\right) = 4.64 \text{ kips/in.}$$

From Equation 20–14, the length of weld required is

$$L = \frac{P}{q} = \frac{48.6 \text{ kips}}{4.63 \text{ kips/in.}} = 10.5 \text{ in.}$$

This length can be pro-
vided by either one of
the two arrangements
shown in Figs. E20–9(2)
and E20–9(3).

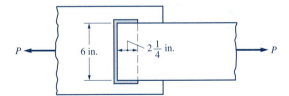

FIGURE E20–9(2)

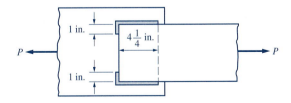

FIGURE E20–9(3)

Note. The *end returns* in Fig. E20–9(3) are required by the AISC specification to reduce the effect of stress concentration. The minimum length of the end return is twice the size of the weld. The end returns are included as part of the effective length of the weld.

───── **EXAMPLE 20–10** ─────

The fillet weld of an E70 electrode connects two 10-in. $\times$ $\frac{3}{4}$-in. plates, as shown in Fig. E20–10(1). The plates are A36 steel, and the fillet weld is along the full width of the plates. Determine the size of the fillet weld if the full strength of the plates is developed.

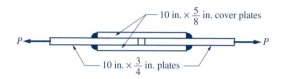

10 in. $\times$ $\frac{5}{8}$ in. cover plates

P

P

10 in. $\times$ $\frac{3}{4}$ in. plates

FIGURE E20–10(1)

Solution. The full strength of the plates is

$$P = \left(10 \times \frac{3}{4} \text{ in.}^2\right)(0.6 \times 36 \text{ kips/in.}^2) = 162 \text{ kips}$$

From the free-body diagram in Fig. E20–10(2), the load P is transmitted by two pieces of fillet weld to the cover plate. Thus, the length of the fillet weld is

$$L = 2 \times 10 = 20 \text{ in.}$$

$\frac{P}{2}$

$\frac{P}{2}$

P

FIGURE E20–10(2)

From Equation 20–14, the required allowable load per in. of weld is

$$q = \frac{P}{L} = \frac{162 \text{ kips}}{20 \text{ in.}} = 8.1 \text{ kips/in.}$$

For an E70 electrode, $\sigma_u = 70$ ksi. The size of the fillet weld may be solved from Equation 20–13. We get

$$\text{Leg size} = \frac{q}{0.212\sigma_u} = \frac{8.1 \text{ kips/in.}}{0.212(70 \text{ kips/in.}^2)} = 0.546 \text{ in.}$$

For the $\frac{5}{8}$-in.-thick cover plates, the maximum size of fillet weld allowed by the AISC specification is $\frac{9}{16}$ in. Thus,

Use a $\frac{9}{16}$-in. (0.563-in.) fillet weld   ⇐ **Ans.**

───── **EXAMPLE 20–11** ─────

The long leg of a steel angle L4 $\times$ $3\frac{1}{2}$ $\times$ $\frac{1}{2}$ is connected by a $\frac{7}{16}$-in. fillet weld of an E90 electrode to the bottom chord of a truss at the joint, as shown in Fig. E20–11. Determine (a) the total length of weld required if the full strength of the angle is developed and (b) the lengths L_1 and L_2 of the weld so that the loading on the angle would have no eccentricity.

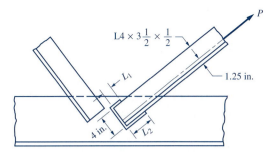

FIGURE E20–11

Solution. The full strength of the angle is

$$P = A(0.60\sigma_y) = (3.50 \text{ in.}^2)(0.60 \times 36 \text{ kips/in.}^2)$$
$$= 75.6 \text{ kips}$$

For an E90 electrode, $\sigma_u = 90$ ksi. The allowable load per unit length of weld is

$$q = 0.212\sigma_u \text{ (leg size)}$$
$$= 0.212(90 \text{ kips/in.}^2)\left(\frac{7}{16} \text{ in.}\right)$$
$$= 8.35 \text{ kips/in.}$$

(a) Total Length. The total length of weld required is

$$L = \frac{P}{q} = \frac{75.6 \text{ kips}}{8.35 \text{ kips/in.}} = 9.05 \text{ in.}$$

$$\text{Use } L = 9\frac{1}{4} \text{ in.} \qquad \Leftarrow \textbf{Ans.}$$

(b) Lengths L_1 and L_2. To avoid eccentric load on the angle, the centroid of the fillet weld must coincide with the centroid of the angle. From the appendix, Table A–4(a), the centroid of the angle L4 $\times$ $3\frac{1}{2}$ $\times$ $\frac{1}{2}$ is located at $y = 1.25$ in. from the outside of the short leg, as indicated in Fig. E20–11. Taking the reference line along L_2, we equate the sum of the first moment of the weld segments to the first moment of the total length located at the centroid of the angle. We write

$$L_1(4) + 4(2) + L_2(0) = L(1.25)$$

From which we get

$$L_1 = \frac{1.25L - 4(2)}{4} = \frac{1.25(9.25) - 8}{4} = 0.891 \text{ in.}$$

$$\text{Use } L_1 = 1 \text{ in.} \qquad \Leftarrow \textbf{Ans.}$$

Then

$$L_2 = 9.25 - 1 - 4 = 4.25 \text{ in.}$$

$$\text{Use } L_2 = 4\frac{1}{4} \text{ in.} \qquad \Leftarrow \textbf{Ans.}$$

20-7
SUMMARY

Structural Connections. Steel structures are almost exclusively connected by using bolts and welding. The use of large rivets requiring hot forming has declined sharply in recent decades. Smaller rivets using cold heading are still used to connect some minor machine components.

Riveted Connections. For a riveted connection, the forces are transmitted through shear forces in the rivets and the bearing force between the rivets and the connected plates. A rivet can be in single or double shear. The bearing stress distribution is approximated on the basis of an average bearing stress acting over the projected area of the rivet's shank onto the cross-section of a plate. Steel rivets used for structural purposes are classified as ASTM A502, grades 1 and 2.

Strength of Riveted Connections. Riveted joints may fail in one of the three modes: (1) failure due to shear in the rivets, (2) failure due to bearing of the rivets on the connected plate, and (3) failure due to tension in the connected plate at the critical section. The allowable load for a riveted connection may be obtained by considering the three modes of failure. The shear strength may be calculated from

$$P_s = nA_s\tau_{\text{allow}} \tag{20-1}$$

The bearing strength may be calculated from

$$P_b = n(td)(\sigma_b)_{\text{allow}} \tag{20-2}$$

The tensile strength may be calculated from

$$P_t = (b_{\text{net}}t)(\sigma_t)_{\text{allow}} \tag{20-3}$$

High-Strength Bolted Connection. Generally speaking, any joint that can be connected by rivets can be connected by bolts. The design considerations of a bolted joint are similar to those of a riveted joint. The two basic types of high-strength steel bolts used in structural joints are the ASTM A325 and A490.

Joints connected by high-strength steel bolts are of two types: the *friction type* and the *bearing type*. In a friction-type joint, the bolt is tightened to a specified minimum initial tension equal to 70% of the bolt tensile strength. The resulting tension in the bolt develops a reliable clamping force and the load is transmitted by friction between the surfaces of the connected plates, which are pressed tightly together. In a bearing-type joint, the load is transmitted by the bearing of the bolts against the joined parts.

Eccentrically Loaded Riveted and Bolted Connection. When the load is applied with eccentricity from the centroid of the connectors, the eccentric load produces direct shear as well as torsion. The eccentric load is equivalent to a direct shear force P and a torque Pe. The direct shear force P is resisted equally by the connectors. To resist the torque Pe, the resisting force on each connector is proportional to its distance to the centroid of the connectors. The horizontal and vertical components H_i and V_i of the force on the ith connector can be computed from

$$H_i = Ky_i \qquad V_i = Kx_i \qquad (20\text{–}10)$$

where

$$K = \frac{Pe}{\Sigma x^2 + \Sigma y^2} \qquad (20\text{–}9)$$

Welded Connection. Welding is the process of connecting metallic parts by heating the surfaces to a plastic or fluid state and allowing the melted parts to join together. The most common welding process is electric-arc welding. Most structural connections are made with fillet welds.

Strength of the Fillet Weld. A fillet weld is designated by its *leg size.* The allowable shear force q per inch of length of the fillet weld is

$$q = 0.212\sigma_u \,(\text{size}) \qquad (20\text{–}13)$$

where σ_u is the ultimate tensile strength of the electrode. The allowable load of a fillet weld of length L (regardless of the direction of the weld with respect to the applied load) is

$$P = qL \qquad (20\text{–}14)$$

PROBLEMS

Section 20–3   Strength of Riveted Connections

20–1   Determine the strength and efficiency of the lap joint in Fig. P20–1. The $\frac{3}{4}$-in.-diameter rivets are made of A502-1 steel, and the plates are made of A36 steel with $\sigma_y = 36$ ksi.

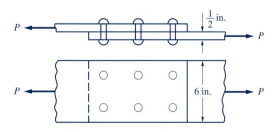

FIGURE P20–1

20–2   Rework Problem 20–1. Assume that the A502-1 rivets in Fig. P20–1 are 20 mm in diameter, and that the plates are 150 mm × 13 mm and are made of A36 steel with $\sigma_y = 250$ MPa.

20–3   Determine the strength and efficiency of the butt joint in Fig. P20–3. The $\frac{7}{8}$-in.-diameter rivets are made of A502-2 steel, and the plates are made of A36 steel with $\sigma_y = 36$ ksi.

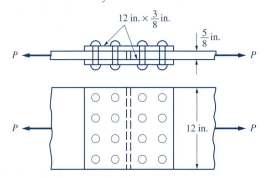

FIGURE P20–3

20–4   Rework Problem 20–3. Assume that the rivets are 22 mm in diameter, the main plates are 300 mm × 16 mm, and the cover plates are 300 mm × 10 mm. The plates are made of A36 steel with $\sigma_y = 250$ MPa.

20–5   Refer to Fig. P20–5. The tension member in a roof truss consists of a single L5 × 3 × $\frac{3}{8}$ connected to the $\frac{1}{2}$-in. gusset plate shown. The angle and the gusset plate are both of A36 steel. The connection is made by four $\frac{7}{8}$-in. -diameter A502-2 rivets. Determine the allowable tensile strength of the member.

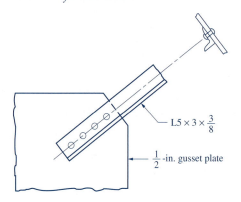

FIGURE P20–5

20–6   In Problem 20–5, determine the number of $\frac{7}{8}$-in.-diameter A502-2 rivets required so that the allowable tensile strength of the member can be developed.

20–7   Refer to Fig. P20–7. The tension member in a bridge truss consisting of a pair of L5 × 3 × $\frac{3}{8}$ angles arranged back to back is connected to the $\frac{5}{8}$-in. gusset plate shown. The angles and gusset plate are made of A36 steel with $\sigma_y = 36$ ksi. The four 1-in. rivets are made of A502-1 steel. Determine the strength and efficiency of the connection.

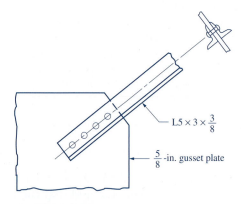

FIGURE P20–7

20–8   Rework Problem 20–7. Assume that a pair of L127 × 76 × 9.5 angles are connected to the 16-mm gusset plate by four 25-mm-diameter A502-1 rivets. The angles and gusset plate are made of A36 steel with σ_y = 250 MPa.

20–9   A W18 × 60 beam is connected to a W12 × 87 column, as shown in Fig. P20–9. The connection consists of two clipped L4 × 4 × $\frac{3}{8}$ angles. Four $\frac{7}{8}$-in. rivets are used to connect the angles to the web of the beam, and eight rivets of the same size are used to connect the angles to the flange of the column. Determine the strength of the joint if the beam, column, and angles are made of A36 steel with σ_y = 36 ksi and the rivets are of A502-1 steel.

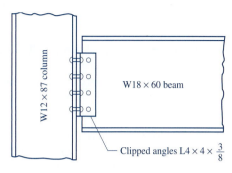

FIGURE P20–9

20–10 Rework Problem 20–9. Assume that a W460 × 0.88 beam is connected to a W300 × 1.27 column. The connection consists of two clipped L102 × 102 × 19.1 angles. Four 22-mm rivets are used to connect the angles to the web of the beam, and eight rivets of the same size are used to connect the angles to the flange of the column. The beam, the column, and the angles are made of A36 steel with σ_y = 250 MPa and the rivets are of A502-1 steel.

20–11 A W18 × 50 beam is connected to the W12 × 65 columns by means of the connection shown in Fig. P20–11. The connection on each end of the beam consists of two clipped L4 × 3$\frac{1}{2}$ × $\frac{5}{16}$ angles and twelve $\frac{3}{4}$-in. rivets. Determine the allowable uniform load that the beam can carry if the beam, the column, and the angles are made of A36 steel with σ_y = 36 ksi and the rivets are made of A502-1 steel. (*Hint:* Consider both the strength of the connection and the flexural strength of the beam using the allowable flexural stress of 0.66σ_y, and consider the beam to be simply supported.)

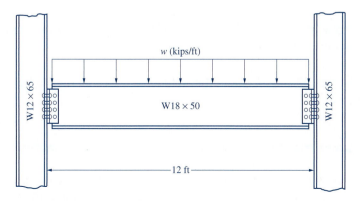

FIGURE P20–11

20–12 Rework Problem 20–11. Assume that a W460 × 0.73 beam is connected to the W300 × 0.95 columns by means of two clipped L102 × 89 × 7.9 angles and twelve 20-mm A502-1 rivets. The beam has a 3.65-m span. The beam, the column, and the angles are made of A36 steel with $\sigma_y = 250$ MPa.

Section 20–4   High-Strength Bolted Connections

20–13 Rework Problem 20–1. Assume that A325 bearing-type, high-strength bolts of $\frac{3}{4}$-in. diameter are used instead of rivets.

20–14 Rework Problem 20–2. Assume that A325 bearing-type, high-strength bolts of 20-mm diameter are used instead of rivets.

20–15 Rework Problem 20–3. Assume that A490 bearing-type, high-strength bolts of $\frac{7}{8}$-in. diameter are used instead of rivets.

20–16 Rework Problem 20–4. Assume that A490 bearing-type, high-strength bolts of 22-mm diameter are used instead of rivets.

20–17 Rework Problem 20–6. Assume that A490 bearing-type, high-strength bolts of $\frac{7}{8}$-in. diameter are used instead of rivets.

20–18 Rework Problem 20–7. Assume that A490 friction-type, high-strength bolts of 1-in. diameter are used instead of rivets.

20–19 Rework Problem 20–8. Assume that A490 friction-type, high-strength bolts of 25-mm diameter are used instead of rivets.

20–20 Rework Problem 20–9. Assume that A325 bearing-type, high-strength bolts of $\frac{5}{8}$-in. diameter are used instead of rivets.

20–21 Rework Problem 20–10. Assume that A490 friction-type, high-strength bolts of 22-mm diameter are used instead of rivets.

20–22 Rework Problem 20–11. Assume that A490 friction-type, high-strength bolts of $\frac{3}{4}$-in. diameter are used instead of rivets.

20–23 Rework Problem 20–12. Assume that A490 friction-type, high-strength bolts of 20-mm diameter are used instead of rivets.

Section 20–5   Eccentrically Loaded Riveted or Bolted Connections

20–24 A steel plate is connected to a machine using five $\frac{7}{8}$-in. rivets, as shown in Fig. P20–24. If the allowable shear stress is 15 ksi, determine the maximum load P that can be applied to the plate.

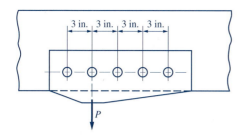

FIGURE P20–24

20–25 Rework Problem 20–24. Assume that the connection is done by using five 20-mm rivets and that the allowable shear stress is 103 MPa. The spacings of the rivets are 75 mm.

20–26 Determine the required size of the bolts for the connection shown in Fig. P20–26 if the allowable shear stress is 20 ksi.

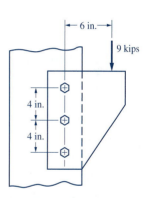

FIGURE P20–26

20–27 The bracket shown in Fig. P20–27 is connected to the flange of a steel column by four $\frac{3}{4}$-in. A490 friction-type bolts. Determine the maximum allowable load P in kips that can be applied to the bracket.

20–28 Rework Problem 20–27. Assume that the connection is made by four 20-mm A490 friction-type bolts. The spacings of the bolts are 100 mm.

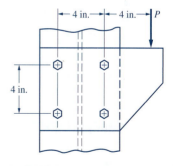

FIGURE P20–27

20–29 The bracket in Fig. P20–29 is connected to the column by six $\frac{3}{4}$-in rivets, arranged as shown. Determine the maximum shear stress in the most critically loaded rivet.

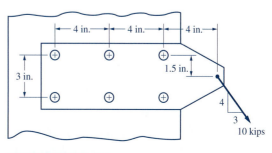

FIGURE P20–29

20–30 The joint shown in Fig. P20–30 connects a $\frac{1}{2}$-in. gusset plate to a W12 × 65 column by eight $\frac{7}{8}$-in. A325 bearing-type bolts. Both the gusset plate and the column are of A36 steel with $\sigma_y = 36$ ksi. Determine the allowable load P that can be applied to the bracket.

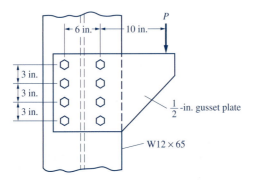

FIGURE P20–30

Section 20–6 Welded Connections

20–31 Determine the strength and efficiency of the welded connection shown in Fig. P20–31. The $\frac{7}{8}$-in. fillet weld of E70 electrode is along the entire width of the plates.

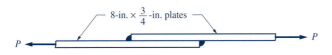

FIGURE P20–31

20–32 In Problem 20–31, determine the size to the nearest $\frac{1}{16}$ in. of the fillet weld of E70 electrode if the full strength of the plates is developed.

20–33 Determine the strength and efficiency of the welded connection shown in Fig. P20–33. The $\frac{1}{2}$-in. fillet weld of E70 electrode is along the entire width of the plates.

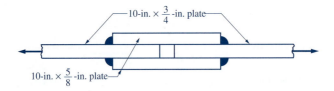

FIGURE P20–33

20–34 For the butt joint shown in Fig. P20–33, determine the required size to the nearest $\frac{1}{16}$ in. of the fillet weld of E70 electrode if the full strength of the plates is developed.

20–35   Refer to Fig. P20–35. Determine the length L_1 required for a $\frac{5}{16}$-in. fillet weld of E70 electrode in the connection shown if the full strength of the plate is developed.

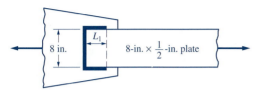

FIGURE P20–35

20–36   Rework Example 20–10. Assume that the angle is L6 × 4 × $\frac{3}{4}$ and that the size of the fillet weld is $\frac{1}{2}$ in.

20–37   The structural joint shown in Fig. P20–37 is connected by fillet welds of E80 electrodes. Assuming that the joint has a strength equal to the full strength of the angle, determine the proper lengths L_1 and L_2 if the centroid of the weld must coincide with the centroid of the angle to avoid eccentric loading. The end returns are required by the AISC code to reduce the effect of stress concentrations, and they can be included as part of the effective length of the weld.

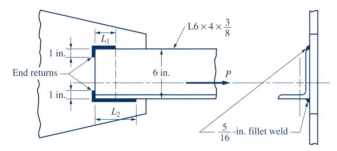

FIGURE P20–37

List of Tables

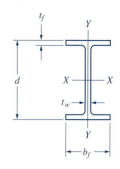

TABLE A–1(α)   Properties of Selected W Shapes (Wide-Flange Sections): U.S. Customary Units

Desig-nation	Area	Depth	Web Thick-ness	Flange Width	Flange Thick-ness	Axis x–x			Axis y–y			Plastic Modulus	
	A	d	t_w	b_f	t_f	I	S	r	I	S	r	Z_X	Z_Y
(in. × lb/ft)	(in.²)	(in.)	(in.)	(in.)	(in.)	(in.⁴)	(in.³)	(in.)	(in.⁴)	(in.³)	(in.)	(in.³)	(in.³)
W36 × 210	61.8	36.69	0.830	12.180	1.360	13200	719	14.6	411	67.5	2.58	833	107
× 150	44.2	35.85	0.625	11.975	0.940	9040	504	14.3	270	45.1	2.47	581	70.9
W33 × 201	59.1	33.68	0.715	15.745	1.150	11500	684	14.0	749	95.2	3.56	772	147
× 130	38.3	33.09	0.580	11.510	0.855	6710	406	13.2	218	37.9	2.39	467	59.5
W30 × 173	50.8	30.44	0.655	14.985	1.065	8200	539	12.7	598	79.8	3.43	605	123
× 108	31.7	29.83	0.545	10.475	0.760	4470	299	11.9	146	27.9	2.15	346	43.9
W27 × 146	42.9	27.38	0.605	13.965	0.975	5630	411	11.4	443	63.5	3.21	461	97.5
× 94	27.7	26.92	0.490	9.990	0.745	3270	243	10.9	124	24.8	2.12	278	38.8
W24 × 131	38.5	24.48	0.605	12.855	0.960	4020	329	10.2	340	53.0	2.97	370	81.5
× 104	30.6	24.06	0.500	12.750	0.750	3100	258	10.1	259	40.7	2.91	289	62.4
× 76	22.4	23.92	0.440	8.990	0.680	2100	176	9.69	82.5	18.4	1.92	200	28.6
W21 × 111	32.7	21.51	0.550	12.340	0.875	2670	249	9.05	274	44.5	2.90	279	68.2
× 83	24.3	21.43	0.515	8.355	0.835	1830	171	8.67	81.4	19.5	1.83	196	30.5
× 62	18.3	20.99	0.400	8.240	0.615	1330	127	8.54	57.5	13.9	1.77	144	21.7
× 50	14.7	20.83	0.380	6.530	0.535	984	94.5	8.18	24.9	7.64	1.30	110	12.2
W18 × 97	28.5	18.59	0.535	11.145	0.870	1750	188	7.82	201	36.1	2.65	211	55.3
× 60	17.6	18.24	0.415	7.555	0.695	984	108	7.47	50.1	13.3	1.69	123	20.6
× 50	14.7	17.99	0.355	7.495	0.570	800	88.9	7.38	40.1	10.7	1.65	101	16.6
× 46	13.5	18.06	0.360	6.060	0.605	712	78.8	7.25	22.5	7.43	1.29	90.7	11.7
× 35	10.3	17.70	0.300	6.000	0.425	510	57.6	7.04	15.3	5.12	1.22	66.5	8.06
W16 × 100	29.4	16.97	0.585	10.425	0.985	1490	175	7.10	186	35.7	2.51	198	54.9
× 89	26.2	16.75	0.525	10.365	0.875	1300	155	7.05	163	31.4	2.49	175	48.1
× 57	16.8	16.43	0.430	7.120	0.715	758	92.2	6.72	43.1	12.1	1.60	105	18.9
× 50	14.7	16.26	0.380	7.070	0.630	659	81.0	6.68	37.2	10.5	1.59	92.0	16.3
× 36	10.6	15.86	0.295	6.985	0.430	448	56.5	6.51	24.5	7.00	1.52	64.0	10.8
× 26	7.68	15.69	0.250	5.500	0.345	301	38.4	6.26	9.59	3.49	1.12	44.2	5.48
W14 × 132	38.8	14.66	0.645	14.725	1.030	1530	209	6.28	548	74.5	3.76	234	113
× 109	32.0	14.32	0.525	14.605	0.860	1240	173	6.22	447	61.2	3.73	192	92.7
× 90	26.5	14.02	0.440	14.520	0.710	999	143	6.14	362	49.9	3.70	157	75.6
× 82	24.1	14.31	0.510	10.130	0.855	882	123	6.05	148	29.3	2.48	139	44.8

(Table continued on page 764)

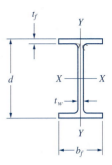

TABLE A–1(b)   Properties of Selected W Shapes (Wide-Flange Sections): SI Units

Desig-nation	Area	Depth	Web Thick-ness	Flange Width	Flange Thick-ness	Elastic Properties Axis x–x			Elastic Properties Axis y–y			Plastic Modulus	
	A $\times 10^{-3}$	d	t_w	b_f	t_f	I $\times 10^{-6}$	S $\times 10^{-3}$	r	I $\times 10^{-6}$	S $\times 10^{-3}$	r	Z_X $\times 10^{-3}$	Z_Y $\times 10^{-3}$
mm × kN/m	(m²)	(mm)	(mm)	(mm)	(mm)	(m⁴)	(m³)	(mm)	(m⁴)	(m³)	(mm)	(m³)	(m³)
W910 × 3.06	39.9	932	21.1	309	34.5	5490	11.8	371	171	1.11	65.5	13.7	1.75
× 2.19	28.5	911	15.9	304	23.9	3760	8.26	363	112	0.739	62.7	9.52	1.16
W840 × 2.93	38.1	855	18.2	400	29.2	4790	11.2	356	312	1.56	90.4	12.7	2.41
× 1.90	24.7	840	14.7	292	21.7	2790	6.65	335	90.7	0.621	60.7	7.65	0.975
W760 × 2.52	32.8	773	16.6	381	27.1	3410	8.83	323	249	1.31	87.1	9.92	2.02
× 1.58	20.5	758	13.8	266	19.3	1860	4.90	302	60.8	0.457	54.6	5.67	0.720
W690 × 2.13	27.7	695	15.4	355	24.8	2340	6.74	290	184	1.04	81.5	7.56	1.60
× 1.37	17.9	684	12.4	254	18.9	1360	3.98	277	51.6	0.406	53.8	4.56	0.636
W610 × 1.91	24.8	622	15.4	327	24.4	1670	5.39	259	142	0.869	75.4	6.06	1.34
× 1.52	19.7	611	12.7	324	19.1	1290	4.23	257	108	0.667	73.9	4.74	1.02
× 1.11	14.5	608	11.2	228	17.3	874	2.88	246	34.3	0.302	48.8	3.28	0.469
W530 × 1.62	21.1	546	14.0	313	22.2	1110	4.08	230	114	0.729	73.7	4.57	1.12
× 1.21	15.7	544	13.1	212	21.2	762	2.80	220	33.9	0.320	46.5	3.21	0.500
× 0.90	11.8	533	10.2	209	15.6	554	2.08	217	23.9	0.228	45.0	2.36	0.356
× 0.73	9.48	529	9.65	166	13.6	410	1.55	208	10.4	0.125	33.0	1.80	0.200
W460 × 1.42	18.4	472	13.6	283	22.1	730	3.08	199	83.7	0.592	67.3	3.46	0.906
× 0.88	11.4	463	10.5	192	17.7	410	1.77	190	20.9	0.218	42.9	2.02	0.338
× 0.73	9.48	457	9.02	190	14.5	333	1.46	187	16.7	0.175	41.9	1.66	0.272
× 0.67	8.71	459	9.14	154	15.4	296	1.29	184	9.36	0.122	32.8	1.49	0.192
× 0.51	6.65	450	7.62	152	10.8	212	0.944	179	6.37	0.084	31.0	1.09	0.132
W410 × 1.46	19.0	431	14.9	265	25.0	620	2.87	180	77.4	0.585	63.8	3.25	0.900
× 1.30	16.9	425	13.3	263	22.2	541	2.54	179	67.8	0.515	63.2	2.87	0.788
× 0.83	10.8	417	10.9	181	18.2	315	1.51	171	17.9	0.198	40.6	1.72	0.310
× 0.73	9.48	413	9.65	180	16.0	274	1.33	170	15.5	0.172	40.4	1.51	0.267
× 0.53	6.84	403	7.49	177	10.9	186	0.926	165	10.2	0.115	38.6	1.05	0.177
× 0.38	4.96	399	6.35	140	8.76	125	0.629	159	3.99	0.057	28.4	0.724	0.090
W360 × 1.93	25.0	372	16.4	374	26.2	637	3.43	160	228	1.22	95.5	3.84	1.85
× 1.59	20.6	364	13.3	371	21.8	516	2.84	158	186	1.00	94.7	3.15	1.52
× 1.31	17.1	356	11.2	369	18.0	416	2.34	156	151	0.818	94.0	2.57	1.24
× 1.20	15.5	363	13.0	257	21.7	367	2.02	154	61.6	0.480	63.0	2.28	0.734

(Table continued on page 765)

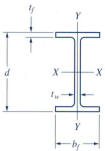

TABLE A–1(a) (Continued)   Properties of Selected W Shapes (Wide-Flange Sections): U.S. Customary Units

			Web	Flange		Elastic Properties						Plastic	
			Thick-	Width	Thick-	Axis x–x			Axis y–y			Modulus	
Desig-nation	Area	Depth	ness		ness								
(in. × lb/ft)	A (in.^{2})	d (in.)	t_w (in.)	b_f (in.)	t_f (in.)	I (in.^{4})	S (in.^{3})	r (in.)	I (in.^{4})	S (in.^{3})	r (in.)	Z_X (in.^{3})	Z_Y (in.^{3})
W14 × 74	21.8	14.17	0.450	10.070	0.785	796	112	6.04	134	26.6	2.48	126	40.6
× 68	20.0	14.04	0.415	10.035	0.720	723	103	6.01	121	24.2	2.46	115	36.9
× 61	17.9	13.89	0.375	9.995	0.645	640	92.2	5.98	107	21.5	2.45	102	32.8
× 53	15.6	13.92	0.370	8.060	0.660	541	77.8	5.89	57.7	14.3	1.92	87.1	22.0
× 43	12.6	13.66	0.305	7.995	0.530	428	62.7	5.82	45.2	11.3	1.89	69.6	17.3
× 38	11.2	14.10	0.310	6.770	0.515	385	54.6	5.87	26.7	7.88	1.55	61.5	12.1
× 34	10.0	13.98	0.285	6.745	0.455	340	48.6	5.83	23.3	6.91	1.53	54.6	10.6
× 30	8.85	13.84	0.270	6.730	0.385	291	42.0	5.73	19.6	5.82	1.49	47.3	8.99
W12 × 87	25.6	12.53	0.515	12.125	0.810	740	118	5.38	241	39.7	3.07	132	60.4
× 65	19.1	12.12	0.390	12.000	0.605	533	87.9	5.28	174	29.1	3.02	96.8	44.1
× 53	15.6	12.06	0.345	9.995	0.575	425	70.6	5.23	95.8	19.2	2.48	77.9	29.1
× 40	11.8	11.94	0.295	8.005	0.515	310	51.9	5.13	44.1	11.0	1.93	57.5	16.8
× 35	10.3	12.50	0.300	6.560	0.520	285	45.6	5.25	24.5	7.47	1.54	51.2	11.5
× 30	8.79	12.34	0.260	6.520	0.440	238	38.6	5.21	20.3	6.24	1.52	43.1	9.56
× 22	6.48	12.31	0.260	4.030	0.425	156	25.4	4.91	4.66	2.31	0.847	29.3	3.66
W10 × 112	32.9	11.36	0.755	10.415	1.250	716	126	4.66	236	45.3	2.68	147	69.2
× 100	29.4	11.10	0.680	10.340	1.120	623	112	4.60	207	40.0	2.65	130	61.0
× 88	25.9	10.84	0.605	10.265	0.990	534	98.5	4.54	179	34.8	2.63	113	53.1
× 77	22.6	10.60	0.530	10.190	0.870	455	85.9	4.49	154	30.1	2.60	97.6	45.9
× 60	17.6	10.22	0.420	10.080	0.680	341	66.7	4.39	116	23.0	2.57	74.6	35.0
× 49	14.4	9.98	0.340	10.000	0.560	272	54.6	4.35	93.4	18.7	2.54	60.4	28.3
× 45	13.3	10.10	0.350	8.020	0.620	248	49.1	4.32	53.4	13.3	2.01	54.9	20.3
× 39	11.5	9.92	0.315	7.985	0.530	209	42.1	4.27	45.0	11.3	1.98	46.8	17.2
× 33	9.71	9.73	0.290	7.960	0.435	170	35.0	4.19	36.6	9.20	1.94	38.8	14.0
× 22	6.49	10.17	0.240	5.750	0.360	118	23.2	4.27	11.4	3.97	1.33	26.0	6.10
W8 × 67	19.7	9.00	0.570	8.280	0.935	272	60.4	3.72	88.6	21.4	2.12	70.2	32.7
× 58	17.1	8.75	0.510	8.220	0.810	228	52.0	3.65	75.1	18.3	2.10	59.8	27.9
× 48	14.1	8.50	0.400	8.110	0.685	184	43.3	3.61	60.9	15.0	2.08	49.0	22.9
× 40	11.7	8.25	0.360	8.070	0.560	146	35.5	3.53	49.1	12.2	2.04	39.8	18.5
× 35	10.3	8.12	0.310	8.020	0.495	127	31.2	3.51	42.6	10.6	2.03	34.7	16.1
× 31	9.13	8.00	0.285	7.995	0.435	110	27.5	3.47	37.1	9.27	2.02	30.4	14.1
× 28	8.25	8.06	0.285	6.535	0.465	98.0	24.3	3.45	21.7	6.63	1.62	27.2	10.1
× 24	7.08	7.93	0.245	6.495	0.400	82.8	20.9	3.42	18.3	5.63	1.61	23.2	8.57
× 21	6.16	8.28	0.250	5.270	0.400	75.3	18.2	3.49	9.77	3.71	1.26	20.4	5.69
× 18	5.26	8.14	0.230	5.250	0.330	61.9	15.2	3.43	7.97	3.04	1.23	17.0	4.66

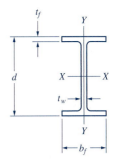

TABLE A–1(b) (Continued)   **Properties of Selected W Shapes (Wide-Flange Sections): SI Units**

Designation	Area	Depth	Web Thickness	Flange Width	Flange Thickness	Elastic Properties						Plastic Modulus	
						Axis x–x			Axis y–y				
	A $\times 10^{-3}$	d	t_w	b_f	t_f	I $\times 10^{-6}$	S $\times 10^{-3}$	r	I $\times 10^{-6}$	S $\times 10^{-3}$	r	Z_X $\times 10^{-3}$	Z_Y $\times 10^{-3}$
mm $\times$ kN/m	(m²)	(mm)	(mm)	(mm)	(mm)	(m⁴)	(m³)	(mm)	(m⁴)	(m³)	(mm)	(m³)	(m³)
W360 $\times$ 1.08	14.1	360	11.4	256	19.9	331	1.84	153	55.8	0.436	63.0	2.07	0.665
$\times$ 0.99	12.9	357	10.5	255	18.3	301	1.69	153	50.4	0.397	62.5	1.88	0.605
$\times$ 0.89	11.5	353	9.53	254	16.4	266	1.51	152	44.5	0.352	62.2	1.67	0.538
$\times$ 0.77	10.1	354	9.40	205	16.8	225	1.28	150	24.0	0.234	48.8	1.43	0.361
$\times$ 0.63	8.13	347	7.75	203	13.5	178	1.03	148	18.8	0.185	48.0	1.14	0.284
$\times$ 0.55	7.23	358	7.87	172	13.1	160	0.895	149	11.1	0.129	39.4	1.01	0.198
$\times$ 0.50	6.45	355	7.24	171	11.6	142	0.797	148	9.70	0.113	38.9	0.895	0.174
$\times$ 0.44	5.71	352	6.86	171	9.78	121	0.688	146	8.16	0.095	37.8	0.775	0.147
W300 $\times$ 1.27	16.5	318	13.1	308	20.6	308	1.93	137	100	0.651	78.0	2.16	0.990
$\times$ 0.95	12.3	308	9.91	305	15.4	222	1.44	134	72.4	0.477	76.7	1.59	0.723
$\times$ 0.77	10.1	306	8.76	254	14.6	177	1.16	133	39.9	0.315	63.0	1.28	0.477
$\times$ 0.58	7.61	303	7.49	203	13.1	129	0.851	130	18.4	0.180	49.0	0.942	0.275
$\times$ 0.51	6.65	318	7.62	167	13.2	119	0.747	133	10.2	0.122	39.1	0.839	0.188
$\times$ 0.44	5.67	313	6.60	166	11.2	99.1	0.633	132	8.45	0.102	38.6	0.706	0.157
$\times$ 0.32	4.18	313	6.60	102	10.8	64.9	0.416	125	1.94	0.038	21.5	0.480	0.060
W250 $\times$ 1.63	21.2	289	19.2	265	31.8	298	2.07	118	98.2	0.742	68.1	2.41	1.13
$\times$ 1.46	19.0	282	17.3	263	28.4	259	1.84	117	86.2	0.656	67.3	2.13	1.000
$\times$ 1.28	16.7	275	15.4	261	25.1	222	1.61	115	74.5	0.570	66.8	1.85	0.870
$\times$ 1.12	14.6	269	13.5	259	22.1	189	1.41	114	64.1	0.493	66.0	1.60	0.752
$\times$ 0.88	11.4	260	10.7	256	17.3	142	1.09	112	48.3	0.377	65.3	1.22	0.574
$\times$ 0.71	9.29	253	8.64	254	14.2	113	0.895	110	38.9	0.306	64.5	0.990	0.464
$\times$ 0.66	8.58	257	8.89	204	15.7	103	0.805	110	22.2	0.218	51.1	0.900	0.333
$\times$ 0.57	7.42	252	8.00	203	13.5	87.0	0.690	108	18.7	0.185	50.3	0.767	0.282
$\times$ 0.48	6.26	247	7.37	202	11.0	70.8	0.574	106	15.2	0.151	49.3	0.636	0.229
$\times$ 0.32	4.19	258	6.10	146	9.14	49.1	0.380	108	4.74	0.065	33.8	0.426	0.100
W200 $\times$ 0.98	12.7	229	14.5	210	23.7	113	0.990	94.5	36.9	0.351	53.8	1.15	0.536
$\times$ 0.85	11.0	222	13.0	209	20.6	94.9	0.852	92.7	31.3	0.300	53.3	0.980	0.457
$\times$ 0.70	9.10	216	10.2	206	17.4	76.6	0.710	91.7	25.3	0.246	52.8	0.803	0.375
$\times$ 0.58	7.55	210	9.14	205	14.2	60.8	0.582	89.7	20.4	0.200	51.8	0.652	0.303
$\times$ 0.51	6.65	206	7.87	204	12.6	52.9	0.511	89.2	17.7	0.174	51.6	0.569	0.264
$\times$ 0.45	5.89	203	7.24	203	11.0	45.8	0.451	88.1	15.4	0.152	51.3	0.498	0.231
$\times$ 0.41	5.32	205	7.24	166	11.8	40.8	0.398	87.6	9.03	0.109	41.1	0.446	0.166
$\times$ 0.35	4.57	201	6.22	165	10.2	34.5	0.343	86.9	7.62	0.092	40.9	0.380	0.140
$\times$ 0.31	3.97	210	6.35	134	10.2	31.3	0.298	88.6	4.07	0.061	32.0	0.334	0.093
$\times$ 0.26	3.39	207	5.84	133	8.38	25.8	0.249	87.1	3.32	0.050	31.2	0.279	0.076

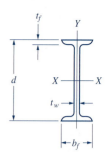

TABLE A–2(α) Properties of S Shapes (American Standard I-Beams): U.S. Customary Units

Designation	Area	Depth	Web Thickness	Flange Width	Flange Thickness	Elastic Properties Axis x–x			Axis y–y			Plastic Modulus	
in. × lb/ft	A (in.²)	d (in.)	t_w (in.)	b_f (in.)	t_f (in.)	I (in.⁴)	S (in.³)	r (in.)	I (in.⁴)	S (in.³)	r (in.)	Z_X (in.³)	Z_Y (in.³)
S24 × 121	35.6	24.50	0.800	8.050	1.090	3160	258	9.43	83.3	20.7	1.53	306	36.2
× 106	31.2	24.50	0.620	7.870	1.090	2940	240	9.71	77.1	19.6	1.57	279	33.2
S24 × 100	29.3	24.00	0.745	7.245	0.870	2390	199	9.02	47.7	13.2	1.27	240	23.9
× 90	26.5	24.00	0.625	7.125	0.870	2250	187	9.21	44.9	12.6	1.30	222	22.3
× 80	23.5	24.00	0.500	7.000	0.870	2100	175	9.47	42.2	12.1	1.34	204	20.7
S20 × 96	28.2	20.30	0.800	7.200	0.920	1670	165	7.71	50.2	13.9	1.33	198	24.9
× 86	25.3	20.30	0.660	7.060	0.920	1580	155	7.89	46.8	13.3	1.36	183	23.0
S20 × 75	22.0	20.00	0.635	6.385	0.795	1280	128	7.62	29.8	9.32	1.16	153	16.7
× 66	19.4	20.00	0.505	6.255	0.795	1190	119	7.83	27.7	8.85	1.19	140	15.3
S18 × 70	20.6	18.00	0.711	6.251	0.691	926	103	6.71	24.1	7.72	1.08	125	14.4
× 54.7	16.1	18.00	0.461	6.001	0.691	804	89.4	7.07	20.8	6.94	1.14	105	12.1
S15 × 50	14.7	15.00	0.550	5.640	0.622	486	64.8	5.75	15.7	5.57	1.03	77.1	9.97
× 42.9	12.6	15.00	0.411	5.501	0.622	447	59.6	5.95	14.4	5.23	1.07	69.3	9.02
S12 × 50	14.7	12.00	0.687	5.477	0.659	305	50.8	4.55	15.7	5.74	1.03	61.2	10.3
× 40.8	12.0	12.00	0.462	5.252	0.659	272	45.4	4.77	13.6	5.16	1.06	53.1	8.85
S12 × 35	10.3	12.00	0.428	5.078	0.544	229	38.2	4.72	9.87	3.89	0.980	44.8	6.79
× 31.8	9.35	12.00	0.350	5.000	0.544	218	36.4	4.83	9.36	3.74	1.00	42.0	6.40
S10 × 35	10.3	10.00	0.594	4.944	0.491	147	29.4	3.78	8.36	3.38	0.901	35.4	6.22
× 25.4	7.46	10.00	0.311	4.661	0.491	124	24.7	4.07	6.79	2.91	0.954	28.4	4.96
S 8 × 23	6.79	8.00	0.441	4.171	0.426	64.9	16.2	3.10	4.31	2.07	0.798	19.3	3.68
× 18.4	5.41	8.00	0.271	4.001	0.426	57.6	14.4	3.26	3.73	1.86	0.831	16.5	3.16
S 7 × 20	5.88	7.00	0.450	3.860	0.392	42.4	12.1	2.69	3.17	1.64	0.734	14.5	2.96
× 15.3	4.50	7.00	0.252	3.662	0.392	36.7	10.5	2.86	2.64	1.44	0.766	12.1	2.44
S 6 × 17.3	5.07	6.00	0.465	3.565	0.359	26.3	8.77	2.28	2.31	1.30	0.675	10.6	2.36
× 12.5	3.67	6.00	0.232	3.332	0.359	22.1	7.37	2.45	1.82	1.09	0.705	8.47	1.85
S 5 × 14.8	4.34	5.00	0.494	3.284	0.326	15.2	6.09	1.87	1.67	1.01	0.620	7.42	1.88
× 10	2.94	5.00	0.214	3.004	0.326	12.3	4.92	2.05	1.22	0.809	0.643	5.67	1.37
S 4 × 9.5	2.79	4.00	0.326	2.796	0.293	6.79	3.39	1.56	0.903	0.646	0.569	4.04	1.13
× 7.7	2.26	4.00	0.193	2.663	0.293	6.08	3.04	1.64	0.764	0.574	0.581	3.51	0.964
S 3 × 7.5	2.21	3.00	0.349	2.509	0.260	2.93	1.95	1.15	0.586	0.468	0.516	2.36	0.826
× 5.7	1.67	3.00	0.170	2.330	0.260	2.52	1.68	1.23	0.455	0.390	0.522	1.95	0.653

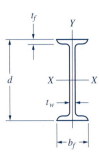

TABLE A–2(b)   Properties of S Shapes (American Standard I-Beams: SI Units

Desig-nation mm × kN/m	Area A $\times 10^{-3}$ (m²)	Depth d (mm)	Web Thickness t_w (mm)	Flange Width b_f (mm)	Flange Thickness t_f (mm)	Elastic Properties Axis x–x I $\times 10^{-6}$ (m⁴)	Axis x–x S $\times 10^{-3}$ (m³)	Axis x–x r (mm)	Axis y–y I $\times 10^{-6}$ (m⁴)	Axis y–y S $\times 10^{-3}$ (m³)	Axis y–y r (mm)	Plastic Modulus Z_X $\times 10^{-3}$ (m³)	Z_Y $\times 10^{-3}$ (m³)
S610 × 1.77	22.97	622	20.3	204	27.7	1320	4.23	240	34.7	0.339	38.9	5.02	0.593
× 1.55	20.13	622	15.7	200	27.7	1220	3.93	247	32.1	0.321	39.9	4.57	0.544
S610 × 1.46	18.90	610	18.9	184	22.1	995	3.26	229	19.9	0.216	32.3	3.93	0.392
× 1.31	17.10	610	15.9	181	22.1	937	3.06	234	18.7	0.207	33.0	3.64	0.365
× 1.17	15.16	610	12.7	178	22.1	874	2.87	241	17.6	0.198	34.0	3.34	0.339
S510 × 1.40	18.19	516	20.3	183	23.4	695	2.70	196	20.9	0.228	33.8	3.25	0.408
× 1.25	16.32	516	16.8	179	23.4	658	2.54	200	19.5	0.218	34.5	3.00	0.377
S510 × 1.09	14.19	508	16.1	162	20.2	533	2.10	194	12.4	0.153	29.5	2.51	0.274
× 0.963	12.52	508	12.8	159	20.2	495	1.95	199	11.5	0.145	30.2	2.29	0.251
S460 × 1.02	13.29	457	18.1	159	17.6	385	1.69	170	10.0	0.127	27.4	2.05	0.236
× 0.798	10.39	457	11.7	152	17.6	335	1.47	180	8.66	0.114	29.0	1.72	0.198
S380 × 0.730	9.48	381	14.0	143	15.8	202	1.06	146	6.53	0.0913	26.2	1.26	0.163
× 0.626	8.13	381	10.4	140	15.8	186	0.977	151	5.99	0.0857	27.2	1.14	0.148
S300 × 0.730	9.48	305	17.4	139	16.7	127	0.833	116	6.53	0.0941	26.2	1.00	0.169
× 0.595	7.74	305	11.7	133	16.7	113	0.744	121	5.66	0.0846	26.9	0.870	0.145
S300 × 0.511	6.65	305	10.9	129	13.8	95.3	0.626	120	4.11	0.0638	24.9	0.734	0.111
× 0.464	6.03	305	8.89	127	13.8	90.7	0.597	123	3.90	0.0613	25.4	0.688	0.105
S250 × 0.511	6.65	254	15.1	126	12.5	61.2	0.482	96.0	3.48	0.0554	22.9	0.580	0.102
× 0.371	4.81	254	7.90	118	12.5	51.6	0.405	103.4	2.83	0.0477	24.2	0.465	0.0813
S200 × 0.336	4.37	203	11.2	106	10.8	27.0	0.266	78.7	1.79	0.0339	20.3	0.316	0.0603
× 0.268	3.49	203	6.88	102	10.8	24.0	0.236	82.8	1.55	0.0305	21.1	0.270	0.0518
S180 × 0.292	3.79	178	11.4	98.0	10.0	17.6	0.198	68.3	1.32	0.0269	18.6	0.238	0.0485
× 0.223	2.90	178	6.40	93.0	10.0	15.3	0.172	72.6	1.10	0.0236	19.5	0.198	0.0400
S150 × 0.252	3.27	152	11.8	90.6	9.12	10.9	0.144	57.9	0.961	0.0213	17.1	0.174	0.0387
× 0.182	2.37	152	5.89	84.6	9.12	9.20	0.121	62.2	0.757	0.0179	17.9	0.139	0.0303
S130 × 0.215	2.80	127	12.5	83.4	8.28	6.33	0.100	47.5	0.695	0.0166	15.7	0.122	0.0308
× 0.146	1.90	127	5.44	76.3	8.28	5.12	0.0806	52.1	0.508	0.0133	16.3	0.0929	0.0225
S100 × 0.139	1.80	102	8.28	71.0	7.44	2.83	0.0556	39.6	0.376	0.0106	14.5	0.0662	0.0185
× 0.112	1.46	102	4.90	67.6	7.44	2.53	0.0498	41.7	0.318	0.0094	14.8	0.0575	0.0158
S 76 × 0.109	1.43	76.2	8.86	63.7	6.60	1.22	0.0320	29.2	0.244	0.0077	13.1	0.0387	0.0135
× 0.0832	1.08	76.2	4.32	59.2	6.60	1.05	0.0275	31.2	0.189	0.0064	13.3	0.0320	0.0107

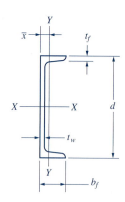

TABLE A–3(α)   Properties of C Shapes (American Standard Channels): U.S. Customary Units

Designation	Area	Depth	Web Thickness	Flange Width	Flange Average Thickness	Axis x–x			Axis y–y			Centroid
in. × lb/ft	A (in.^{2})	d (in.)	t_w (in.)	b_f (in.)	t_f (in.)	I (in.^{4})	S (in.^{3})	r (in.)	I (in.^{4})	S (in.^{3})	r (in.)	$\bar{x}$ (in.)
C15 × 50	14.7	15.00	0.716	3.716	0.650	404	53.8	5.24	11.0	3.78	0.867	0.798
× 40	11.8	15.00	0.520	3.520	0.650	349	46.5	5.44	9.23	3.37	0.886	0.777
× 33.9	9.96	15.00	0.400	3.400	0.650	315	42.0	5.62	8.13	3.11	0.904	0.787
C12 × 30	8.82	12.00	0.510	3.170	0.501	162	27.0	4.29	5.14	2.06	0.763	0.674
× 25	7.35	12.00	0.387	3.047	0.501	144	24.1	4.43	4.47	1.88	0.780	0.674
× 20.7	6.09	12.00	0.282	2.942	0.501	129	21.5	4.61	3.88	1.73	0.799	0.698
C10 × 30	8.82	10.00	0.673	3.033	0.436	103	20.7	3.42	3.94	1.65	0.669	0.649
× 25	7.35	10.00	0.526	2.886	0.436	91.2	18.2	3.52	3.36	1.48	0.676	0.617
× 20	5.88	10.00	0.379	2.739	0.436	78.9	15.8	3.66	2.81	1.32	0.692	0.606
× 15.3	4.49	10.00	0.240	2.600	0.436	67.4	13.5	3.87	2.28	1.16	0.713	0.634
C 9 × 20	5.88	9.00	0.448	2.648	0.413	60.9	13.5	3.22	2.42	1.17	0.642	0.583
× 15	4.41	9.00	0.285	2.485	0.413	51.0	11.3	3.40	1.93	1.01	0.661	0.586
× 13.4	3.94	9.00	0.233	2.433	0.413	47.9	10.6	3.48	1.76	0.962	0.669	0.601
C 8 × 18.75	5.51	8.00	0.487	2.527	0.390	44.0	11.0	2.82	1.98	1.01	0.599	0.565
× 13.75	4.04	8.00	0.303	2.343	0.390	36.1	9.03	2.99	1.53	0.854	0.615	0.553
× 11.5	3.38	8.00	0.220	2.260	0.390	32.6	8.14	3.11	1.32	0.781	0.625	0.571
C 7 × 14.75	4.33	7.00	0.419	2.299	0.366	27.2	7.78	2.51	1.38	0.779	0.564	0.532
× 12.25	3.60	7.00	0.314	2.194	0.366	24.2	6.93	2.60	1.17	0.703	0.571	0.525
× 9.8	2.87	7.00	0.210	2.090	0.366	21.3	6.08	2.72	0.968	0.625	0.581	0.540
C 6 × 13	3.83	6.00	0.437	2.157	0.343	17.4	5.80	2.13	1.05	0.642	0.525	0.514
× 10.5	3.09	6.00	0.314	2.034	0.343	15.2	5.06	2.22	0.866	0.564	0.529	0.499
× 8.2	2.40	6.00	0.200	1.920	0.343	13.1	4.38	2.34	0.693	0.492	0.537	0.511
C 5 × 9	2.64	5.00	0.325	1.885	0.320	8.96	3.56	1.83	0.632	0.450	0.489	0.478
× 6.7	1.97	5.00	0.190	1.750	0.320	7.49	3.00	1.95	0.479	0.378	0.493	0.484
C 4 × 7.25	2.13	4.00	0.321	1.721	0.296	4.59	2.29	1.47	0.433	0.343	0.450	0.459
× 5.4	1.59	4.00	0.184	1.584	0.296	3.85	1.93	1.56	0.319	0.283	0.449	0.457
C 3 × 6	1.76	3.00	0.356	1.596	0.273	2.07	1.38	1.08	0.305	0.268	0.416	0.455
× 5	1.47	3.00	0.258	1.498	0.273	1.85	1.24	1.12	0.247	0.233	0.410	0.438
× 4.1	1.21	3.00	0.170	1.410	0.273	1.66	1.10	1.17	0.197	0.202	0.404	0.436

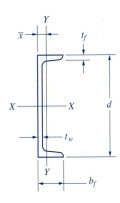

TABLE A–3(b) Properties of C Shapes (American Standard Channels): SI Units

Desig-nation	Area	Depth	Web Thick-ness	Flange		Axis x–x			Axis y–y			Cen-troid
				Width	Average Thick-ness							
	A $\times 10^{-3}$	d	t_w	b_f	t_f	I $\times 10^{-6}$	S $\times 10^{-3}$	r	I $\times 10^{-6}$	S $\times 10^{-3}$	r	$\bar{x}$
mm × kN/m	(m²)	(mm)	(mm)	(mm)	(mm)	(m⁴)	(m³)	(mm)	(m⁴)	(m³)	(mm)	(m³)
C380 × 0.730	9.48	381.0	18.2	94.4	16.5	168	0.882	133	4.58	0.0620	22.0	20.3
× 0.584	7.61	381.0	13.2	89.4	16.5	145	0.762	138	3.84	0.0552	22.5	19.7
× 0.495	6.43	381.0	10.2	86.4	16.5	131	0.688	143	3.38	0.0510	23.0	20.0
C300 × 0.438	5.69	304.8	13.0	80.5	12.7	67.4	0.443	109	2.14	0.0338	19.4	17.1
× 0.365	4.74	304.8	9.83	77.4	12.7	59.9	0.395	113	1.86	0.0308	19.8	17.1
× 0.302	3.93	304.8	7.16	74.7	12.7	53.7	0.352	117	1.61	0.0284	20.3	17.7
C250 × 0.438	5.69	254.0	17.1	77.0	11.1	42.9	0.339	86.9	1.64	0.0270	17.0	16.5
× 0.365	4.74	254.0	13.4	73.3	11.1	38.0	0.298	89.4	1.40	0.0243	17.2	15.7
× 0.292	3.79	254.0	9.63	69.6	11.1	32.8	0.259	93.0	1.17	0.0216	17.6	15.4
× 0.223	2.90	254.0	6.10	66.0	11.1	28.1	0.221	98.3	0.949	0.0190	18.1	16.1
C230 × 0.292	3.79	228.6	11.4	67.3	10.5	25.3	0.221	81.8	1.01	0.0192	16.3	14.8
× 0.219	2.85	228.6	7.24	63.1	10.5	21.2	0.185	86.4	0.803	0.0166	16.8	14.9
× 0.196	2.54	228.6	5.92	61.8	10.5	19.9	0.174	88.4	0.733	0.0158	17.0	15.3
C200 × 0.274	3.56	203.2	12.4	64.2	9.91	18.3	0.180	71.6	0.824	0.0166	15.2	14.4
× 0.201	2.61	203.2	7.70	59.5	9.91	15.0	0.148	75.9	0.637	0.0140	15.6	14.0
× 0.168	2.18	203.2	5.59	57.4	9.91	13.6	0.133	79.0	0.549	0.0128	15.9	14.5
C180 × 0.215	2.79	177.8	10.6	58.4	9.30	11.3	0.128	63.8	0.574	0.0128	14.3	13.5
× 0.179	2.32	177.8	7.98	55.7	9.30	10.1	0.114	66.0	0.487	0.0115	14.5	13.3
× 0.143	1.85	177.8	5.33	53.1	9.30	8.87	0.100	69.1	0.403	0.0102	14.8	13.7
C150 × 0.190	2.47	152.4	11.1	54.8	8.71	7.24	0.0951	54.1	0.437	0.0105	13.3	13.1
× 0.153	1.99	152.4	7.98	51.7	8.71	6.33	0.0829	56.4	0.360	0.00924	13.4	12.7
× 0.120	1.55	152.4	5.08	48.8	8.71	5.45	0.0718	59.4	0.288	0.00806	13.6	13.0
C130 × 0.131	1.70	127.0	8.26	47.9	8.13	3.70	0.0583	46.5	0.263	0.00738	12.4	12.1
× 0.098	1.27	127.0	4.83	44.5	8.13	3.12	0.0492	49.5	0.199	0.00620	12.5	12.3
C100 × 0.106	1.37	101.6	8.15	43.7	7.52	1.91	0.0375	37.3	0.180	0.00562	11.4	11.7
× 0.079	1.03	101.6	4.67	40.2	7.52	1.60	0.0316	39.6	0.133	0.00464	11.4	11.6
C 76 × 0.088	1.14	76.2	9.04	40.5	6.93	0.862	0.0226	27.4	0.127	0.00439	10.6	11.6
× 0.073	0.948	76.2	6.55	38.0	6.93	0.770	0.0203	28.4	0.103	0.00382	10.4	11.1
× 0.060	0.781	76.2	4.32	35.8	6.93	0.691	0.0180	29.7	0.0820	0.00331	10.3	11.1

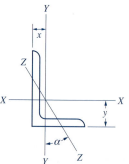

TABLE A–4(α)　Properties of Selected L Shapes (Steel Angles): U.S. Customary Units

Size and Thickness	Weight per ft	Area	Axis x–x				Axis y–y				Axis z–z	
		A	I	S	r	y	I	S	r	x	r	$\tan \alpha$
(in.)	(lb/ft)	(in.²)	(in.⁴)	(in.³)	(in.)	(in.)	(in.⁴)	(in.³)	(in.)	(in.)	(in.)	
L 8 × 6　× 1	44.2	13.0	80.8	15.1	2.49	2.65	38.8	8.92	1.73	1.65	1.28	0.543
× $\frac{3}{4}$	33.8	9.94	63.4	11.7	2.53	2.56	30.7	6.92	1.76	1.56	1.29	0.551
× $\frac{1}{2}$	23.0	6.75	44.3	8.02	2.56	2.47	21.7	4.79	1.79	1.47	1.30	0.558
L 8 × 4　× 1	37.4	11.0	69.6	14.1	2.52	3.05	11.6	3.94	1.03	1.05	0.846	0.247
× $\frac{3}{4}$	28.7	8.44	54.9	10.9	2.55	2.95	9.36	3.07	1.05	0.953	0.852	0.258
× $\frac{1}{2}$	19.6	5.75	38.5	7.49	2.59	2.86	6.74	2.15	1.08	0.859	0.865	0.267
L 7 × 4　× $\frac{3}{4}$	26.2	7.69	37.8	8.42	2.22	2.51	9.05	3.03	1.09	1.01	0.860	0.324
× $\frac{1}{2}$	17.9	5.25	26.7	5.81	2.25	2.42	6.53	2.12	1.11	0.917	0.872	0.335
× $\frac{3}{8}$	13.6	3.98	20.6	4.44	2.27	2.37	5.10	1.63	1.13	0.870	0.880	0.340
L 6 × 6　× 1	37.4	11.0	35.5	8.57	1.80	1.86	35.5	8.57	1.80	1.86	1.17	1.000
× $\frac{7}{8}$	33.1	9.73	31.9	7.63	1.81	1.82	31.9	7.63	1.81	1.82	1.17	1.000
× $\frac{3}{4}$	28.7	8.44	28.2	6.66	1.83	1.78	28.2	6.66	1.83	1.78	1.17	1.000
× $\frac{5}{8}$	24.2	7.11	24.2	5.66	1.84	1.73	24.2	5.66	1.84	1.73	1.18	1.000
× $\frac{1}{2}$	19.6	5.75	19.9	4.61	1.86	1.68	19.9	4.61	1.86	1.68	1.18	1.000
× $\frac{3}{8}$	14.9	4.36	15.4	3.53	1.88	1.64	15.4	3.53	1.88	1.64	1.19	1.000
L 6 × 4　× $\frac{3}{4}$	23.6	6.94	24.5	6.25	1.88	2.08	8.68	2.97	1.12	1.08	0.860	0.428
× $\frac{5}{8}$	20.0	5.86	21.1	5.31	1.90	2.03	7.52	2.54	1.13	1.03	0.864	0.435
× $\frac{1}{2}$	16.2	4.75	17.4	4.33	1.91	1.99	6.27	2.08	1.15	0.987	0.870	0.440
× $\frac{3}{8}$	12.3	3.61	13.5	3.32	1.93	1.94	4.90	1.60	1.17	0.941	0.877	0.446
L 6 × 3$\frac{1}{2}$ × $\frac{3}{8}$	11.7	3.42	12.9	3.24	1.94	2.04	3.34	1.23	0.988	0.787	0.767	0.350
× $\frac{5}{16}$	9.8	2.87	10.9	2.73	1.95	2.01	2.85	1.04	0.996	0.763	0.772	0.352
L 5 × 5　× $\frac{7}{8}$	27.2	7.98	17.8	5.17	1.49	1.57	17.8	5.17	1.49	1.57	0.973	1.000
× $\frac{3}{4}$	23.6	6.94	15.7	4.53	1.51	1.52	15.7	4.53	1.51	1.52	0.975	1.000
× $\frac{1}{2}$	16.2	4.75	11.3	3.16	1.54	1.43	11.3	3.16	1.54	1.43	0.983	1.000
× $\frac{3}{8}$	12.3	3.61	8.74	2.42	1.56	1.39	8.74	2.42	1.56	1.39	0.990	1.000
× $\frac{5}{16}$	10.3	3.03	7.42	2.04	1.57	1.37	7.42	2.04	1.57	1.37	0.994	1.000
L 5 × 3$\frac{1}{2}$ × $\frac{3}{4}$	19.8	5.81	13.9	4.28	1.55	1.75	5.55	2.22	0.977	0.996	0.748	0.464
× $\frac{1}{2}$	13.6	4.00	9.99	2.99	1.58	1.66	4.05	1.56	1.01	0.906	0.755	0.479
× $\frac{3}{8}$	10.4	3.05	7.78	2.29	1.60	1.61	3.18	1.21	1.02	0.861	0.762	0.486
× $\frac{5}{16}$	8.7	2.56	6.60	1.94	1.61	1.59	2.72	1.02	1.03	0.838	0.766	0.489
L 5 × 3　× $\frac{1}{2}$	12.8	3.75	9.45	2.91	1.59	1.75	2.58	1.15	0.829	0.750	0.648	0.357
× $\frac{3}{8}$	9.8	2.86	7.37	2.24	1.61	1.70	2.04	0.888	0.845	0.704	0.654	0.364
× $\frac{5}{16}$	8.2	2.40	6.26	1.89	1.61	1.68	1.75	0.753	0.853	0.681	0.658	0.368
× $\frac{1}{4}$	6.6	1.94	5.11	1.53	1.62	1.66	1.44	0.614	0.861	0.657	0.663	0.371

(Table continued on page 772)

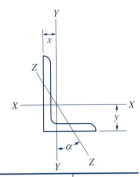

TABLE A–4(b) Properties of Selected L Shapes (Steel Angles): SI Units

Size and Thickness	Weight per ft	Area	Axis x–x					Axis y–y				Axis z–z	
		A	I	S	r	y	I	S	r	x	r	$\tan \alpha$
	$\times 10^{-3}$	$\times 10^{-3}$	$\times 10^{-6}$	$\times 10^{-3}$			$\times 10^{-6}$	$\times 10^{-3}$				
(mm)	(kN/m)	(m²)	(m⁴)	(m³)	(mm)	(mm)	(m⁴)	(m³)	(mm)	(mm)	(mm)	
L203 × 152 × 25.4	0.645	8.39	33.6	0.247	63.2	67.3	16.1	0.146	43.9	41.9	32.5	0.543
× 19.1	0.493	6.41	26.4	0.192	64.3	65.0	12.8	0.113	44.7	39.6	32.8	0.551
× 12.7	0.336	4.36	18.4	0.131	65.0	62.7	9.03	0.0785	45.5	37.3	33.0	0.558
L203 × 102 × 25.4	0.546	7.10	29.0	0.231	64.0	77.5	4.83	0.0646	26.2	26.7	21.5	0.247
× 19.1	0.419	5.45	22.8	0.179	64.8	74.9	3.90	0.0503	26.7	24.2	21.6	0.258
× 12.7	0.286	3.71	16.0	0.123	65.8	72.6	2.81	0.0352	27.4	21.8	22.0	0.267
L178 × 102 × 19.1	0.382	4.96	15.7	0.138	56.4	63.8	3.77	0.0497	27.7	25.7	21.8	0.324
× 12.7	0.261	3.39	11.1	0.0952	57.2	61.5	2.72	0.0347	28.2	23.3	22.1	0.335
× 9.5	0.193	2.57	8.57	0.0728	57.7	60.2	2.12	0.0267	28.7	22.1	22.4	0.340
L152 × 152 × 25.4	0.546	7.10	14.8	0.140	45.7	47.2	14.8	0.140	45.7	47.2	29.7	1.000
× 22.2	0.483	6.28	13.3	0.125	46.0	46.2	13.3	0.125	46.0	46.2	29.7	1.000
× 19.1	0.419	5.45	11.7	0.109	46.5	45.2	11.7	0.109	46.5	45.2	29.7	1.000
× 15.9	0.353	4.59	10.1	0.0928	46.7	43.9	10.1	0.0928	46.7	43.9	30.0	1.000
× 12.7	0.286	3.71	8.28	0.0756	47.2	42.7	8.28	0.0756	47.2	42.7	30.0	1.000
× 9.5	0.217	2.81	6.41	0.0579	47.8	41.7	6.41	0.0579	47.8	41.7	30.2	1.000
L152 × 102 × 19.1	0.344	4.48	10.2	0.102	47.8	52.8	3.61	0.0487	28.4	27.4	21.8	0.428
× 15.9	0.292	3.78	8.78	0.0870	48.3	51.6	3.13	0.0416	28.7	26.2	21.9	0.435
× 12.7	0.236	3.06	7.24	0.0710	48.5	50.5	2.61	0.0341	29.2	25.1	22.1	0.440
× 9.5	0.179	2.33	5.62	0.0544	49.0	49.3	2.04	0.0262	29.7	23.9	22.3	0.446
L152 × 89 × 9.5	0.171	2.21	5.37	0.0531	49.3	51.8	1.39	0.0202	25.1	20.0	19.5	0.350
× 7.9	0.143	1.85	4.54	0.0447	49.5	51.1	1.19	0.0170	25.3	19.4	19.6	0.352
L127 × 127 × 22.2	0.397	5.15	7.41	0.0847	37.8	39.9	7.41	0.0847	37.8	39.9	24.7	1.000
× 19.1	0.344	4.48	6.53	0.0742	38.4	38.6	6.53	0.0742	38.4	38.6	24.8	1.000
× 12.7	0.236	3.06	4.70	0.0518	39.1	36.3	4.70	0.0518	39.1	36.3	25.0	1.000
× 9.5	0.179	2.33	3.64	0.0397	39.6	35.3	3.64	0.0397	39.6	35.3	25.1	1.000
× 7.9	0.150	1.95	3.09	0.0334	39.9	34.8	3.09	0.0334	39.9	34.8	25.2	1.000
L127 × 89 × 19.1	0.289	3.75	5.79	0.0701	39.4	44.5	2.31	0.0364	24.8	25.3	19.0	0.464
× 12.7	0.198	2.58	4.16	0.0490	40.1	42.2	1.69	0.0256	25.7	23.0	19.2	0.479
× 9.5	0.152	1.97	3.24	0.0375	40.6	40.9	1.32	0.0198	25.9	21.9	19.4	0.486
× 7.9	0.127	1.65	2.75	0.0318	40.9	40.4	1.13	0.0167	26.2	21.3	19.5	0.489
L127 × 76 × 12.7	0.187	2.42	3.93	0.0477	40.4	44.5	1.07	0.0188	21.1	19.1	16.5	0.357
× 9.5	0.143	1.85	3.07	0.0367	40.9	43.2	0.849	0.0146	21.5	17.9	16.6	0.364
× 7.9	0.120	1.55	2.61	0.0310	40.9	42.7	0.728	0.0123	21.7	17.3	16.7	0.368
× 6.4	0.0963	1.25	2.13	0.0251	41.1	42.2	0.599	0.0101	21.9	16.7	16.8	0.371

(Table continued on page 773)

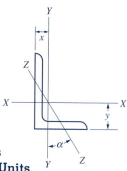

TABLE A–4(a) (Continued) **Properties of Selected L Shapes (Steel Angles): U.S. Customary Units**

Size and Thickness (in.)		Weight per ft (lb/ft)	Area A (in.²)	Axis x–x				Axis y–y				Axis z–z	
				I (in.⁴)	S (in.³)	r (in.)	y (in.)	I (in.⁴)	S (in.³)	r (in.)	x (in.)	r (in.)	$\tan \alpha$
L 4 × 4	× $\frac{3}{4}$	18.5	5.44	7.67	2.81	1.19	1.27	7.67	2.81	1.19	1.27	0.778	1.000
	× $\frac{5}{8}$	15.7	4.61	6.66	2.40	1.20	1.23	6.66	2.40	1.20	1.23	0.779	1.000
	× $\frac{1}{2}$	12.8	3.75	5.56	1.97	1.22	1.18	5.56	1.97	1.22	1.18	0.782	1.000
	× $\frac{3}{8}$	9.8	2.86	4.36	1.52	1.23	1.14	4.36	1.52	1.23	1.14	0.788	1.000
	× $\frac{5}{16}$	8.2	2.40	3.71	1.29	1.24	1.12	3.71	1.29	1.24	1.12	0.791	1.000
	× $\frac{1}{4}$	6.6	1.94	3.04	1.05	1.25	1.09	3.04	1.05	1.25	1.09	0.795	1.000
L 4 × 3½	× $\frac{1}{2}$	11.9	3.50	5.32	1.94	1.23	1.25	3.79	1.52	1.04	1.00	0.722	0.750
	× $\frac{3}{8}$	9.1	2.67	4.18	1.49	1.25	1.21	2.95	1.17	1.06	0.955	0.727	0.755
	× $\frac{5}{16}$	7.7	2.25	3.56	1.26	1.26	1.18	2.55	0.994	1.07	0.932	0.730	0.757
	× $\frac{1}{4}$	6.2	1.81	2.91	1.03	1.27	1.16	2.09	0.808	1.07	0.909	0.734	0.579
L 4 × 3	× $\frac{1}{2}$	11.1	3.25	5.05	1.89	1.25	1.33	2.42	1.12	0.864	0.827	0.639	0.543
	× $\frac{3}{8}$	8.5	2.48	3.96	1.46	1.26	1.28	1.92	0.866	0.879	0.782	0.644	0.551
	× $\frac{5}{16}$	7.2	2.09	3.38	1.23	1.27	1.26	1.65	0.734	0.887	0.759	0.647	0.554
	× $\frac{1}{4}$	5.8	1.69	2.77	1.00	1.28	1.24	1.36	0.599	0.896	0.736	0.651	0.558
L 3½ × 3½	× $\frac{3}{8}$	8.5	2.48	2.87	1.15	1.07	1.01	2.87	1.15	1.07	1.01	0.687	1.000
	× $\frac{5}{16}$	7.2	2.09	2.45	0.976	1.08	0.990	2.45	0.976	1.08	0.990	0.690	1.000
	× $\frac{1}{4}$	5.8	1.69	2.01	0.794	1.09	0.968	2.01	0.794	1.09	0.968	0.694	1.000
L 3½ × 3	× $\frac{3}{8}$	7.9	2.30	2.72	1.13	1.09	1.08	1.85	0.851	0.897	0.830	0.625	0.721
	× $\frac{5}{16}$	6.6	1.93	2.33	0.954	1.10	1.06	1.58	0.722	0.905	0.808	0.627	0.724
	× $\frac{1}{4}$	5.4	1.56	1.91	0.776	1.11	1.04	1.30	0.589	0.914	0.785	0.631	0.727
L 3½ × 2½	× $\frac{3}{8}$	7.2	2.11	2.56	1.09	1.10	1.16	1.09	0.592	0.719	0.660	0.537	0.496
	× $\frac{5}{16}$	6.1	1.78	2.19	0.927	1.11	1.14	0.939	0.504	0.727	0.637	0.540	0.501
	× $\frac{1}{4}$	4.9	1.44	1.80	0.755	1.12	1.11	0.777	0.412	0.735	0.614	0.544	0.506
L 3 × 3	× $\frac{1}{2}$	9.4	2.75	2.22	1.07	0.898	0.932	2.22	1.07	0.898	0.932	0.584	1.000
	× $\frac{3}{8}$	7.2	2.11	1.76	0.833	0.913	0.888	1.76	0.833	0.913	0.888	0.587	1.000
	× $\frac{5}{16}$	6.1	1.78	1.51	0.707	0.922	0.865	1.51	0.707	0.922	0.865	0.589	1.000
	× $\frac{1}{4}$	4.9	1.44	1.24	0.577	0.930	0.842	1.24	0.577	0.930	0.842	0.592	1.000
	× $\frac{3}{16}$	3.71	1.09	0.962	0.441	0.939	0.820	0.962	0.441	0.939	0.820	0.596	1.000
L 3 × 2½	× $\frac{3}{8}$	6.6	1.92	1.66	0.810	0.928	0.956	1.04	0.581	0.736	0.706	0.522	0.676
	× $\frac{1}{4}$	4.5	1.31	1.17	0.561	0.945	0.911	0.743	0.404	0.753	0.661	0.528	0.684
	× $\frac{3}{16}$	3.39	0.996	0.907	0.430	0.954	0.888	0.577	0.310	0.761	0.638	0.533	0.688
L 3 × 2	× $\frac{3}{8}$	5.9	1.73	1.53	0.781	0.940	1.04	0.543	0.371	0.559	0.539	0.430	0.428
	× $\frac{5}{16}$	5.0	1.46	1.32	0.664	0.948	1.02	0.470	0.317	0.567	0.516	0.432	0.435
	× $\frac{1}{4}$	4.1	1.19	1.09	0.542	0.957	0.993	0.392	0.260	0.574	0.493	0.435	0.440
	× $\frac{3}{16}$	3.07	0.902	0.842	0.415	0.966	0.970	0.307	0.200	0.583	0.470	0.439	0.446

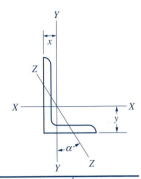

TABLE A–4(b) (Continued)   **Properties of Selected L Shapes (Steel Angles): SI Units**

Size and Thickness	Weight per ft	Area	Axis x–x				Axis y–y				Axis z–z	
		A	I	S	r	y	I	S	r	x	r	$\tan \alpha$
	$\times 10^{-3}$	$\times 10^{-3}$	$\times 10^{-6}$	$\times 10^{-3}$			$\times 10^{-6}$	$\times 10^{-3}$				
(mm)	(kN/m)	(m²)	(m⁴)	(m³)	(mm)	(mm)	(m⁴)	(m³)	(mm)	(mm)	(mm)	
L102 × 102 × 19.1	0.270	3.51	3.19	0.0461	30.2	32.3	3.19	0.0461	30.2	32.3	19.8	1.000
× 15.9	0.229	2.97	2.77	0.0393	30.5	31.2	2.77	0.0393	30.5	31.2	19.8	1.000
× 12.7	0.187	2.42	2.31	0.0323	31.0	30.0	2.31	0.0323	31.0	30.0	19.9	1.000
× 9.5	0.143	1.85	1.81	0.0249	31.2	29.0	1.81	0.0249	31.2	29.0	20.0	1.000
× 7.9	0.120	1.55	1.54	0.0211	31.5	28.4	1.54	0.0211	31.5	28.4	20.1	1.000
× 6.4	0.0963	1.25	1.27	0.0172	31.8	27.7	1.27	0.0172	31.8	27.7	20.2	1.000
L102 × 89 × 12.7	0.174	2.26	2.21	0.0318	31.2	31.8	1.58	0.0249	26.4	25.4	18.3	0.750
× 9.5	0.133	1.72	1.74	0.0244	31.8	30.7	1.23	0.0192	26.9	24.3	18.5	0.755
× 7.9	0.112	1.45	1.48	0.0207	32.0	30.0	1.06	0.0163	27.2	23.7	18.5	0.757
× 6.4	0.0905	1.17	1.21	0.0169	32.3	29.5	0.870	0.0132	27.2	23.1	18.6	0.579
L102 × 76 × 12.7	0.162	2.10	2.10	0.0310	31.8	33.8	1.01	0.0184	21.9	21.0	16.2	0.543
× 9.5	0.124	1.60	1.65	0.0239	32.0	32.5	0.799	0.0142	22.3	19.9	16.4	0.551
× 7.9	0.105	1.35	1.41	0.0202	32.3	32.0	0.687	0.0120	22.5	19.3	16.4	0.554
× 6.35	0.0846	1.09	1.15	0.0164	32.5	31.5	0.566	0.00982	22.8	18.7	16.5	0.558
L89 × 89 × 9.5	0.124	1.60	1.19	0.0188	27.2	25.7	1.19	0.0188	27.2	25.7	17.4	1.000
× 7.9	0.105	1.35	1.02	0.0160	27.4	25.1	1.02	0.0160	27.4	25.1	17.5	1.000
× 6.4	0.0846	1.09	0.837	0.0130	27.7	24.6	0.837	0.0130	27.7	24.6	17.6	1.000
L89 × 76 × 9.5	0.115	1.48	1.13	0.0185	27.7	27.4	0.770	0.0139	22.8	21.1	15.9	0.721
× 7.9	0.0963	1.25	0.970	0.0156	27.9	26.9	0.658	0.0118	23.0	20.5	15.9	0.724
× 6.4	0.0788	1.01	0.795	0.0127	28.2	26.4	0.541	0.00965	23.2	19.9	16.0	0.727
L89 × 64 × 9.5	0.105	1.36	1.07	0.0179	27.9	29.5	0.454	0.00970	18.3	16.8	13.6	0.496
× 7.9	0.0890	1.15	0.911	0.0152	28.2	29.0	0.391	0.00826	18.5	16.2	13.7	0.501
× 6.4	0.0715	0.929	0.749	0.0124	28.4	28.2	0.323	0.00675	18.7	15.6	13.8	0.506
L76 × 76 × 12.7	0.137	1.77	0.924	0.0175	22.8	23.7	0.924	0.0175	22.8	23.7	14.8	1.000
× 9.5	0.105	1.36	0.733	0.0137	23.2	22.6	0.733	0.0137	23.2	22.6	14.9	1.000
× 7.9	0.0890	1.15	0.628	0.0116	23.4	22.0	0.628	0.0116	23.4	22.0	15.0	1.000
× 6.4	0.0715	0.929	0.516	0.00946	23.6	21.4	0.516	0.00946	23.6	21.4	15.0	1.000
× 4.8	0.0541	0.703	0.400	0.00723	23.9	20.8	0.400	0.00723	23.9	20.8	15.1	1.000
L76 × 64 × 9.5	0.0963	1.24	0.691	0.0133	23.6	24.3	0.433	0.00952	18.7	17.9	13.3	0.676
× 6.4	0.0657	0.845	0.487	0.00919	24.0	23.1	0.309	0.00662	19.1	16.8	13.4	0.684
× 4.8	0.0495	0.643	0.377	0.00705	24.2	22.6	0.240	0.00508	19.3	16.2	13.5	0.688
L76 × 51 × 9.5	0.0861	1.12	0.637	0.0128	23.9	26.4	0.226	0.00608	14.2	13.7	10.9	0.428
× 7.9	0.0730	0.942	0.549	0.0109	24.1	25.9	0.196	0.00520	14.4	13.1	11.0	0.435
× 6.4	0.0598	0.768	0.454	0.00888	24.3	25.2	0.163	0.00426	14.6	12.5	11.0	0.440
× 4.8	0.0448	0.582	0.350	0.00680	24.5	24.6	0.128	0.00328	14.8	11.9	11.2	0.446

TABLE A–5(a)   Properties of Structural Steel Pipes: U.S. Customary Units

Nominal Diameter	Outside Diameter	Inside Diameter	Wall Thickness	Weight per ft	Properties			
(in.)	d_o (in.)	d_i (in.)	t (in.)	w (lb/ft)	A (in.^{2})	I (in.^{4})	S (in.^{3})	r (in.)
Standard Weight								
$\frac{1}{2}$	0.840	0.622	0.109	0.85	0.250	0.017	0.041	0.261
$\frac{3}{4}$	1.050	0.824	0.113	1.13	0.333	0.037	0.071	0.334
1	1.315	1.049	0.133	1.68	0.494	0.087	0.133	0.421
$1\frac{1}{4}$	1.660	1.380	0.140	2.27	0.669	0.195	0.235	0.540
$1\frac{1}{2}$	1.900	1.610	0.145	2.72	0.799	0.310	0.326	0.623
2	2.375	2.067	0.154	3.65	1.07	0.666	0.561	0.787
$2\frac{1}{2}$	2.875	2.469	0.203	5.79	1.70	1.53	1.06	0.947
3	3.500	3.068	0.216	7.58	2.23	3.02	1.72	1.16
$3\frac{1}{2}$	4.000	3.548	0.226	9.11	2.68	4.79	2.39	1.34
4	4.500	4.026	0.237	10.79	3.17	7.23	3.21	1.51
5	5.563	5.047	0.258	14.62	4.30	15.2	5.45	1.88
6	6.625	6.065	0.280	18.97	5.58	28.1	8.50	2.25
8	8.625	7.981	0.322	28.55	8.40	72.5	16.8	2.94
10	10.750	10.020	0.365	40.48	11.9	161	29.9	3.67
12	12.750	12.000	0.375	49.56	14.6	279	43.8	4.38
Extra Strong								
$\frac{1}{2}$	0.840	0.546	0.147	1.09	0.320	0.020	0.048	0.250
$\frac{3}{4}$	1.050	0.742	0.154	1.47	0.433	0.045	0.085	0.321
1	1.315	0.957	0.179	2.17	0.639	0.106	0.161	0.407
$1\frac{1}{4}$	1.660	1.278	0.191	3.00	0.881	0.242	0.291	0.524
$1\frac{1}{2}$	1.900	1.500	0.200	3.63	1.07	0.391	0.412	0.605
2	2.375	1.939	0.218	5.02	1.48	0.868	0.731	0.766
$2\frac{1}{2}$	2.875	2.323	0.276	7.66	2.25	1.92	1.34	0.934
3	3.500	2.900	0.300	10.25	3.02	3.89	2.23	1.14
$3\frac{1}{2}$	4.000	3.364	0.318	12.50	3.68	6.28	3.14	1.31
4	4.500	3.826	0.337	14.98	4.41	9.61	4.27	1.48
5	5.563	4.813	0.375	20.78	6.11	20.7	7.43	1.84
6	6.625	5.761	0.432	28.57	8.40	40.5	12.2	2.19
8	8.625	7.625	0.500	43.39	12.8	106	24.5	2.88
10	10.750	9.750	0.500	54.74	16.1	212	39.4	3.63
12	12.750	11.750	0.500	65.42	19.2	362	56.7	4.33

TABLE A–5(b)   Properties of Structural Steel Pipes: SI Units

Nominal Diameter	Outside Diameter	Inside Diameter	Wall Thickness	Weight per ft	Properties			
	d_o	d_i	t	w $\times 10^{-3}$	A $\times 10^{-3}$	I $\times 10^{-6}$	S $\times 10^{-6}$	r
(mm)	(mm)	(mm)	(mm)	(kN/m)	(m²)	(m⁴)	(m³)	(mm)
Standard Weight								
15	21.33	15.80	2.77	0.0124	0.161	0.00708	0.672	6.63
20	26.67	20.93	2.87	0.0165	0.215	0.0154	1.16	8.48
25	33.40	26.64	3.38	0.0245	0.319	0.0362	2.18	10.7
35	42.16	35.05	3.56	0.0331	0.432	0.0812	3.85	13.7
40	48.26	40.89	3.68	0.0397	0.516	0.129	5.34	15.8
50	60.33	52.50	3.91	0.0533	0.690	0.277	9.19	20.0
65	73.03	62.71	5.16	0.0845	1.10	0.637	17.4	24.1
80	88.90	77.93	5.49	0.111	1.44	1.26	28.2	29.5
90	101.60	90.12	5.74	0.133	1.73	1.99	39.2	34.0
100	114.30	102.26	6.02	0.157	2.05	3.01	52.6	38.4
125	141.30	128.19	6.55	0.213	2.77	6.33	89.3	47.8
150	168.28	154.05	7.11	0.277	3.60	11.7	139	57.2
200	219.08	202.72	8.18	0.417	5.42	30.2	275	74.7
250	273.05	254.51	9.27	0.591	7.68	67.0	490	93.2
300	323.85	304.80	9.53	0.723	9.42	116	718	111
Extra Strong								
15	21.34	13.87	3.73	0.0159	0.206	0.00832	0.787	6.4
20	26.67	18.85	3.91	0.0214	0.279	0.0187	1.39	8.2
25	33.40	24.31	4.55	0.0317	0.412	0.0441	2.64	10.3
35	42.16	32.46	4.85	0.0438	0.568	0.101	4.77	13.3
40	48.26	38.10	5.08	0.0530	0.690	0.163	6.75	15.4
50	60.33	49.25	5.54	0.0732	0.955	0.361	12.0	19.5
65	73.03	59.00	7.01	0.112	1.45	0.799	22.0	23.7
80	88.90	73.66	7.62	0.150	1.95	1.62	36.5	29.0
90	101.60	85.45	8.08	0.182	2.37	2.61	51.5	33.3
100	114.30	97.18	8.56	0.219	2.85	4.00	70.0	37.6
125	141.30	122.25	9.53	0.303	3.94	8.62	122	46.7
150	168.28	146.33	10.98	0.417	5.42	16.9	200	55.6
200	219.08	193.68	12.70	0.633	8.26	44.1	402	73.2
250	273.05	247.65	12.70	0.799	10.4	88.2	646	92.2
300	323.85	298.45	12.70	0.954	12.4	151	929	110

TABLE A–6(α) Properties of Structural Timber: U.S. Customary Units

Nominal Size (in.)	Standard Dressed Size (in.)	Area of Section A (in.^{2})	Moment of Inertia I (in.^{4})	Section Modulus S (in.^{3})	Weight per ft w (lb/ft)
2×4	$1\frac{1}{2} \times 3\frac{1}{2}$	5.25	5.36	3.06	1.46
$\times 6$	$\times 5\frac{1}{2}$	8.25	20.8	7.56	2.29
$\times 8$	$\times 7\frac{1}{4}$	10.9	47.6	13.14	3.02
$\times 10$	$\times 9\frac{1}{4}$	13.9	98.9	21.4	3.85
3×4	$2\frac{1}{2} \times 3\frac{1}{2}$	8.75	8.93	5.10	2.43
$\times 6$	$\times 5\frac{1}{2}$	13.8	34.7	12.6	3.82
$\times 8$	$\times 7\frac{1}{4}$	18.1	79.4	21.9	5.04
$\times 10$	$\times 9\frac{1}{4}$	23.1	165	35.7	6.42
$\times 12$	$\times 11\frac{1}{4}$	28.1	297	52.7	7.81
4×4	$3\frac{1}{2} \times 3\frac{1}{2}$	12.3	12.5	7.15	3.40
$\times 6$	$\times 5\frac{1}{2}$	19.3	48.5	17.6	5.35
$\times 8$	$\times 7\frac{1}{4}$	25.4	111	30.7	7.05
$\times 10$	$\times 9\frac{1}{4}$	32.4	231	49.9	8.93
$\times 12$	$\times 11\frac{1}{4}$	39.4	415	73.8	10.9
$\times 14$	$\times 13\frac{1}{4}$	46.4	678	102	12.9
6×6	$5\frac{1}{2} \times 5\frac{1}{2}$	30.3	76.3	27.7	8.40
$\times 8$	$\times 7\frac{1}{2}$	41.3	193	51.6	11.5
$\times 10$	$\times 9\frac{1}{2}$	52.3	393	82.7	14.5
$\times 12$	$\times 11\frac{1}{2}$	63.3	697	121	17.6
$\times 14$	$\times 13\frac{1}{2}$	74.3	1128	167	20.6
$\times 16$	$\times 15\frac{1}{2}$	85.3	1707	220	23.7
$\times 18$	$\times 17\frac{1}{2}$	96.3	2456	281	26.7
8×8	$7\frac{1}{2} \times 7\frac{1}{2}$	56.3	264	70.3	15.6
$\times 10$	$\times 9\frac{1}{2}$	71.3	536	113	19.8
$\times 12$	$\times 11\frac{1}{2}$	86.3	951	165	24.0
$\times 14$	$\times 13\frac{1}{2}$	101	1538	228	28.1
$\times 16$	$\times 15\frac{1}{2}$	116	2327	300	32.3
$\times 18$	$\times 17\frac{1}{2}$	131	3350	383	36.5
$\times 20$	$\times 19\frac{1}{2}$	146	4634	475	40.6
10×10	$9\frac{1}{2} \times 9\frac{1}{2}$	90.3	679	143	25.1
$\times 12$	$\times 11\frac{1}{2}$	109	1204	209	30.3
$\times 14$	$\times 13\frac{1}{2}$	128	1948	289	35.6
$\times 16$	$\times 15\frac{1}{2}$	147	2948	380	40.9
$\times 18$	$\times 17\frac{1}{2}$	166	4243	485	46.2
$\times 20$	$\times 19\frac{1}{2}$	185	5870	602	51.5
$\times 22$	$\times 21\frac{1}{2}$	204	7868	732	56.7

Note: Properties and weights are for dressed sizes. Weight per unit foot is based on an assumed average weight of 40 lb/ft^3. Moment of inertia and section modulus are about the strong axis.

TABLE A–6(b) Properties of Structural Timber: SI Units

Nominal Size (mm)	Standard Dressed Size (mm)	Area of Section A ($\times 10^{-3}$ m^2)	Moment of Inertia I ($\times 10^{-6}$ m^4)	Section Modulus S ($\times 10^{-3}$ m^3)	Weight per ft w (kN/m)
50 × 100	38.1 × 88.9	3.39	2.23	0.0502	0.0213
× 150	× 140	5.32	8.66	0.124	0.0334
× 200	× 184	7.03	19.8	0.215	0.0441
× 260	× 235	8.97	41.2	0.351	0.0562
80 × 100	63.5 × 88.9	5.65	3.72	0.0836	0.0355
× 150	× 140	8.90	14.4	0.207	0.0557
× 200	× 184	11.7	33.0	0.359	0.0735
× 250	× 235	14.9	68.7	0.585	0.0937
× 300	× 286	18.1	124	0.864	0.114
100 × 100	88.9 × 88.9	7.94	5.20	0.117	0.0496
× 150	× 140	12.5	20.2	0.288	0.0781
× 200	× 184	16.4	46.2	0.503	0.103
× 250	× 235	20.9	96.1	0.818	0.130
× 300	× 286	25.4	173	1.21	0.159
× 360	× 337	29.9	282	1.67	0.188
150 × 150	140 × 140	19.5	31.8	0.454	0.123
× 200	× 191	26.6	80.3	0.851	0.168
× 250	× 241	33.7	164	1.36	0.212
× 300	× 292	40.8	290	1.98	0.257
× 360	× 343	47.9	469	2.74	0.301
× 410	× 394	55.0	710	3.61	0.346
× 460	× 445	62.1	1022	4.601	0.390
200 × 200	191 × 191	36.3	110	1.15	0.228
× 250	× 241	46.0	223	1.85	0.289
× 300	× 292	55.7	396	2.70	0.350
× 360	× 343	65.2	640	3.74	0.410
× 410	× 394	74.8	968	4.92	0.471
× 460	× 445	84.5	1390	6.28	0.533
× 510	× 495	94.2	1929	7.79	0.592
250 × 250	241 × 241	58.3	283	2.34	0.366
× 300	× 292	70.3	501	3.43	0.442
× 360	× 343	82.6	811	4.74	0.519
× 410	× 394	94.8	1230	6.23	0.597
× 460	× 445	107	1770	7.95	0.674
× 510	× 495	119	2440	9.87	0.751
× 560	× 546	132	3270	12.0	0.827

Note: Properties and weights are for dressed sizes. Weight per unit foot is based on an assumed average weight of 6.28 kN/m^3. Moment of inertia and section modulus are about the strong axis.

TABLE A–7(α) Typical Mechanical Properties of Common Materials: U.S. Customary Units

Material	Specific Weight γ (lb/ft^3)	Modulus of Elasticity E ($\times 10^3$ ksi)	Modulus of Rigidity G ($\times 10^3$ ksi)	Yield Strength Tension σ_y (ksi)	Yield Strength Shear τ_y (ksi)	Ultimate Strength Tension $(\sigma_u)_t$ (ksi)	Ultimate Strength Compression $(\sigma_u)_c$ (ksi)	Ultimate Strength Shear τ_u (ksi)	Coefficient of Thermal Expansion α ($\times 10^{-6}$/°F)
Steel:									
ASTM-A36 (carbon)	490	30	12	36	21	58			6.5
ASTM-A441 (alloy)	490	30	12	46		67			6.5
AISI 1020 (hot rolled)	490	30	11.5	30		55			6.5
AISI 1040 (hot rolled)	490	30	11.5	42		76			6.5
Stainless steel (annealed)	490	30	11.6	38	22	85			9.6
Cast Iron:									
Gray cast iron	450	13	6			25	90	32	5.8
Malleable cast iron	460	25	12	33		50	90	48	6.7
Aluminum:									
Alloy 2014-T6	173	10.9	3.9	58	33	66		40	12.8
Alloy 2024-T4	173	10.6		47		68		41	12.9
Alloy 6061-T6	169	10.1	3.7	35	20	38		24	13.1
Copper:									
Annealed	556	17	6.4	10		32		22	9.4
Hard-drawn	556	17	6.4	53		57		29	9.4
Alloys:									
Magnesium alloy	110	6.5	2.4	22		40		21	14.0
Titanium alloy	275	16.5	6.5	120		130		100	5.3
Timber:									
Douglas fir	30	1.9				15	7.2	1.1	
Westen white pine	31	1.7					5.0	1.4	
Southern pine	36	1.8					8.4	1.5	
White oak	43	1.8					7.4	2.0	
Red oak	41	1.8					6.8	1.8	
Western hemlock	28	1.6				13	7.2	1.3	
California redwood	26	1.3				9.6	6.1	0.9	
Concrete:									
Medium strength	150	3.6					4.0		5.5
High strength	150	4.5					6.0		5.5

TABLE A–7(b)   Typical Mechanical Properties of Common Materials: SI Units

Material	Specific Weight γ (kN/m³)	Modulus of Elasticity E (GPa)	Modulus of Rigidity G (GPa)	Yield Strength Tension σ_y (MPa)	Yield Strength Shear τ_y (MPa)	Ultimate Strength Tension $(\sigma_u)_t$ (MPa)	Ultimate Strength Compression $(\sigma_u)_c$ (MPa)	Ultimate Strength Shear τ_u (MPa)	Coefficient of Thermal Expansion α ($\times 10^{-6}$/°C)
Steel:									
ASTM-A36 (carbon)	77	210	83	250	145	400			12
ASTM-A441 (alloy)	77	210	83	320		460			12
AISI 1020 (hot rolled)	77	210	79	210		380			12
AISI 1040 (hot rolled)	77	210	79	290		520			12
Stainless steel (annealed)	77	210	80	260	152	590			17
Cast Iron:									
Gray cast iron	71	90	41			170	620	221	10
Malleable cast iron	72	170	83	230		350	620	331	12
Aluminum:									
Alloy 2014-T6	27	75	27	400	228	460		276	23.0
Alloy 2024-T4	27	73		320		470		283	23.2
Alloy 6061-T6	27	70	26	240	138	260		165	23.6
Copper:									
Annealed	87	120	44	69		220		152	17
Hard-drawn	87	120	44	370		390		200	17
Alloys:									
Magnesium alloy	17	45	17	150		280		145	25.2
Titanium alloy	43	110	45	830		900		690	9.5
Timber:									
Douglas-fir	4.7	13				100	50	7.6	
Westen white pine	4.9	12					35	9.7	
Southern pine	5.7	12					58	10	
White oak	6.8	12					51	14	
Red oak	6.4	12					47	12	
Western hemlock	4.4	11				90	50	9.0	
California redwood	4.1	9.0				66	42	6.2	
Concrete:									
Medium strength	24	25					28		9.9
High strength	24	31					41		9.9

CHAPTER 1

1–1 (a) A rigid body is a solid body in which all the points in the body remain in fixed positions relative to each other; that is, a rigid body does not deform.

(b) In statics and dynamics, all bodies are considered rigid because a small deformation, if any, is negligible in the static and dynamic analyses.

(c) Deformations of bodies are important in the study of strength of materials for two reasons: (1) the amount of deformation (even if small) is required in designing a structural member, (2) the deformation condition is needed in solving statically indeterminate problems.

1–3 The characteristics of a vector quantity are magnitude, direction, line of action, and point of application.

1–5 A concurrent coplanar force system

1–7 Mass is a measure of a particle's (or a body's) inertia, i.e., its resistance to a change of motion. A body of greater mass has greater resistance to a change of motion, and hence less acceleration is caused by a given force.

1–9 No, a force acting on a rigid body can be considered to act anywhere along its line of action, according to the principle of transmissibility.

1–11 In a gravitational system, the unit of force (or weight) that is dependent on the gravitational attraction is chosen as one of the base units. In an absolute system, the unit of mass that is independent of the gravitational attraction is chosen as one of the base units.

1–13 15.53 slug

1–15 101.9 kg

1–17 10.19 kg

1–19 (a) 6.38×10^6 kg   (b) 9×10^5 m   (c) 37.6 Mg   (d) 0.070 m   (e) 23.4 kN

1–21 22.78 mph

1–23 3960 mi

1–25 (a) 271 N · m   (b) 96.5 km/h   (c) 74.6 kW

1–27 1.30 acres

1–29 4000 ft

1–31 25 lb

1–33 4.72 mm

1–35 28.0 m/s

1–37 $a = 402$ mm, $b = 573$ mm

1–39 468 ft

1–41 $a = 13.1$ m, $b = 28.9$ m

1–43 $C = 80°, b = 158$ mm, $c = 171.7$ mm

1–45 $a = 10.7$ m, $B = 52.5°, C = 82.5°$

1–47 $A = 24.8°, B = 55.1°, C = 100.1°$

1–49   138.9 mi

1–51 $x = 2.04, y = 1.61$

1–53 $P = 405.3$ lb, $Q = 9064$ lb, $R = -12\ 560$ lb

1–55 $x = 1.5, y = -2.5$

1–57 $T = 257$ lb, $P = 69.5$ lb

1–59 $x = 9.00$ kN, $y = 7.00$ kN, $z = 15.00$ kN

CHAPTER 2

2–1   **(a)** 46 N ∠ 64°   **(b)** 46 N ∠ 64°

2–3   **(a)** 88 N ⦣ 121°   **(b)** 87.4 N ⦣ 120.9°

2–5   500 lb

2–7   2.45 kN ⬈ 40°

2–9 $T_{AB} = 516$ lb, $R = 1202$ lb

2–11   The resultant passes through the point 1.70 ft to the left of D and is within the middle third of the base.

2–13   5.8 kN ⦣ 99°

2–15 $F_x = 352$ lb ⟵ , $F_y = 296$ lb ↓

2–17 $F_x = 143.4$ N ⟶ , $F_y = 205$ N ↓

2–19 $F_x = 8$ kips ⟶ , $F_y = 6$ kips ↑

2–21 $P_x = 13.42$ kips ⟶ , $P_y = 6.71$ kips ↑

  $Q_x = 4.47$ kips ⟶ , $Q_y = 8.94$ kips ↑

2–23 $W_x = 48.6$ lb ↘ , $W_y = -87.4$ lb ↗

2–25   26.0 lb ⟻ 14.5°

2–27   156.1 kN ∠ 75.1°

2–29   20 kN · m ↺

2–31   64.3 lb · ft ↻

2–33   740 N · m ↺

2–35 $a = 90°, (M_B)_{max} = 100$ N · m

2–37   0

2–39   **(a)** 50.0 kN · m ↺   **(b)** 100.0 kN · m ↺

2–41   500 lb · ft ↺

2–43 0.6 kN · m $\circlearrowleft$

2–45   0

2–47   0 (The system of forces is balanced.)

2–49   10 kips $\downarrow$, 10 kip · ft $\circlearrowright$

2–51   600 lb $\downarrow$, 400 lb · ft $\circlearrowright$

2–53   20 lb $\rotatebox{70}{\downarrow}$ 70°, 14.1 lb · ft $\circlearrowright$

2–55 $\mathbf{R} = 800$ lb $\angle$ 30°, $d = 2$ in. to the left of O.

2–57 $\mathbf{R} = 4$ kN $\downarrow$ at 150 mm to the left of B.

2–59 52.5 N · m $\circlearrowleft$

2–61 $\mathbf{R} = 100$ lb $\longrightarrow$ at 10.5 ft (above B)

2–63   The resultant passes through A.

2–65   19.9 kN

2–67   8.94 kN $\nwarrow$ 26.6° at 0.913 m above A

2–69   287 N $\searrow$ 141.7° at $x = 0.112$ m and $y = 0.141$ m

2–71   114 kN $\swarrow$ 51.6° at 0.812 m above point C

2–73   The resultant passes through the point 1.81 ft to the right of A and is within the middle third of the base.

2–75   94 lb $\downarrow$ at 3.21 ft to the right of A

2–77   16 kips $\downarrow$ at 7.13 ft to the right of A

2–79   2150 lb $\downarrow$ at 7.98 ft to the right of A

2–81   19.55 kips $\downarrow$ at 6.81 ft to the right of A

2–83   14.15 kN $\swarrow$ 238.0° at 6.75 m to the right of A

2–85 $a = 5.56$ ft, $b = 1.22$ ft

2–87   The resultant passes through the point 1.61 ft to the right of A and is within the middle third of the base.

CHAPTER 3

3–11   18.8 N

3–13 $R_A = 647$ N, $R_B = 256$ N

3–15 $T_{AB} = 272$ N, $T_{AC} = 254$ N, $T_{AD} = 294$ N

3–17 $T_{AB} = 4.11$ kN, $T_{AC} = 3.23$ kN

3–19 $T_{AB} = 1430$ N, $F_{AC} = 2200$ N

3–21   60 lb

3–23   491 N

3–25 $R_A = 13.32$ lb, $R_B = 30.0$ lb

3–27 $\alpha = 14.48°$, $\beta = 61.0°$

3–29 $x_{AB} = 0.32$ m, $x_{AC} = 0.3$ m

3–31 (a) $R_A = 30.5$ lb ⬐ $79.7°$, $R_B = 5.45$ lb ←

(b) $B_x = 5.47$ lb ← , $A_x = 5.47$ lb → , $A_y = 30$ lb ↑

3–33 (a) **A** $= 10.0$ kips ↑ , **C** $= 17.9$ kips ⬊ $63.4°$

(b) $A_y = 10$ kips ↑ , $C_x = 8$ kips → , $C_y = 16$ kips ↓

3–35 (a) **P** $= 14.4$ lb ← , $\mathbf{R}_A = 52.0$ lb ⬐ $73.9°$

(b) **P** $= 14.4$ lb ← , $A_x = 14.4$ lb → , $A_y = 50$ lb ↑

3–37 (a) $\mathbf{R}_A = 750$ N ↑ , $\mathbf{R}_B = 661$ N ⬊ $49.1°$

(b) $A_y = 750$ N ↑ , $B_x = 433$ N → , $B_y = 500$ N ↓

3–39 (a) $\mathbf{R}_A = 896$ lb ⬈ $26.6°$, $\mathbf{R}_D = 1133$ lb ⬋ $45°$

(b) $A_x = 800$ lb ← , $A_y = 400$ lb ↓ , $\mathbf{R}_D = 1130$ lb ⬋ $45°$

3–41 (a) $\mathbf{R}_A = 3.62$ kN ⬊³⁄₂ , $\mathbf{R}_D = 5.00$ kN ⬊³⁄₄

(b) $\mathbf{R}_D = 5.00$ kN ⬊³⁄₄ , $A_x = 2.00$ kN → , $A_y = 3.00$ kN ↓

3–43 $\mathbf{R}_D = 19.8$ kips ⬈¹⁄₁ , $A_x = 14.0$ kips ← , $A_y = 4.00$ kips ↓

3–45 $T = 28.33$ kN, $B_x = 22.7$ kN ← , $B_y = 7$ kN ↑

3–47 $B_x = 48.0$ kips ← , $B_y = 12.0$ kips ↑ , $A_x = 38.0$ kips → , $A_y = 15.3$ kips ↑

3–49 $B_x = 0$, $B_y = 6.5$ kips ↑ , $A_y = 8.5$ kips ↑

3–51 $A_x = 98.0$ kN → , $B_x = 98.0$ kN ← , $B_y = 39.2$ kN ↑

3–53 $A_x = 0$, $A_y = 6.84$ kips ↑ , $B_y = 13.17$ kips ↑

3–55 $D_x = 30$ lb → , $D_y = 25$ lb ↓ , $B_y = 185$ lb ↑

3–57 $A_x = 0$, $A_y = 3.51$ kN ↑ , $B_y = 6.29$ kN ↑

3–59 $\mathbf{R}_D = 6870$N ⬋ $30°$, $A_x = 5950$ N ← , $A_y = 491$ N ↓

3–61 200 mm

3–63 $T = 143$ N, $A_x = 71.5$ N ← , $A_y = 464$ N ↑

CHAPTER 4

4–1 $F_{AB} = 120$ lb (T), $F_{AC} = 179$ lb (T), $F_{BC} = 200$ lb (C)

4–3 $F_{AB} = 250$ lb (C), $F_{AC} = 600$ lb (T), $F_{BC} = 600$ lb (T), $F_{CD} = 600$ lb (T), $F_{BD} = 750$ lb (C)

4–5 $F_{AB} = 5$ kips (T), $F_{AC} = 4$ kips (C), $F_{BC} = 3$ kips (T), $F_{CE} = 4$ kips (C), $F_{BE} = 10$ kips (C), $F_{BD} = 12$ kips (T), $F_{DE} = 6$ kips (T)

4–7 $F_{AB} = 0$, $F_{AC} = 20.6$ kips (T), $F_{AD} = 35.4$ kips (T), $F_{BD} = 45$ kips (C), $F_{CD} = 10$ kips (C), $F_{CE} = 0$, $F_{CF} = 25$ kips (T), $F_{DF} = 20$ kips (C), $F_{EF} = 0$

4–9 $F_{AB} = 15$ kN (C), $F_{AC} = 9$ kN (T), $F_{BC} = 0$, $F_{BD} = 18$ kN (C), $F_{BE} = 15$ kN (T), $F_{CE} = 9$ kN (T), $F_{DF} = 18$ kN (C), $F_{DE} = 0$, $F_{EF} = 5$ kN (T), $F_{EG} = 15$ kN (T), $F_{FG} = 16$ kN (T), $F_{FH} = 25$ kN (C), $F_{GH} = 15$ kN (T)

4–11 $F_{AB} = F_{CE} = F_{EF} = 0$

4–13 $F_{AC} = F_{CD} = F_{GH} = F_{GI} = 0$

4–15 $F_{BC} = F_{CD} = F_{DG} = F_{FG} = F_{IJ} = F_{JK} = F_{KL} = F_{KN} = F_{MN} = 0$

4–17 $F_{FH} = 28.5$ kips (C), $F_{FI} = 0$, $F_{GI} = 27$ kips (T)

4–19 $F_{BD} = 11.2$ kips (T), $F_{CD} = 14.1$ kips (C), $F_{CE} = 0$

4–21 $F_{DF} = 76.5$ kN (C), $F_{DG} = 26.0$ kN (T), $F_{FG} = 58.3$ kN (T)

4–23 $F_{DF} = 510$ kN (C), $F_{DG} = 80.2$ kN (T), $F_{EG} = 467$ kN (T)

4–25 $F_{AB} = 63.6$ kN (C), $F_{AC} = 45$ kN (T), $F_{BC} = 0$, $F_{BD} = 67.1$ kN (C), $F_{BE} = 21.2$ kN (T), $F_{CE} = 45$ kN (T), $F_{DE} = 60$ kN (T), $F_{DF} = 67.1$ kN (C), $F_{EF} = 21.2$ kN (C), $F_{EG} = 75$ kN (T), $F_{FG} = 60$ kN (T), $F_{FH} = 106.1$ kN (C), $F_{GH} = 75$ kN (T)

4–27 $F_{AB} = 33.3$ kips (C), $F_{AD} = 27.4$ kips (T), $F_{BC} = 51$ kips (C), $F_{BD} = 30.0$ kips (T), $F_{CD} = 20$ kips (C), $F_{CE} = 51$ kips (C), $F_{DE} = 34.9$ kips (T), $F_{DF} = 23.2$ kips (T), $F_{EF} = 36.6$ kips (C)

4–29 $F_{BD} = 287$ lb (C), $R_{Cx} = 25$ lb $\longleftarrow$, $R_{Cy} = 143.3$ lb $\downarrow$

4–31 $R_{Ax} = 1.05$ kN $\longleftarrow$, $R_{Ay} = 2$ kN $\uparrow$, $M_A = 3.50$ kN $\cdot$ m $\circlearrowright$, $F_{BD} = 5$ kN (C), $C_y = 1$ kN, $C_x = 4$ kN

4–33 $R_{Ay} = 7.67$ kN $\uparrow$, $R_{By} = +32.33$ kN $\uparrow$, $R_{Ax} = 0.28$ kN $\longrightarrow$, $R_{Bx} = 10.28$ kN $\longleftarrow$

4–35 $T = 50$ lb, $C_x = 43.3$ lb, $C_y = 35$ lb, $B_y = 92.4$ lb $\uparrow$, $M_B = 440$ lb $\cdot$ ft $\circlearrowright$

4–37 $A_x = \frac{9}{8}$ kip $\longleftarrow$, $A_y = \frac{1}{6}$ kip $\uparrow$, $B_x = 1$ kip, $B_y = 1$ kip, $D_x = 1$ kip, $D_y = 1$ kip, $C_x = \frac{9}{8}$ kip, $C_y = \frac{11}{6}$ kip

4–39 1.47 kN

4–41 $P = 11.25$ lb, $E_x = 30$ lb, $E_y = 63.2$ lb

4–43 $M = 0.2$ kN $\cdot$ m $\circlearrowright$

4–45 1613 lb (T)

CHAPTER 5

5–1   The block is at rest, $F = 141$ N.

5–3   182 N

5–5   170.7 lb

5–7   16.7°

5–9   54.1 N $\angle$ 16°

5–11   64.2 kg

5–13   883 N

5–15   12.9 ft

5–17   The file cabinet will tip at $P = 15.0$ lb.

5–19   **(a)** $P = 335$ N   **(b)** $h = 934$ mm

5–21   369 N

5–23   1928 N

5–25 $R_1' = R_2' = 628$ lb

5–27   99.5 lb

5–29   **(a)** $W = 323$ lb   **(b)** $M' = 7.51$ lb · in.

5–31   603 lb · in.

5–33   205 lb · in.

5–35   62.4 lb

5–37   0.228

5–39   7.90 lb

5–41   88.7 lb · ft $\circlearrowright$

5–43   185 N · m $\circlearrowright$

5–45   3050 lb · in.

5–47   Yes. A car with fully inflated tires would have better gas mileage because fully inflated tires deform less and have a smaller coefficient of rolling resistance.

5–49   **(a)** $P_1 = 0.035$ W   **(b)** $P_2 = 0.0025$ W   **(c)** $P_3 = 0.0015$ W

5–51   471 N

CHAPTER 6

6–1 $F_x = +137$ lb, $F_y = +257$ lb, $F_z = -274$ lb

6–3 $F_x = 86.6$ lb, $F_y = 42.8$ lb, $F_z = 25.9$ lb

6–5 $F = 2600$ N, $\theta_x = 76.7°$, $\theta_y = 72.1°$, $\theta_z = 157.4°$

6–7 $F = 481$ N, $\theta_x = 109.4°$, $\theta_y = 51.4°$, $\theta_z = 135.0°$

6–9 $(T_{AC})_x = +700$ N, $(T_{AC})_y = -400$ N, $(T_{AC})_z = +400$ N

6–11 $T_x = -136.8$ lb, $T_y = 171.0$ lb, $T_z = -205.2$ lb

6–13 $T_x = -20.2$ lb, $T_y = 28.5$ lb, $T_z = 35.8$ lb

6–15 $R_x = -6$ kN, $R_y = 6$ kN, $R_z = -2$ kN

6–17 $R_x = 3$ kN, $R_y = 24$ kN, $R_z = 4$ kN

6–19 $F_{AB} = 1020$ lb (C), $T_{BC} = 540$ lb (T), $T_{BD} = 480$ lb (T)

6–21 $F_{AB} = 20$ kN (C), $T_{BD} = 11.0$ kN (T), $T_{BC} = 21.0$ kN (T)

6–23 $T_{BC} = 9.00$ kN (T), $T_{BD} = 7.00$ kN (T), $F_{AB} = 15.0$ kN (C)

6–25 $T_{AD} = 422.5$ lb (T), $T_{AC} = 366.2$ lb (T), $T_{AB} = 211.3$ lb (T)

6–27 $F_{AB} = 123$ lb (C), $T_{AC} = 12.2$ lb (T), $T_{AD} = 80.8$ lb (T)

CHAPTER 7

7–1 2.62 in.

7–3 53.5 mm

7–5 46.4 mm

7–7 $\bar{x} = 201$ mm, $\bar{y} = 98.5$ mm, $\bar{z} = 20$ mm

7–9 $\bar{x} = 1.50$ m, $\bar{y} = 0.520$ m

7–11 $\bar{x} = 1.68$ m, $\bar{y} = 0.659$ m

7–13 $\bar{x} = 762$ mm, $\bar{y} = 308$ mm

7–15 $\bar{x} = 0, \bar{y} = -0.750$ in.

7–17 $\bar{x} = 0, \bar{y} = 4.30$ in.

7–19 $\bar{x} = 0, \bar{y} = 4.84$ in.

7–21 $\bar{x} = 0, \bar{y} = 140$ mm

7–23 **(a)** $P = 13.88$ N   **(b)** $(\mu_s)_{min} = 0.141$

7–25 $R_{Ax} = 0, R_{Ay} = 4$ kN $\uparrow$, $M_A = 4$ kN $\cdot$ m $\circlearrowright$

7–27 $R_B = 4250$ N $\uparrow$, $R_C = 1750$ N $\uparrow$

7–29 $d = 0.921$ m. The resultant does not act through the middle third of the base and the dam is not safe.

7–31 117.6 kN

7–33 5.20 m

7–35 $T_{min} = 5014$ kips, $T_{max} = 5362$ kips, length = 972 ft

7–37 2230 kips $\searrow$ 33.1°

7–39 **(a)** $T_{max} = 1364$ kN   **(b)** $s_{AB} = 100.5$ m

7–41 0.5 ft

CHAPTER 8

8–1 $\bar{r}_x = \dfrac{h}{\sqrt{12}}, \bar{r}_y = \dfrac{b}{\sqrt{12}}$

8–3 $I_x = 8340$ in.^{4}, $r_x = 10.3$ in.

8–5 8.33×10^{-5} m^4

8–7 30 920 in.^{4}

8–9 $\bar{r}_x = 3.00$ in., $r_y = 3.80$ in.

8–11 136 in.^{4}

8–13 820 in.^{4}

8–15 7.39×10^{-4} m^4

8–17 2.57×10^{-5} m^4

8–19 $I_y = 832$ in.^{4}, $r_y = 3.79$ in.

8–21 5.11×10^{-3} m^4

8–23 689 in.^{4}

8–25 7.971×10^{-5} m^4

8–27 $\bar{I}_x = 3.11 \times 10^{-4}$ m^4, $\bar{r}_x = 115.3$ mm

8–29 $\bar{I}_x = 2.33 \times 10^{-3}$ m^4, $\bar{r}_x = 221$ mm

8–31 $\bar{I}_x = 2840$ in.^{4}, $\bar{r}_x = 7.21$ in.

8–33 $\bar{I}_y = 936$ in.^{4}, $\bar{r}_y = 5.11$ in.

CHAPTER 9

9–1 20 370 psi

9–3 $\sigma_{AB} = 15.92$ ksi (T), $\sigma_{BC} = 3.18$ ksi (T), $\sigma_{CD} = 6.37$ ksi (C)

9–5 $\sigma_{AB} = 49.5$ MPa (C), $\sigma_{BC} = 47.7$ MPa (T), $\sigma_{CD} = 63.7$ MPa (T)

9–7 Use $d = 1\frac{9}{16}$ in.

9–9 318 MPa (T)

9–11 $\sigma_{AB} = 1000$ psi (C), $\sigma_{BC} = 20\,000$ psi (T)

9–13 2010 kN

9–15 $(A_{BD})_{req} = 383$ mm^2, $(A_{BE})_{req} = 383$ mm^2, $(A_{CE})_{req} = 343$ mm^2

9–17 (a) $\tau = 95.5$ MPa   (b) $\sigma_b = 125$ MPa

9–19 (a) $\tau = 135.7$ MPa   (b) $\sigma_b = 272$ MPa

9–21 (a) $\tau = 14.4$ ksi   (b) $\sigma_b = 28.8$ ksi

9–23 (a) $P = 37.7$ kips   (b) $\tau = 12.0$ ksi

9–25 Use $d = 1$ in.

9–27 Use $d = 2\frac{1}{16}$ in.

9–29 (a) Use $d_{bar} = \frac{13}{16}$ in.   (b) Use $d_{pin} = \frac{3}{4}$ in.

9–31 $\sigma_\theta = 5$ ksi, $\tau_\theta = 8.67$ ksi

9–33 57.8 kN

9–35 3.55 kN

9–37 8 mm

9–39 4 mm

CHAPTER 10

10–1 (a) $\epsilon = 6.67 \times 10^{-4}$ (b) $\delta = 0.08$ in. (elongation)

10–3 $\sigma = 15.3$ ksi (T), $\epsilon = 0.000\,527$, elongation length $= 20.011$ ft

10–5 100.003 ft

10–7   0.040 in. (elongation)

10–9   0.0529 in. (elongation)

10–11 0.782 mm (elongation)

10–13 0.530 in.

10–15 $\sigma_{AC} = 10.25$ ksi (C), $\sigma_{CB} = 12.83$ ksi (T)

10–17 $\sigma_{al} = 170$ MPa, $\sigma_{br} = 36.0$ MPa

10–19 $\sigma_1 = 13.0$ ksi (T), $\sigma_2 = 7.00$ ksi (T)

10–21 $T_{CD} = 24$ kN, $T_{EF} = 48$ kN

10–23 3890 psi

10–25 $\sigma_{br} = 28.8$ MPa (T), $\sigma_{st} = 115$ MPa (T)

10–27 $\sigma_{st} = 5160$ psi (T), $\sigma_{al} = 4130$ psi (C)

10–29 2.07×10^{-4} in. (contraction)

CHAPTER 11

11–1   Elastic deformation is the deformation that can be recovered fully once the load is removed. Plastic deformation is permanent and cannot be recovered after the load is removed.

11–3   Necking is a drastic decrease in diameter at a localized area. Necking usually occurs a little beyond the ultimate strength of ductile materials.

11–5   For bars of the same length and subjected to the same stress, the one with a lower value of modulus of elasticity stretches more. Hence, the aluminum bar, which has a lower value of modulus of elasticity than that of copper, stretches more.

11–7   For ductile materials, the modulus of elasticity E and the yield strength σ_y obtained from the tension test and compression test are the same. The tension test provides more information, such as ultimate strength, after the material has yielded. Tensile strength for ductile materials is utilized more than compressive strength. The tensile test is therefore more important than the compression test for ductile materials.

11–9   **(a)** $\sigma_u = 77\,400$ psi   **(b)** % elongation = 29.5%
          (c) % reduction in area = 28.4%

11–11 **(a)** $\sigma_p = 30$ ksi   **(b)** $E = 15 \times 10^3$ ksi   **(c)** $(\sigma_y)_{0.2\%} = 42$ ksi
          (d) $\sigma_u = 56$ ksi   **(e)** % elongation = 39%
          (f) % reduction in area = 32.6%

11–13 Use $d = \frac{7}{16}$ in.

11–15 Use $1\frac{3}{8}$ in. $\times$ $1\frac{3}{8}$ in. section

11–17 130.8°

11–19 Due to the lack of yield point in brittle materials, additional load causes continuous increase of the maximum stress at the point of stress concentration. The material will start to crack when σ_u is reached. Therefore, stress concentration is an important design factor for brittle materials, even when the member is subjected to static load.

11–21 (a) 11.9 ksi (b) 12.5 ksi (c) 17.0 ksi

11–23 15.8 kips

11–25 The straight link is 60% stronger than the link with an enlarged section.

11–27 (a) $(P_{allow})_{elastic} = 31.9$ kN (b) $(P_{allow})_{plastic} = 54.1$ kN

CHAPTER 12

12–1 $T_{AB} = +2$ kN · m, $T_{BC} = -3$ kN · m

12–3 $T_{AB} = -500$ N · m, $T_{BC} = -800$ N · m

12–5 $T_{AB} = +4500$ lb · in., $T_{BC} = +6500$ lb · in., $T_{CD} = +1500$ lb · in.

12–7 7.96 ksi

12–9 9.31 ksi

12–11 $\tau_{min} = 6000$ psi

12–13 6.52 MPa

12–15 24.5 kip · in.

12–17 % reduction of torsional strength = 19.8%, % reduction of weight = 44.4%

12–19 Use $d = 54$ mm

12–21 Use $d_o = 2\frac{13}{16}$ in., $d_i = 2\frac{1}{16}$ in.

12–25 210 lb · in.

12–27 5990 psi

12–29 25 hp

12–31 24.4 kW

12–33 46.9 MPa

12–35 8210 psi

12–37 0.267°

12–39 +1.14°

12–41 −1.59°

12–43 Use $d = 73$ mm

12–45 Use $d = 73$ mm

CHAPTER 13

13–1 $R_A = 7.5$ kips $\uparrow$, $R_B = 10.5$ kips $\uparrow$

13–3 $R_A = 8.5$ kN $\uparrow$, $R_B = 0.5$ kN $\uparrow$

13–5 $M_B = 15$ kN · m $\circlearrowright$, $R_B = 15$ kN $\uparrow$

13–7 $V_{1-1} = -10$ kN, $V_{2-2} = -20$ kN, $V_{3-3} = -20$ kN
$M_{1-1} = -5$ kN · m, $M_{2-2} = -20$ kN · m, $M_{3-3} = -40$ kN · m

13–9 $V_{1-1} = 10$ kN, $V_{2-2} = 7$ kN, $V_{3-3} = 7$ kN
$M_{1-1} = M_{2-2} = -19$ kN · m, $M_{3-3} = -12$ kN · m

13–11 $V_{1-1} = 4800$ lb, $V_{2-2} = 1800$ lb, $V_{3-3} = 450$ lb
      $M_{1-1} = -20\ 100$ lb · ft, $M_{2-2} = -3600$ lb · ft, $M_{3-3} = -450$ lb · ft

13–13 $V_{A^-} = 0$, $V_{A^+} = 6.8$ kips, $V_{B^-} = +6.8$ kips, $V_{B^+} = -1.2$ kips,
      $V_C = V_D = -1.2$ kips, $V_E = -2.2$ kips, $V_{F^-} = -3.2$ kips, $V_{F^+} = 0$

13–15 $V_A = 0$, $V_B = -1$ kip, $V_C = -4$ kips, $V_D = V_{E^-} = -9$ kips,
      $V_{E^+} = V_{F^-} = -14$ kips, $V_{F^+} = 0$
      $M_A = 0$, $M_B = -0.667$ kip · ft, $M_C = -5.33$ kip · ft, $M_D = -18$ kip · ft,
      $M_E = -36$ kip · ft, $M_F = -64$ kip · ft

13–17 $V_{max} = 12$ kN, $V_{min} = -11$ kN, $M_{max} = 3.125$ kN · m, $M_{min} = -12$ kN · m

13–19 $V_{max} = 25$ kN, $V_{min} = -55$ kN, $M_{max} = 37.8$ kN · m, $M_{min} = -20$ kN · m

13–21 $V_{max} = 24$ kN, $V_{min} = -18$ kN, $M_{max} = 6$ kN · m, $M_{min} = -18$ kN · m

13–23 $V_{max} = 10$ kN, $V_{min} = -8.4$ kN, $M_{max} = 7.22$ kN · m, $M_{min} = -10$ kN · m

13–25 $V_{max} = 26$ kN, $V_{min} = -26$ kN, $M_{max} = 24.5$ kN · m

13–27 $V_{max} = \frac{1}{2}wL$, $V_{min} = -\frac{1}{2}wL$, $M_{max} = \frac{1}{8}wL^2$

13–29 $V_{max} = 40$ kN, $V_{min} = -32$ kN, $M_{max} = 20$ kN · m, $M_{min} = -40$ kN · m

13–31 $V_{max} = 20$ kN, $V_{min} = -25$ kN, $M_{max} = 5.63$ kN · m, $M_{min} = -10$ kN · m

13–33 $V_{max} = 99$ kN, $V_{min} = -41$ kN, $M_{max} = 123$ kN · m, $M_{min} = -15$ kN · m

13–35 $V_{max} = 50$ kN, $V_{min} = -50$ kN, $M_{max} = 50$ kN · m, $M_{min} = -100$ kN · m

13–37 $V_{max} = 10.2$ kips, $V_{min} = -14.8$ kips, $M_{max} = 18.0$ kip · ft, $M_{min} = -16$ kip · ft

13–39 $|V_{max}| = 12$ kips, $|M_{max}| = 56$ kip · ft

13–41 $V_{max} = 5$ kips, $M_{max} = 19$ kip · ft

13–43 $|V_{max}| = 270$ N, $|M_{max}| = 230$ N · m

13–45 $V_{max} = 1080$ N, $M_{max} = 621$ N · m

CHAPTER 14

14–1   1000 psi

14–3   153 MPa

14–5   23.7 ksi

14–7   87 MPa

14–9   24.5 ksi

14–11   8.26 MPa

14–13 $\sigma_{max}^{(T)} = 2550$ psi, $\sigma_{max}^{(C)} = 2310$ psi

14–15   162 kip · ft

14–17   18.2 kN · m

14–19   135.3 kips

14–21 $\tau_A = 0$, $\tau_B = 66.0$ psi, $\tau_C = 118.8$ psi

14–23 $\tau_A = 294$ psi, $\tau_B = 919$ psi, $\tau_C = 588$ psi

14–25   696 lb/ft

14–27   6.24 MPa

14–29 $\tau_1 = 0$, $\tau_2 = 29.4$ psi, $\tau_2' = 88.2$ psi, $\tau_3 = 91.9$ psi, $\tau_4 = 68.9$ psi, $\tau_5 = 0$

14–31 $(\sigma_{st})_{max} = 116$ MPa, $(\sigma_{wd})_{max} = 5.38$ MPa

14–33 $M_{allow} = 44.3$ kN · m

14–35 $w_{allow} = 1.34$ kip/ft

14–37 $(\sigma_{cn})_{max} = 1.30$ ksi (C),$(\sigma_{st})_{max} = 20.2$ ksi (T)

14–39 **(a)** $x = 6$ in.   **(b)** $A_{st} = 1.80$ in.²   **(c)** $w_{allow} = 1503$ lb/ft

14–41 $\sigma_{max} = 627$ psi, $\tau_{max} = 305$ psi

14–43 $\sigma_A = 0$, $\tau_A = 1.52$ ksi, $\sigma_B = 8.42$ ksi (C), $\tau_B = 1.45$ ksi

14–45 124 kN/m

14–47 **(a)** Allowable linear load = 147 lb/ft   **(b)** Allowable live load = 66 lb/ft²

14–49 **(a)** $M_y = 498$ kip · ft   **(b)** $M_y = 633$ kip · ft
(c) $w_y = 4.43$ kip/ft, $w_p = 5.63$ kip/ft

14–51 **(a)** $M_y = 95.5$ kN · m   **(b)** $M_y = 172.5$ kN · m
(c) $w_y = 5.31$ kN/m, $w_p = 9.58$ kN/m

14–53 149 kN

CHAPTER 15

15–1 Use W18 × 60

15–3 Use W12 × 22

15–5 Use W14 × 43

15–7 Use W460 × 0.88

15–9 Use W12 × 22

15–11 Use a 10 × 14 rectangular section

15–13 Use a 150 × 410 rectangular section

15–15 Use a 6 × 12 rectangular section

15–17 Use a 200 × 250 rectangular section

15–19 Use a 4 × 14 Southern pine section

15–21 $w_{allow} = 278$ lb/ft

15–23 Use $\frac{3}{4}$ in. pitch

15–25 Use 6 in. pitch for the AB portion and 3 in. pitch for the BC portion

15–27 Use W18 × 46

15–29 Use W410 × 0.38

15–31 Use W300 × 0.32

CHAPTER 16

16–1 31.25 ksi

16–3 26.0 ksi

16–5 3.26 ft

16–7 0.868 in.

16–9 4.46 mm

16–11 $\delta_{max} = 27.0$ mm, $\theta_{max} = 0.688°$

16–13 1.73 kips/ft

16–15 0.431 in. $\downarrow$

16–17 $\delta_B = 3.40$ mm $\downarrow$, $\delta_C = 9.95$ mm $\downarrow$

16–19 11.3 mm $\downarrow$

16–21 0.216 in. $\downarrow$

16–23 9.36 mm $\downarrow$

16–25 6.62 mm $\downarrow$

16–27 0.337 in. $\downarrow$

16–29 $\delta_C = 0.209$ in. $\downarrow$, $\delta_D = 0.0725$ in. $\uparrow$

16–31 $\theta_{max} = \dfrac{Pa^2}{2EI}$, $\delta_{max} = \dfrac{Pa^2}{6EI}(3L - a)$

16–33 $\theta_{max} = \dfrac{PL^2}{16EI}$, $\delta_{max} = \dfrac{PL^3}{48EI}$

16–35 0.543 in.

16–37 7.14 mm

16–39 $\delta_C = 3.92$ mm $\uparrow$, $\delta_D = 10.5$ mm $\downarrow$

16–41 $M_B = -80$ kN · m, $M_A = -290$ kN · m

16–43 $M_C = 4$ kip · ft

16–45 $M_A = -10$ kN · m

16–47 $M_A = -30$ kip · ft

16–49 0.826 in. $\downarrow$

16–51 8.54 mm $\downarrow$

16–53 0.216 in. $\downarrow$

16–55 0.320 in. $\downarrow$

16–57 0.186 in. $\downarrow$

16–59 $\delta_C = 0.142$ in. $\uparrow$, $\delta_D = 0.337$ in. $\downarrow$

16–61 $\delta_C = 0.209$ in. $\downarrow$, $\delta_D = 0.0730$ in. $\uparrow$

CHAPTER 17

17–1 $R_B = 3.38$ kips $\uparrow$, $R_A = 6.63$ kips $\uparrow$, $M_A = 14.63$ kip · ft $\circlearrowright$

17–3 $R_B = 32.25$ kN $\uparrow$, $R_A = 1.25$ kN $\downarrow$, $M_A = 3.5$ kN · m $\circlearrowright$

17–5 $R_B = \frac{5}{16}P \uparrow$, $R_A = \frac{11}{16}P \uparrow$, $M_A = \frac{3}{16}PL$

17–7 $R_B = 6.80$ kips $\uparrow$, $R_A = 2.80$ kips $\uparrow$, $M_A = 1.60$ kip · ft $\circlearrowright$

17–9 Use W14 × 43

17–11 $M_A = 2$ kip · ft, $M_B = 4$ kip · ft, $R_A = 2.33$ kips $\uparrow$, $R_B = 6.67$ kips $\uparrow$

17–13 $R_A = R_B = 11$ kN $\uparrow$, $M_A = M_B = 13.5$ kN · m

17–15 Use W200 × 0.31

17–17 $R_B = 7.33$ kips $\uparrow$, $R_A = 2.50$ kips $\downarrow$, $R_C = 1.17$ kips $\uparrow$

17–19 $R_B = R_C = 22.0$ kN $\uparrow$, $R_A = R_D = 8$ kN $\uparrow$

17–21 Use W12 × 22

17–23 $M_B = -18.13$ kip · ft, $R_A = 1.19$ kips $\uparrow$, $R_B = 16.62$ kips $\uparrow$, $R_C = 8.19$ kips $\uparrow$

17–25 $M_B = -21.25$ kN · m, $R_A = 20.75$ kN $\uparrow$, $R_B = 50.5$ kN $\uparrow$, $R_C = 48.75$ kN $\uparrow$

17–27 Use W530 × 0.90

CHAPTER 18

18–1 $\sigma_{max}^{(T)} = 15.5$ ksi, $\sigma_{max}^{(C)} = -14.14$ ksi

18–3 **(a)** $F = 52.8$ kN **(b)** $\sigma_B = 9.39$ MPa (C)

18–5 Use two C9 × 13.4

18–7 $h = 5.06$ ft

18–9 $\sigma_{max}^{(T)} = 10.53$ MPa, $\sigma_{max}^{(C)} = -10.53$ MPa

18–11 $a_{req} = 91.5$ mm

18–13 Use W12 × 30

18–15 $\sigma_A = 600$ psi (C), $\sigma_B = 1200$ psi (T)

18–17 $\sigma_A = 20.7$ ksi (T), $\sigma_B = 17.5$ ksi (C)

18–19 $\sigma_A = 89.1$ MPa (C), $\sigma_B = 114.5$ MPa (T)

18–21 5050 lb

18–23 $\sigma_A = 1.33$ MPa (C), $\sigma_B = 9.330$ MPa (C), $\sigma_C = 1.33$ MPa (C), $\sigma_D = 6.67$ MPa (T)

18–25 49.4 kips

18–27 Diameter of kern = d/4

18–29 $\sigma_\theta = 0$, $\tau_\theta = +50$ MPa

18–31 $\sigma_\theta = +2.83$ ksi, $\tau_\theta = -6.83$ ksi

18–33 $\sigma_\theta = +5.09$ ksi, $\tau_\theta = +9.49$ ksi

18–35 $\sigma_\theta = 0$, $\tau_\theta = +50$ MPa

18–37 $\sigma_\theta = +2.9$ ksi, $\tau_\theta = -6.8$ ksi

18–39 $\sigma_\theta = +5.1$ ksi, $\tau_\theta = +9.5$ ksi

18–41   **(a)** $\sigma_x = +450$ psi, $\sigma_y = 0$, $\tau_x = -14.1$ psi
        (b) $\sigma_\theta = +100.3$ psi, $\tau_\theta = -187.8$ psi

18–43 $\sigma_1 = 4$ ksi, $\sigma_2 = -6$ ksi, $\theta_1 = -18.4°$, $\theta_2 = 71.6°$,
        $\tau_{max} = 5$ ksi, $\tau_{min} = -5$ ksi, $\sigma' = -1$ ksi, $\theta_s = -63.4°$, $\theta_s' = 26.6°$

18–45 $\sigma_1 = 3$ MPa, $\sigma_2 = -23$ MPa, $\theta_1 = 56.3°$, $\theta_2 = -33.7°$,
        $\tau_{max} = 13$ MPa, $\tau_{min} = -13$ MPa, $\sigma' = -10$ MPa, $\theta_s = 11.3°$, $\theta_s' = -78.7°$

18–47 $\sigma_1 = 52.4$ MPa, $\sigma_2 = -32.4$ MPa, $\theta_1 = -67.5°$, $\theta_2 = 22.5°$,
        $\tau_{max} = 42.4$ MPa, $\tau_{min} = -42.4$ MPa, $\sigma' = 10$ MPa, $\theta_s = 67.5°$, $\theta_s' = -22.5°$

18–49 $\sigma_1 = 24$ ksi, $\sigma_2 = -6$ ksi, $\theta_1 = -26.6°$, $\theta_2 = 63.4°$,
        $\tau_{max} = 15$ ksi, $\tau_{min} = -15$ ksi, $\sigma' = 9$ ksi, $\theta_s = -71.6°$, $\theta_s' = 18.4°$

18–51   At point A: $\sigma_1 = 25.3$ MPa (T), $\sigma_2 = 0$
        At point B: $\sigma_1 = 0.28$ MPa (T), $\sigma_2 = -19.28$ MPa (C)
        At point C: $\sigma_1 = 0$, $\sigma_2 = -50.6$ MPa (C)

18–53 $\sigma_1 = 26.5$ MPa (T), $\sigma_2 = -0.65$ MPa (C),
        $\tau_{max} = 13.58$ MPa, $\tau_{min} = -13.58$ MPa, $\sigma' = 12.93$ MPa

CHAPTER 19

19–1   Slenderness ratio = 166

19–3   For the same cross-sectional area, a square column is better than a circular column because the radius of gyration of a square section is greater than that of a circular section.

19–5   158 kN

19–7   28.8 kips

19–9   384 kips

19–11   24.3 mm

19–13   9.61 ft

19–15   Use $2\frac{1}{2}$-in. standard steel pipe

19–17   1.34 in.

19–19   3.82 MN

19–21   160 kips

19–23   2235 kN

19–25   14 kips

19–27   26.3 kips

19–29   230 kips

19–31   1800 kN

19–33   410 kN

19–35 178 kips

19–37 Use W12 $\times$ 87

19–39 Use W300 $\times$ 0.95

19–41 Use W10 $\times$ 49

19–43 Use 250 mm diameter standard steel pipe

CHAPTER 20

20–1 Joint strength = 39.8 kips, joint efficiency = 61.4%

20–3 Joint strength = 108 kips, joint efficiency = 66.7%

20–5 Allowable tensile strength = 53.8 kips

20–7 Joint strength = 94.2 kips, joint efficiency = 76.2%

20–9 Joint strength = 70.6 kips

20–11 w_{allow} = 8.63 kips/ft

20–13 Joint strength = 45.9 kips, joint efficiency = 70.8%

20–15 Joint strength = 108 kips, joint efficiency = 66.7%

20–17 Use 3 bolts

20–19 Joint strength = 475 kN, joint efficiency = 85.6%

20–21 Joint strength = 420 kN

20–23 w_{allow} = 145 kN/m

20–25 P = 80.9 kN

20–27 P = 12.1 kips

20–29 τ_{max} = 11.6 ksi

20–31 Strength of the weld = 119 kips, joint efficiency = 91.8%

20–33 Strength of the weld = 148 kips, joint efficiency = 91.4%

20–35 Use L = 19 in., L_1 = $5\frac{1}{2}$ in.

20–37 Use L = 15 in., L_1 = $3\frac{7}{8}$ in., L_2 = $9\frac{1}{8}$ in.

U.S. Customary Units and Their SI Equivalents

Quantity	U.S. Customary Unit	SI Equivalent
Length	ft	0.3048 m
	in.	25.40 mm
	mi	1.609 km
Mass	slug	14.59 kg
Force	lb	4.448 N
	kip	4.448 kN
Area	ft^2	$0.0929 \ m^2$
	$in.^2$	$0.6452 \times 10^{-3} \ m^2$
Volume	ft^3	$0.02832 \ m^3$
	$in.^3$	$16.39 \times 10^{-6} \ m^3$
Velocity	ft/s	0.3048 m/s
	mi/h (mph)	0.4470 m/s
	mi/h (mph)	1.609 km/h
Acceleration	ft/s^2	$0.3048 \ m/s^2$
Moment of a force	lb · ft	1.356 N · m
	lb · in.	0.1130 N · m
Pressure or stress	lb/ft^2 (psf)	47.88 Pa (pascal or N/m^2)
	$lb/in.^2$ (psi)	6.895 kPa (kN/m^2)
Spring constant	lb/ft	14.59 N/m
	lb/in.	175.1 N/m
Load intensity	lb/ft	14.59 N/m
	kip/ft	14.59 kN/m
Area moment of inertia	$in.^4$	$0.4162 \times 10^{-6} \ m^4$
Work or energy	lb · ft	1.356 J (joule or N · m)
Power	lb · ft/s	1.356 W (watt or N · m/s)
	hp (1 horsepower = 550 ft · lb/s)	745.7 W (watt or N · m/s)